Atlas of Clinically Important Fungi

Atlas of Clinically Important Fungi

Carmen V. Sciortino, Jr.

Division of Infectious Diseases
University of Louisville School of Medicine, and
Robley Rex Veterans Healthcare Medical Center
Louisville, Kentucky, USA

WILEY Blackwell

Published by John Wiley & Sons, Inc., Hoboken, New Jersey
Published simultaneously in Canada

Library of Congress Cataloging-in-Publication Data

Names: Sciortino, Carmen V., Jr., 1953–
Title: Atlas of clinically important fungi / Carmen V. Sciortino, Jr.
Description: Hoboken, New Jersey : John Wiley & Sons, Inc., [2017] |
Includes bibliographical references and index.
Identifiers: LCCN 2016036788 (print) | LCCN 2016046904 (ebook) | ISBN 9781119069669 (cloth) |
ISBN 9781119069676 (pdf) | ISBN 9781119069683 (epub)
Subjects: LCSH: Fungi–Identification. | Fungi–Atlases. | Materia medica, Vegetable–Atlases.
Classification: LCC QK603 .S344 2017 (print) | LCC QK603 (ebook) | DDC 579.5–dc23
LC record available at https://lccn.loc.gov/2016036788

Set in 10/12pt Warnock by SPi Global, Pondicherry, India

10 9 8 7 6 5 4 3 2 1

This book is dedicated in loving memory of my Father and to the many other War Veterans that served our country.

Contents

Preface

Although there are many texts that provide quality information for the identification of fungi, I have found that medical technologists rarely have time to read the text. Hence, most are rushed and seek morphological information that helps guide them to the identification of fungi. In other words, one picture is worth a thousand words. This atlas is a reference for those who have a basic understanding of mycology and wish to identify an organism with just culture morphology and a lacto-phenol cotton blue (LPCB) stain. It is written in simplified, common terms rather than in strict mycological terms to avoid the need to consult a glossary.

I have attempted to show many photographs without extensive morphological descriptions. Photographs show most structures that the technologist can correlate with their own preparations. For this reason, numerous color photographs are included so that technologists will have sufficient reference photos for identifying the various morphologies of a single organism. Organism photographs begin with the macroscopic colony views followed by the microscopic views. Also included for some microorganisms are clinical pathology photographs demonstrating how the organism appears in human tissues.

Acknowledgments

Special thanks go to James "Jim" Lay, Mary Kemper, Karen Petach, Gerald "Doug" Thompson, Derli Yung, Mark Constant, Kimberly Pittman, Mark Heckman, and Charles "Chuck" Cornett, a great team to work with over the years. James "Jim" Snyder is much appreciated for providing several of the fungal isolates and his valuable consultations. A special thanks to the pathologists, Larry Rubel and Saad Shaheen, for their consultations. The University of Texas Health Science Center Fungus Testing Laboratory, San Antonio, TX, kindly provided the *Conidiobolus* isolate from Dr M. Kimura's laboratory.

Abbreviations

BA	Blood agar
BHI	Brain heart infusion agar
CMA	Corn meal agar
GMS	Grocott's methenamine silver stain
H&E	Hematoxylin and eosin
LPCB	Lacto-phenol cotton blue
PAS	Periodic acid–Schiff stain
PDA	Potato dextrose agar
SABDX	Sabouraud dextrose agar – Emmons modification

Part I

Divisions of the Fungi

1

Morphological Growth Index

1.1 Filamentous Bacteria

Growth: slow, beginning glabrous, developing into fuzzy appearance in later culture. May also appear folded and convoluted, giving a "molar tooth-like" colony.

1.1.1 Gram stain shows Gram-positive branching rods

Nocardia (white, pink to orange) (see figures in Chapter 4)

Rhodococcus equi (coral) (see figures in Chapter 4)

Streptomyces (white, cream, brown, black) (see figures in Chapter 4)

1.1.2 Modified Acid-fast Stain Shows Partial Acid Fastness

Rhodococcus equi (see figures in Chapter 4)

Nocardia (see figures in Chapter 4)

1.2 Yeasts and Yeast-like Organisms

1.2.1 Growth (30 °C): Rapid, White to Cream, or Tan

Candida species (see figures in Chapter 4)

Cryptococcus (see figures in Chapter 4) (for further information see reference 64 in Additional Reading)

Pichia (Hansenula) anomala (see figures in Chapter 4)

Malassezia (see figures in Chapter 4) (for further information see reference 65 in Additional Reading)

Trichosporon (see figures in Chapter 4) (for further information see reference 63 in Additional Reading)

Prototheca… Algae (see figures in Chapter 4)

Atlas of Clinically Important Fungi, First Edition. Carmen V. Sciortino.

1.2.2 Growth (30 °C): Rapid, Pigmented

Cryptococcus (yellow pigment on Sabouraud dextrose) (see figures in Chapter 4) (for further information see reference 64 in Additional Reading)

Rhodotorula (pink or coral pigment) (see figures in Chapter 4)

1.3 Yeast/Yeast-like Organisms That Produce Hyphae

Coccidioides immitis (see figures in Chapter 5) (for further information see references 10 and 11 in Additional Reading)

Blastoschizomyces (see figures in Chapter 4) (for further information see reference 61 in Additional Reading)

Geotrichum (see figures in Chapter 4)

Malbranchea (see figures in Chapter 8)

Trichosporon (see figures in Chapter 4) (for further information see reference 63 in Additional Reading)

Arthrographis (see figures in Chapter 4)

1.4 Thermally Dimorphic Fungi

Blastomyces dermatitidis (see figures in Chapter 5) (for further information see references 1–4 in Additional Reading)

Histoplasma capsulatum (see figures in Chapter 5) (for further information see references 5–9 in Additional Reading)

Paracoccidioides brasiliensis (for further information see references 12 and 13 in Additional Reading)

Penicillium marneffei (for further information see reference 26 in Additional Reading)

Sporothrix schenckii (see figures in Chapter 8)

1.5 Thermally Monomorphic Molds

1.5.1 Having Microconidia, Macroconidia or Phialoconidia

Acremonium (see figures in Chapter 9) (for further information see reference 38 in Additional Reading)

Aspergillus (see figures in Chapter 8) (for further information see reference 35 in Additional Reading)

Beauveria (see figures in Chapter 8)

Chrysosporium (see figures in Chapter 8) (for further information see reference 30 in Additional Reading)

Emmonsia (see figures in Chapter 8)

Epidermophyton (see figures in Chapter 7)

Exophiala (see figures in Chapter 9) (for further information see reference 51 in Additional Reading)

Fonsecaea (see figures in Chapter 9) (for further information see reference 54 in Additional Reading)

Fusarium (see figures in Chapter 8) (for further information see reference 37 in Additional Reading)

Gliocladium (see figures in Chapter 8)

Graphium (for further information see reference 36 in Additional Reading)

Lasiodiplodia theobromae (for further information see reference 57 in Additional Reading)

Metarrhizium anisopliae (for further information see reference 39 in Additional Reading)

Microsporum (see figures in Chapter 7)

Paecilomyces (see figures in Chapter 8) (for further information see references 33 and 34 in Additional Reading)

Paraconiothyrium cyclothyrioides (for further information see reference 29 in Additional Reading)

Penicillium (see figures in Chapter 8) (for further information see references 31 and 32 in Additional Reading)

Penicillium marneffei (for further information see references 26 and 28 in Additional Reading)

Phaeoacremonium parasiticum (see figures in Chapter 9) (for further information see reference 54 in Additional Reading)

Phaeoannellomyces (Hortaea) werneckii (see figures in Chapter 9)

Phialemonium (see figures in Chapter 9)

Phialophora (see figures in Chapter 9) (for further information see references 53 and 54 in Additional Reading)

Pseudallescheria boydii (see figures in Chapter 9) (for further information see reference 48 in Additional Reading)

(Also *Scedosporium apiospermum*)

Ramichloridium (see figures in Chapter 8) (for further information see references 41 and 42 in Additional Reading)

Rhinocladiella, Veronaea (see figures in Chapter 8) (for further information see reference 27 in Additional Reading)

Scopulariopsis (see figures in Chapter 8)

Sepedonium (see figures in Chapter 8)

Sporotrichum (see figures in Chapter 8)

Stachybotrys (see figures in Chapter 9)

Syncephalastrum (see figures in Chapter 6)

Trichoderma (see figures in Chapter 8)

Trichophyton (see figures in Chapter 7)

Trichothecium roseum (see figures in Chapter 8)

1.5.2 Having Sporangia

Absidia (Lichtheimia) (see figures in Chapter 6) (for further information see reference 22 in Additional Reading)

Alternaria (see figures in Chapter 9)

Apophysomyces (for further information see references 20, 22 and 25 in Additional Reading)

Basidiobolus (for further information see references 17 and 22 in Additional Reading)

Cokeromyces (for further information see references 16, 19 and 22 in Additional Reading)

Conidiobolus (see figures in Chapter 6) (for further information see references 18 and 22 in Additional Reading)

Cunninghamella (see figures in Chapter 6) (for further information see reference 22 in Additional Reading)

Mucor (see figures in Chapter 6) (for further information see references 22 and 23 in Additional Reading)

Rhizopus (see figures in Chapter 6) (for further information see reference 22 in Additional Reading)

Saksenaea (for further information see references 20 and 22 in Additional Reading)

Stemphylium (see figures in Chapter 9) (for further information see reference 22 in Additional Reading)

1.5.3 Having Arthroconidia

Arthrographis (see figures in Chapter 4)

Coccidioides (see figures in Chapter 5) (for further information see references 10 and 11 in Additional Reading)

Geotrichum (see figures in Chapter 4)

Malbranchea (see figures in Chapter 8)

Neoscytalidium (see figures in Chapter 9)

1.5.4 Having Only Hyphae with Chlamydoconidia

Madurella grisea (see figures in Chapter 9)

Madurella mycetomatis (see figures in Chapter 9)

Microsporum ferrugineum (see figures in Chapter 7)

Trichophyton schoenleinii (see figures in Chapter 7)

Trichophyton verrucosum (see figures in Chapter 7)

Trichophyton violaceum (see figures in Chapter 7)

1.6 Dematiaceous (Brown or Black Colored Fungi)

Alternaria (see figures in Chapter 9) (for further information see reference 60 in Additional Reading)

Aureobasidium (see figures in Chapter 9) (for further information see reference 59 in Additional Reading)

Bipolaris (see figures in Chapter 9) (for further information see references 52, 54 and 60 in Additional Reading)

Chaetomium (see figures in Chapter 9)

Chrysosporium (see figures in Chapter 8)

Cladophialophora carrionii (see figures in Chapter 9) (for further information see references 44 and 50 in Additional Reading)

Cladosporium species (see figures in Chapter 9) (for further information see references 45, 46, 49, 55 and 58 in Additional Reading)

Curvularia (see figures in Chapter 9) (for further information see reference 54 in Additional Reading)

Ochroconis (Dactylaria) (see figures in Chapter 9) (for further information see references 52 and 54 in Additional Reading)

Emmonsia (see figures in Chapter 8)

Epicoccum (see figures in Chapter 9)

Exophiala jeanselmei (see figures in Chapter 9) (for further information see reference 52 in Additional Reading)

Exophiala dermatitidis (see figures in Chapter 9) (for further information see reference 52 in Additional Reading)

Exserohilum (see figures in Chapter 9) (for further information see reference 54 in Additional Reading)

Graphium (for further information see reference 36 in Additional Reading)

Hormonema dematioides (see figures in Chapter 9) (for further information see reference 56 in Additional Reading)

Madurella grisea (see figures in Chapter 9)

Microascus trigonosporus (for further information see reference 54 in Additional Reading)

Nigrospora (see figures in Chapter 9)

Phaeoacremonium parasiticum (see figures in Chapter 9) (for further information see reference 54 in Additional Reading)

Phaeoannellomyces (Hortaea) werneckii (see figures in Chapter 9) (for further information see reference 47 in Additional Reading)

Phialophora verrucosum (see figures in Chapter 9) (for further information see references 52 and 54 in Additional Reading)

Phialophora richardsiae (for further information see reference 54 in Additional Reading)

Pseudallescheria boydii (Scedosporium) (see figures in Chapter 9) (for further information see reference 48 in Additional Reading)

Phoma (see figures in Chapter 9) (for further information see reference 43 in Additional Reading)

Pithomyces (see figures in Chapter 9)

Neoscytalidium (see figures in Chapter 9)
Sporotrichum (see figures in Chapter 8)
Stachybotrys (see figures in Chapter 9)
Stemphylium (see figures in Chapter 9)
Ulocladium (see figures in Chapter 9)

1.7 Other Fungi for Which No Photographs are Available

Actinomadura (white, red, pink, orange)
Botrytis
Drechlsera
Geomyces
Helminthosporium
Lecythophora
Monilia sitophilia
Nocardiopsis
Piedraia hortae
Saccharomyces
Sporobolomyces
Verticillium

2

Division of Fungi by Type of Sporulation

2.1 Annelloconidia

Blastoschizomyces
Hortaea

2.2 Arthroconidia

Arthrographis
Geotrichum
Malbranchea
Neoscytalidium

2.3 Arthroconidia and Blastoconidia

Trichosporon

2.4 Arthroconidia; Spherule in Tissue

Coccidioides

2.5 Basidiospores

Schizophyllum

2.6 Phialoconidia

Aspergillus
Acremonium
Cladophialophora

Atlas of Clinically Important Fungi, First Edition. Carmen V. Sciortino.

Exophiala
Paecilomyces
Penicillium
Phaeoacremonium
Phialemonium
Phialophora

2.7 Produce Only Macroconidia

Alternaria – poroconidia
Bipolaris – poroconidia
Curvularia – poroconidia
Epicoccum
Exserohilum – poroconidia
Ochroconis
Pithomyces
Sepedonium
Stemphylium – poroconidia
Ulocladium – poroconidia

2.8 Macroconidia and Microconidia

Emmonsia
Epidermophyton
Fusarium
Histoplasma
Microsporum
Trichophyton

2.9 Perithecium and Ascospores

Chaetomium
Phoma

2.10 Sporangia or Sporangiospores

Conidiobolus
Mucor
Prototheca

2.11 Sporangia with Rhizoids

Absidia (Lichtheimia)
Rhizomucor
Rhizopus
Syncephalastrum

2.12 Produce Only Microconidia

Acremonium
Beauveria
Chrysosporium
Cladophialophora
Cladosporium
Fonsecaea
Gliocladium
Nigrospora
Paecilomyces
Penicillium
Phaeoacremonium
Phialemonium
Phialophora
Scedosporium
Stachybotrys
Trichothecium

2.13 Microconidia and Chlamydoconidia

Madurella

2.14 Microconidia and Arthroconidia

Aureobasidium
Hormonema
Sporotrichum

2.15 Microconidia and Cleistothecium

Aspergillus
Pseudallescheria

2.16 Yeast Only

Cryptococcus
Malassezia
Rhodotorula
Yarrowia

2.17 Yeast and Hyphae and Phialides

Exophiala

2.18 Yeast and Pseudohyphae

Candida
Pichia

2.19 Yeast and Conidia

Blastomyces
Paracoccidioides
Sporothrix

3

Division of Fungi, Yeast, and Bacteria by Morphological Group

3.1 Actinomycetes, Yeast, and Yeast-like Organisms

Actinomycetes
Arthrographis
Blastoschizomyces
Candida
Cryptococcus
Exophiala
Geotrichum
Malassezia
Nocardia
Pichia
Pneumocystis
Prototheca
Rhodococcus
Rhodotorula
Streptomyces
Trichosporon
Yarrowia

3.2 Dimorphic Organisms

Blastomyces
Coccidioides
Histoplasma
Paracoccidioides
Sporothrix

3.3 Dermatophytes

Epidermophyton
Microsporum
Trichophyton

Atlas of Clinically Important Fungi, First Edition. Carmen V. Sciortino.

3.4 Zygomycetes (Aseptate)

Basidiobolus
Conidiobolus
Cunninghamella
Lichtheimia (Absidia)
Mucor
Rhizomucor
Rhizopus
Syncephalastrum

3.5 Hyaline Molds (Septate)

Aspergillus
Beauveria
Chrysosporium
Emmonsia
Fusarium
Gliocladium
Malbranchea
Paecilomyces
Penicillium
Ramichloridium
Schizophyllum
Scopulariopsis
Sepedonium
Sporotrichum
Trichoderma
Trichothecium

3.6 Dematiaceous Fungi (Septate)

Acremonium
Alternaria
Aureobasidium
Bipolaris
Chaetomium
Cladophialophora
Cladosporium
Curvularia
Epicoccum
Exserohilum
Fonsecaea
Hormonema

Hortaea
Madurella
Neoscytalidium
Nigrospora
Ochroconis
Phaeoacremonium
Phialemonium
Phialophora
Pithomyces
Pseudallescheria
Scedosporium
Stachybotrys
Stemphylium
Ulocladium

Part II

Descriptions and Photographs of Microorganisms

4

Actinomycetes, Yeast (Thermally Monomorphic), and Yeast-like Organisms

4.1 Microorganisms with Fungal-like Morphology Occurring in Fungal Cultures

4.1.1 Actinomycetales

The order *Actinomycetales* contains the aerobic genera of bacteria known as the 'Actinomycetes': *Nocardia*, *Rhodococcus*, and *Streptomyces*. Because these microorganisms are Gram-positive branching rods and grow on fungal media, they are included here for comparison with fungi. *Nocardia* and *Rhodococcus* exhibit yeast-like growth and both are partially acid fast whereas *Streptomyces* are not. The anaerobic Gram-positive rods of *Actinomycetales* belong to the genus *Actinomyces* and are not included in this book.

4.1.2 Actinomycetes

The *Actinomycetes* are Gram-positive branching rods often confused with fungi in microscopic exams. They may form heaped, molar tooth-like colonies on SABDX or other fungal media. For that reason, photos of these common organisms are presented.

4.2 Yeast and Yeast-like Fungi

A true fungus, *Pneumocystis jirovecii*, phylum *Ascomycota*, subphylum *Taphrinomycotina*, class *Pneumocystidomycetes*, order *Pneumocystidales*, is included because it is a pathogen of man. *Pneumocystis* is non-cultivable, therefore allowing only the demonstration of the pathogen in tissue.

Other yeasts that occur in fungal cultures include *Pichia anomala* (formerly *Hansenula anomala*), *Rhodotorula*, *Cryptococcus*, *Yarrowia*, *Malassezia*, *Trichosporon*, *Prototheca wickerhamii* (algae), and *Blastoschizomyces*.

Colonial morphology: yeast cells, smooth to rough, moist and cream colored, demonstrating a "fried egg" morphology.

Microscopic morphology: budding yeast cells with different morphologies of various size and shapes. Pseudohyphae may be present. Some cells are round and contain several nuclei. Some species, when grown on ascospore media, or *in vivo*,

Atlas of Clinically Important Fungi, First Edition. Carmen V. Sciortino.

ascospores are produced. Ascospores are half-moon shaped with a brim on each end, giving the morphology of a hat or helmet. Yeast biochemical testing is performed for the identification of most yeasts.

Photos of the above microorganisms are presented on the following pages.

Actinomycetes

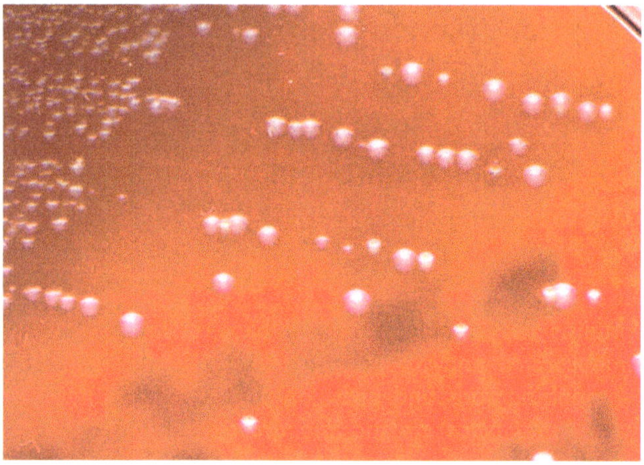

Fig. 4.2.1 *Actinomycetes* on blood agar.

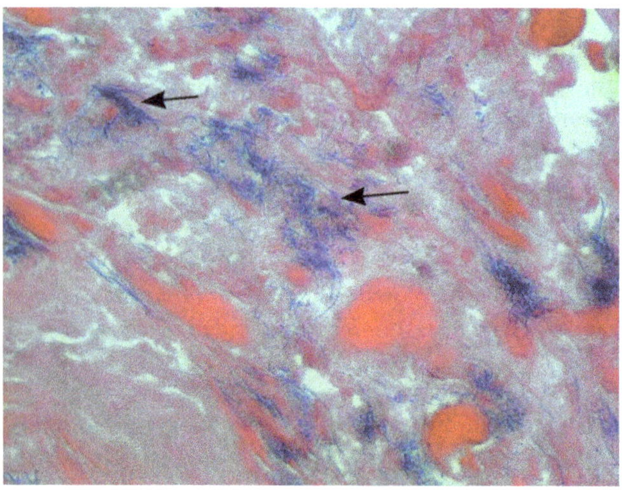

Fig. 4.2.2 *Actinomycetes* in tissue, H&E stain.

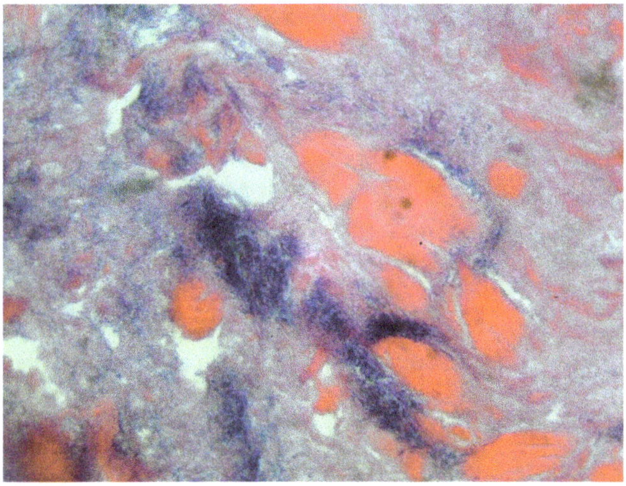

Fig. 4.2.3 *Actinomycetes* in tissue, H&E stain.

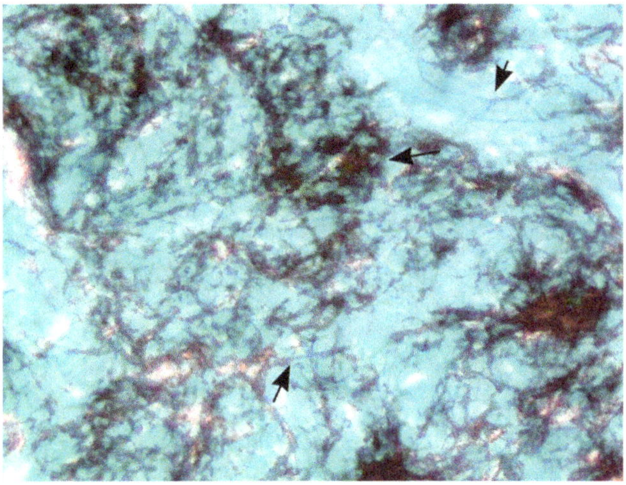

Fig. 4.2.4 *Actinomycetes* in tissue, GMS silver stain.

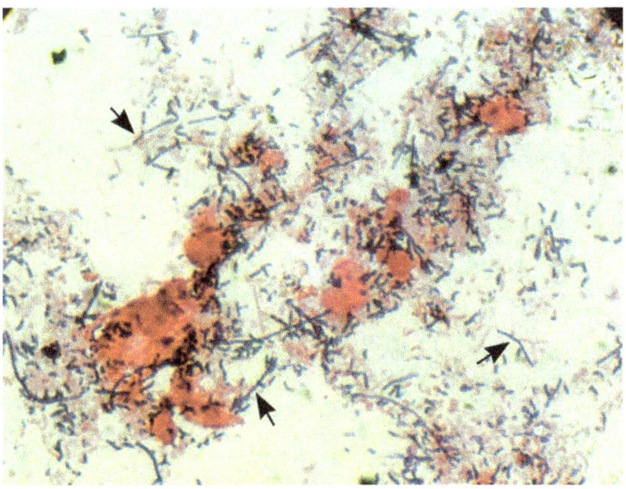

Fig. 4.2.5 *Actinomycetes*, Gram stain from abscess.

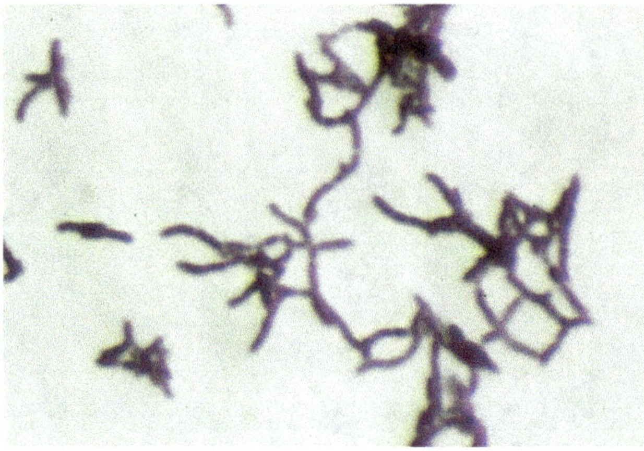

Fig. 4.2.6 *Actinomycetes*, Gram stain.

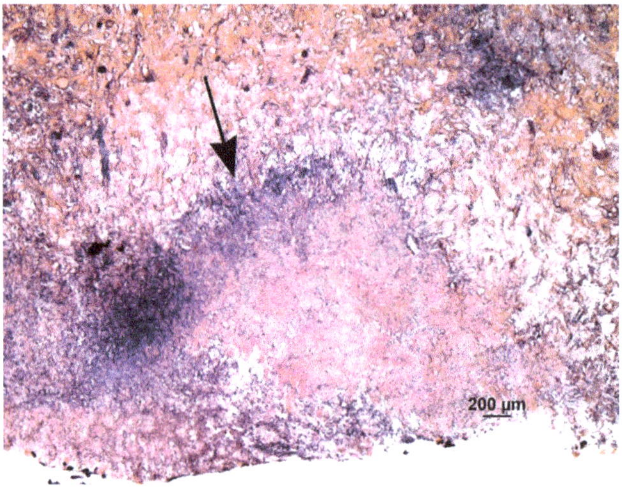

Fig. 4.2.7 Sulfur granule in human sinus tissue.

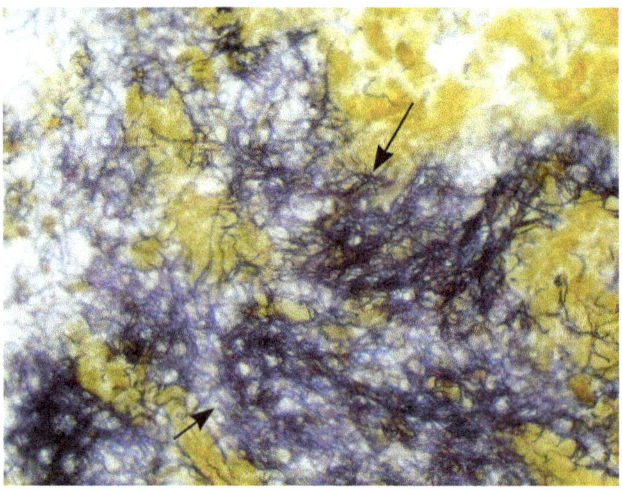

Fig. 4.2.8 *Actinomycetes* in tissue, PAS stain.

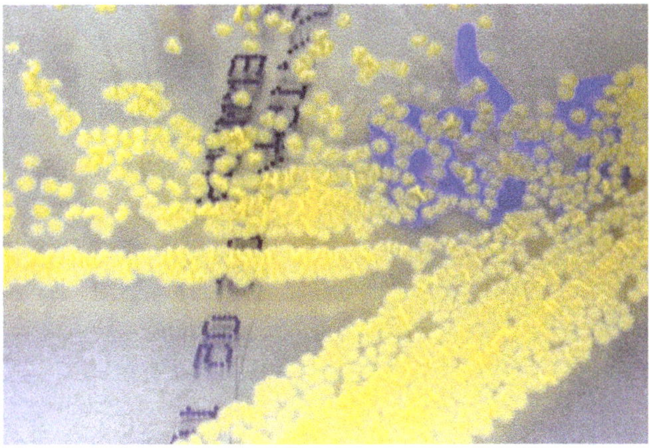

Fig. 4.2.9 *Nocardia asteroides*, close colonies SABDX (top).

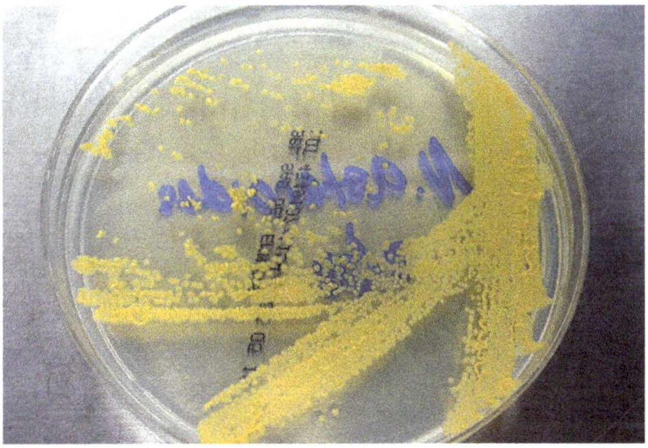

Fig. 4.2.10 *N. asteroides*, SABDX (top).

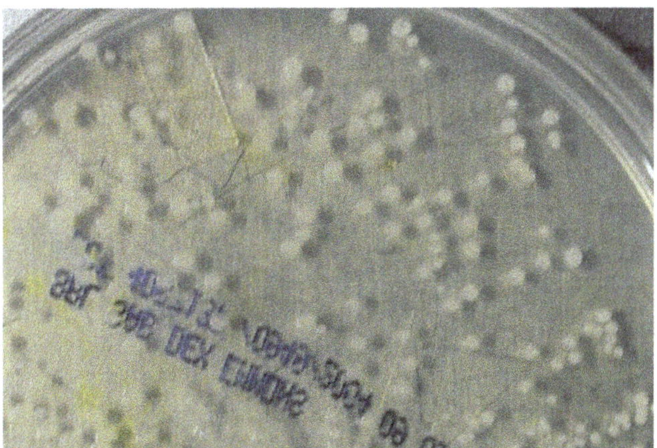

Fig. 4.2.11 *N. brasiliensis*, SABDX (top).

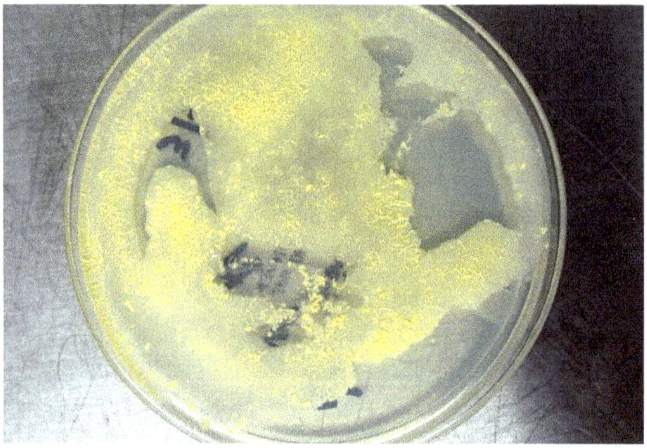

Fig. 4.2.12 *N. brasiliensis*, PDA (top).

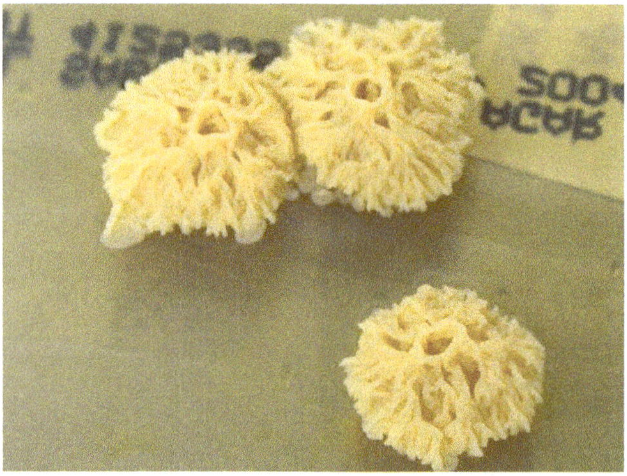

Fig. 4.2.13 *N. brasiliensis*, SAB BHI with CC.

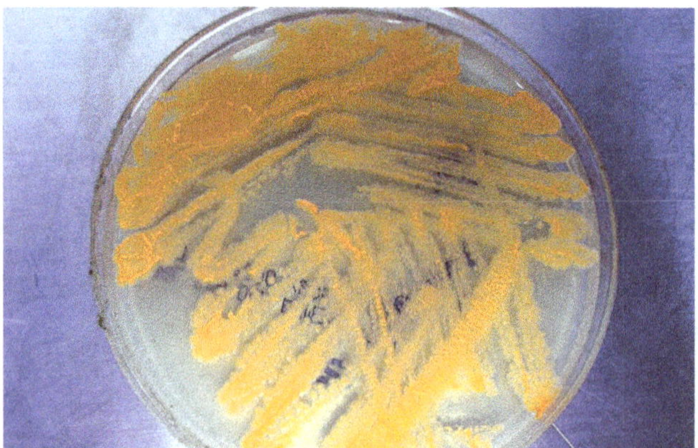

Fig. 4.2.14 *N. caviae* on SABDX (top).

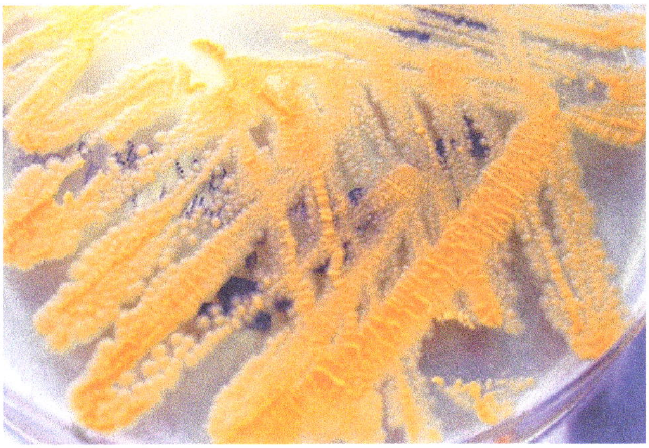

Fig. 4.2.15 *N. caviae* close colonies on SABDX (top).

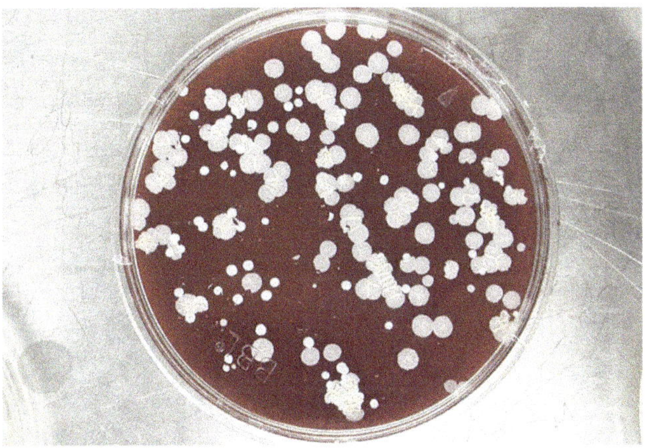

Fig. 4.2.16 *N. farcinica* on BHI blood agar.

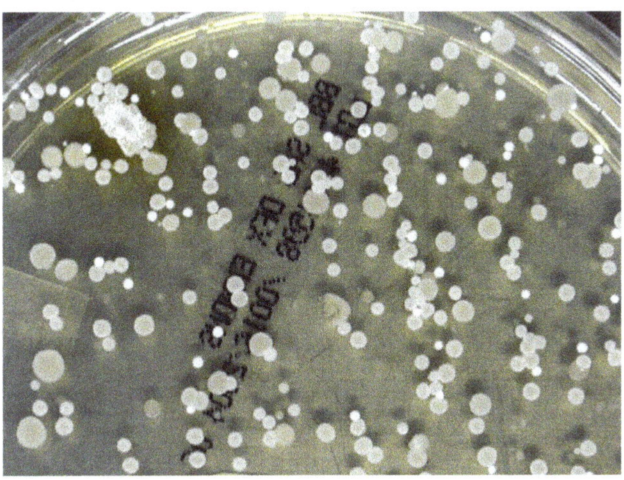

Fig. 4.2.17 *N. farcinica*, SABDX (top).

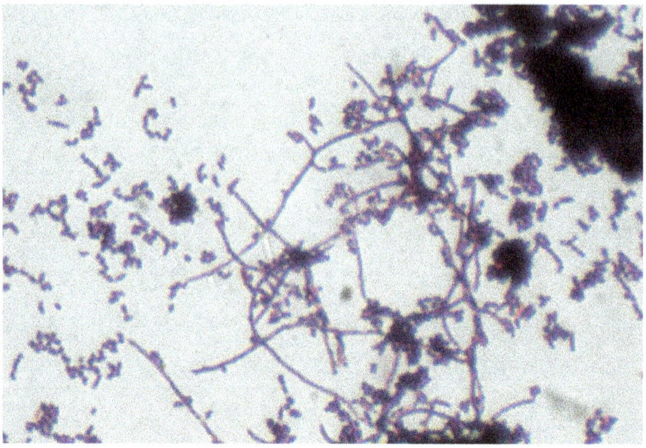

Fig. 4.2.18 *N. farcinica*, Gram stain from culture.

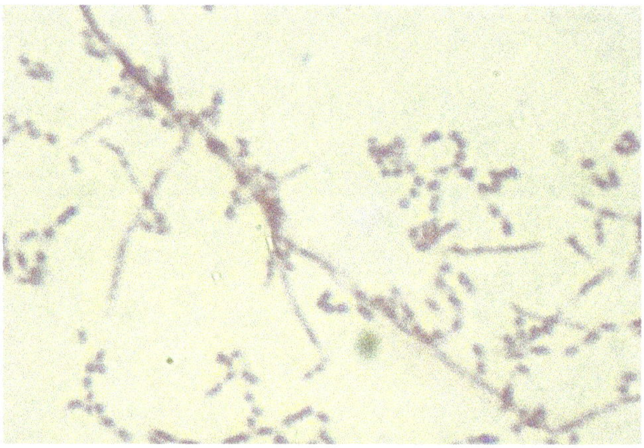

Fig. 4.2.19 *N. farcinica*, modified acid-fast stain from culture.

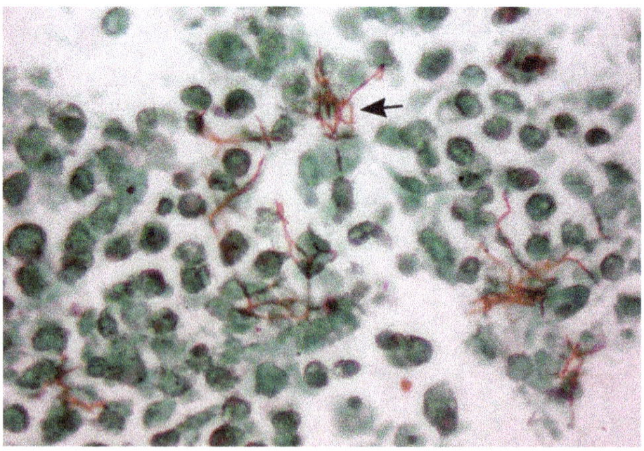

Fig. 4.2.20 *Nocardia* acid-fast stain from tissue.

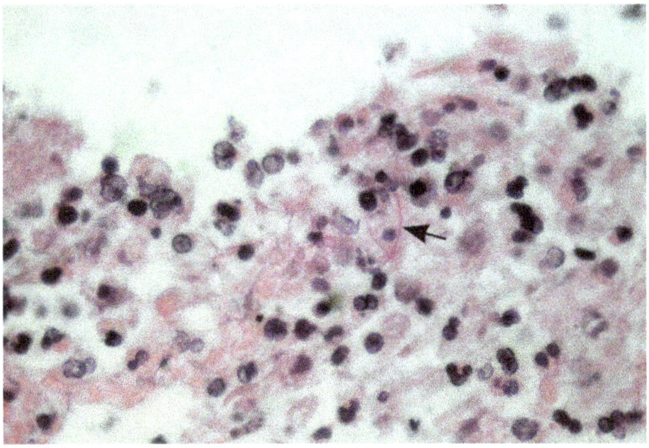

Fig. 4.2.21 *Nocardia* H&E stain from tissue.

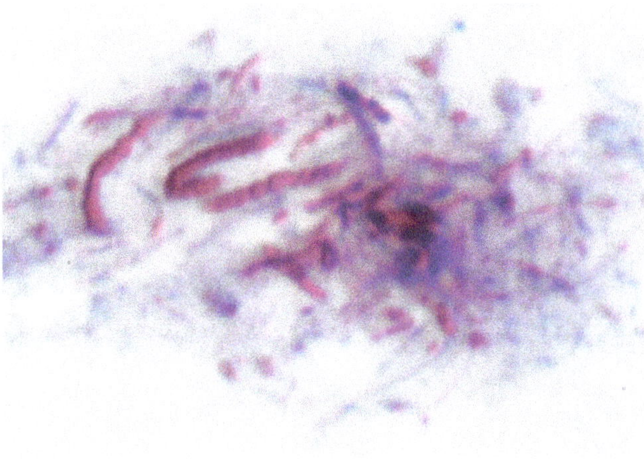

Fig. 4.2.22 *Nocardia* acid-fast stain, beaded appearance.

Streptomyces spp

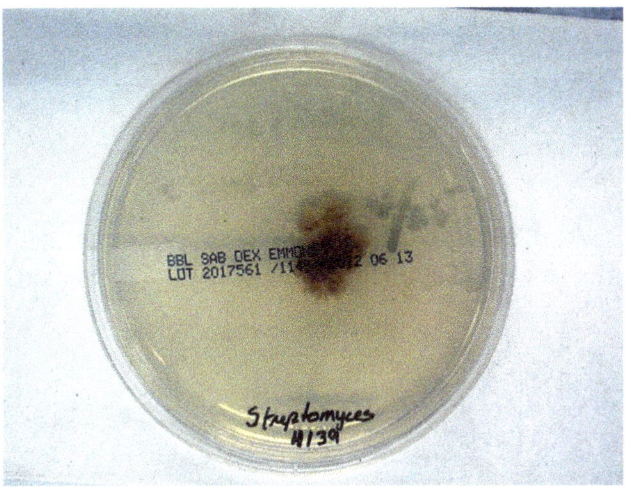

Fig. 4.2.23 *Streptomyces* (bottom).

Fig. 4.2.24 *Streptomyces*, SABDX (top).

Fig. 4.2.25 *Streptomyces*, SABDX (side).

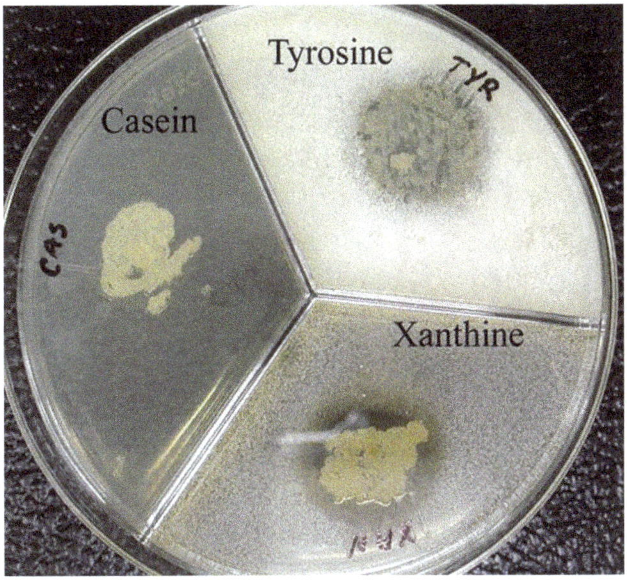

Fig. 4.2.26 *Streptomyces* casein, xanthine, tyrosine, assimilation agar (bottom).

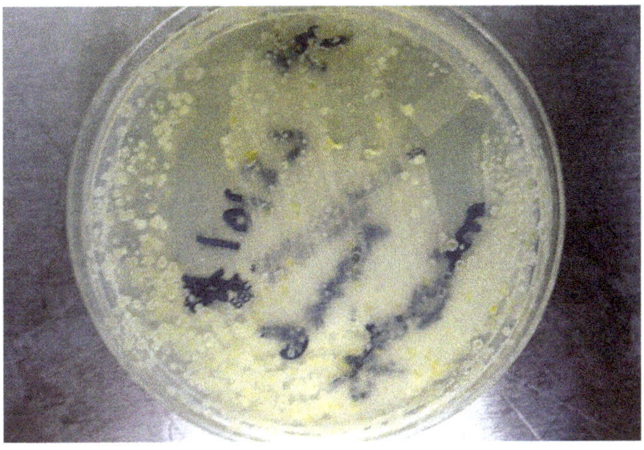

Fig. 4.2.27 *S. griseus*, SABDX (top).

Rhodococcus equi

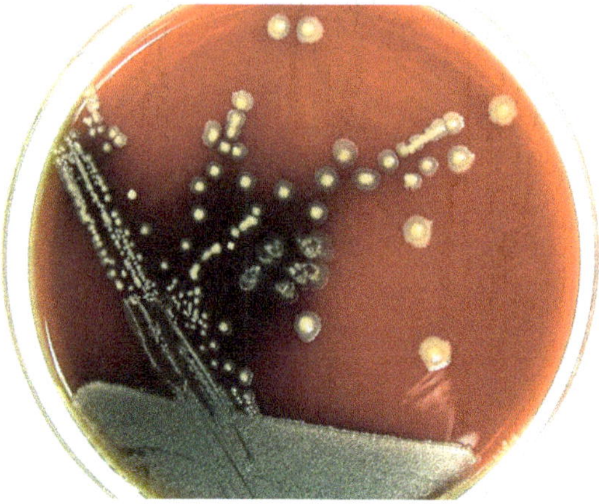

Fig. 4.2.28 *Rhodococcus equi*, sheep blood agar (top).

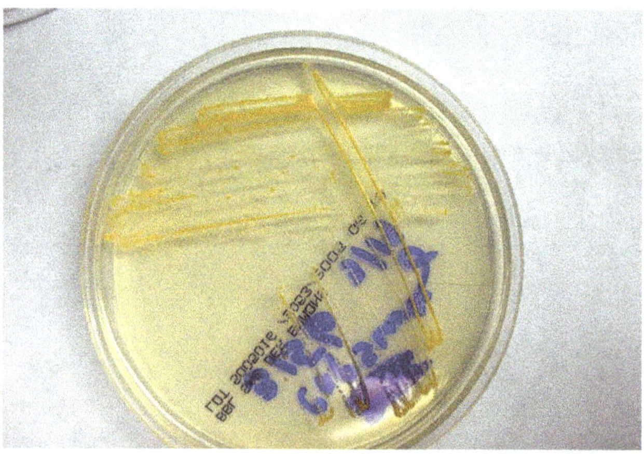

Fig. 4.2.29 *Rhodococcus equi*, SABDX (top).

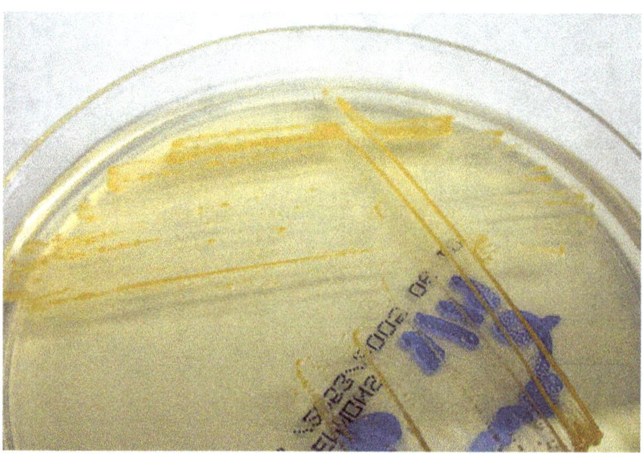

Fig. 4.2.30 *Rhodococcus equi*, SABDX close (top).

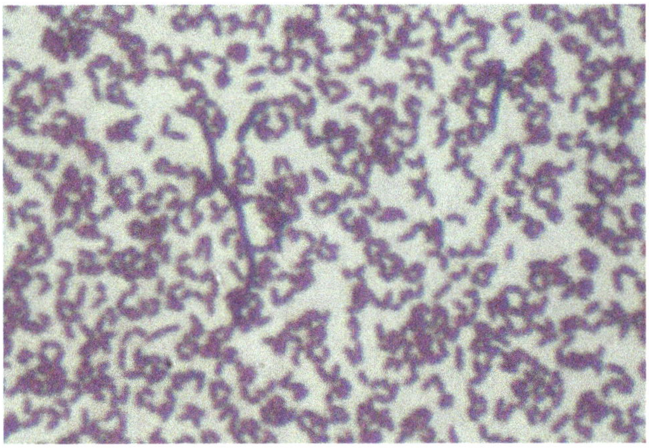

Fig. 4.2.31 *R. equi*, Gram stain.

Pichia anomala

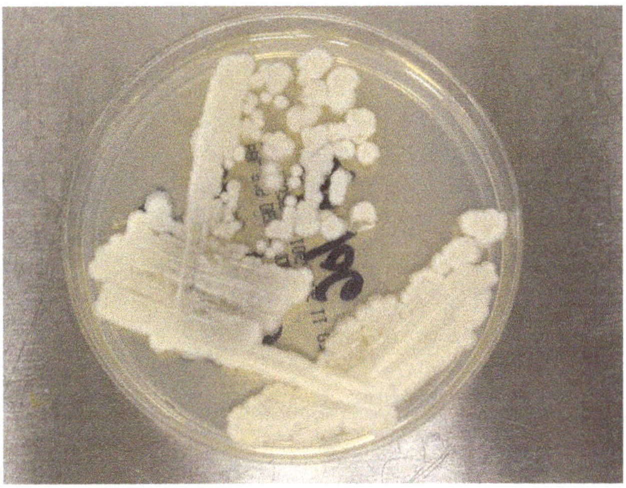

Fig. 4.2.32 *Pichia*, SABDX (top).

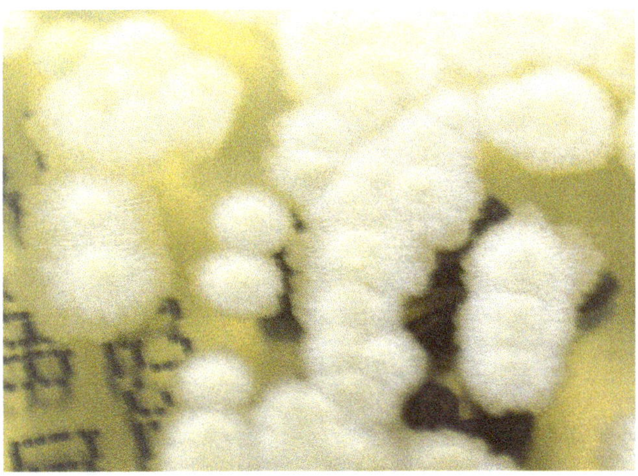

Fig. 4.2.33 *Pichia*, SABDX (side).

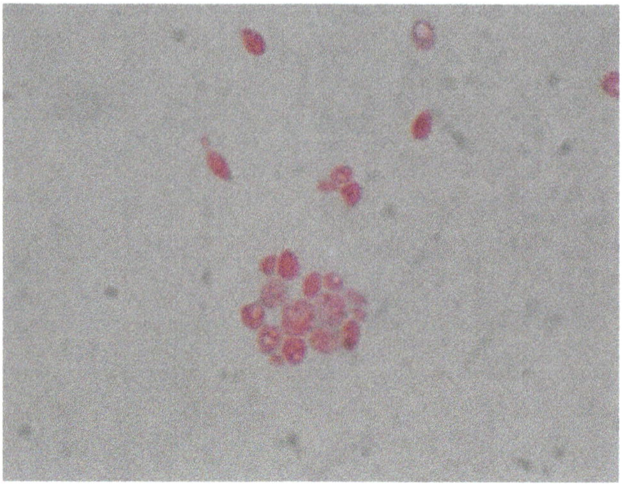

Fig. 4.2.34 *Pichia*, Gram stain.

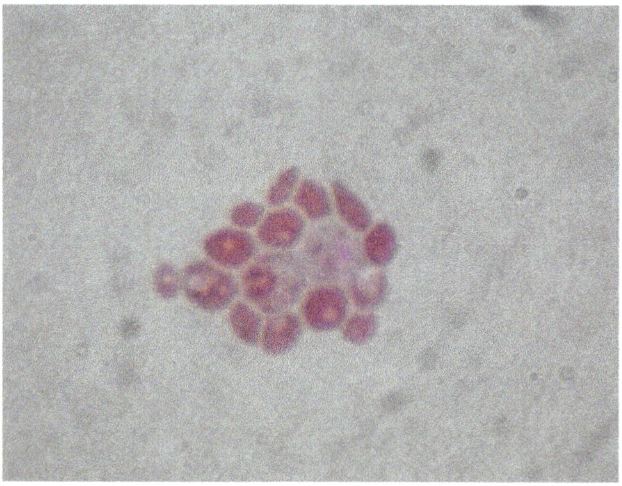

Fig. 4.2.35 *Pichia*, Gram stain.

Pneumocystis

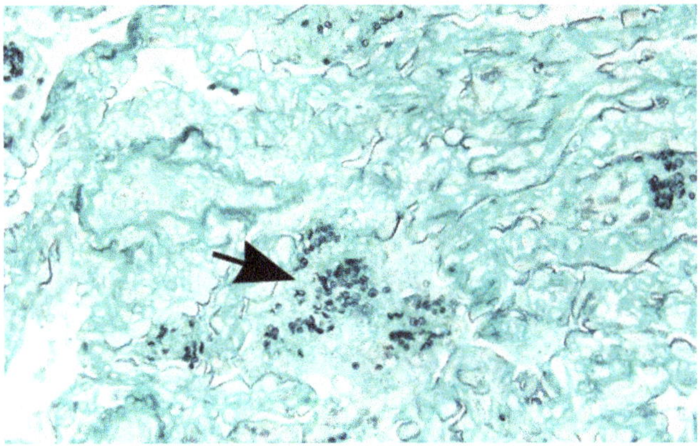

Fig. 4.2.36 *Pneumocystis*, GMS silver stain.

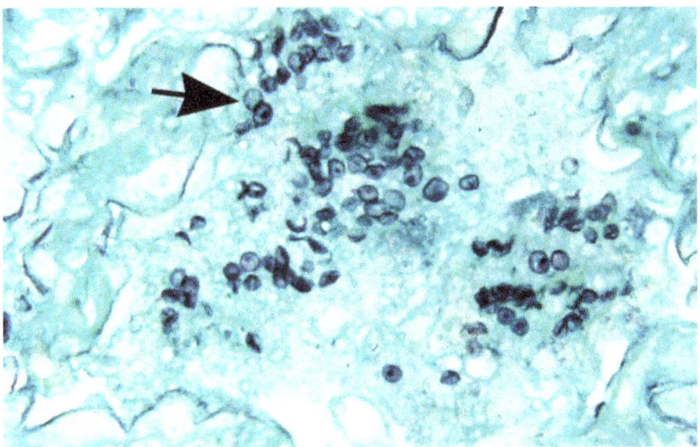

Fig. 4.2.37 *Pneumocystis*, GMS silver stain.

Yeasts culture

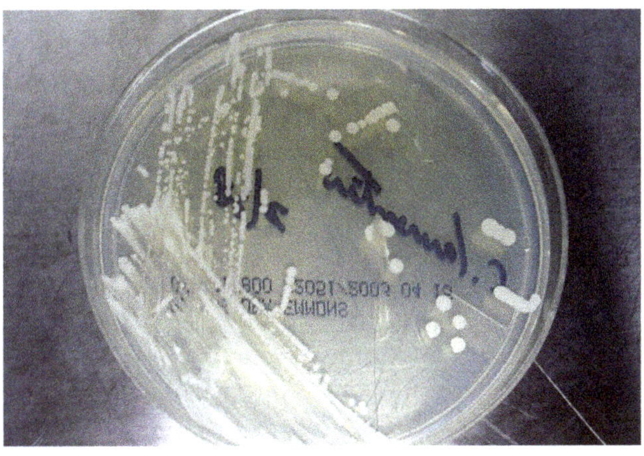

Fig. 4.2.38 *Cryptococcus laurentii*, SABDX (top).

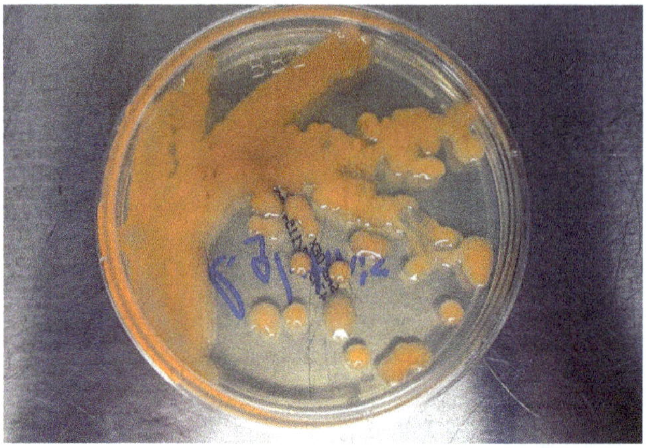

Fig. 4.2.39 *Rhodotorula glutinis*, SABDX (top).

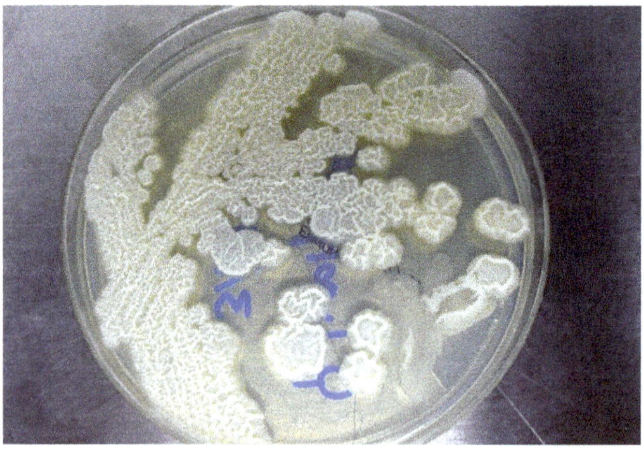

Fig. 4.2.40 *Yarrowia lipolytica*, SABDX (top).

4.3 *Candida* spp

Colonial morphology: colonies are white to cream colored, 2–5 mm in diameter, shiny or rough, and some may have a starburst-like appearance. *Candida* colonies may take on odd shapes and appearances on some culture media. Some cultures may show whorls of yeast covering the surface of the plate while others may look like spiked asteroids that have crashed into the agar.

Microscopic morphology: yeast cells are round, oval or cigar shaped. All *Candida* bud; some form pseudohyphae and blastoconidia. The diameter of an average yeast cell is 3–7 μm except for *C. glabrata* which are 2–5 μm. In the past, culture of *Candida* on cornmeal agar, which stimulates the production of chlamydospores and pseudohyphae, was used for identification. Presently, biochemical testing is the method of choice for species differentiation.

Candida in tissue

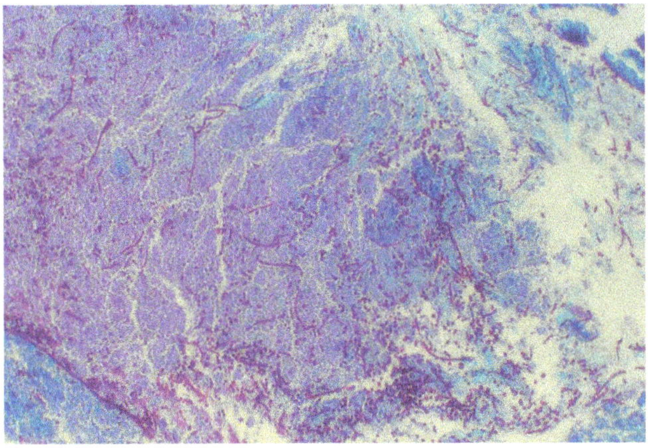

Fig. 4.3.1 *Candida*.

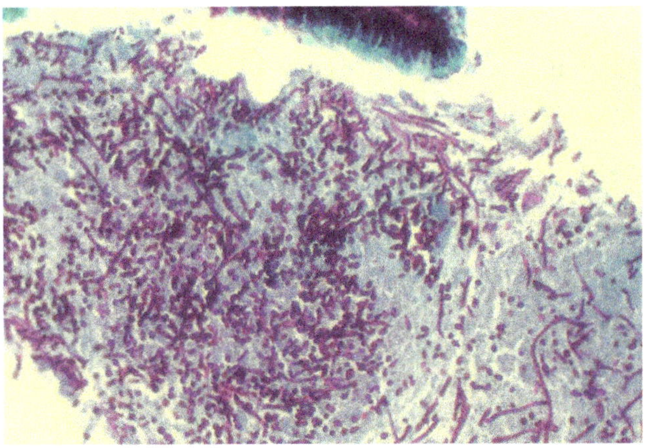

Fig. 4.3.2 *Candida*.

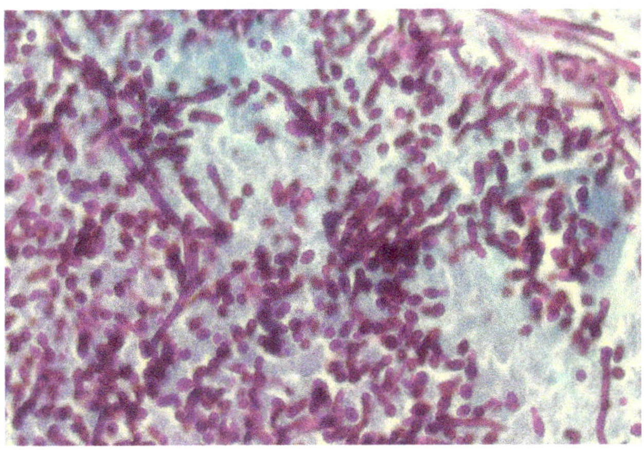

Fig. 4.3.3 *Candida*.

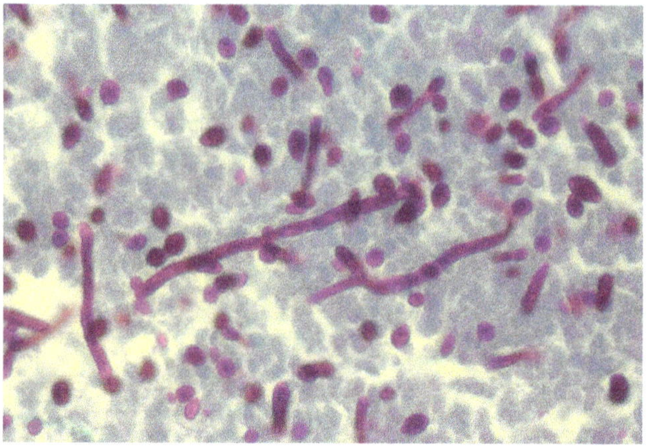

Fig. 4.3.4 *Candida*.

4.4 *Prototheca* spp

Colonial morphology: an alga that gives the appearance of a yeast. Colonies are white, cream or yellow and grow on fungal media.

Microscopic morphology: sporangia are produced that contain sporangiospores. Sporangia give the appearance of a "wagon wheel" and are 7–25 μm in diameter. Sporangia do not bud and no hyphae are produced. Sporangia may contain one or many sporangiospores. Sporangia have a thick cell wall, and *P. stagnora* produces a capsule. Sporangiospores are best viewed using phase-contrast microscopy. Common species are *P. wickerhamii*, and *P. zopfii*. Species are separated on biochemical utilization and common growth patterns.

Prototheca

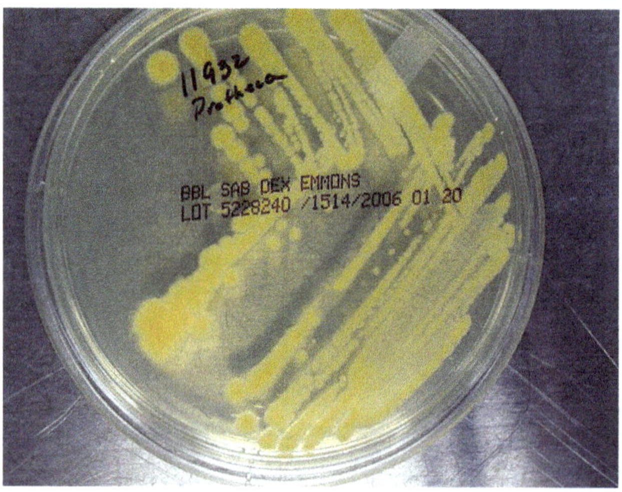

Fig. 4.4.1 *Prototheca* (bottom).

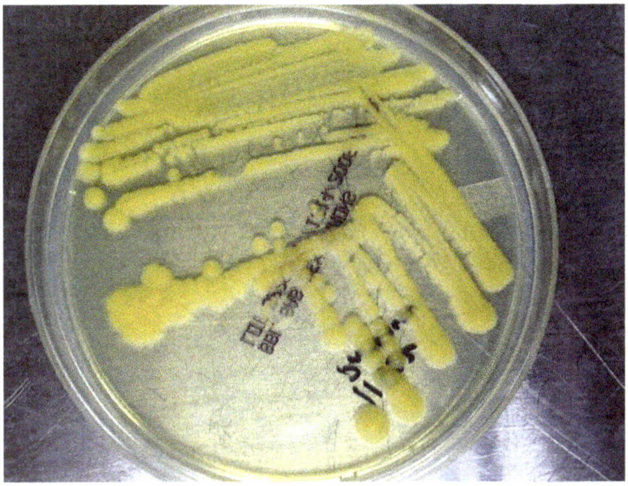

Fig. 4.4.2 *Prototheca* (top).

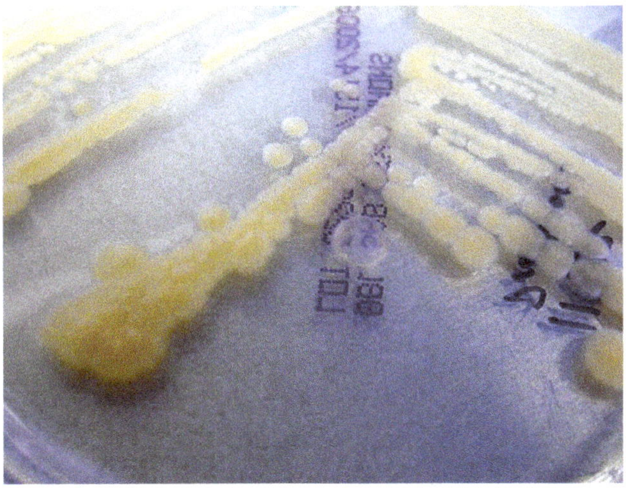

Fig. 4.4.3 *Prototheca* (side).

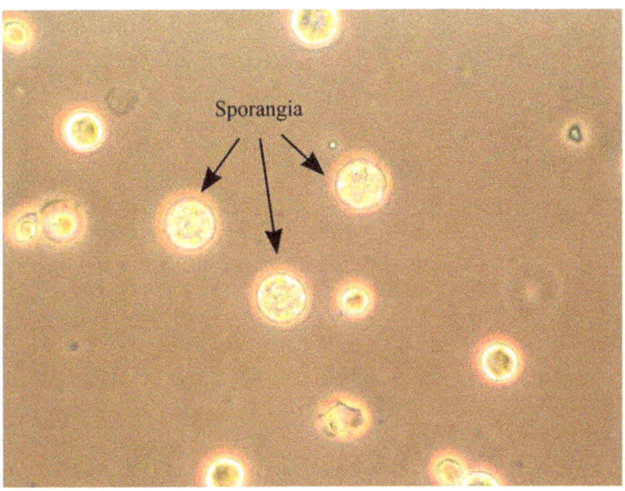

Fig. 4.4.4 *Prototheca*, phase contrast microscopy.

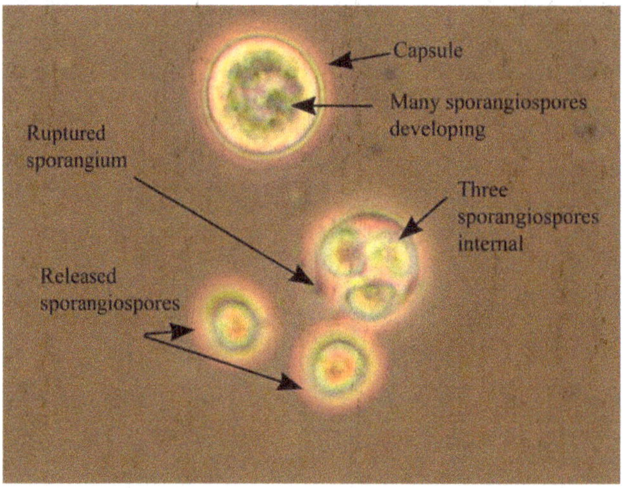

Fig. 4.4.5 *Prototheca* sporangiospores, phase contrast microscopy, 400×.

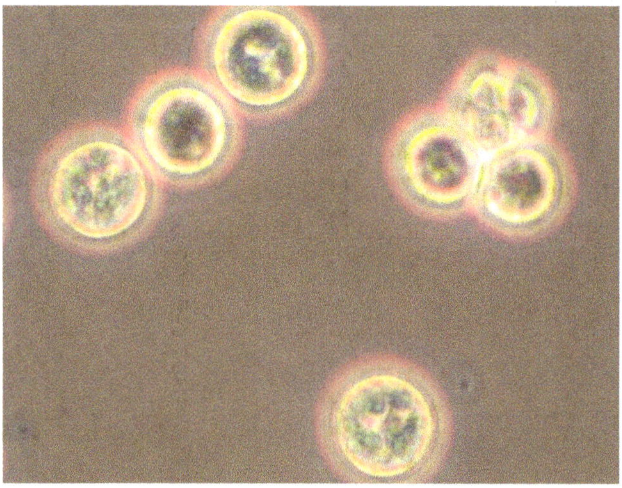

Fig. 4.4.6 *Prototheca*, wagon wheel morphology, phase contrast microscopy, 400×.

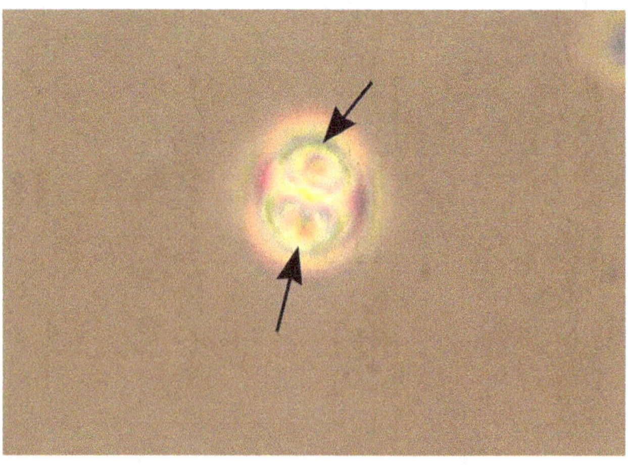

Fig. 4.4.7 *Prototheca*, showing two sporangiospores.

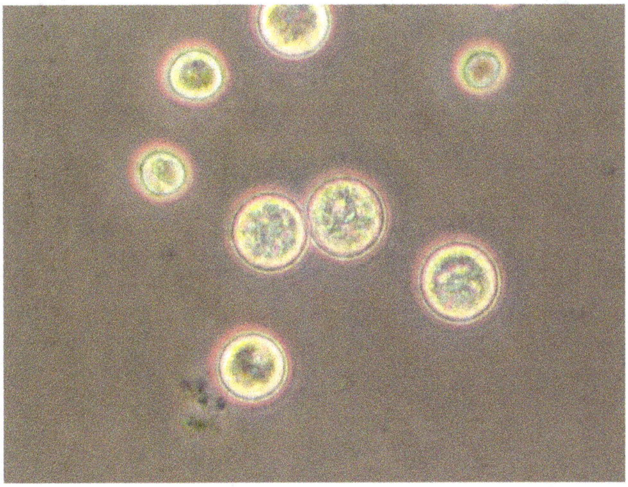

Fig. 4.4.8 *Prototheca* – sporangia with endospores.

4.5 *Arthrographis kalrae*

Colonial morphology: colonies develop slowly at first, then rapidly. At first they are smooth and yeast-like, and later (1 week) become velvety and floccose. They are white to pale yellow on top, and white to tan on the reverse.

Microscopic morphology: yeast may appear in young cultures. In older cultures, hyphae are septate, hyaline, and produce hyaline conidiophores that are both simple and branched. Conidiophores produce chains of arthroconidia at the tips, or intercalary. Arthroconidia (2–3.5 × 3–7 µm) are mostly boxcar shaped, but may be round. They are consecutive (without spaces), not barrel shaped, and may be easily confused with *Geotrichum*. *Geotrichum* lacks branching conidiophores. *Arthrographis* may produce blastoconidia. The non-pathogenic *A. cuboidea* will not grow at 45 °C whereas *A. kalrae* does. *A. cuboidea* also produces a pink-violet diffusible pigment at 36 °C.

Arthrographis kalrae

Fig. 4.5.1 *Arthrographis* (bottom) 3 days.

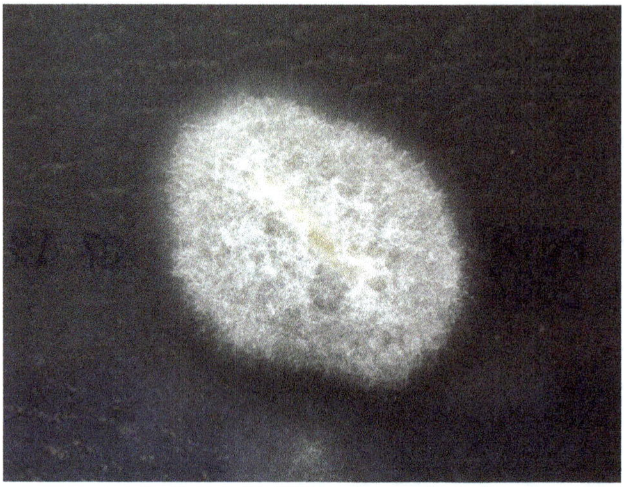

Fig. 4.5.2 *Arthrographis* (top) 3 days.

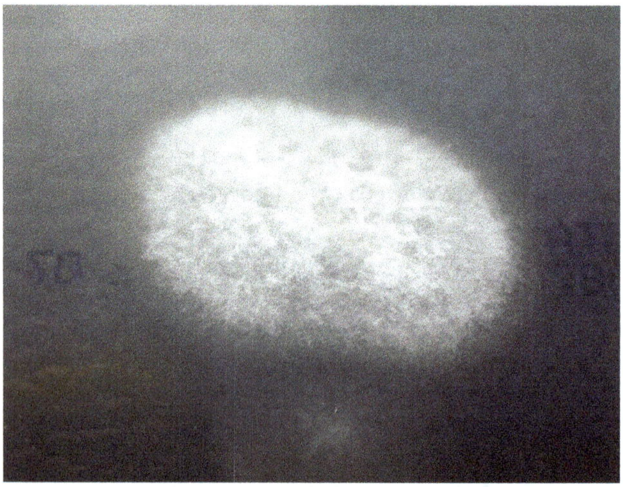

Fig. 4.5.3 *Arthrographis* (side) 3 days.

Fig. 4.5.4 *Arthrographis* (bottom) 12 days.

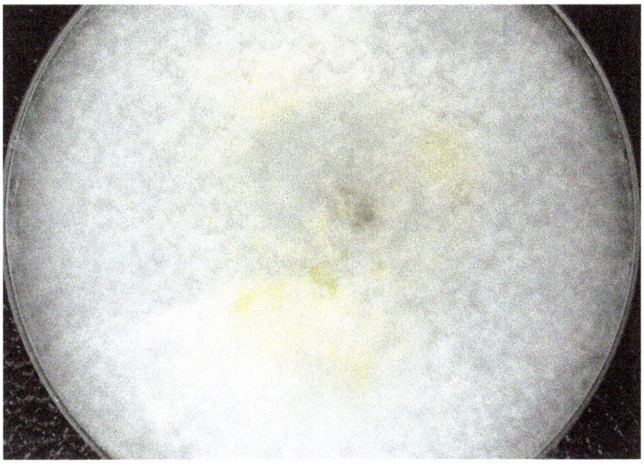

Fig. 4.5.5 *Arthrographis* (top) 12 days.

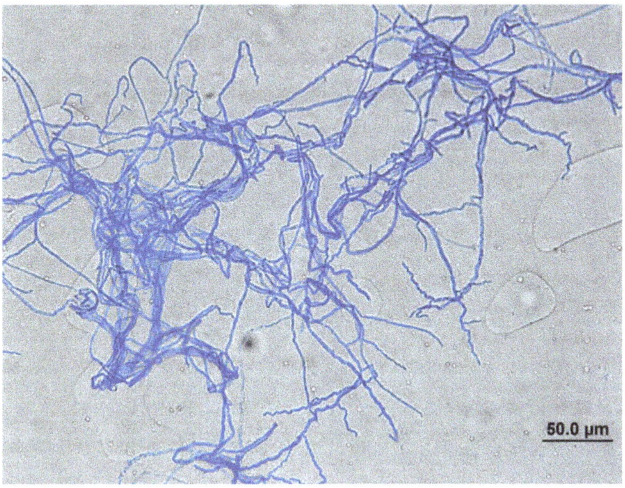

Fig. 4.5.6 Rough hyphae.

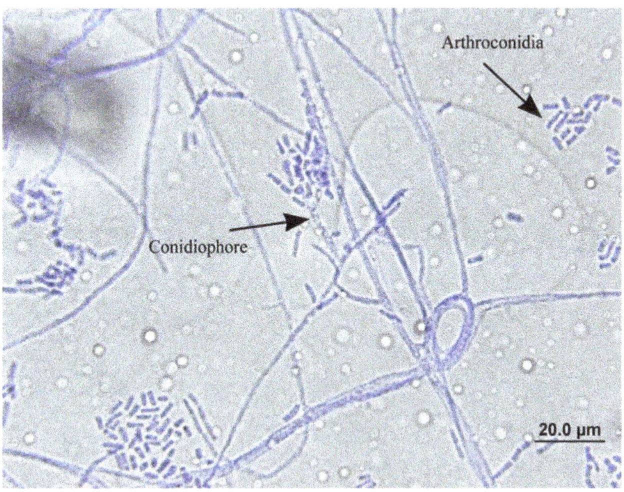

Fig. 4.5.7 Branching conidiophore.

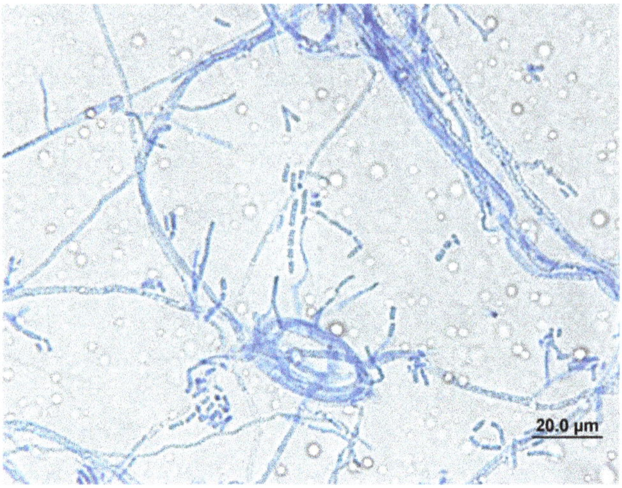

Fig. 4.5.8 Hyphae in whorls and stuck together.

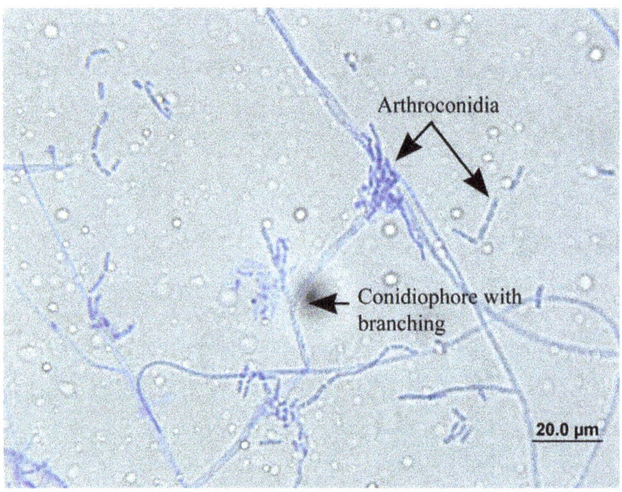

Fig. 4.5.9 Arthroconidia produced in chains at tip of conidiophore.

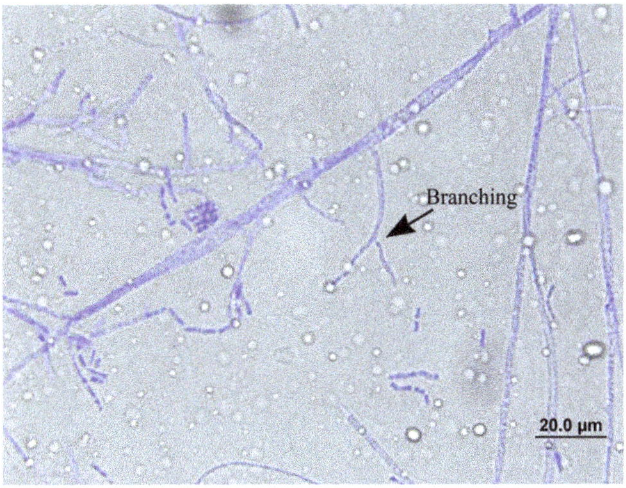

Fig. 4.5.10 Branching conidiophore.

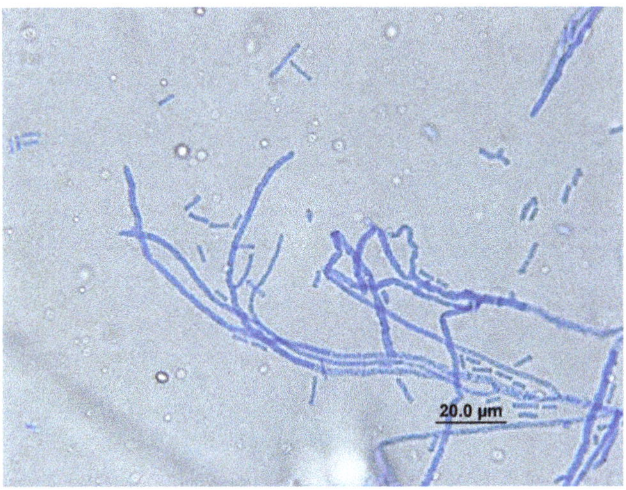

Fig. 4.5.11 Rough hyphae produce arthroconidia.

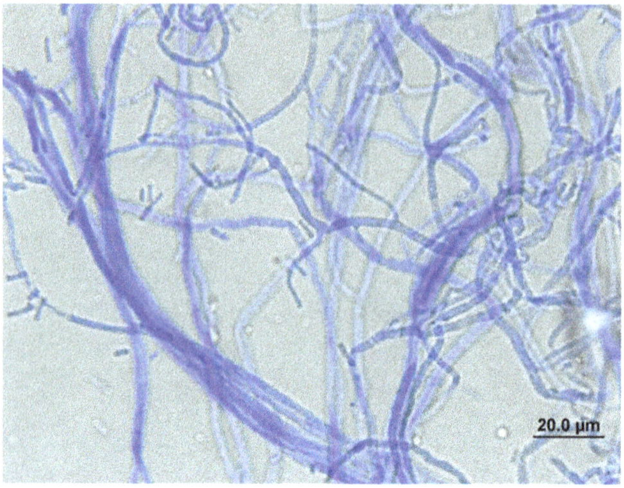

Fig. 4.5.12 Hyphae stick together.

4.6 *Cryptococcus neoformans*

Colonial morphology: *C. neoformans* grows on Sabouraud dextrose agar colonies as cream colored, smooth, mucoid yeast.

Microscopic morphology: round to elongated budding yeast-like cells or blastoconidia, 3.0–7.0 × 3.3–7.9 µm in size.

The teleomorph stage, *Filobasidiella neoformans* var. *bacillispora,* grows as a mycelia colony. A basidium is produced at the end of hyphae. Basidiospores are produced in chains at four equidistant positions (four corners) at the apical end of the basidium. This stage is rarely seen so photographs of this stage are scarce.

Cryptococcus

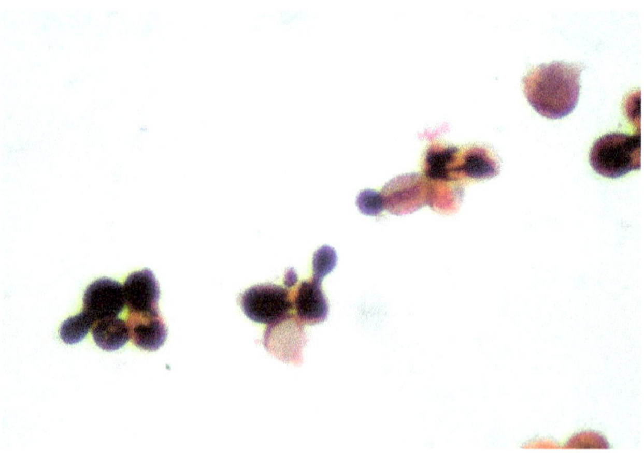

Fig. 4.6.1 Gram stain.

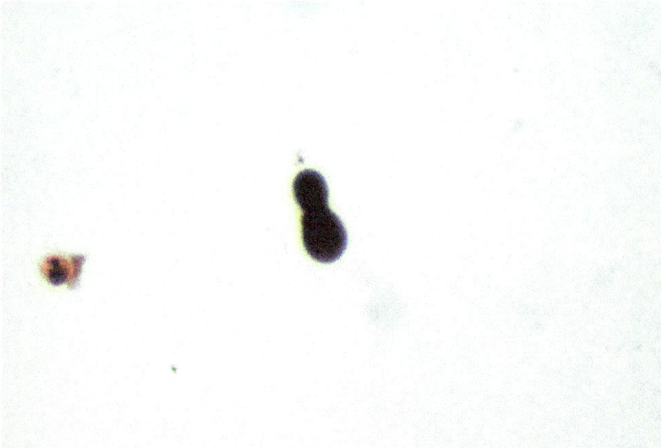

Fig. 4.6.2 Gram stain.

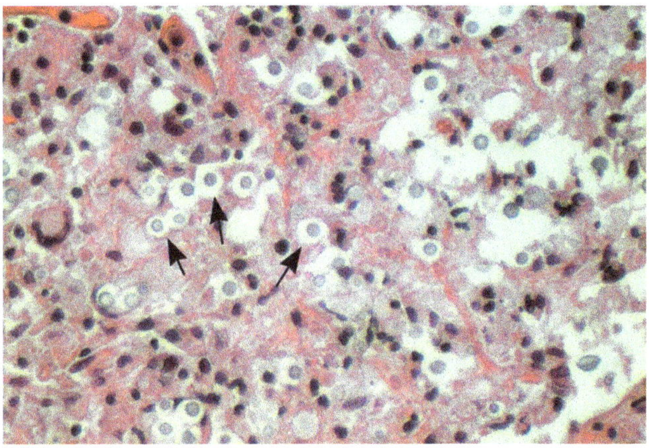

Fig. 4.6.3 *Cryptococcus* in tissue H&E stain.

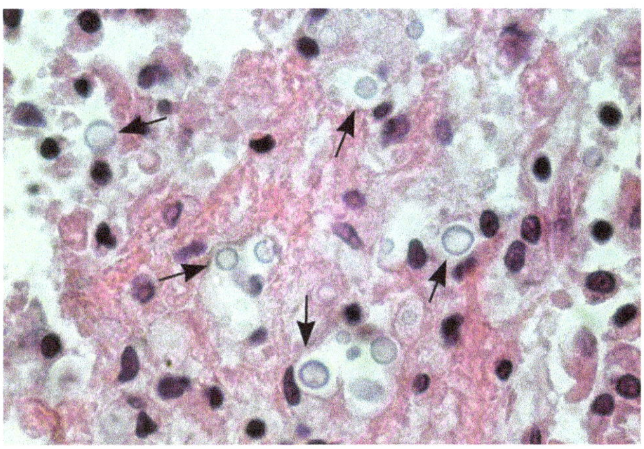

Fig. 4.6.4 *Cryptococcus* in tissue H&E stain.

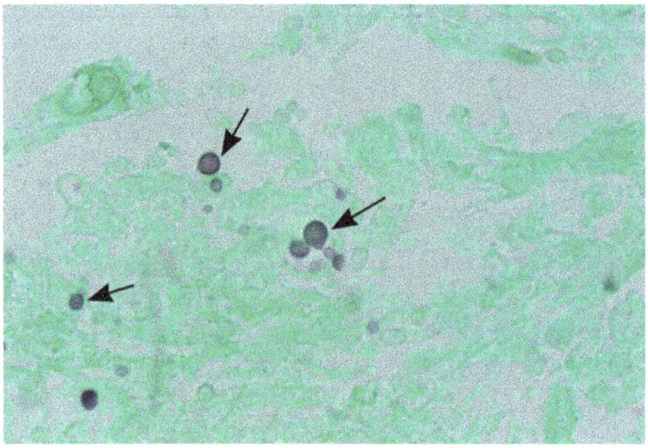

Fig. 4.6.5 *Cryptococcus* in tissue silver stain.

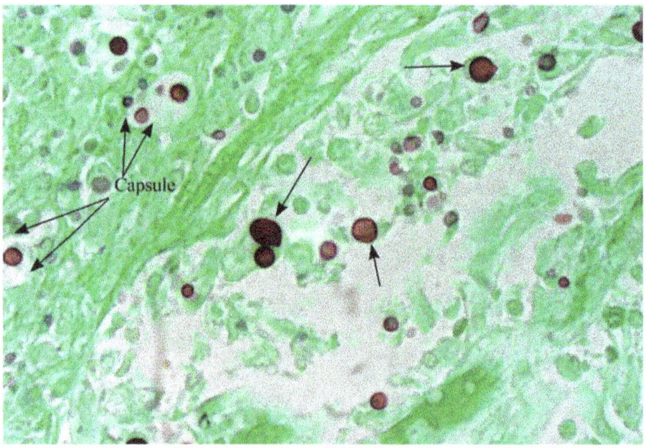

Fig. 4.6.6 *Cryptococcus* in tissue silver stain (computer-enhanced contrast).

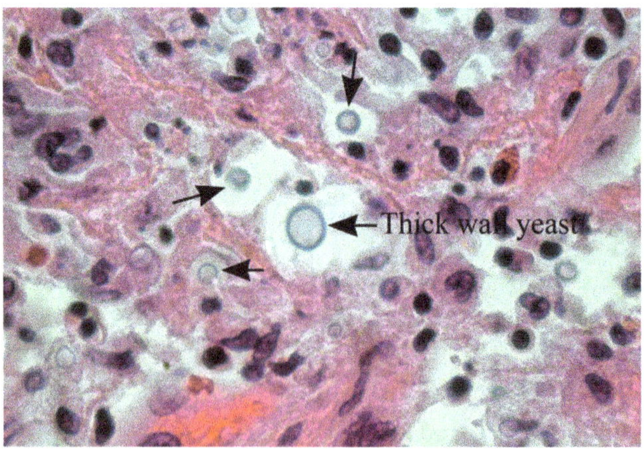

Fig. 4.6.7 *Cryptococcus* in tissue H&E stain.

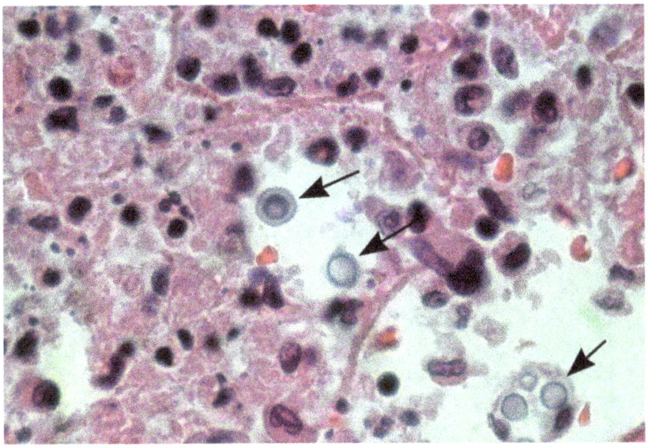

Fig. 4.6.8 *Cryptococcus* in tissue H&E stain.

4.7 *Geotrichum candidum*

Colonial morphology: the colony appears yeast-like and can spread across the culture plate within days. Growth is beige or tan. When incubated at 25 °C, the growth is above the agar; at 30 °C, some growth is above the agar and some is below the surface; at 37 °C, growth is below the agar surface or there is no growth. Growth appears as ground glass but can be displaced with an inoculating loop, enough for a wet preparation. Scotch tape preparations are often negative unless some growth is displaced first. Some cultures may form aerial mycelia.

Microscopic morphology: hyphae are septate, coarse, and true hyphae (not pseudohyphae). With age, hyphae separate into arthroconidia. Arthroconidia are consecutively attached in chains, not separated by a space, and they are rectangular in shape with thick walls. They are 6–12 × 3–6 µm in size and break apart by the separation of a double septum. Arthroconidia bud from the corner of the cell, but the bud forms true hyphae and not another yeast cell or pseudohyphae.

Geotrichum must be differentiated from the following:

- *Coccidioides immitis* – alternating barrel shaped (swollen) arthroconidia with empty cells between
- *Malbrancheae* – alternating arthroconidia with parallel sides and empty cells between
- *Trichosporon* – also forms blastoconidia and pseudohyphae
- *Arthrographis* – arthrospores are formed similar to *Geotrichum*, but branching conidiophores are present
- *Geomyces* – conidiophores are present; alternating arthroconidia.

Geotrichum candidum

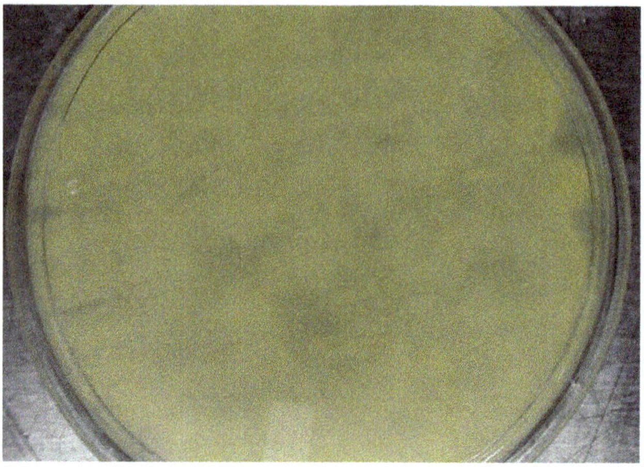

Fig. 4.7.1 *Geotrichum* (bottom) CMA.

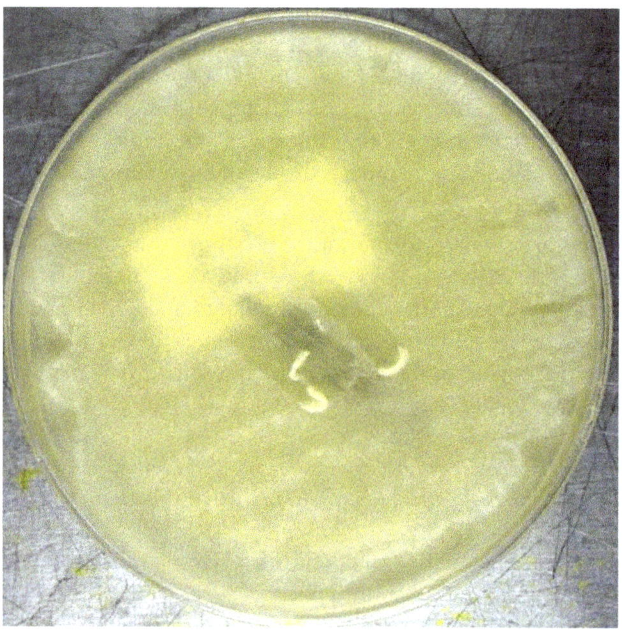

Fig. 4.7.2 *Geotrichum candidum* (top) CMA.

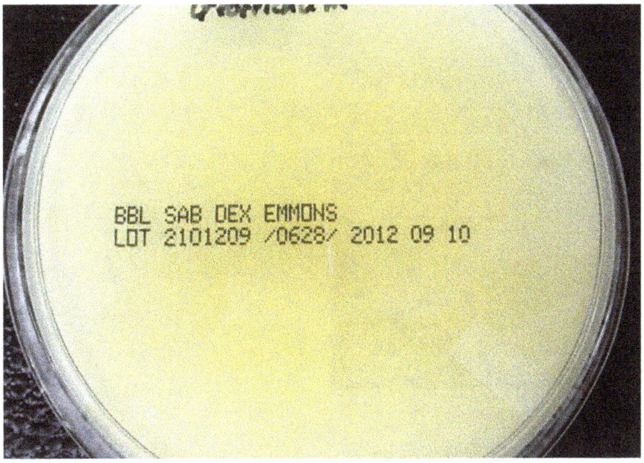

Fig. 4.7.3 *Geotrichum* (bottom) SABDX.

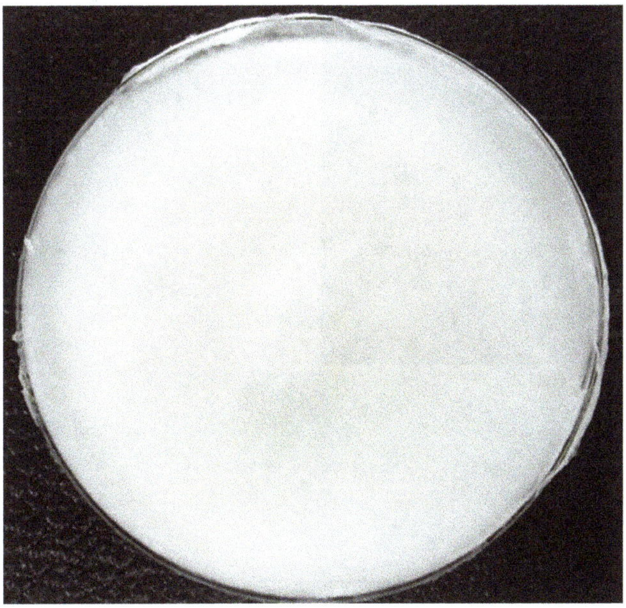

Fig. 4.7.4 *Geotrichum* (top) SABDX.

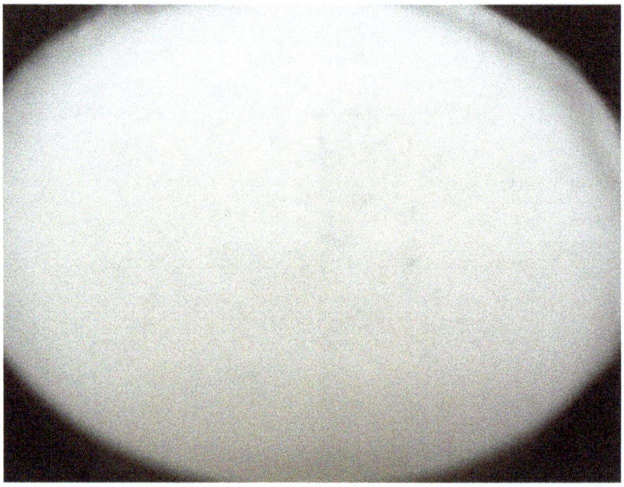

Fig. 4.7.5 *Geotrichum* (side) SABDX.

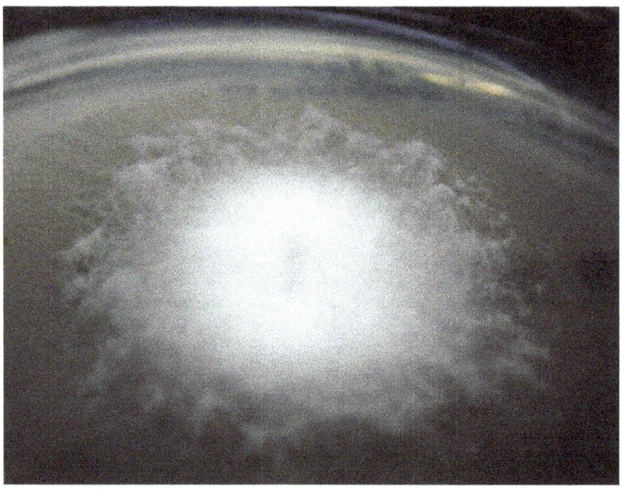

Fig. 4.7.6 *Geotrichum* (side) 5 days SABDX.

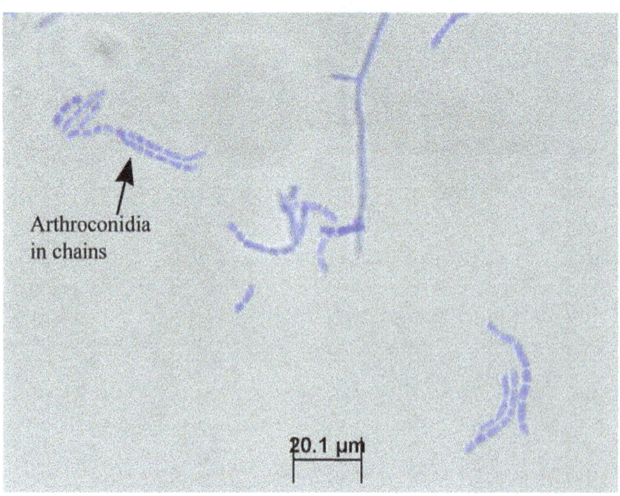

Fig. 4.7.7 *Geotrichum* with arthoconidia in chains.

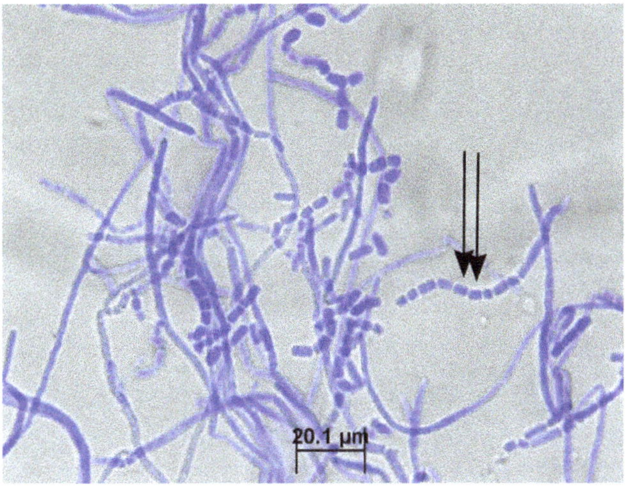

Fig. 4.7.8 Arthroconidia without spaces.

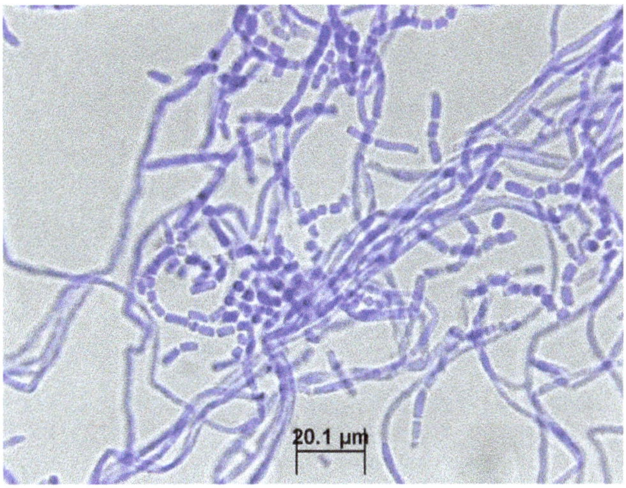

Fig. 4.7.9 Box-shaped arthroconidia with parallel sides.

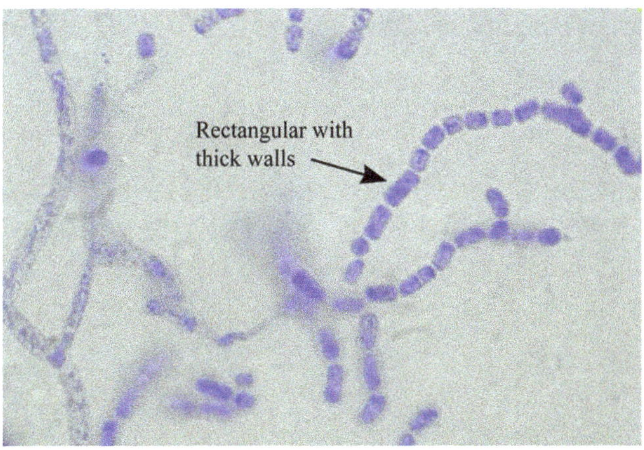

Fig. 4.7.10 Thick walled arthroconidia, 1000×.

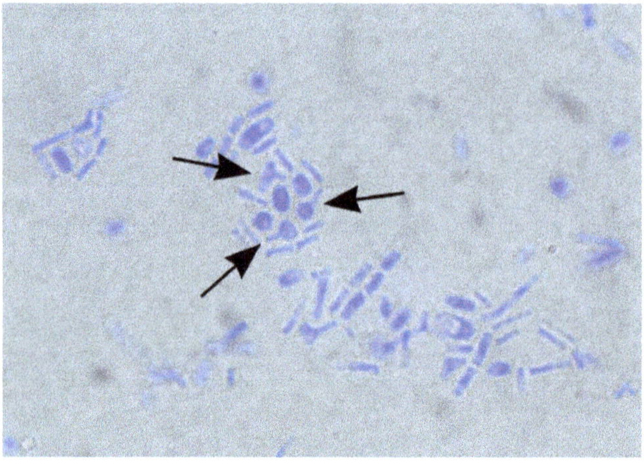

Fig. 4.7.11 Germinating arthroconidia form new hyphae.

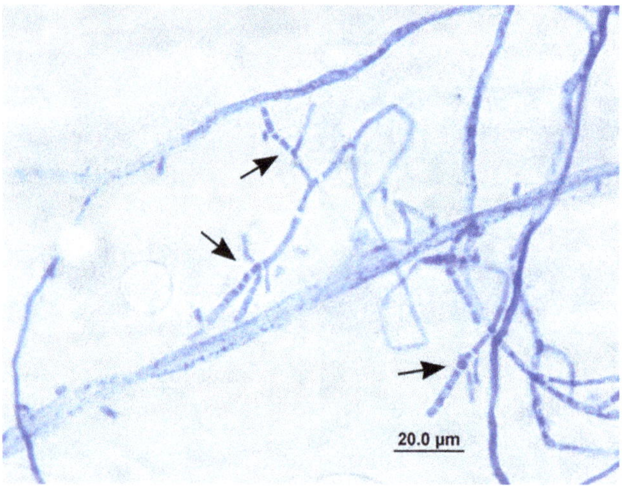

Fig. 4.7.12 False-branching showing new chains of arthroconidia.

Blastoschizomyces - Geotricum capitatum

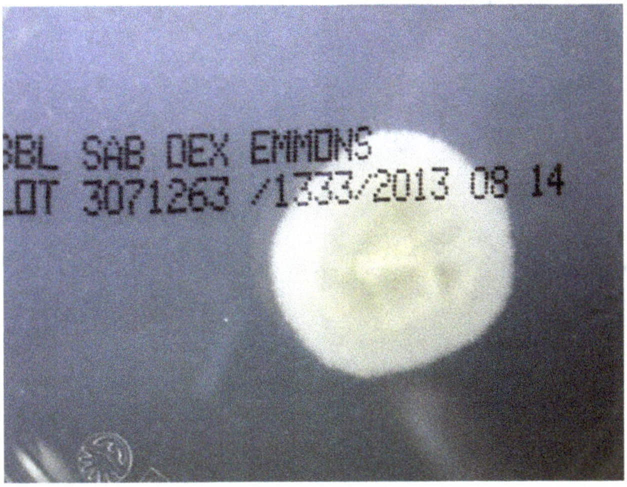

Fig. 4.7.13 *Geotricum capitatum* (bottom).

Fig. 4.7.14 *Geotricum capitatum* (top).

Fig. 4.7.15 *Geotricum capitatum* (side).

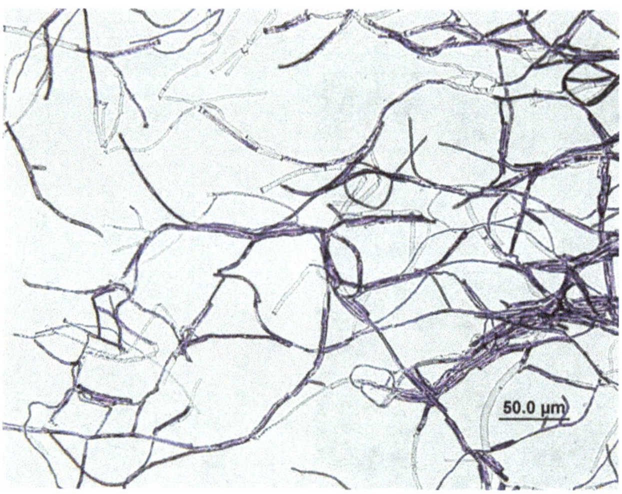

Fig. 4.7.16 *Geotricum capitatum*.

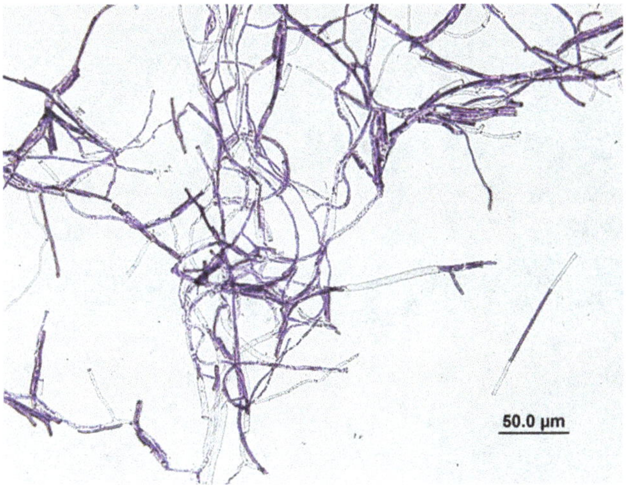

Fig. 4.7.17 *Geotricum capitatum.*

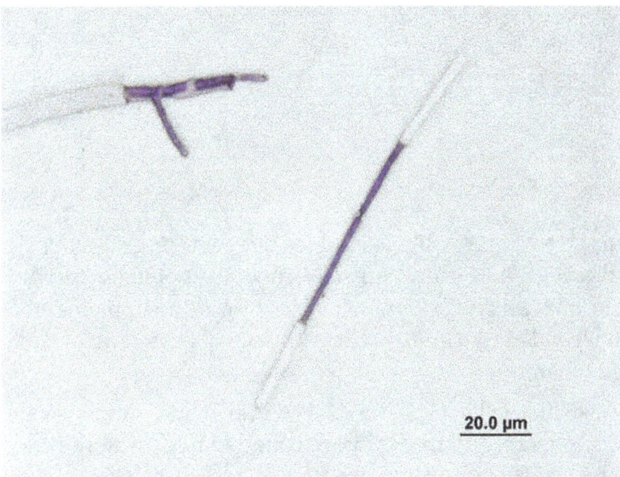

Fig. 4.7.18 *Geotricum capitatum.*

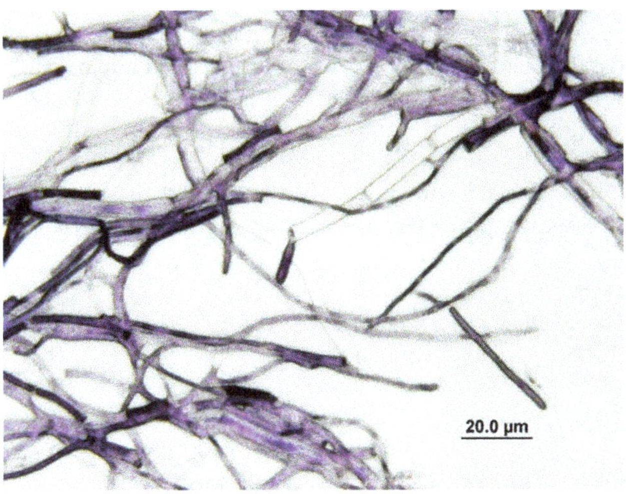

Fig. 4.7.19 *Geotricum capitatum.*

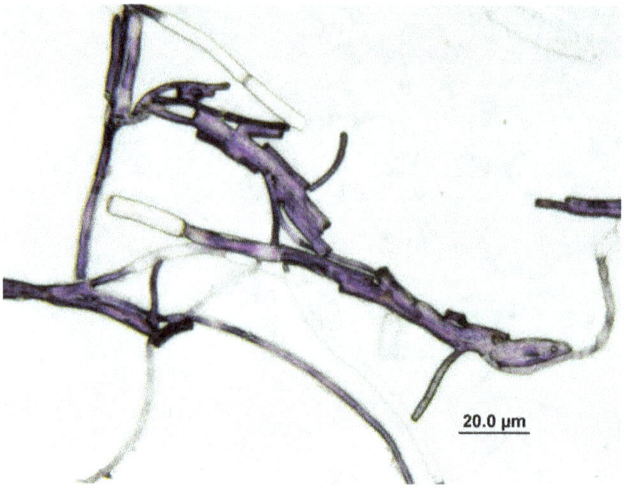

Fig. 4.7.20 *Geotricum capitatum.*

4.8 *Malassezia* spp

Colonial morphology: grows as a yeast at 30–37 °C and does not grow or grows poorly at 25 °C. *M. furfur* is a lipophilic yeast and grows on solid media supplemented with olive oil, or glycerol monooleate. *M. pachydermatis* will grow on non-supplemented media. Colonies are smooth to wrinkled, cream to yellowish-brown, glistening or dull, with the borders being entire or lobate.

Microscopic morphology: yeast-like cells (1.5–4.5 × 2.0–6.0 µm) that demonstrate collarettes. Cells have a unique appearance in that they are rounded to oval at one end, have a septum (collarette) and then a protruding bud occurs at the septum. These yeast-like cells are actually small phialides with small collarettes. Cells can be stained with safranin or calcofluor white. *M. furfur* does not identify with biochemical tests because of its lipid growth requirements, whereas *M. pachydermatis* often misidentifies as *Candida lipolytica*. Therefore, microscopic exam is the best means for identification of these microorganisms. Note that the collarette of *M. furfur* is more distinct and broader than that of *M. pachydermatis*.

Sometimes, early positive blood cultures may contain sufficient lipids to promote yeast growth, but as lipids become depleted, yeast cells die and autolyze. Therefore, Gram stains of blood cultures showing early yeast cells that are non-cultivable to solid media and, later, negative Gram stains are highly suggestive of *M. furfur*.

Malassezia furfur

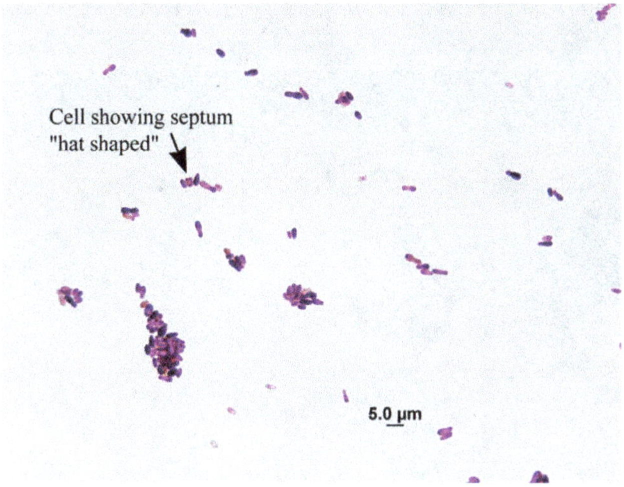

Fig. 4.8.1 Gram stain.

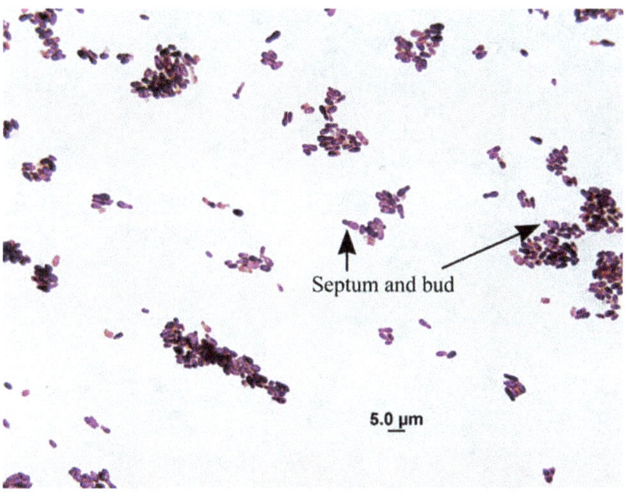

Fig. 4.8.2 Gram stain.

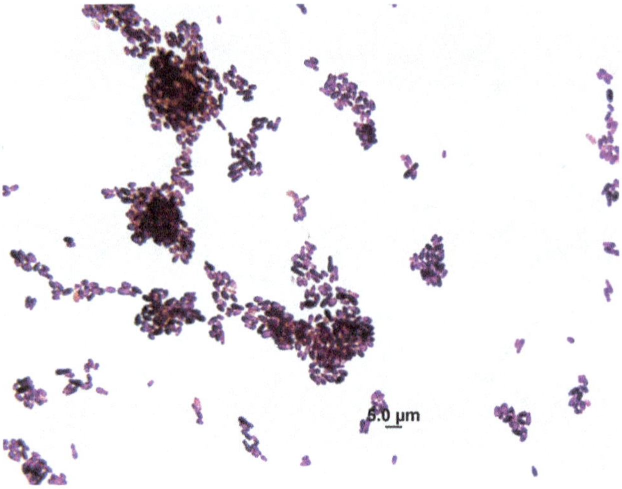

Fig. 4.8.3 Gram stain.

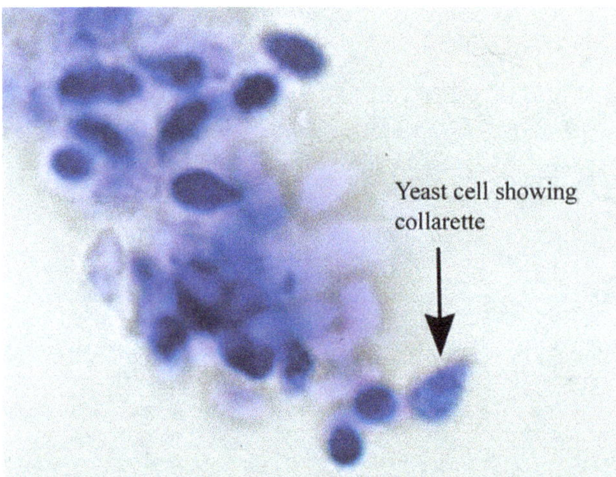

Fig. 4.8.4 LPCB stain.

4.9 *Trichosporon* spp

Colonial morphology: colonies on Sabouraud dextrose agar are yeast-like, white, yellow or cream, with a smooth, wrinkled or rough surface and an irregular edge. Splashes of white may occur over the surface of the colony.

Microscopic morphology: hyphae are hyaline and septate, producing arthroconidia. Pseudohyphae occur frequently and help differentiate this genus from other genera that form arthrospores. Numerous blastoconidia (3.5–7.0 × 3.5–14.0 µm) are present. The blastoconidia undergo budding and their presence on the hyphae separates this genus from *Geotrichum*. Note that blastoconidia are best seen on corn meal Tween 80 agar incubated at 25 °C. Species characterization is by physiological growth tests and biochemical assimilation tests. All species are urease positive.

The following table may help differentiate similar-appearing organisms.

Organism	Urease	Arthroconidia	Blastoconidia	Hyphae	Pseudohyphae	Sugar hydrolysis	Chlamydospores	Germ tube	Growth at 37 °C	Growth at 42 °C	CC resistant
Candida spp	Neg (±)	No	Yes	Few	Yes	Many	Yes alb/dubl	Yes alb/dubl	Yes	Varies	Most No
Trichosporon	Pos	Yes	Yes	Yes	Yes	Many	No	No	Yes	No (±)	Most No
Blastoschizomyces	Neg (±)	Yes (older)	Yes	Yes	Yes	Glu/Gal	No	No	Yes	Yes	Yes

alb, albicans; CC, chloramphenicol and cycloheximide; dubl, dubliniensis; Gal, galactose; Glu, glucose.

Trichosporon sp

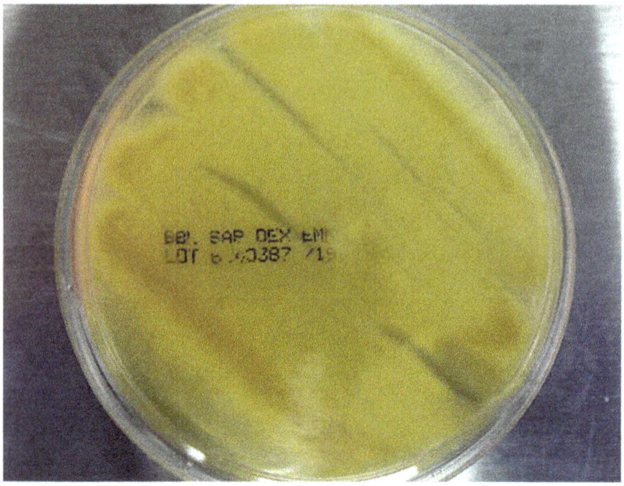

Fig. 4.9.1 *Trichosporon asahii* (bottom).

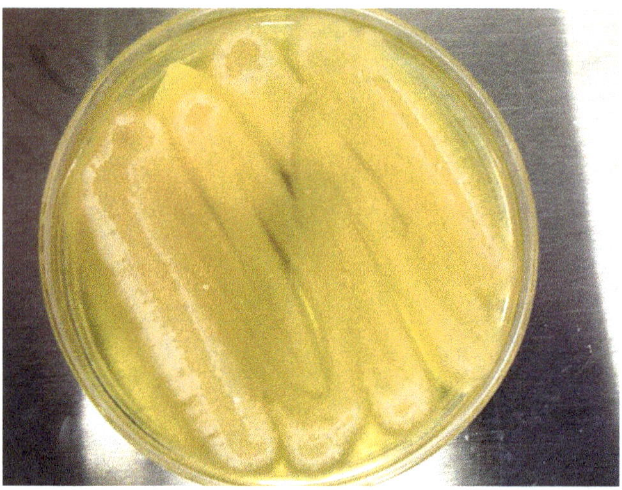

Fig. 4.9.2 *Trichosporon asahii* (top).

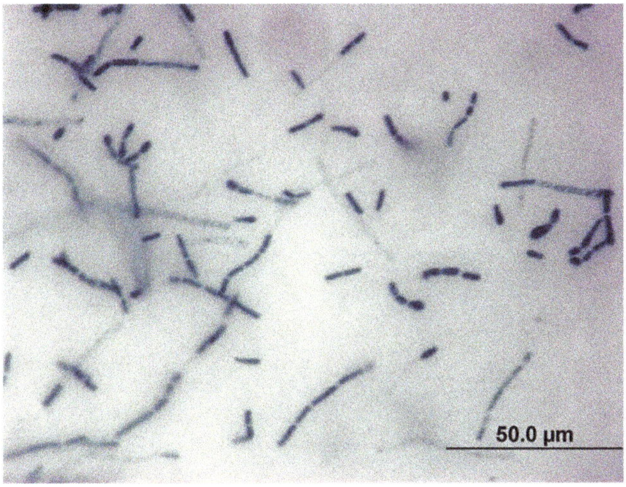

Fig. 4.9.3 *T. asahii.*

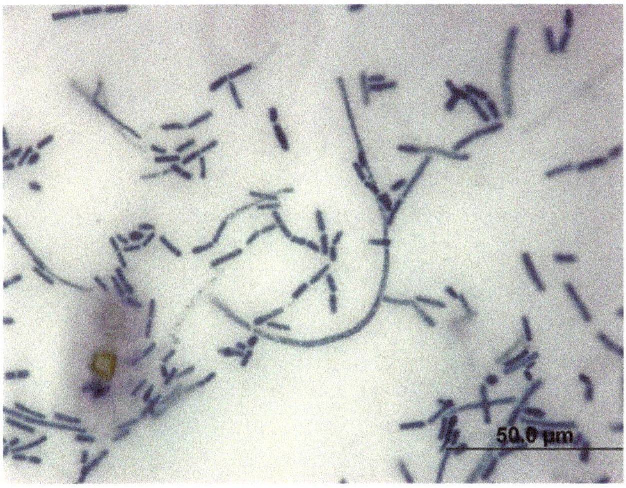

Fig. 4.9.4 *T. asahii.*

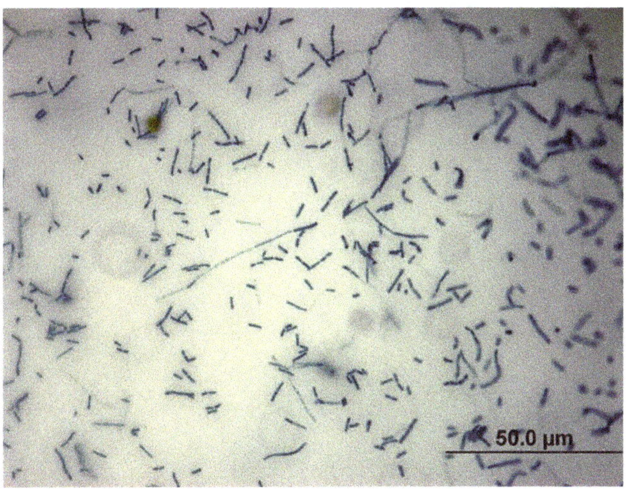

Fig. 4.9.5 *T. asahii.*

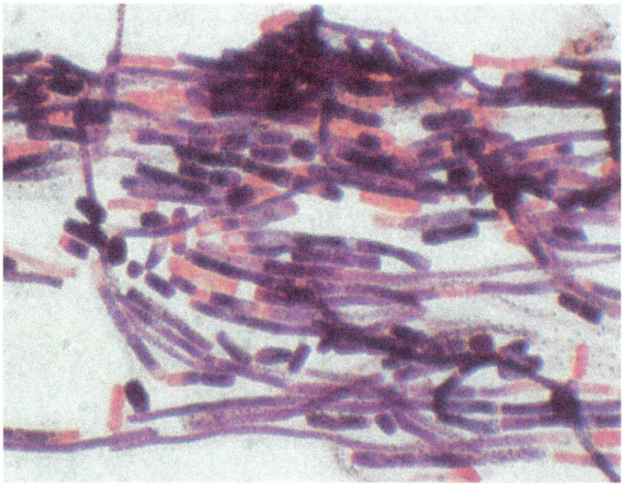

Fig. 4.9.6 *T. asahii* Gram stain.

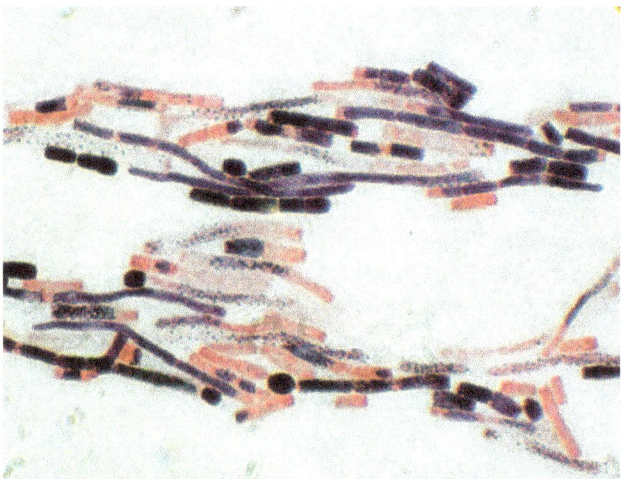

Fig. 4.9.7 *T. asahii* Gram stain.

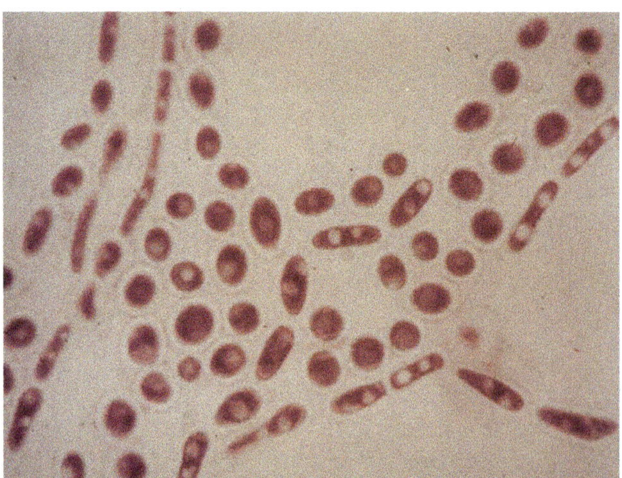

Fig. 4.9.8 *Trichosporon beigelii*.

5

Yeast (Thermally Dimorphic)

5.1 *Blastomyces dermatitidis*

Colonial morphology: *Blastomyces* is a thermally dimorphic microorganism, that grows as a mycelia form at 25–30 °C and a yeast at 35–37 °C. Fungal colonies begin as moist, tan colonies and mature to white fluffy, cottony colonies. Organisms grown on blood agar may stay moist, tan, and wrinkled. Aerial mycelia are produced. Some strains may turn from white to brown with age. Once colonies begin to grow, they mature in 14 days, but growth is slow and cultures suspected for blastomycosis should be held for 6–8 weeks. It is common that cultures remain negative for weeks before colonies appear. This organism is a BSL-3 pathogen and should be grown and handled only in laboratories equipped with BSL-3 or -4 containment facilities. Slide cultures should not be performed!

Microscopic morphology: hyphae are septate and consist of two forms: hyaline-thin (1–2 μm) and hyaline-wide (5–8 μm). Septations are not prominent and may be difficult to see. Conidia are produced from the thin hyphae in three ways.

- The most commonly seen are pear-shaped conidia that form at the end of hyphae and swell the hyphae at the point of attachment. Some pear-shaped conidia may directly arise along the hyphae with no conidiophore. There are usually visible septa at the base of the conidia and as the conidia separate from the hyphae, they become truncated with flat ends. These closely resemble the conidia of *Chrysosporium*.
- The least common conidia are the lollipop type, that appear as tiny lollipops along the hyphae. These are round (2–6 μm) and are produced perpendicular to the hyphal axis on short, thin conidiophores. Some may chain together, giving the appearance of a strand of pearls. Some cultures may have both types of conidia present.
- Older cultures may contain chlamydoconidia.

Final identification of this fungus is by nucleic acid probes and not by growth and morphology studies. Thermal conversion of isolates to the yeast phase can aid in the identification of this fungus when nucleic acid probes are unavailable. Yeasts are 8–15 μm in diameter and demonstrate broad-base budding. Recognition of large, round, broad-base, budding yeast in tissue samples can be diagnostic for blastomycosis.

Atlas of Clinically Important Fungi, First Edition. Carmen V. Sciortino.

Blastomyces dermatitidis

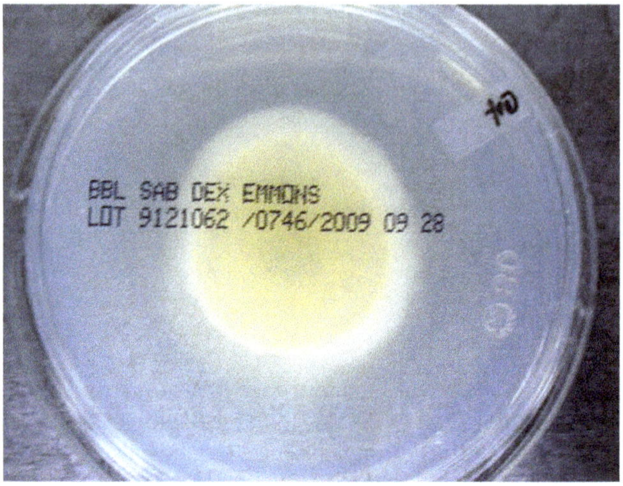

Fig. 5.1.1 *Blastomyces dermititidis* (bottom).

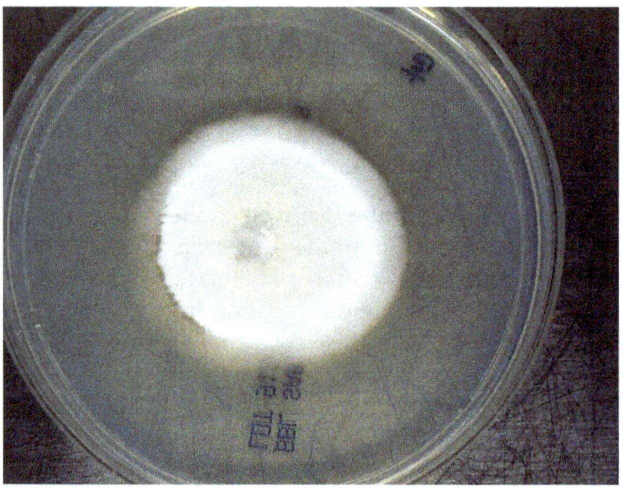

Fig. 5.1.2 *Blastomyces dermititidis*.

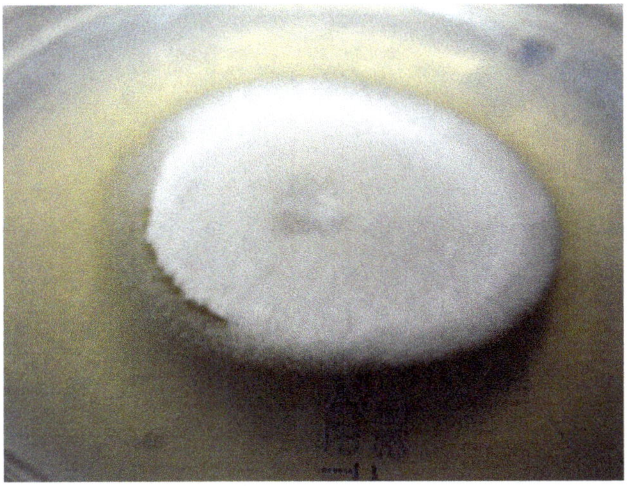

Fig. 5.1.3 *Blastomyces dermititidis* (side).

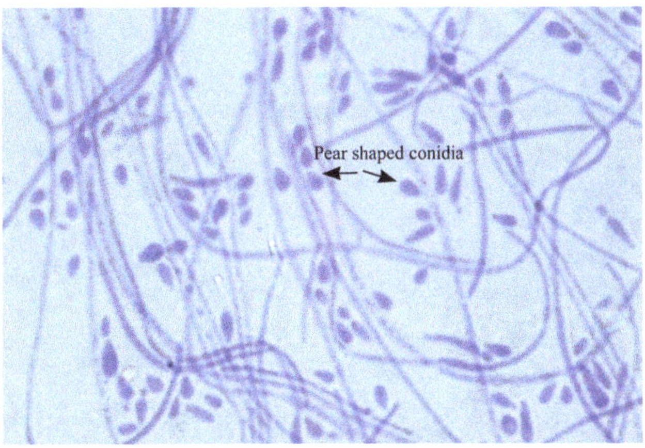

Fig. 5.1.4 Commonly seen pear shaped conidia.

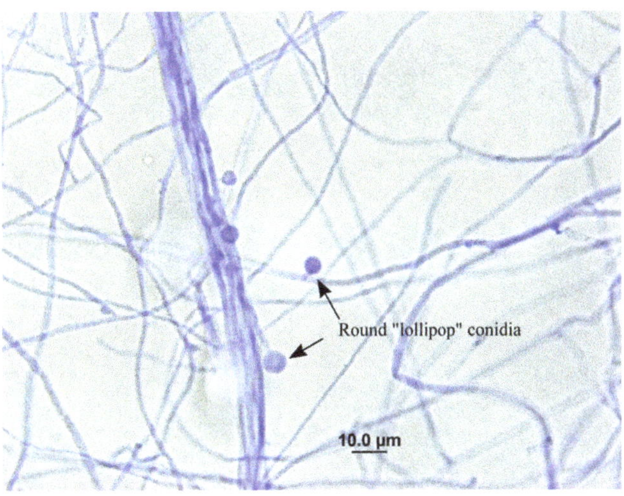

Fig. 5.1.5 Rare round conidia.

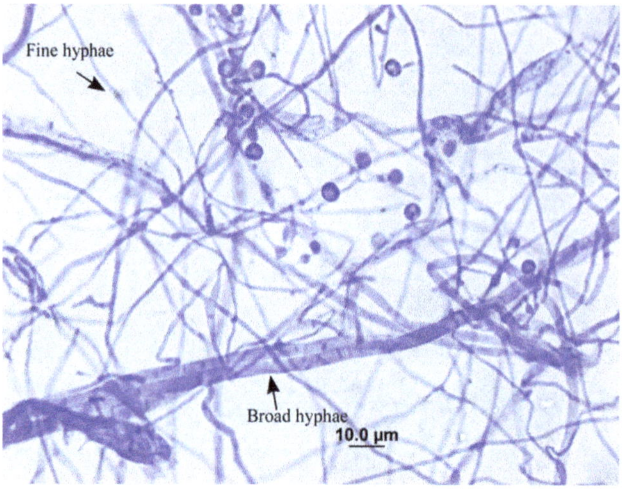

Fig. 5.1.6 Both broad and fine hyphae are present.

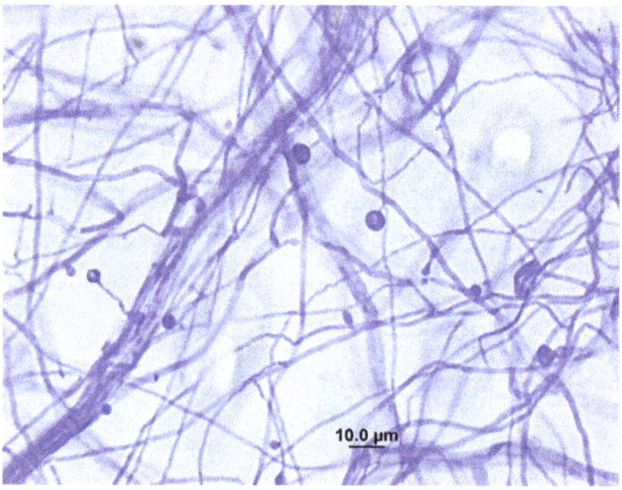

Fig. 5.1.7 Conidiophores vary in width and length.

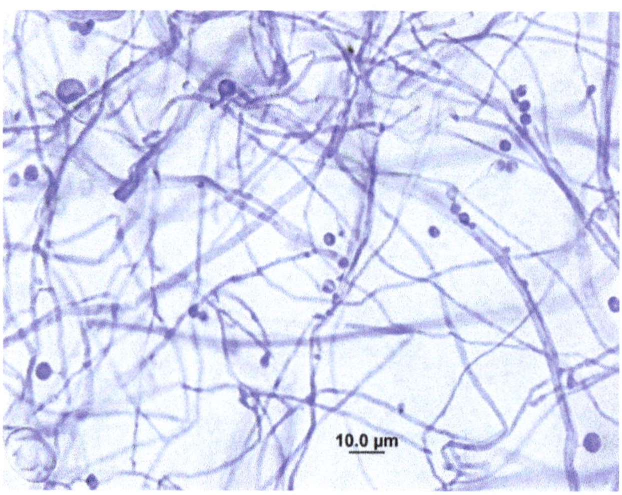

Fig. 5.1.8 Some conidia may seem to form chains.

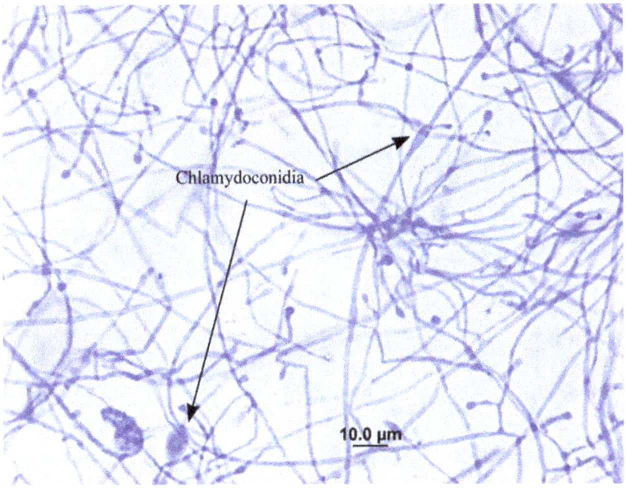

Fig. 5.1.9 Intercalary chlamydoconidia.

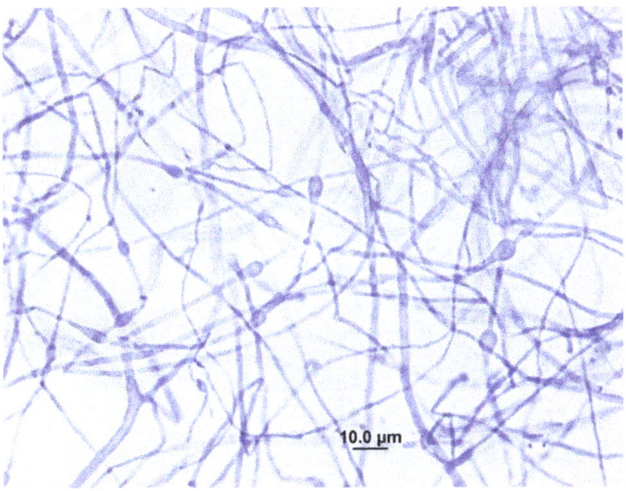

Fig. 5.1.10 Abundant chlamydoconidia.

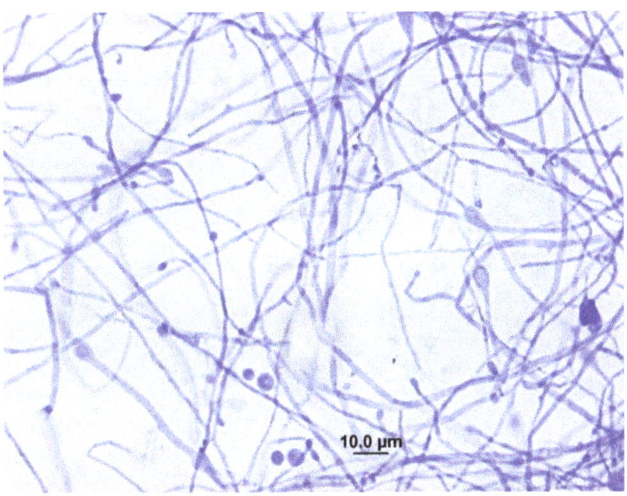

Fig. 5.1.11 Showing a variety of conidia.

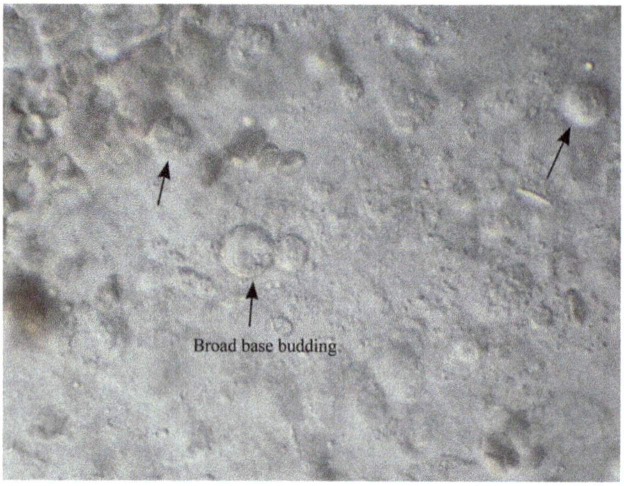

Fig. 5.1.12 Yeast, Nomarski optics.

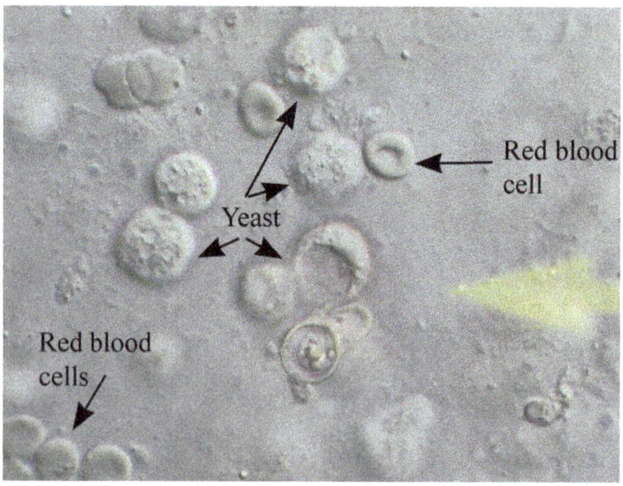

Fig. 5.1.13 Yeast, Nomarski optics.

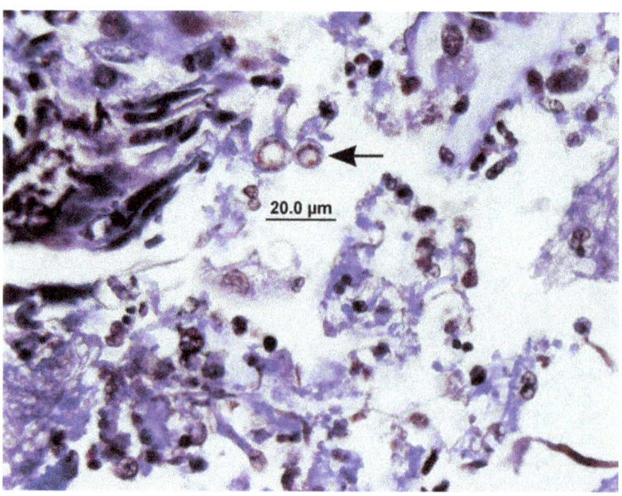

Fig. 5.1.14 In tissue, trichrome stain.

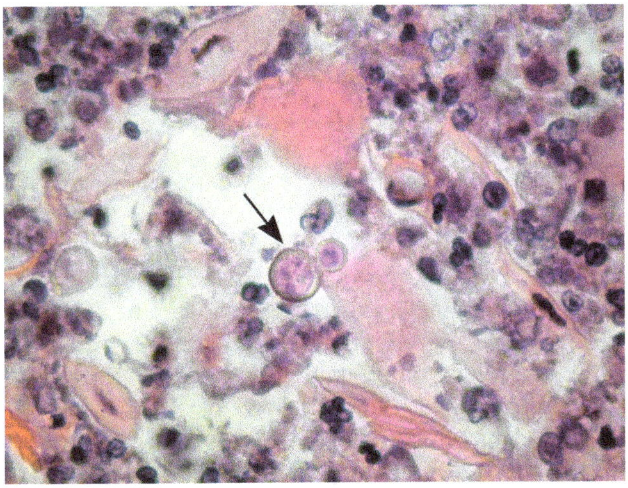

Fig. 5.1.15 In tissue, H&E stain.

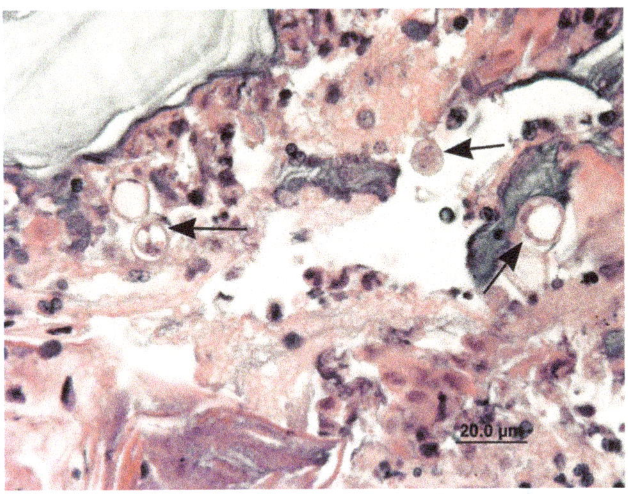

Fig. 5.1.16 In tissue, H&E stain.

5.2 *Coccidioides immitis*

Colonial morphology: cultures of *C. immitis* are highly infectious and should only be handled in a BSL-3 laboratory under a biosafety hood. Extra caution must be maintained for lacto-phenol cotton blue preparations. Slide cultures should not be done. Coccidiomycosis is spread by the respiratory route and a single arthroconidium is infectious.

Colonies of C. *immitis* begin as white suede-like colonies that may become cottony. Some cultures may develop first as tan and moist, resembling bacterial colonies, but later become white and fluffy. As colonies mature, they may become gray-white, with a tan-to-brown reverse.

Microscopic morphology: hyphae are hyaline, fine, delicate, and tiny. They develop into chains of arthroconidia. Chains of arthroconidia (2.5–4 × 3–6 μm) alternate with definitive spaces between the barrel-shaped arthroconidia. The arthroconidia appear swollen, or larger than the hyphae. Chains of arthroconidia terminate in a single conidium that is oval to round and thick walled. It is often difficult to see the arthroconidia within these fine hyphae, so phase-contrast microscopy can be a helpful tool. When arthroconidia separate from hyphae, they often carry pieces of the hyphae with them, giving them a flanged appearance. Laboratory technicians often refer to these as arthroconidia with "ears.".

Coccidioides is dimorphic but not so in culture. In the host, it grows at 37 °C where it develops into round spherules as small as 5 μm and as large as 120 μm. Spherules are not produced *in vitro*. Spherules stain with H&E. Inside the spherules are round endospores 2–5 μm in diameter. Endospores do not bud, but can be mistaken for yeast cells of *Blastomyces dermatitidis* because they can bunch together. Mature spherules do not readily stain with GMS or PAS stains. Some have reported hyphal growth in tissue, but this is extremely rare.

Coccidioides immitis

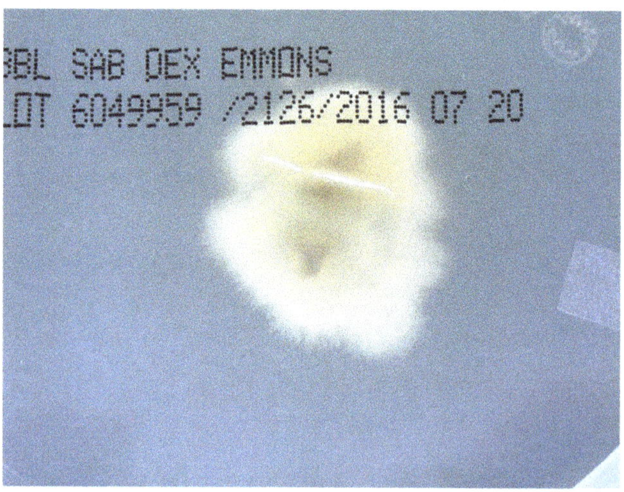

Fig. 5.2.1 *Coccidioides immitis*, Sab DX (bottom).

Fig. 5.2.2 *Coccidioides immitis*, Sab DX (top).

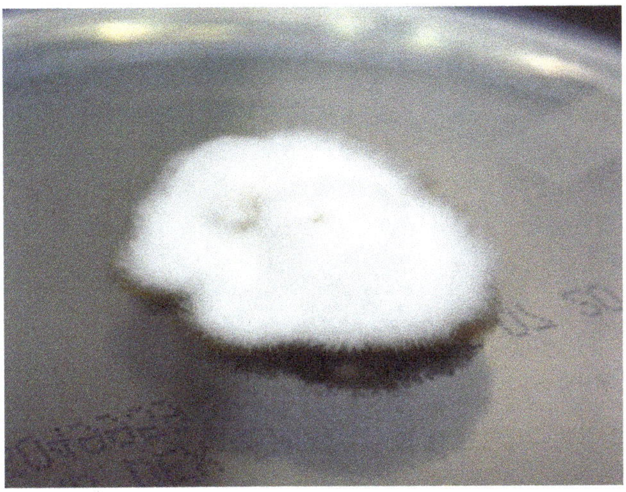

Fig. 5.2.3 *Coccidioides immitis*, Sab DX (side).

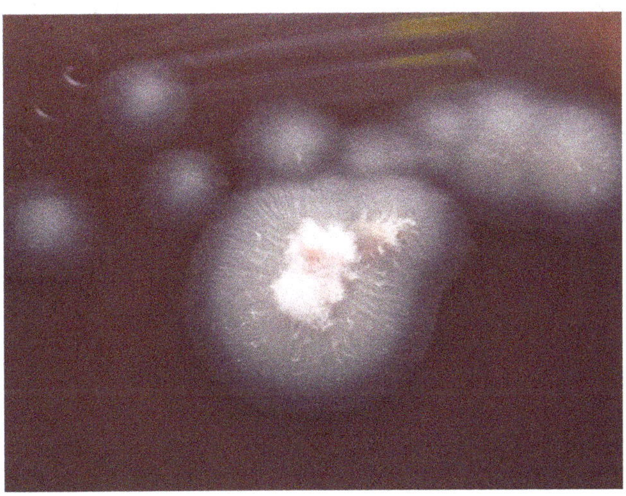

Fig. 5.2.4 *Coccidioides immitis*, BHI BA (top).

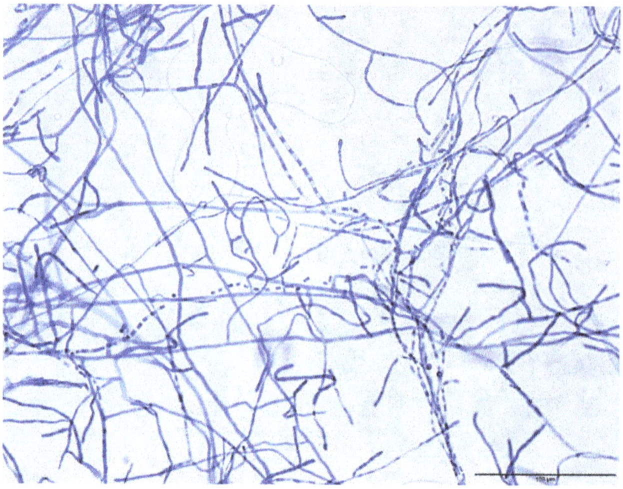

Fig. 5.2.5 *C. immitis*, 400×.

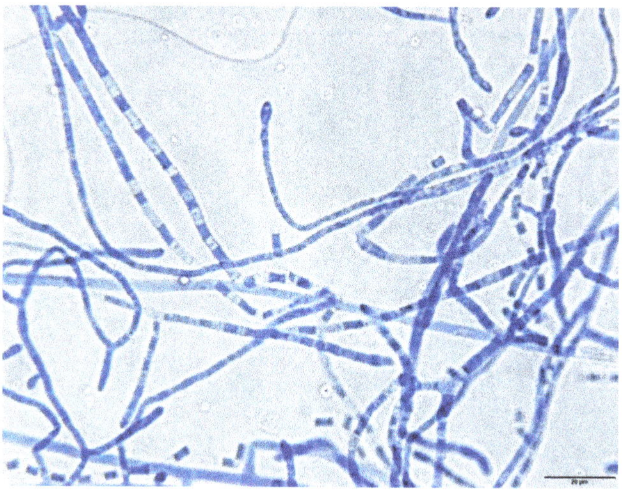

Fig. 5.2.6 *C. immitis*, 1000×.

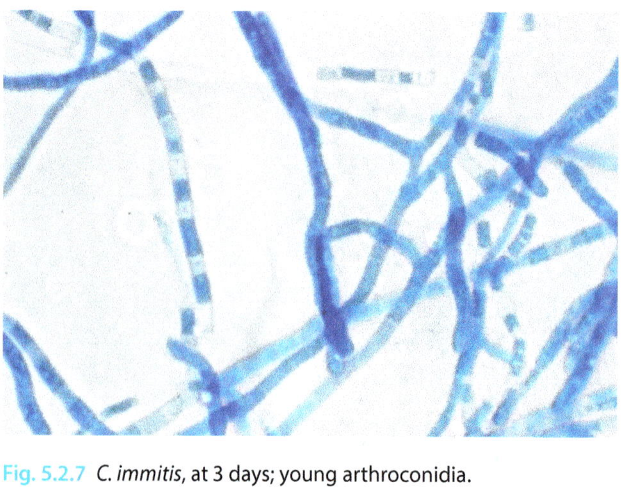

Fig. 5.2.7 *C. immitis*, at 3 days; young arthroconidia.

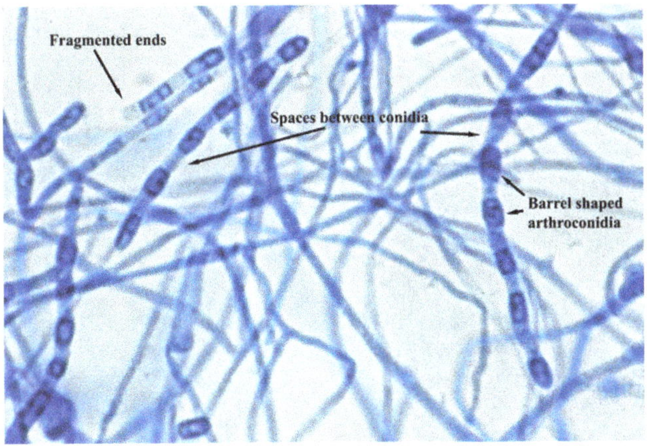

Fig. 5.2.8 Seven days with mature arthroconidia.

5.3 *Histoplasma capsulatum*

Colonial morphology: grown on blood agar at 30 °C, mycelia colonies appear yeast-like, cream to white, smooth, and later become wrinkled and folded, with a yellow center. Grown on blood agar at 37 °C, yeast colonies form which are small (5–15 mm), and cream to yellow. Grown at 25–30 °C on Sabouraud dextrose agar, colonies grow slowly, become white and cottony and may spread over the surface of the agar. With age, some strains may have a brownish tint in the center. The reverse is tan. Grown at 25-30 °C on potato dextrose agar, colonies are tan-brown, entire, small (1–2.5 cm in diameter), and develop a white tuft in the center of the colony. At first, the texture is suede-like but with age, colonies become fluffy. The reverse is gray-brown. This organism is a BSL-3 pathogen and should be grown and handled only in laboratories equipped with BSL-3 or -4 containment facilities. Slide cultures should not be performed!

Microscopic morphology: cultures grown at 25–30 °C show septate, hyaline hyphae that are fine and indistinct. Characteristic large, tuberculate macroconidia (8–14 µm in diameter; macroaleurioconidia) are produced that are rounded, single-celled and warty, formed on short, hyaline, undifferentiated conidiophores. Cultures may produce microconidia that are small (2–4 µm in diameter), round to pyriform and occur on short branches or directly on the sides of the hyphae. Cultures grown at 37 °C produce budding yeasts. Yeasts are small, 2–3 × 4–5 µm, round to tear-shaped and stain Gram negative from culture.

In tissue, silver stains are best for revealing small tear-shaped yeasts. Yeasts (5–6) are often seen inside phagocytic cells. *Histoplasma capsulatum* var. *duboisii* may also cause human infection, but the yeast cells are larger, 3–4 × 6–8 µm.

Histoplasma capsulatum

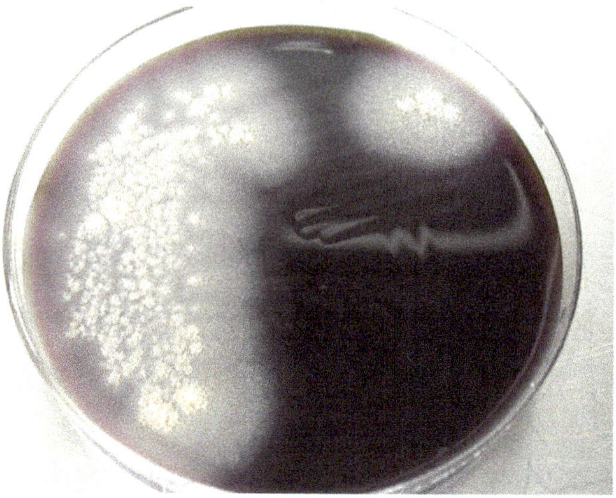

Fig. 5.3.1 *Histoplasma*, blood agar, mature culture (top).

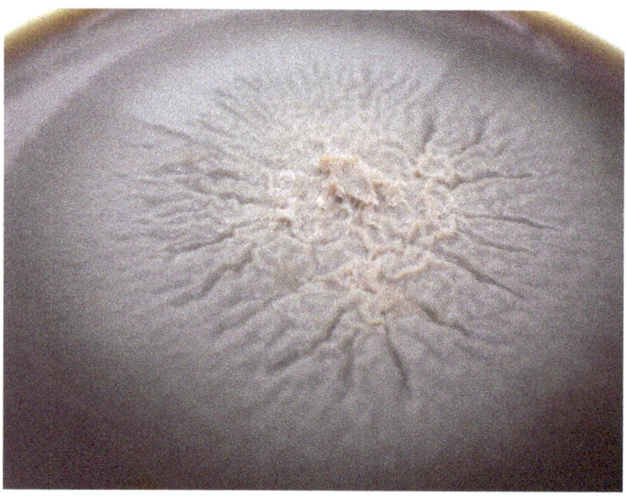

Fig. 5.3.2 *Histoplasma*, blood agar, mature colony (side).

Fig. 5.3.3 *Histoplasma*, LPCB stain from blood agar.

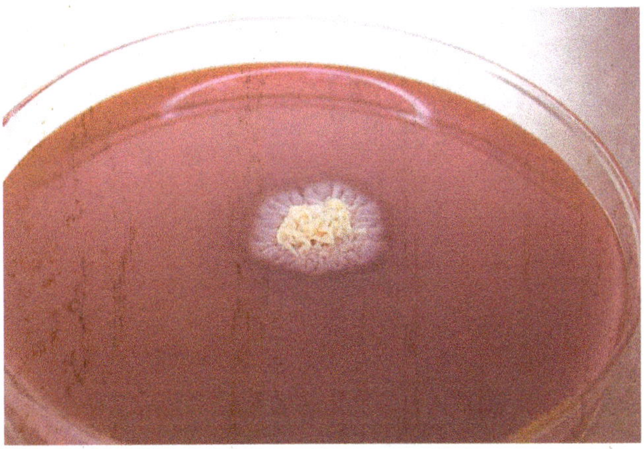

Fig. 5.3.4 *Histoplasma*, blood agar, young colony (side).

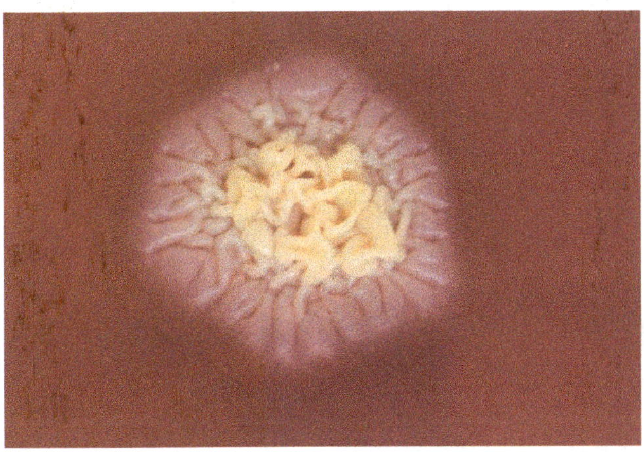

Fig. 5.3.5 *Histoplasma*, blood agar, young colony (top).

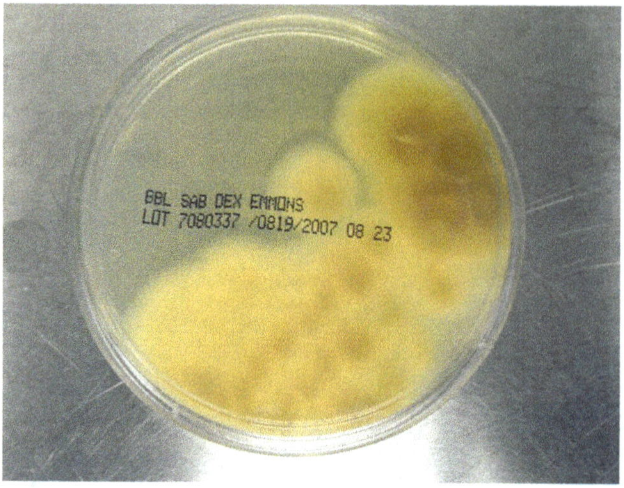

Fig. 5.3.6 *Histoplasma*, SABDX (bottom).

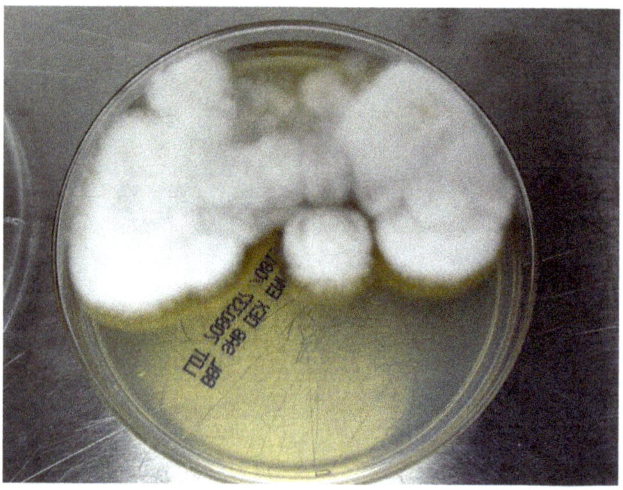

Fig. 5.3.7 *Histoplasma*, SABDX (top).

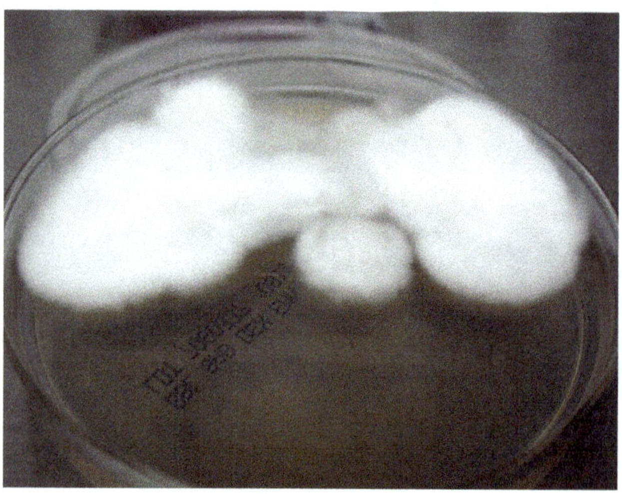

Fig. 5.3.8 *Histoplasma*, SABDX (side).

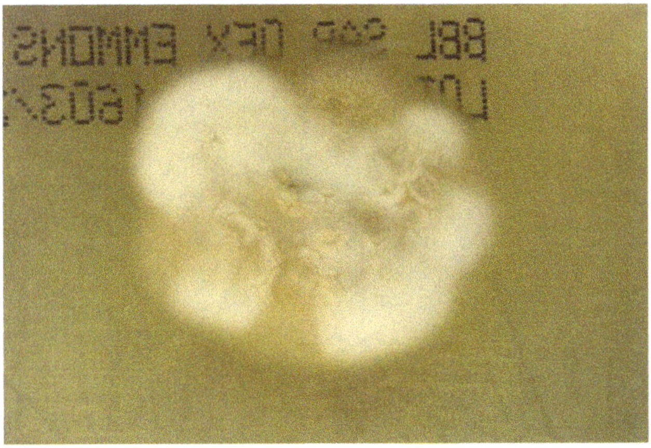

Fig. 5.3.9 *Histoplasma*, SABDX, young colony (top).

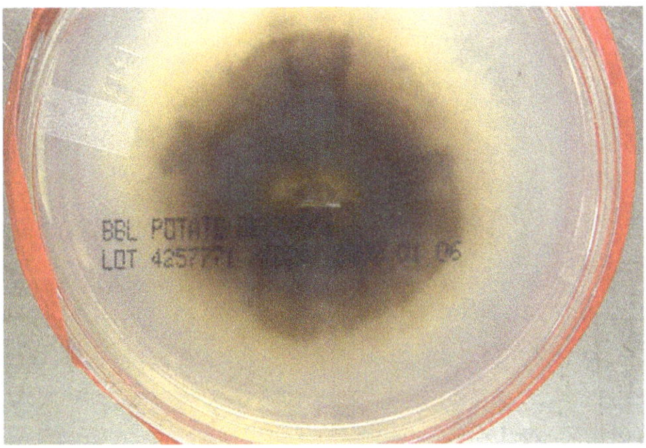

Fig. 5.3.10 *Histoplasma*, PDA, (bottom).

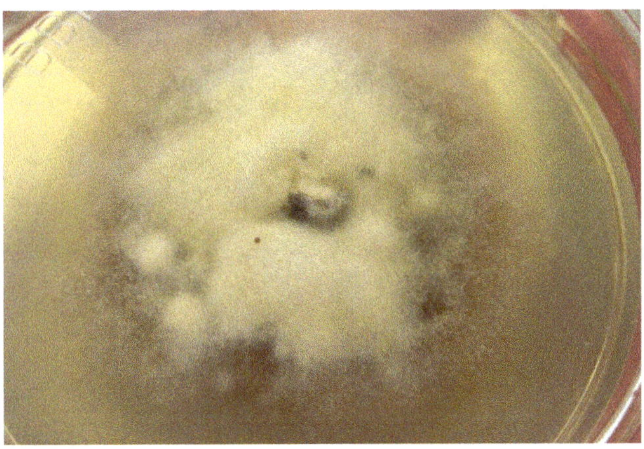

Fig. 5.3.11 *Histoplasma*, PDA (top).

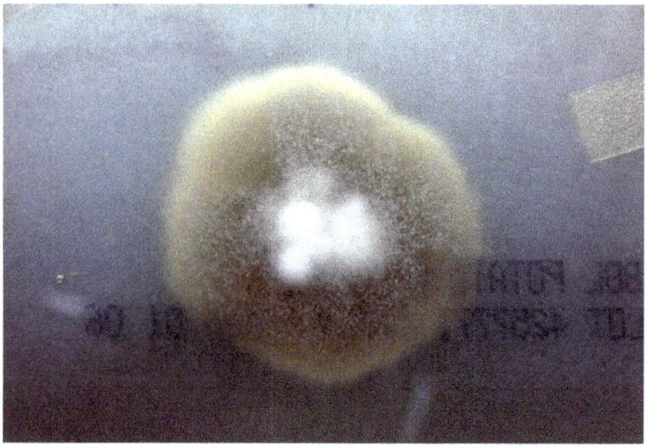

Fig. 5.3.12 *Histoplasma*, PDA, young colony (top).

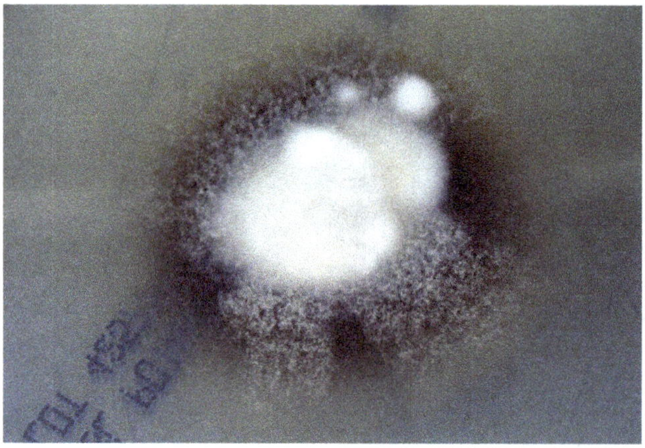

Fig. 5.3.13 *Histoplasma*, PDA, young colony (side).

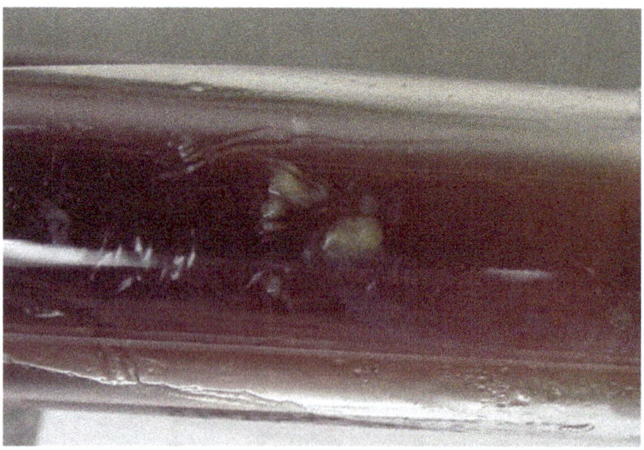

Fig. 5.3.14 Yeast phase.

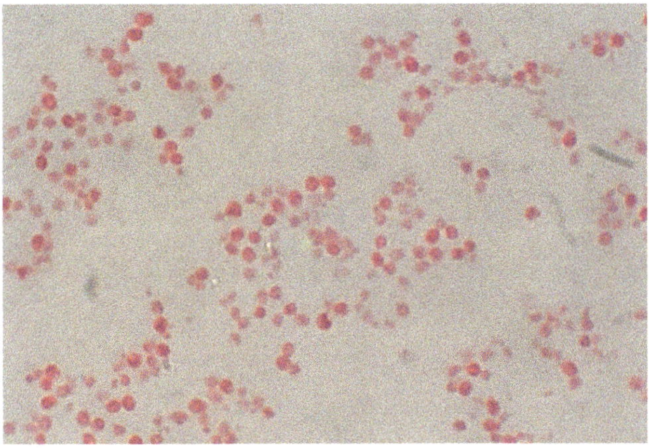

Fig. 5.3.15 Yeast Gram stain.

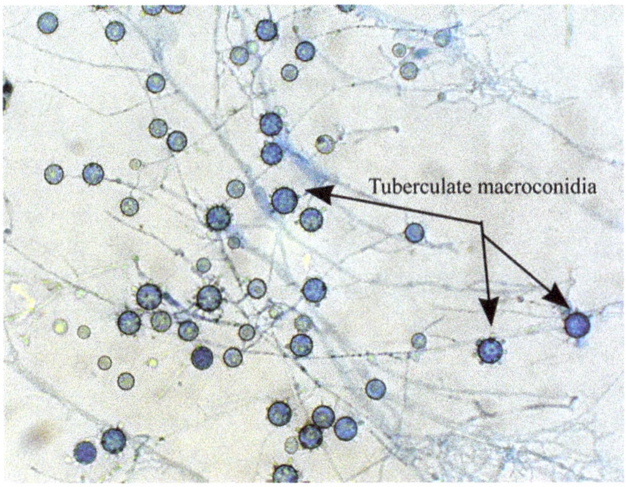

Fig. 5.3.16 Macroconidia.

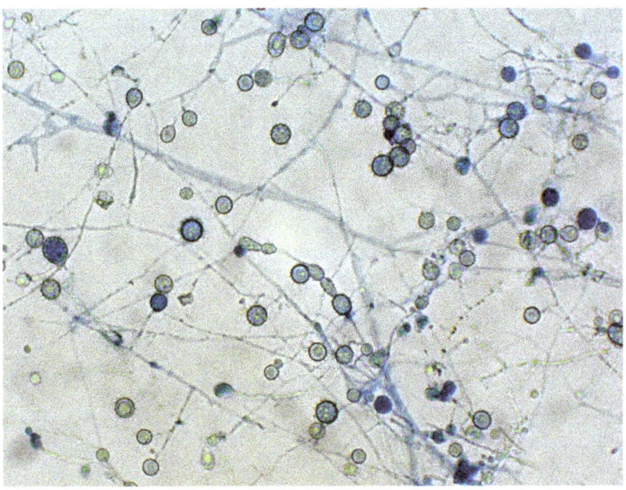

Fig. 5.3.17 Macroconidia vary in shape and size.

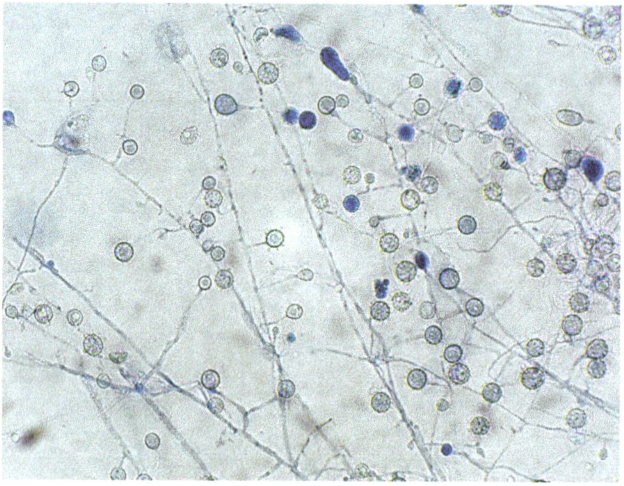

Fig. 5.3.18 Macroconidia produced on conidiophores at right angles from hyphae.

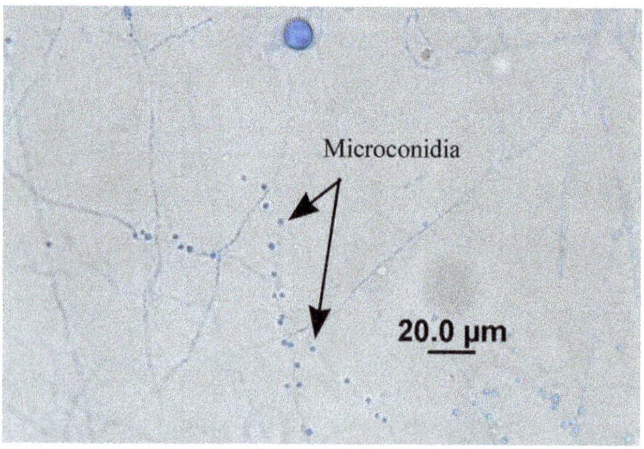

Fig. 5.3.19 Microconidia.

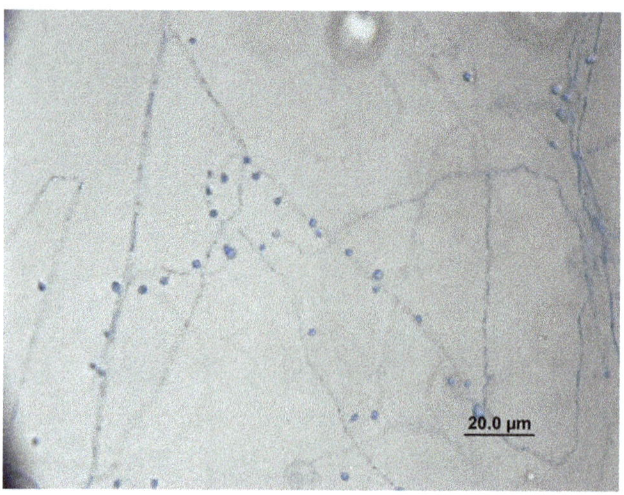

Fig. 5.3.20 Microconidia along hyphae.

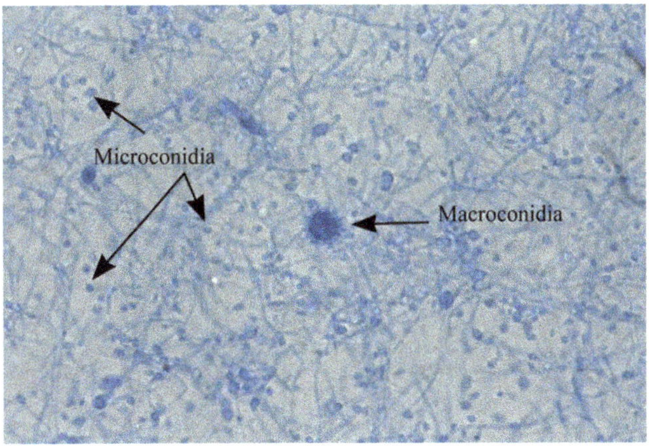

Fig. 5.3.21 Microconidia and tuberculate macroconidia.

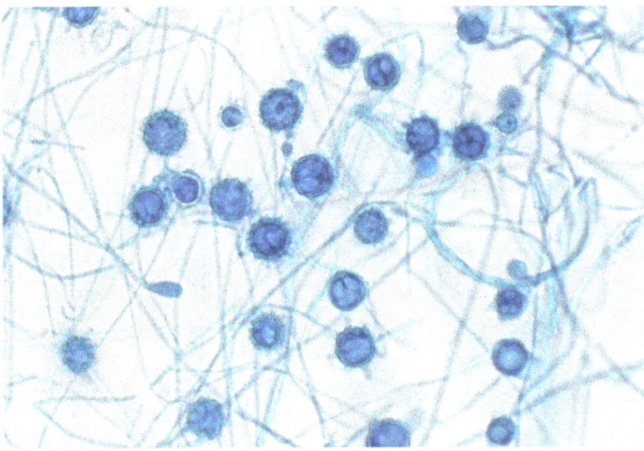

Fig. 5.3.22 Macroconidia, 400×.

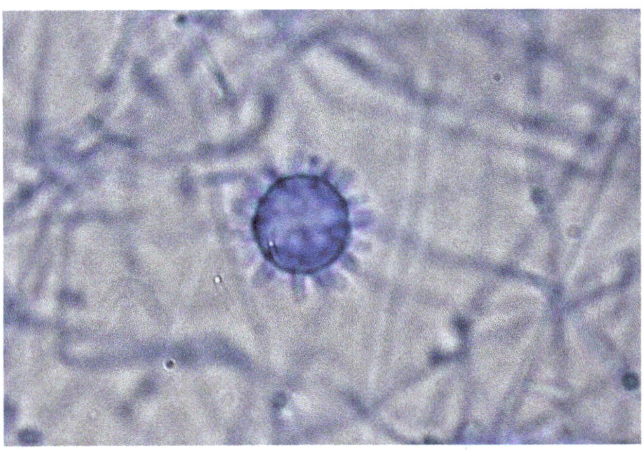

Fig. 5.3.23 Tuberculate macroconidium, 1000×.

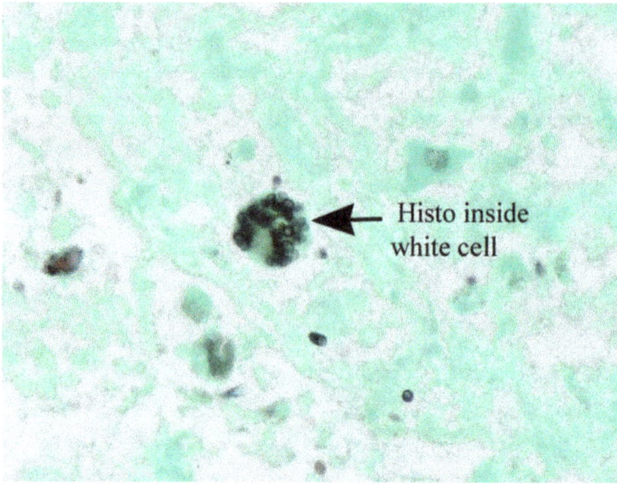

Fig. 5.3.24 *H. capsulatum* in lung tissue, silver stain, 1000×.

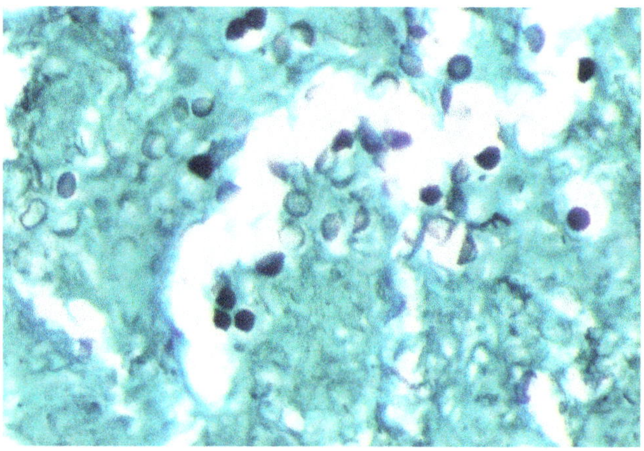

Fig. 5.3.25 *H. capsulatum* in lung tissue, silver stain, 1000×.

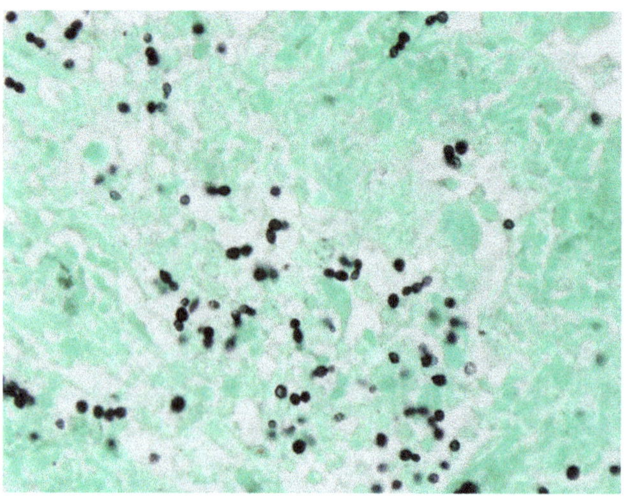

Fig. 5.3.26 *H. capsulatum* in lung tissue, silver stain, 1000×.

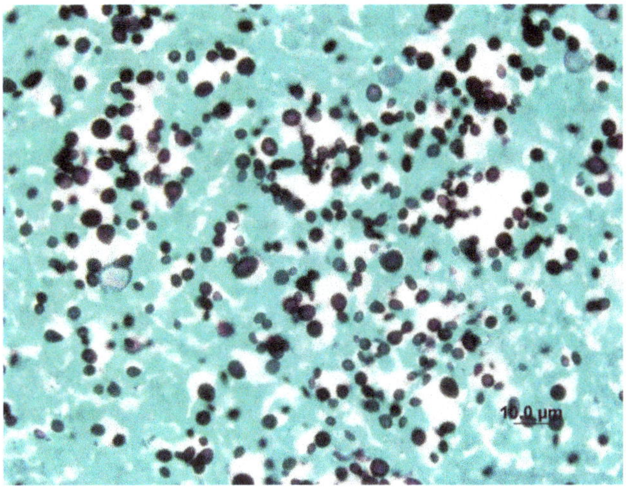

Fig. 5.3.27 *H. capsulatum* var *dubosii* showing large cells, silver stain, 1000×.

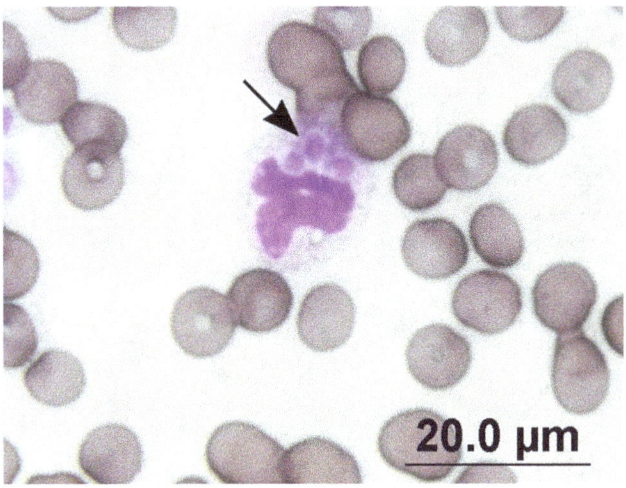

Fig. 5.3.28 Peripheral smear with *Histoplasma* inside leukocyte.

5.4 *Sporothrix schenckii*

Colonial morphology: this organism is thermally dimorphic, exhibiting a mold phase at 25–30 °C and a yeast phase at 37 °C. Colonial morphology is variable depending upon growth medium and organism strain. On Sabouraud dextrose agar at 25–30 °C, colonies are moist and smooth, developing a folded surface with a starfish-like wrinkling pattern. Colonies may exhibit variable colors of black and white, or white, gray and brown, intermingled. The reverse is cream to white. Subcultures may grow only black colonies with a dark reverse. On blood agar at 25–30 °C, colonies are first white and cottony, becoming wrinkled and cracked with a black border. Subcultures on blood agar may become dark, wrinkled, and leathery with age. On BHI blood agar at 37 °C, colonies grow as round or elongate yeasts exhibiting a white-to-tan color.

Microscopic morphology: hyphae are hyaline, fine (1–2 μm in diameter) and septate. Conidiophores are produced perpendicular to the hyphae and are mostly single, vertical, and narrowing towards the top. They are spaced individually along the hyphae at broad intervals 10–35 μm apart. Conidia sometimes form on thread-like denticles displaying sympodial production but most often conidia cluster into rosettes, giving a daisy-like appearance. Hyaline, non-clavate, smooth conidia (3–6 × 2–3 μm) are abundant. Infrequently, single, dark, thick-walled, non-clavate, oval to elongated conidia may be produced along the hyphae. Elongated conidia may chain together with parallel walls adjacent (not end to end), much like a picket fence. The morphological variability of this fungus makes it difficult to match with photographs. The key most common morphological feature is the appearance of long-stemmed daisies sprouting along a hyphal path.

Sporothrix shenckii

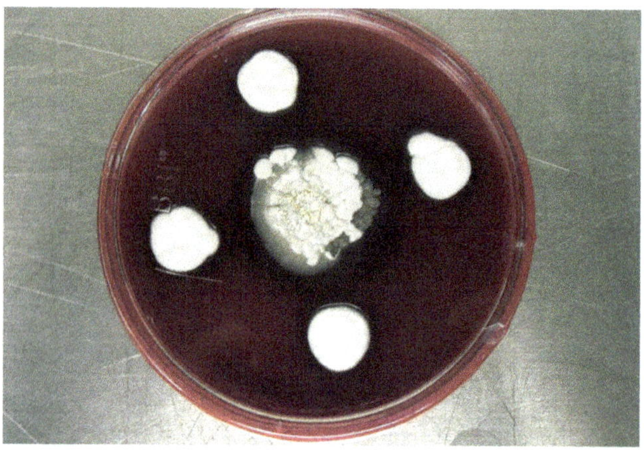

Fig. 5.4.1 *Sporothrix schenckii* on blood agar (top).

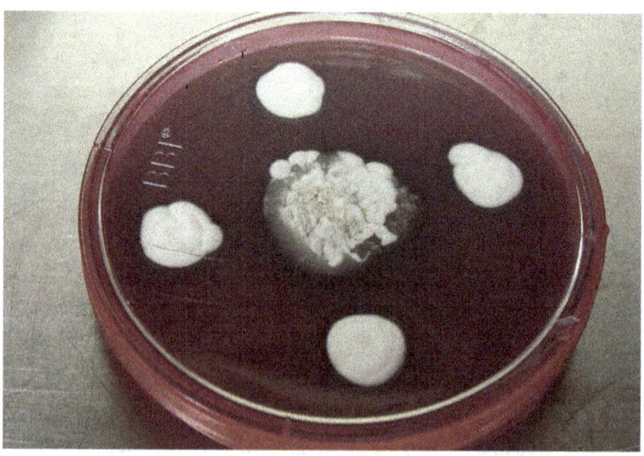

Fig. 5.4.2 *S. schenckii* (side).

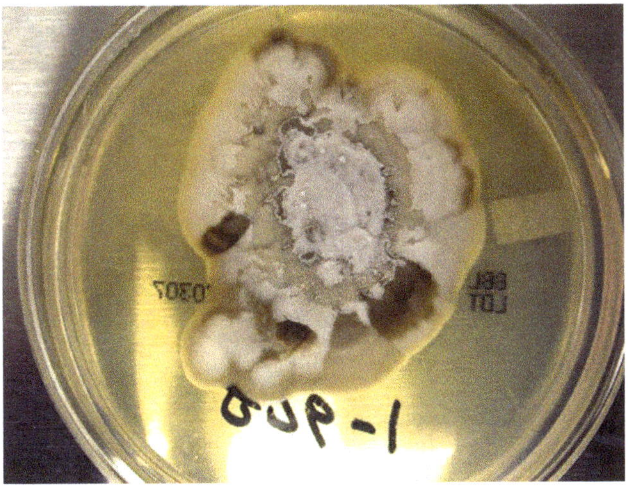

Fig. 5.4.3 *S. schenckii* on SABDX (top).

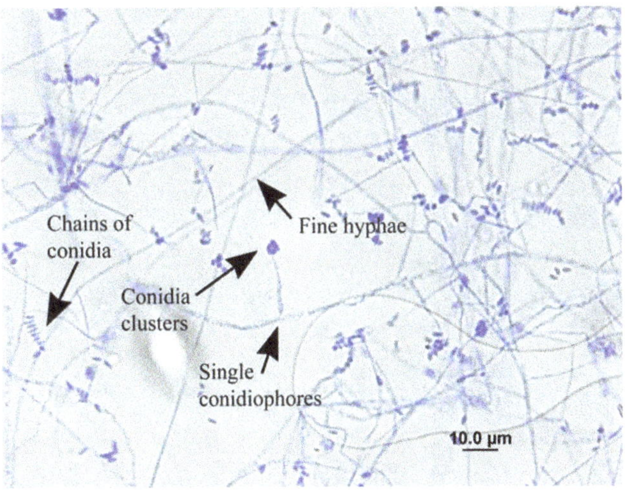

Fig. 5.4.4 Palisading chains of conidia at tips of conidiophores.

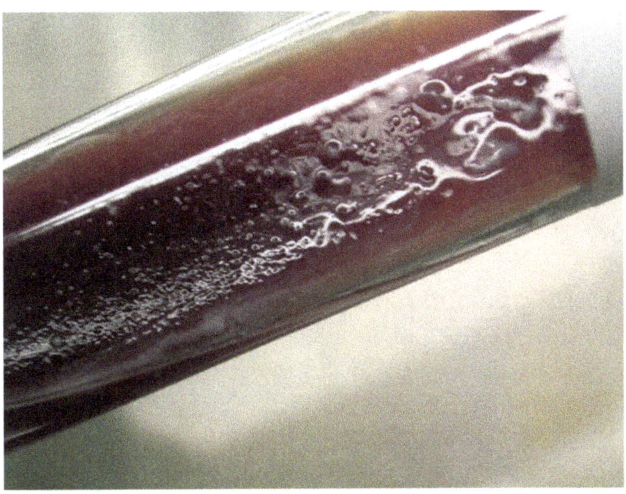

Fig. 5.4.5 *S. schenckii* on BHI BA, yeast at 37 °C.

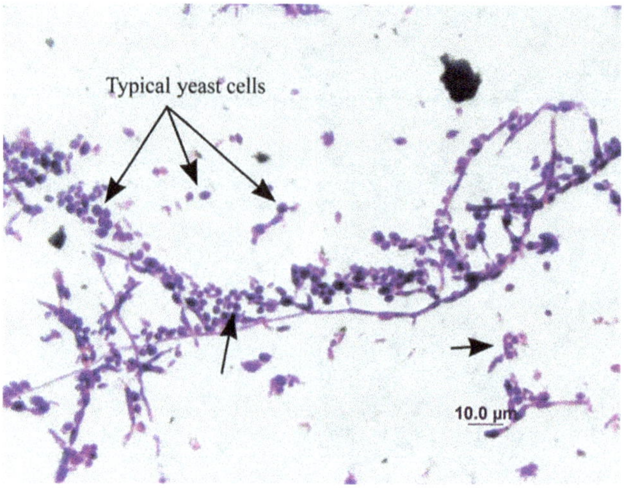

Fig. 5.4.6 Gram stain during yeast conversion.

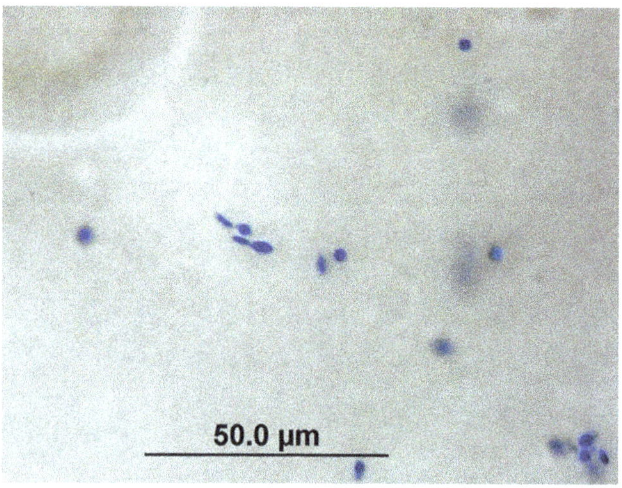

Fig. 5.4.7 Yeast phase.

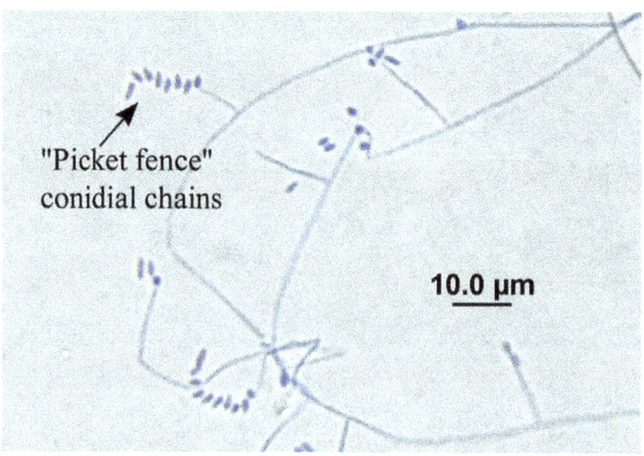

Fig. 5.4.8 Palisading conidia show "picket fence" appearance.

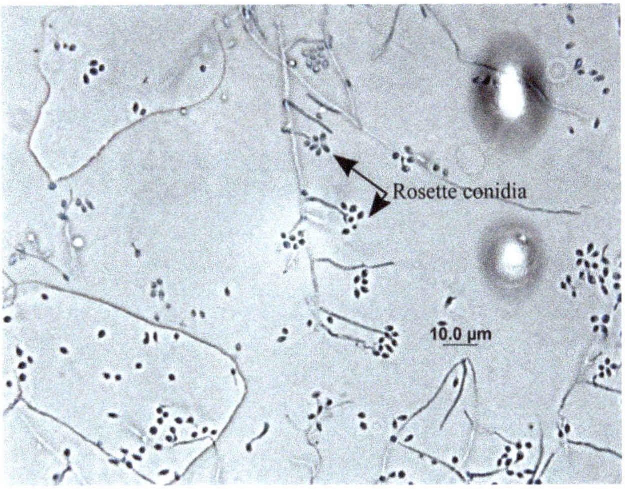

Fig. 5.4.9 Conidia form rosettes at the tip of slender denticles.

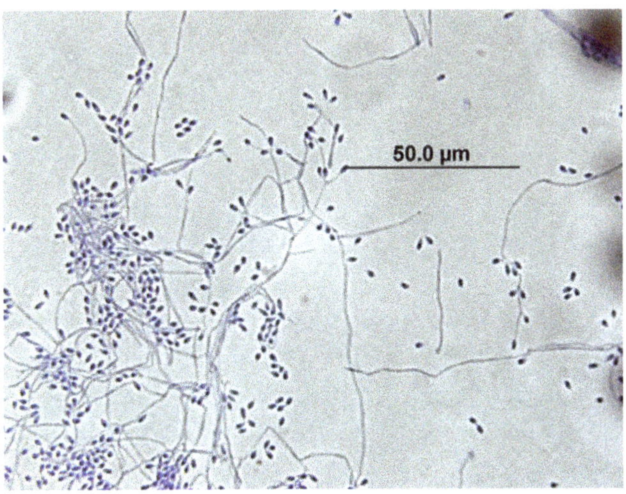

Fig. 5.4.10 Abundant conidia.

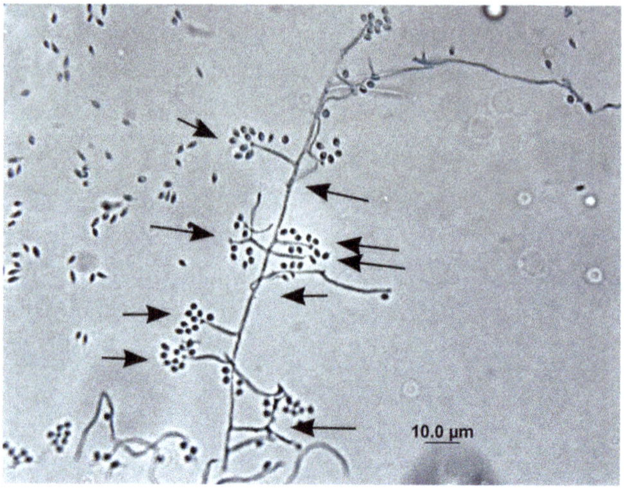

Fig. 5.4.11 Denticle arrangement along hyphae.

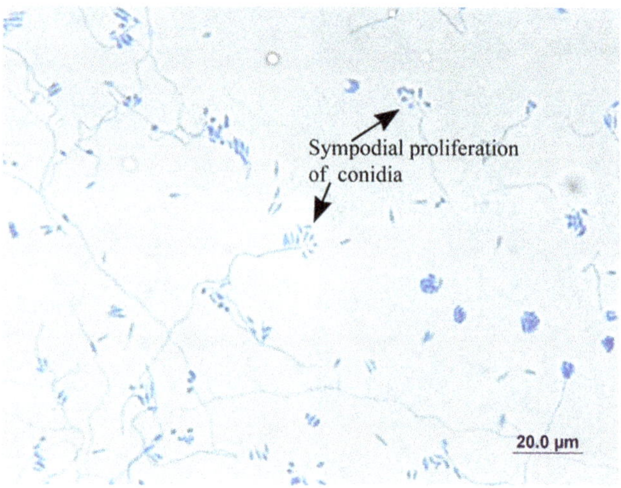

Fig. 5.4.12 *Rhinocladiella* type conidia formation (center).

6

Zygomycetes (Aseptate)

6.1 *Lictheimia (Absidia) corymbifera*

Colonial morphology: white surface, fluffy, and may become light gray with age. Reverse is white.

Microscopic morphology: mycelia are non-septate, hyaline, wide (6–16 μm), and ribbon-like. Rhizoids and aerial mycelia (stolons) are produced. Sporangiophores (up to 400 μm long) are simple but may be branched and arise from the stolons. They occur singly or may be seen in clusters of three to seven. Sporangia are 10–40 μm in diameter and have a pyriform shape. There is a conical columella with a semi-circular top extending beyond the apophysis. At the top of the columella, a small projection is often seen. As the sporangium separates from the columella, a collarette is often seen. Sporangiospores are hyaline, vary in size and shape, and are smooth walled. Zygospores may form between mating strains. These are round to somewhat flattened, 50–80 μm in diameter, with rough walls and several divisions. Giant cells are produced by some strains. They are large, pleomorphic and may clump together in a mass. Giant cells are forms of chlamydoconidia.

Lictheimia differs from *Apophysomyces elegans* in having a conical apophysis rather than a bell-shaped one. *Apophysomyces* often has a foot cell at the base of the columella. *Lictheimia* develops rhizoids at the base of the columella near the node. *Rhizopus* has a sporangiophore arising from the nodes, sometimes appearing as foot cells that are found opposite the rhizoids. *Rhizomucor* develops rhizoids along the stolon and small rhizoids can often be seen at the base of the sporangiophore. *L. corymbifera* grows faster at 37 °C than at 25 °C, and will also grow within 48–52 °C, whereas other species do not.

Atlas of Clinically Important Fungi, First Edition. Carmen V. Sciortino.

Absidia (Lictheimia) corymbifera

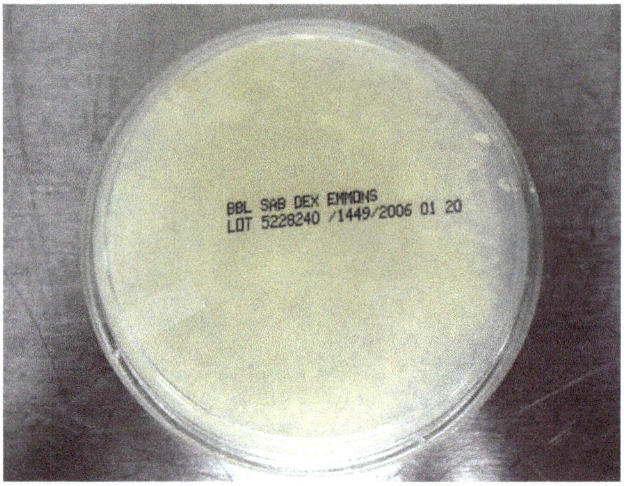

Fig. 6.1.1 *Lictheimia* (bottom).

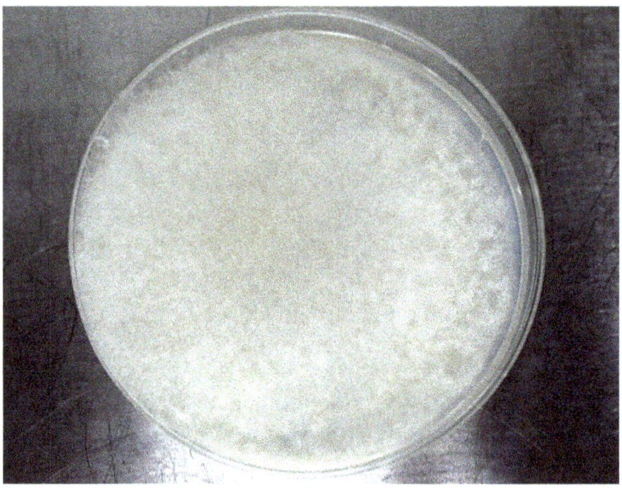

Fig. 6.1.2 *Lictheimia* (top).

Fig. 6.1.3 *Lictheimia* (side).

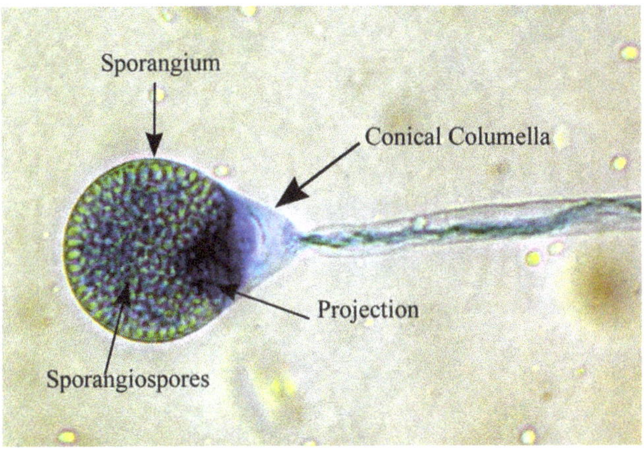

Fig. 6.1.4 Sporangium with sporangiospores, morphology at 1000×.

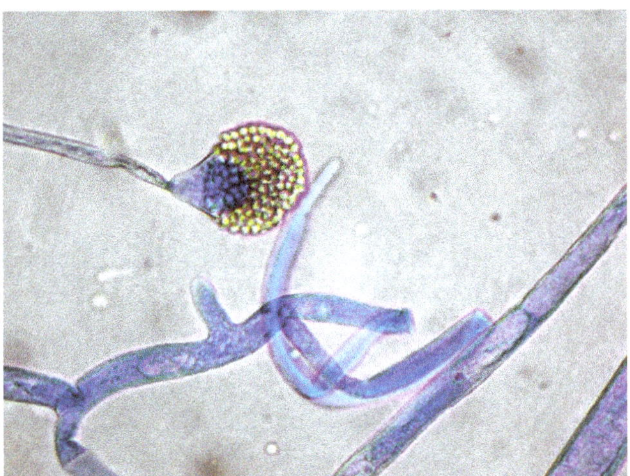

Fig. 6.1.5 Mature sporangiospores are golden brown.

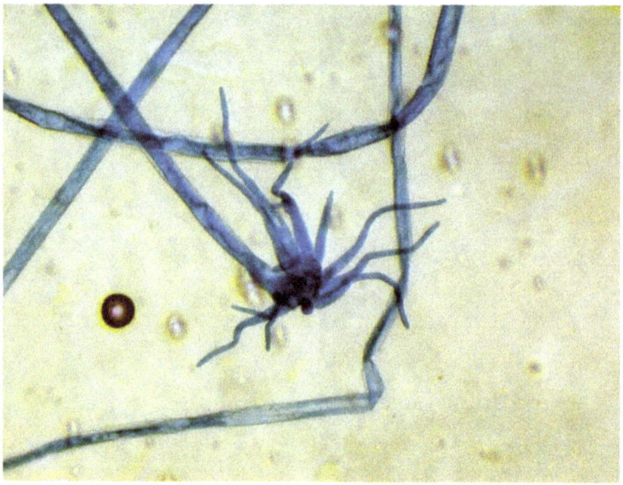

Fig. 6.1.6 Rhizoids.

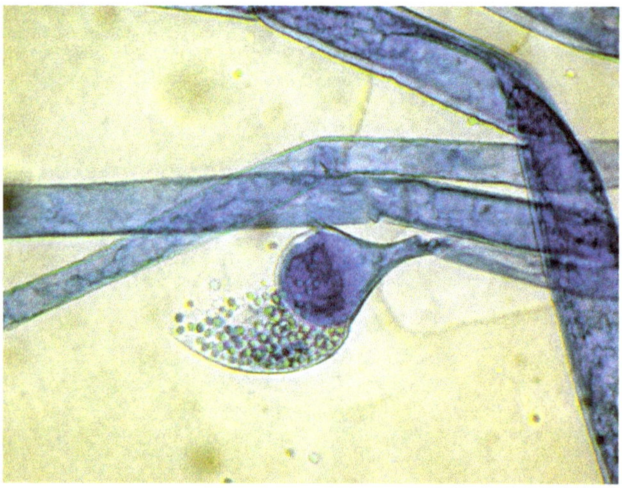

Fig. 6.1.7 Conical columella with semi-circular top; collarette present.

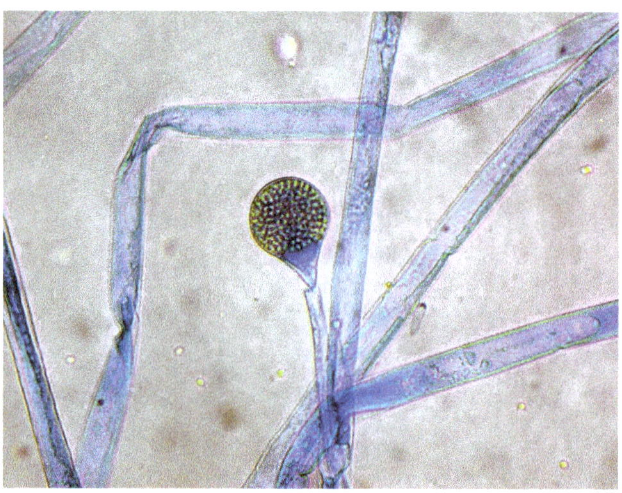

Fig. 6.1.8 *Lictheimia* with ribbon like sporangiophores.

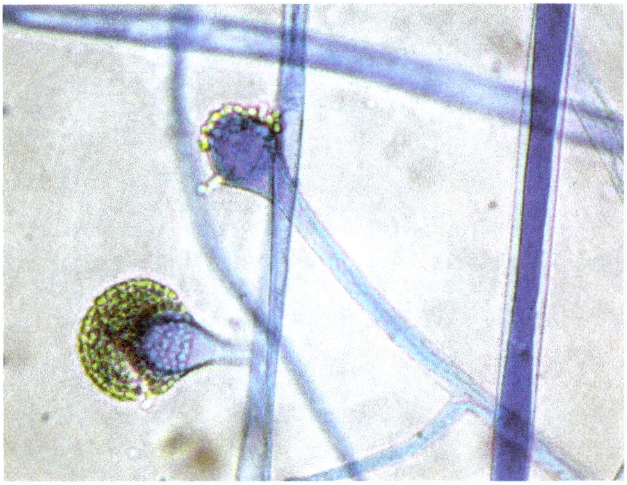

Fig. 6.1.9 Older sporangia.

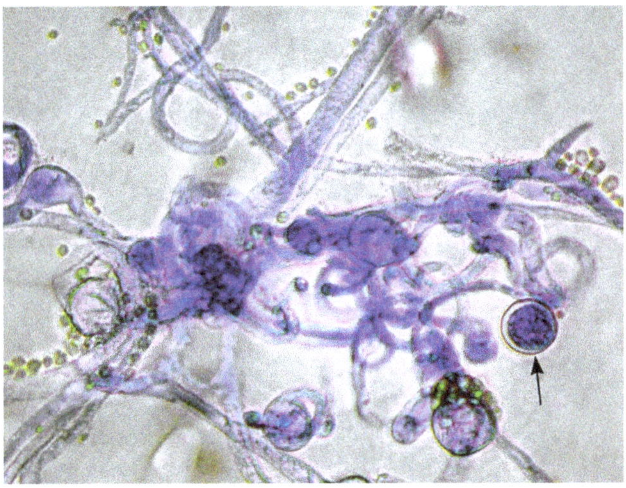

Fig. 6.1.10 Zygospore on right (*arrow*).

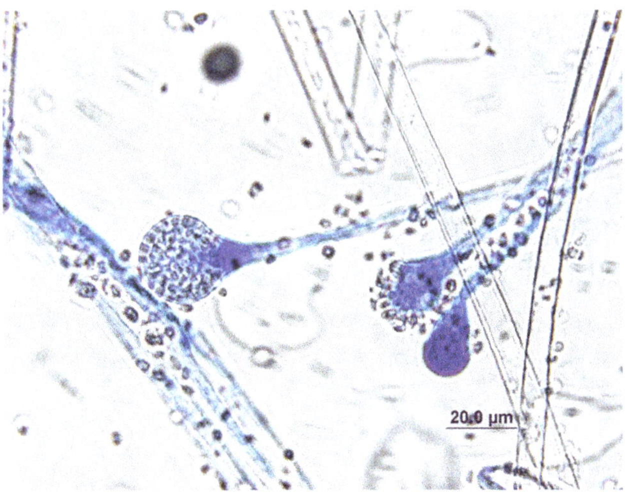

Fig. 6.1.11 *Lichtheimia*.

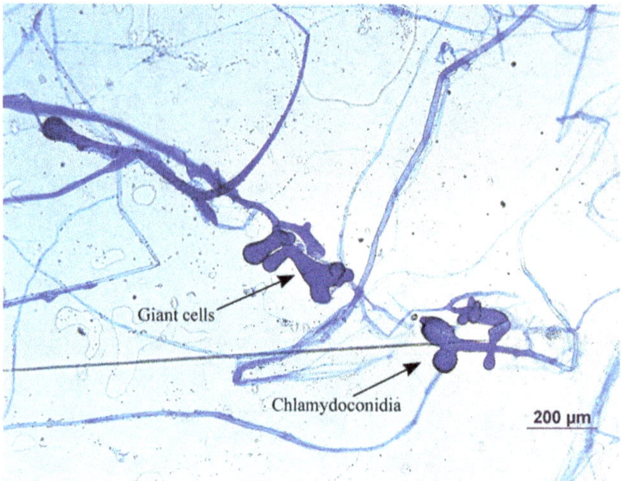

Fig. 6.1.12 Giant cells.

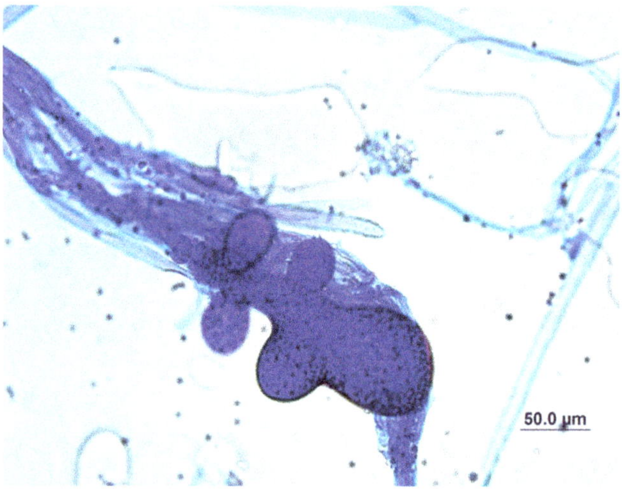

Fig. 6.1.13 Giant cells, 400×.

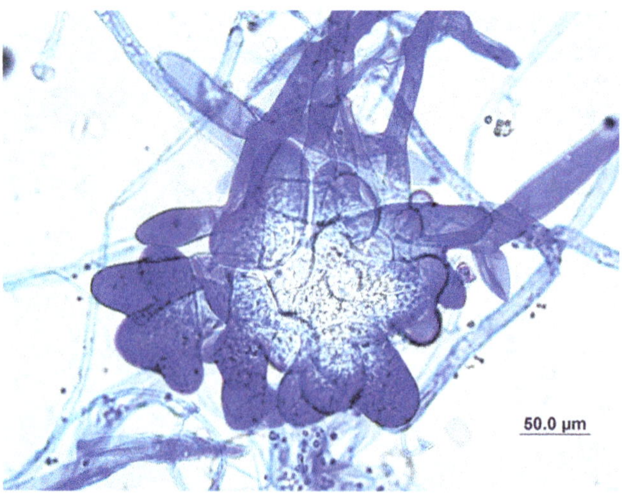

Fig. 6.1.14 Clumps of giant cells.

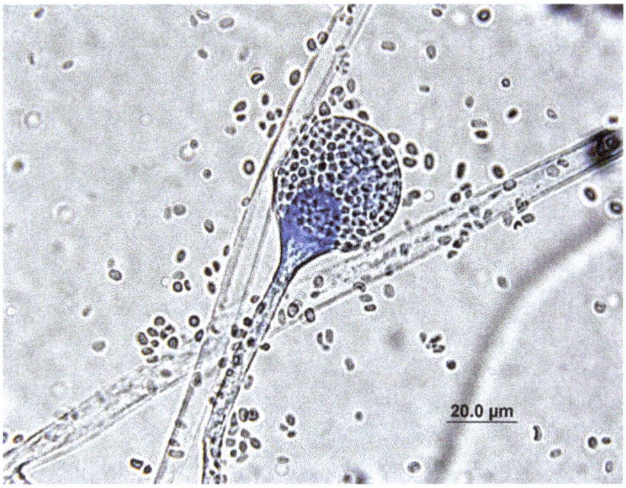

Fig. 6.1.15 *Lictheimia*, sporangiospores oval to round.

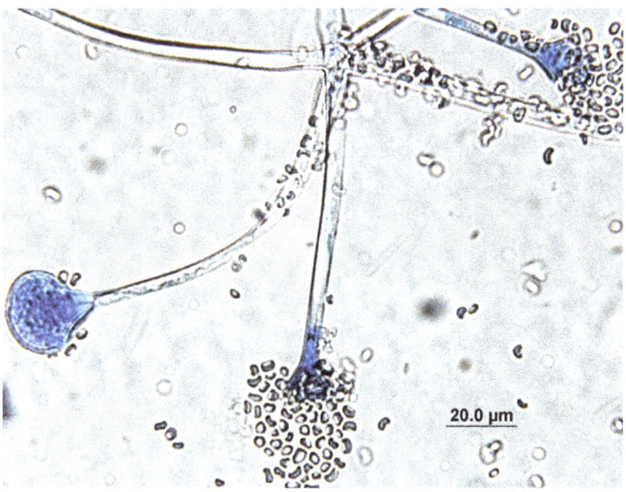

Fig. 6.1.16 *Lictheimia*, sporangiospores crescent shaped.

6.2 *Conidiobolus lamprauges, coronatus, incongruus*

Colonial morphology: on Sabouraud dextrose agar or potato flake agar, colonies are white to tan and yeast-like at first. On corn meal agar, colonies are flat, transparent, and spreading. With age, the colony becomes beige to tan and rapidly covers the entire plate. When cultures are grown at 36 °C, they develop aerial mycelia or sporophores that will launch spores (ballistospores) outward and against the lid of the petri dish. This results in an imprint of the colony appearing on the lid of the dish. Growth at 36 °C is optimal and fast, whereas at 27 °C it is slow. There is no growth at 45 °C.

Note: Petri plates should be handled with caution under a biosafety hood and sealed, as ballistospores are continuously released. *Conidiobolus* is known to cause respiratory mucosal infection with dissemination to all organs resulting in a fatal disease.

Microscopic morphology: hyphae are broad, partially septate to aseptate, and hyaline. Hyphae may contain swellings that are intercalary produced zygospores. Zygospores may also be produced at the ends of hyphae or on sporophores. Zygospores are 10–25 µm in diameter. Those that are produced at the end of sporophores show a narrow constriction at the base of the spore. At the point of attachment to the sporophore, zygospores have a flat, truncated end. As the zygospores are released, the flat end becomes conical and protrudes outward to form a papilla. Zygospores later germinate from the papilla and form new hyphae or new sporophores that may give rise to secondary zygospores, giving a dumbbell arrangement. In some instances, this sporulation may be multiple, resulting in a corona of secondary spores. Zygospores of *C. coronatus* become villose (spiked) which distinguishes the species. *C. incongruus* and *C. lamprauges* produce zygospores without conjugation breaks but not villose zygospores. *C. lamprauges* also produces Hülle cells whereas the other two species do not. In tissues, hyphae are broad, septate, with 90° dichotomous branching similar to *Aspergillus.* However, monilliaceous hyphae are also seen in tissues, but can be distinguished from *Mucor* by bulbous swellings at the ends. *C. lamprauges* is resistant to cycloheximide (0.04%).

Conidiobolus lamprauges

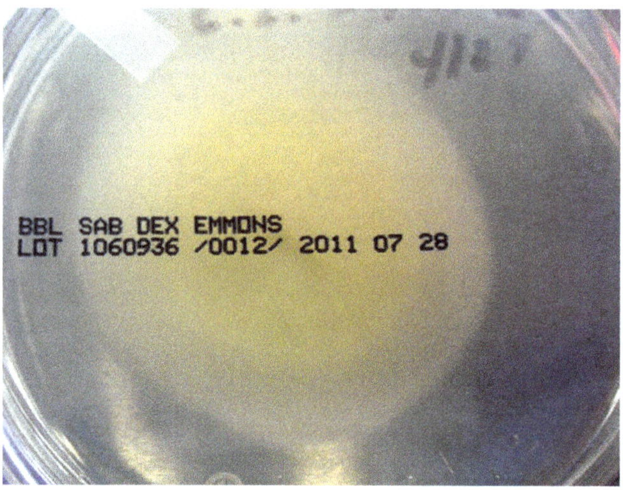

Fig. 6.2.1 *Conidiobolus lamprauges*, Sab DX (bottom).

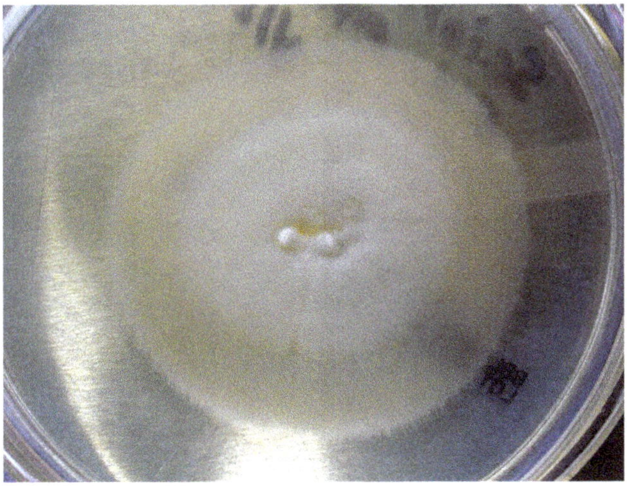

Fig. 6.2.2 *Conidiobolus lamprauges*, Sab DX.

Fig. 6.2.3 *Conidiobolus lamprauges*, potato DX (bottom).

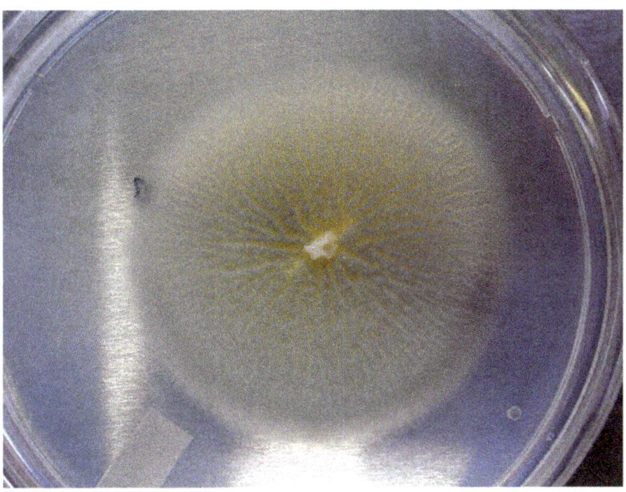

Fig. 6.2.4 *Conidiobolus lamprauges*, potato DX.

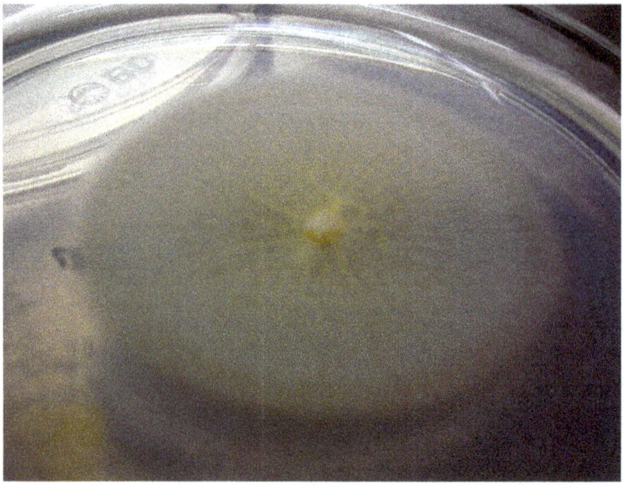

Fig. 6.2.5 *Conidiobolus lamprauges*, potato DX (side).

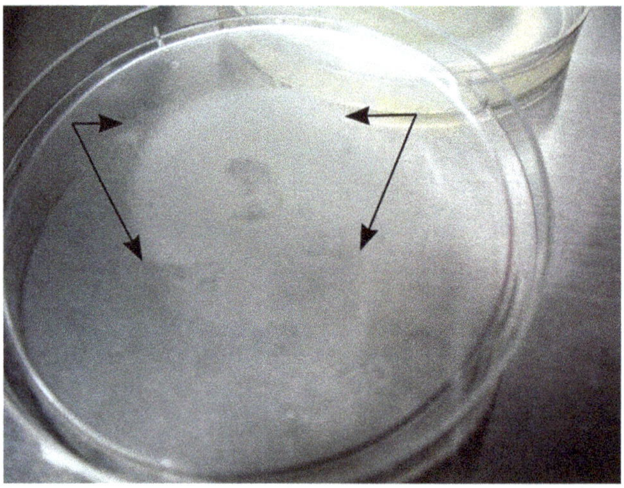

Fig. 6.2.6 Spores on petri dish lid.

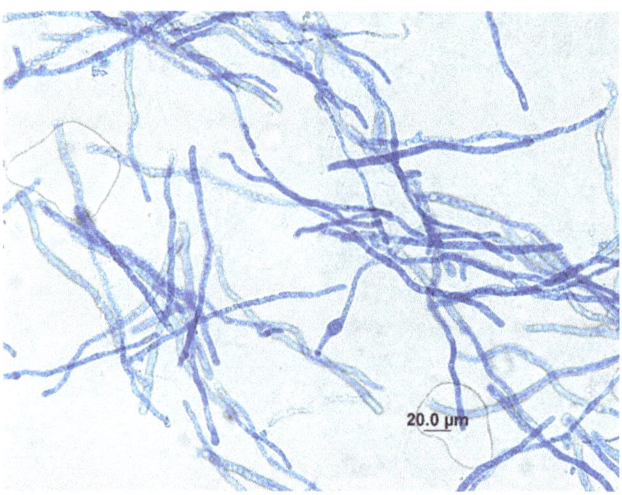

Fig. 6.2.7 Hyphae.

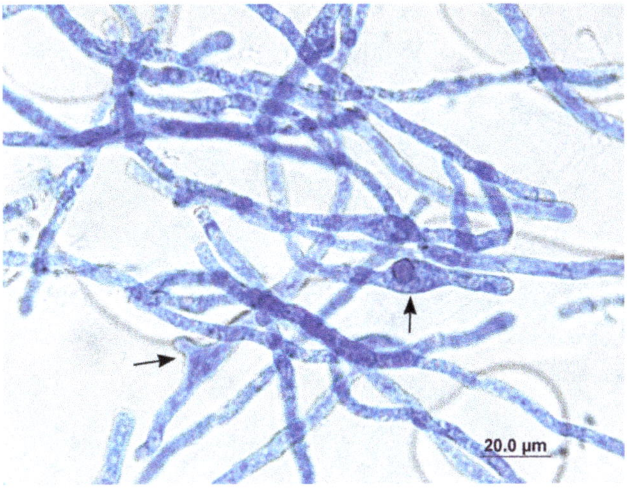

Fig. 6.2.8 Developing zygospores.

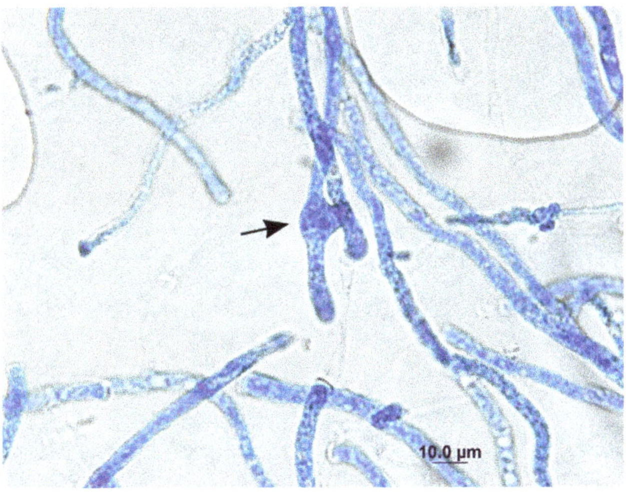

Fig. 6.2.9 Mating cells.

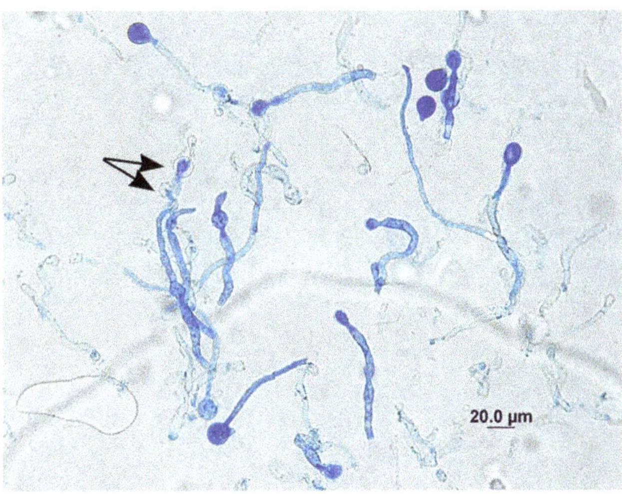

Fig. 6.2.10 Hülle cell.

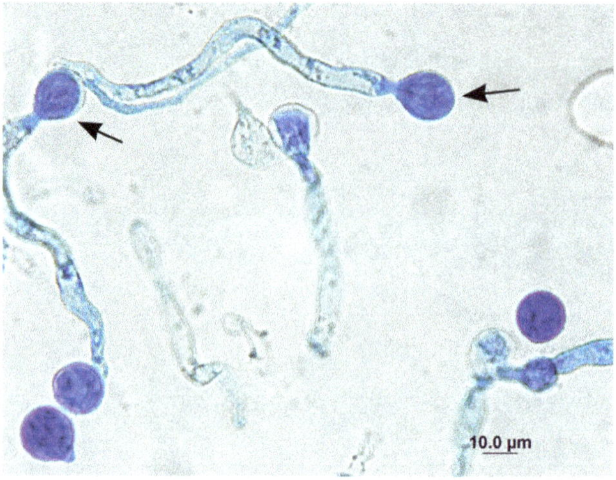

Fig. 6.2.11 Zygospore at end of sporophore.

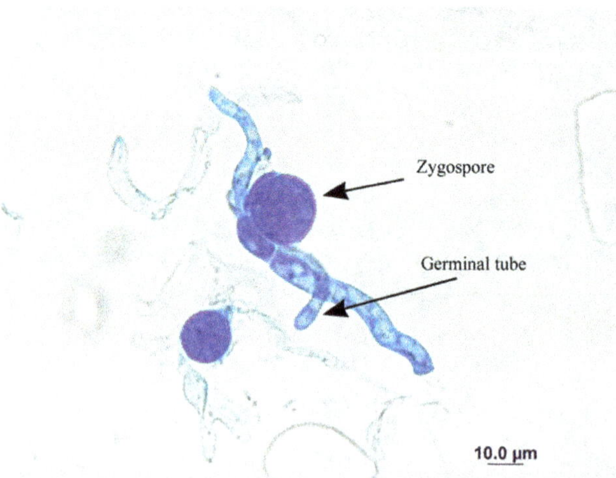

Fig. 6.2.12 Germinating zygospore.

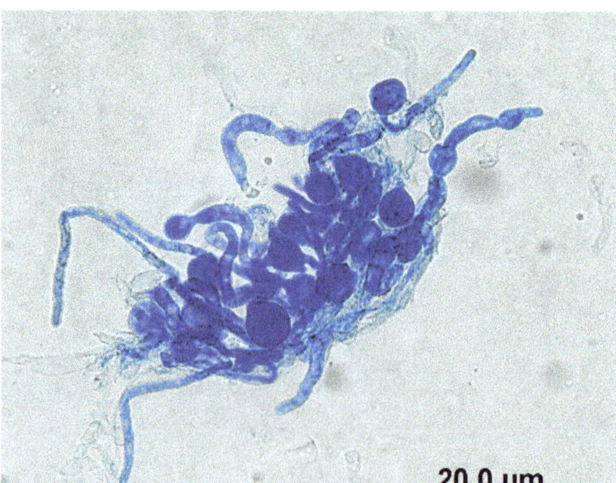

Fig. 6.2.13 Clumps of germinating zygospores.

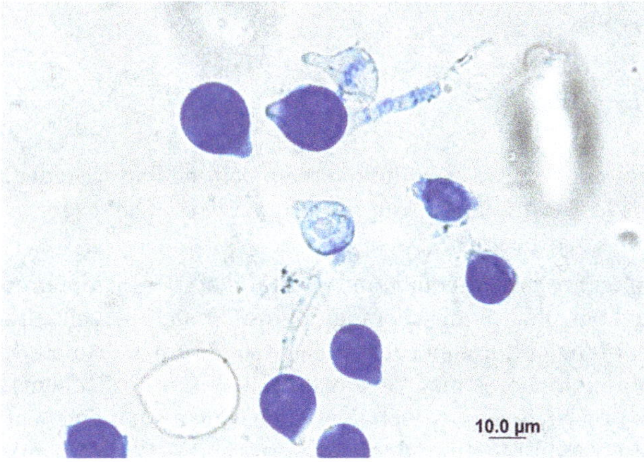

Fig. 6.2.14 Ballistospores from lid of plate.

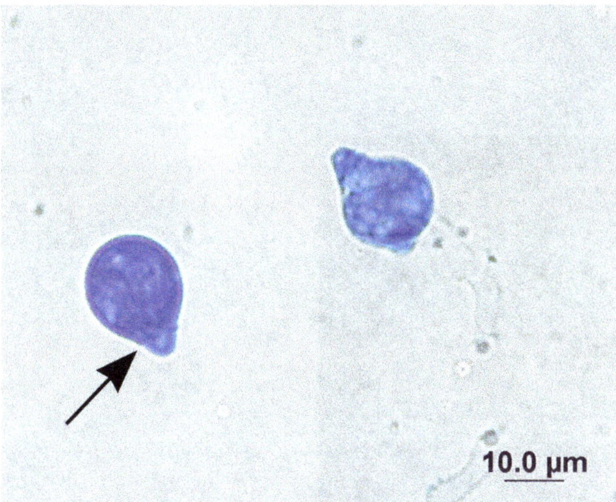

Fig. 6.2.15 Ballistospores showing papilla.

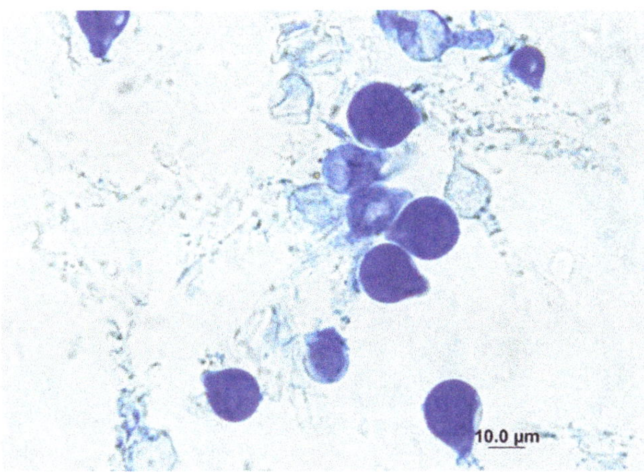

Fig. 6.2.16 Ballistospores with variable staining.

6.3 *Cunninghamella bertholletiae*

Colonial morphology: colonies appear first as white to cream colored and wrinkled, but mature quickly in a few days to produce fluffy white or gray surfaces. The reverse is cream to white, but some variants may be gray.

Microscopic morphology: hyphae are large, aseptate, and ribbon-like. Sporangiophores are produced perpendicular to the hyphae. Some sporangiophores branch. Vesicles are formed at the terminal end of sporangiophores and are swollen (30–40 μm in diameter), while vesicles on branched sporangiophores may be smaller (10–30 μm in diameter). Multiple denticles are produced on each vesicle, each denticle having a sporangiola at the end. Hyaline sporangiola are round (7–10 μm diameter) or oval (9–12 × 6–10 μm), and often appear dark when bunched together. They are often warty, but sometimes exhibit a spiny surface. Chlamydoconidia, zygospores, and synnema may also be seen. *C. bertholletiae* is a zygomycete.

Cunninghamella bertholletiae

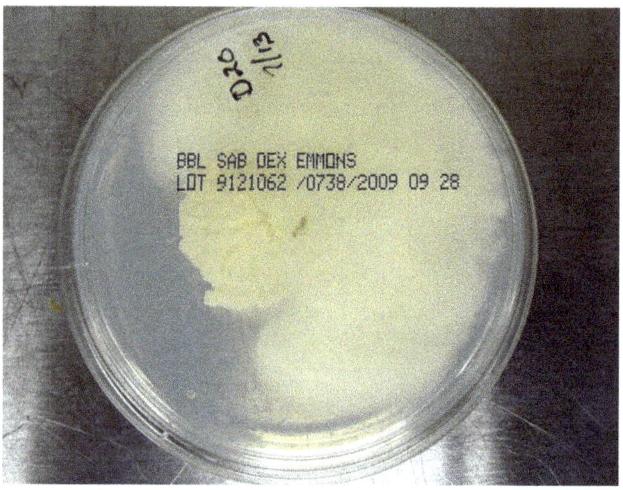

Fig. 6.3.1 *Cunninghamella bertholletiae* (bottom).

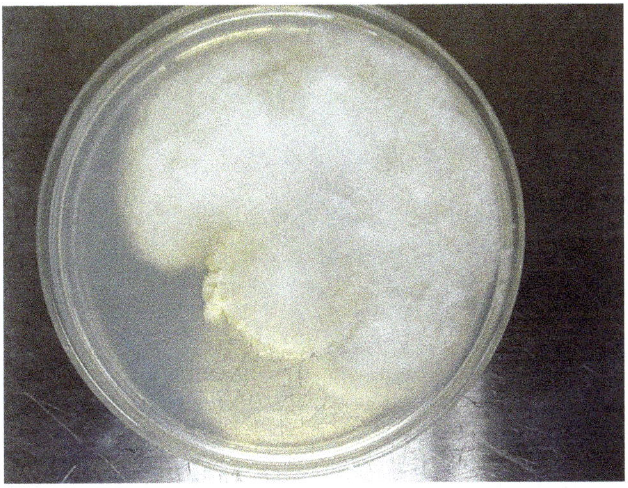

Fig. 6.3.2 *Cunninghamella bertholletiae*.

Fig. 6.3.3 *Cunninghamella bertholletiae* (side).

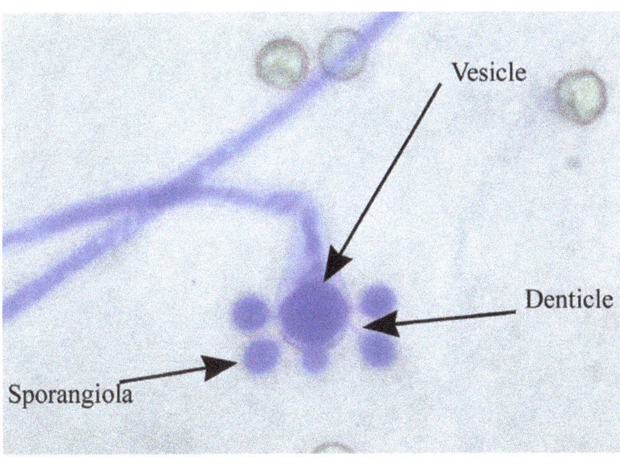

Fig. 6.3.4 Young fruiting head showing fine denticles.

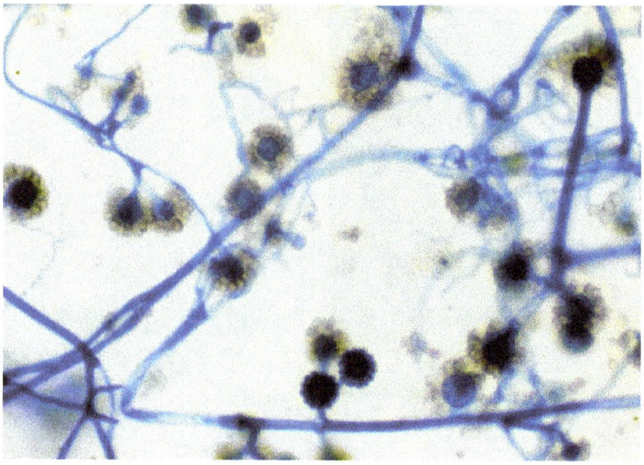

Fig. 6.3.5 Older fruiting heads with dark sporangiola.

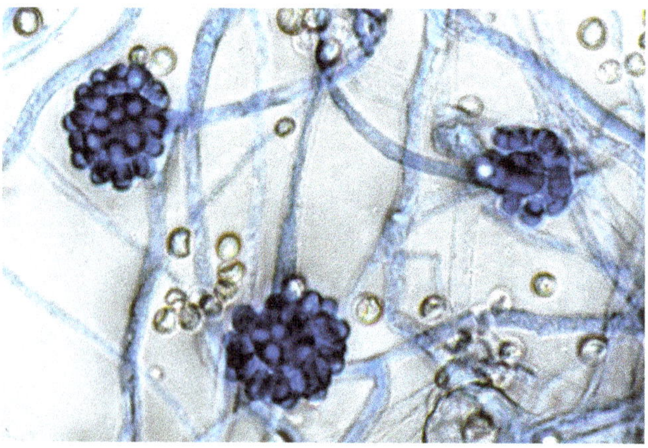

Fig. 6.3.6 Sporangiola are round and pitted.

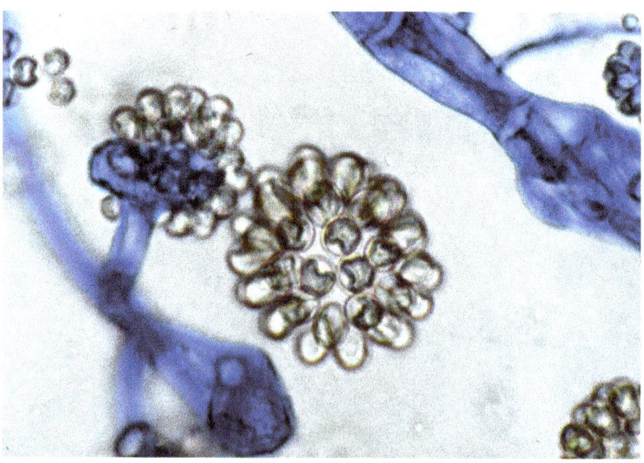

Fig. 6.3.7 Dark mature sporangiola.

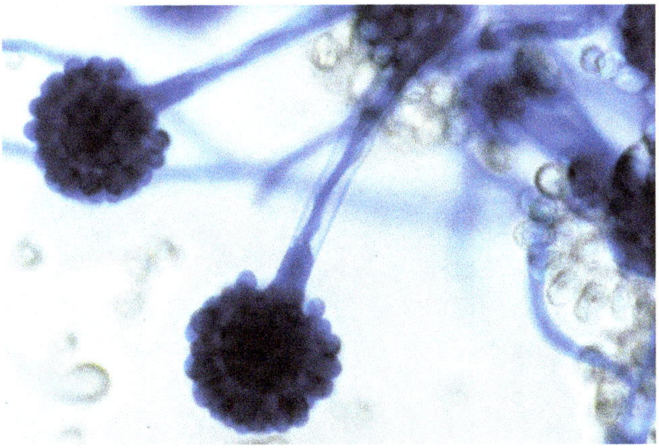

Fig. 6.3.8 Immature fruiting heads stain dark with LPCB.

Fig. 6.3.9 Swollen vesicle on right later develops into those shown on left with sporangiola.

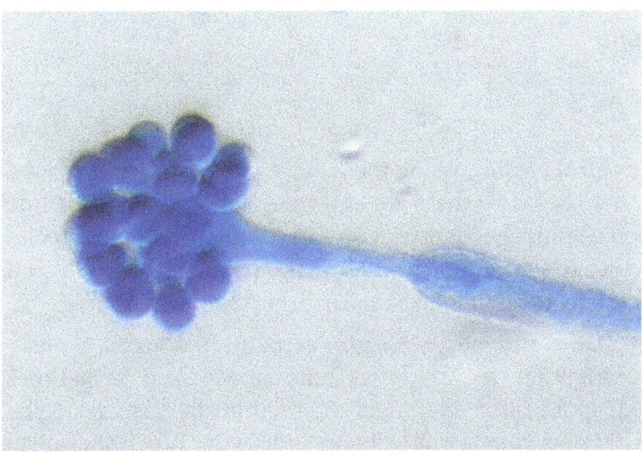

Fig. 6.3.10 Sporangiophores narrow or collapse at the apices.

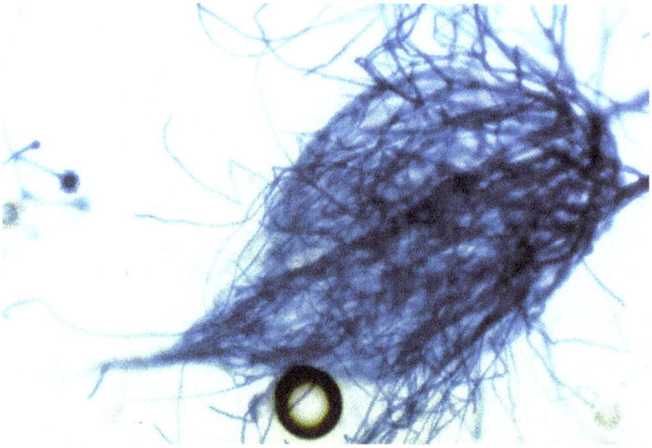

Fig. 6.3.11 Synnema.

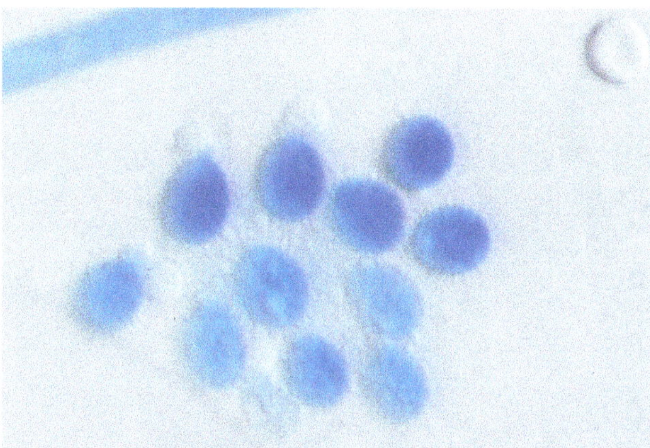

Fig. 6.3.12 Spiny sporangiola.

6.4 *Mucor* spp

There are 365 species of *Mucor*.

Colonial morphology: colonies grow rapidly, filling the culture plate within a few days. Aerial growth of hyphae may touch the lid, so these are known as "lid-lifter" fungi. White growth is first observed and then colonies become grayish to brownish and some species may have a yellow center. Reverse is white to yellow. Growth is cottony or wooly and inhibited by cycloheximide. Most species do not grow at 37 °C, but grow best at 25–30 °C.

Microscopic morphology: hyphae are broad (6–15 μm), aseptate to scarcely septate, hyaline and ribbon-like. Sporangiophores are long, unbranched and sometimes branched with terminal, round, spore-filled sporangia (50–300 μm in diameter). The rounded columella extends into the sporangia but lacks an apophysis. When sporangia rupture releasing spores, the sporangium adheres to the columella, revealing a collarette at the base of the sporangium. Sporangiospores (4–8 µm) are round to oval, hyaline, gray or brownish, and smooth walled or rough. Chlamydospores and zygospores are sometimes produced, but there are no true rhizoids or stolons.

Mucor circinelloides shows elaborate branching of the sporangiophores. Characteristic elongated zygospores have star-like spines. Hair-like, fine hyphae not developing from stolons or foot cells (pseudorhizoids) may be produced.

Mucor plumbeus also shows elaborate branching of sporangiophores. The sporangium is verticullate, having a rough, spiny appearance. Pseudorhizoids may be seen at the ends of sporangiophores opposite the sporangium, often resembling *Rhizomucor.* These can be differentiated from *Rhizomucor* by temperature growth studies. *Rhizomucor* will grow at temperatures >40 °C <58 °C whereas *Mucor* will not.

Mucor species

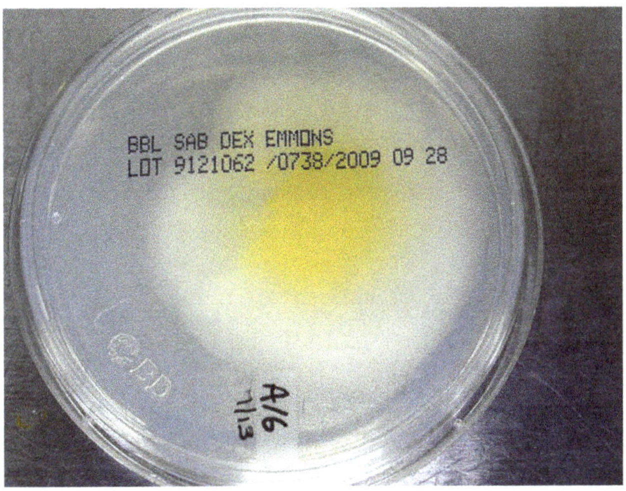

Fig. 6.4.1 *Mucor circinelloides* (bottom).

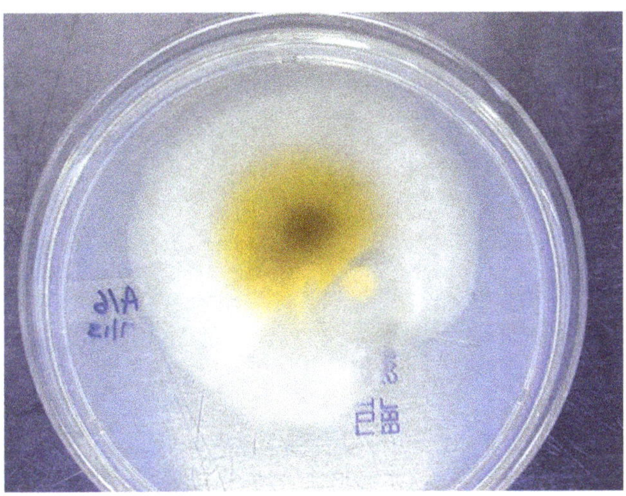

Fig. 6.4.2 *Mucor circinelloides* (top).

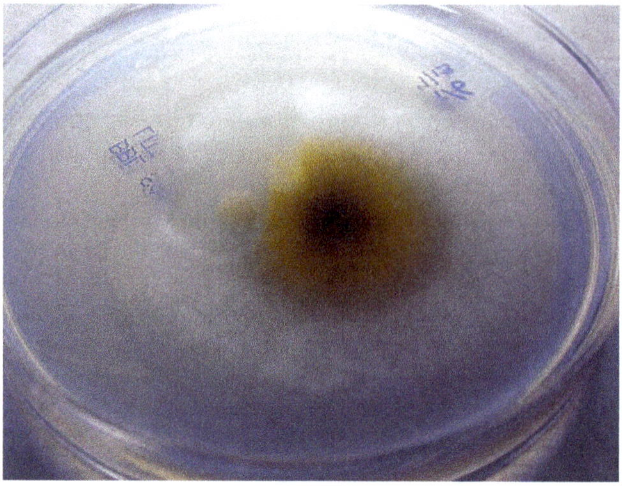

Fig. 6.4.3 *Mucor circinelloides* (side).

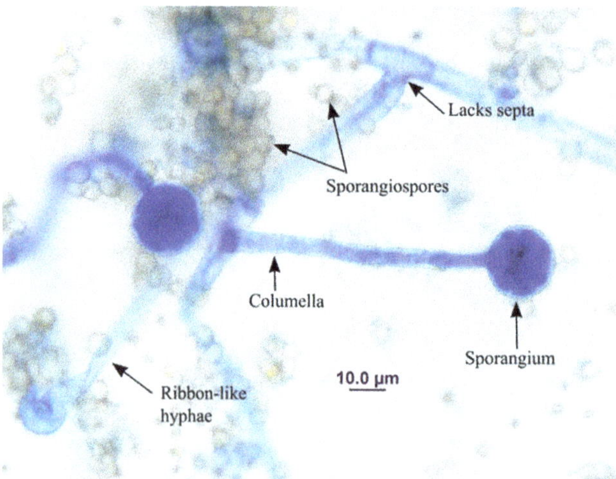

Fig. 6.4.4 *M. circinelloides*.

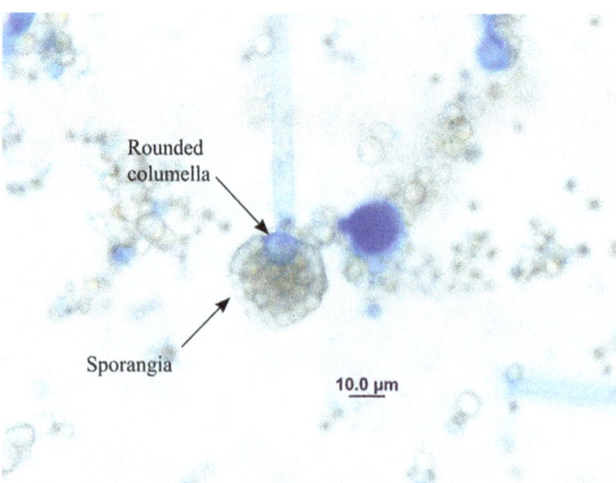

Fig. 6.4.5 *M. circinelloides*.

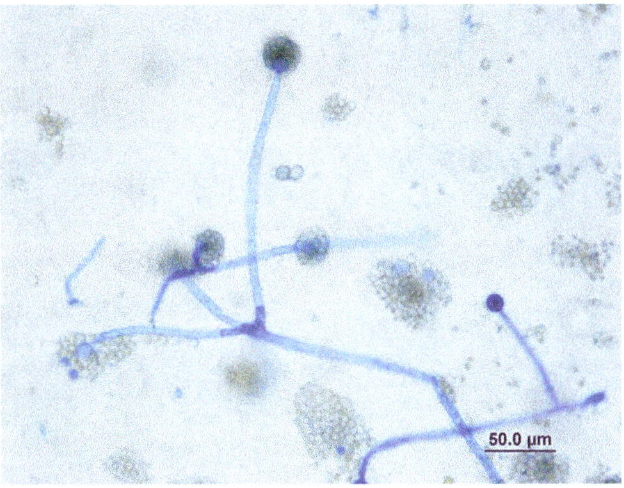

Fig. 6.4.6 *M. circinelloides*.

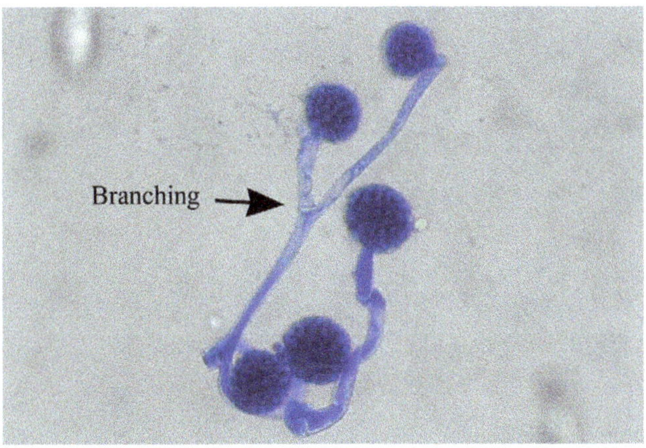

Fig. 6.4.7 *M. circinelloides*, conidiophores bend and curl.

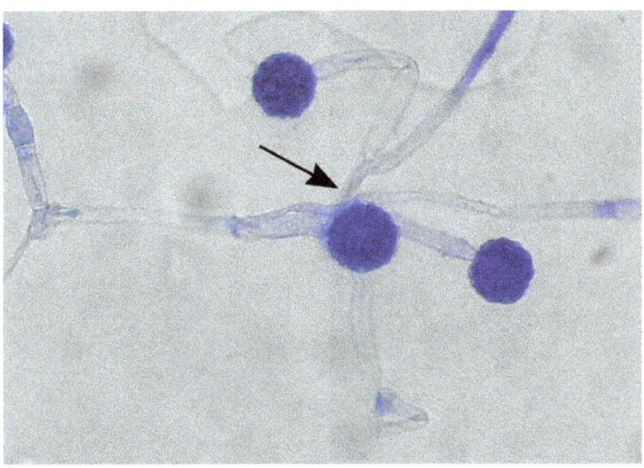

Fig. 6.4.8 *M. circinelloides*, conidiophores branch from single point.

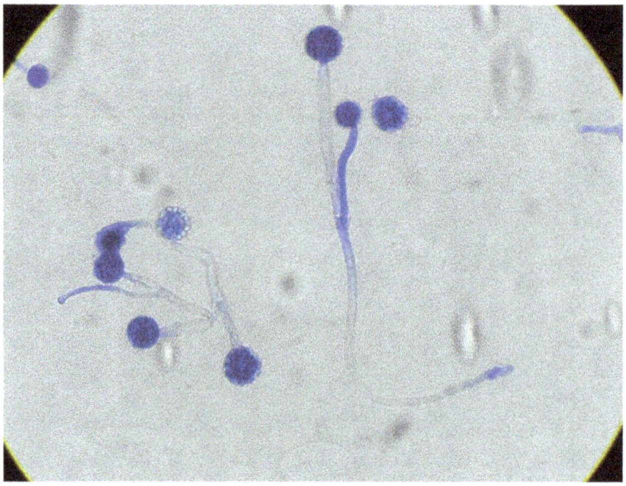

Fig. 6.4.9 *M. circinelloides*.

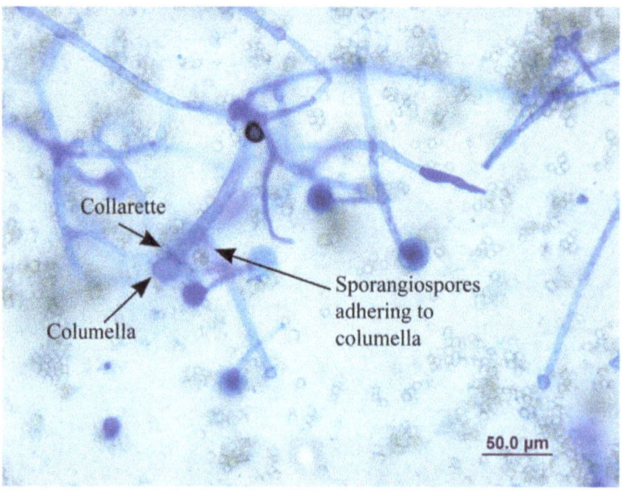

Fig. 6.4.10 *M. circinelloides* with ruptured sporangium.

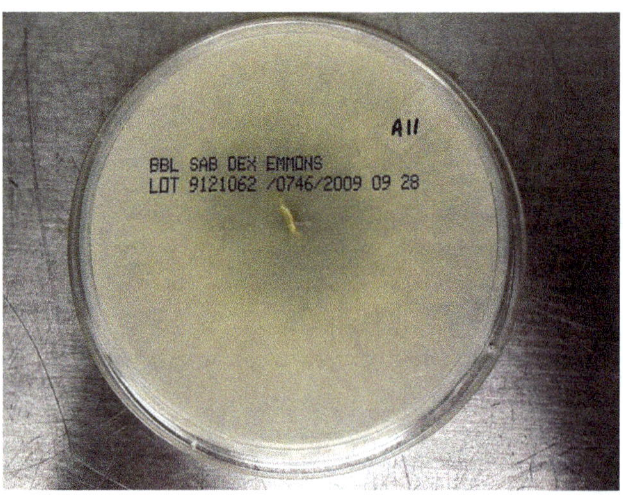

Fig. 6.4.11 *Mucor plumbeus* (bottom).

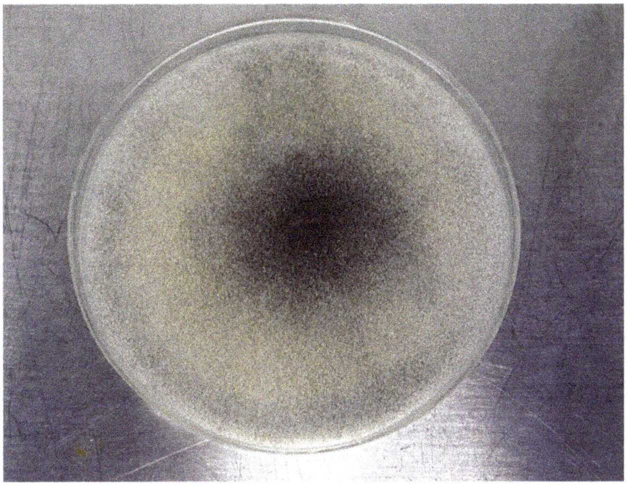

Fig. 6.4.12 *Mucor plumbeus* (top).

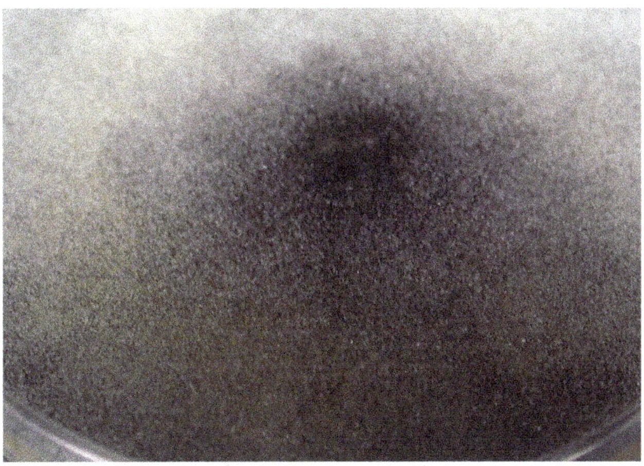

Fig. 6.4.13 *Mucor plumbeus* (side).

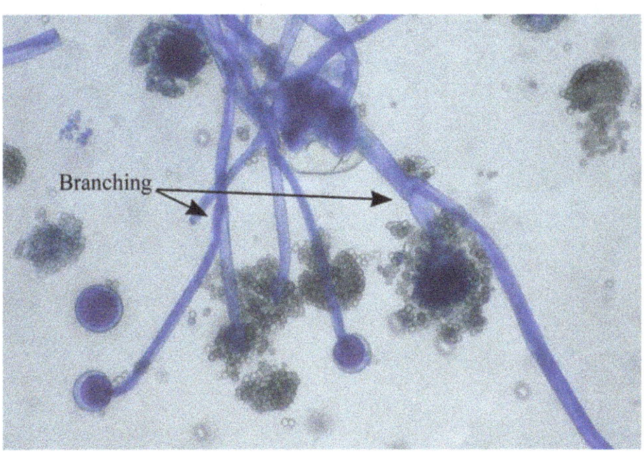

Fig. 6.4.14 *M. plumbeus*.

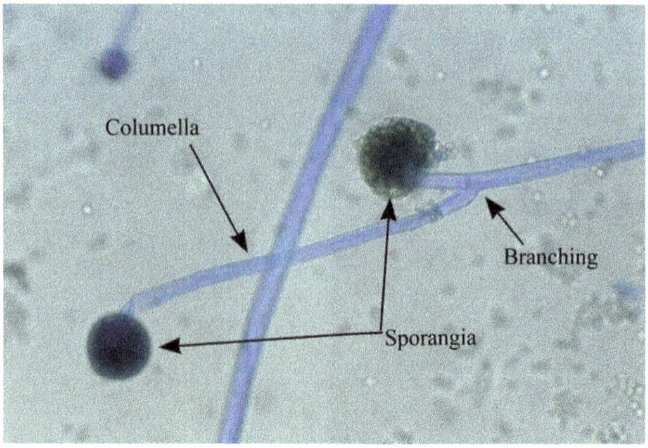

Fig. 6.4.15 *M. plumbeus*.

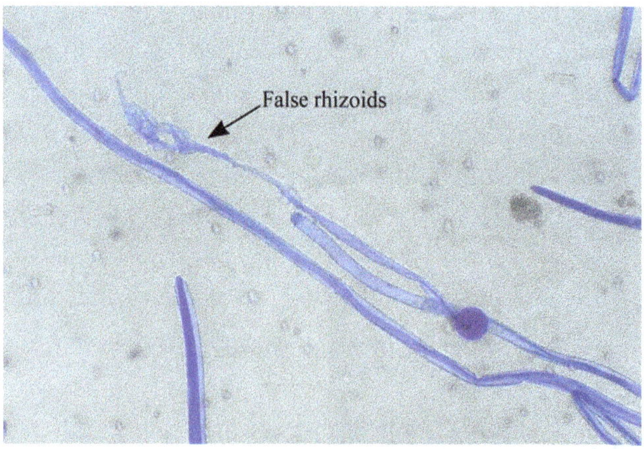

Fig. 6.4.16 *M. plumbeus* – false rhizoids.

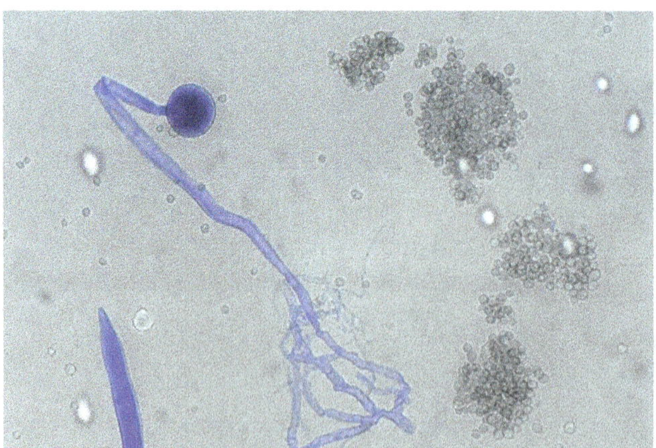

Fig. 6.4.17 *M. plumbeus* – false rhizoids.

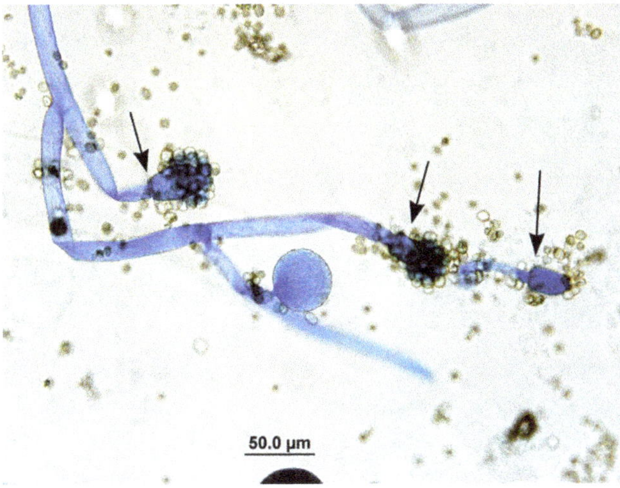

Fig. 6.4.18 *M. plumbeus*, oblong columnella with basal collarette.

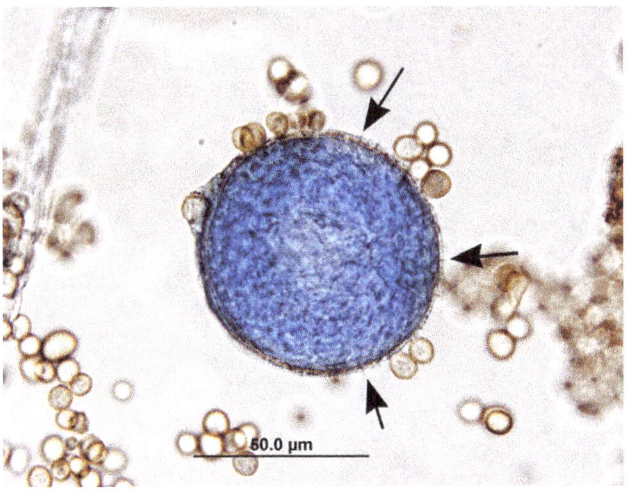

Fig. 6.4.19 *M. plumbeus*, finely echinulate sporangia.

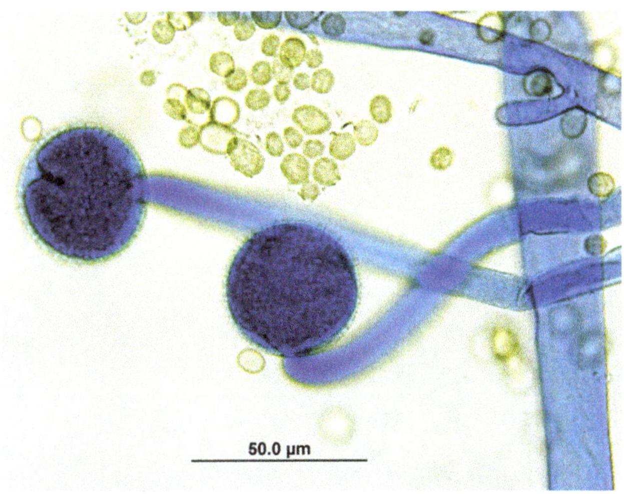

Fig. 6.4.20 *M. plumbeus*.

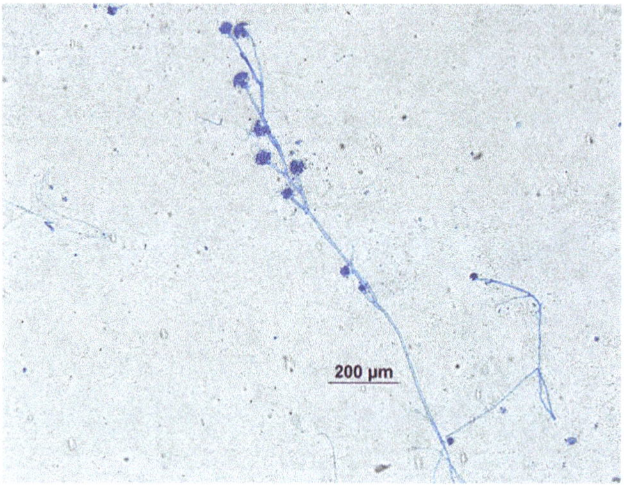

Fig. 6.4.21 *Mucor racemosus*, multiple branching conidiophores.

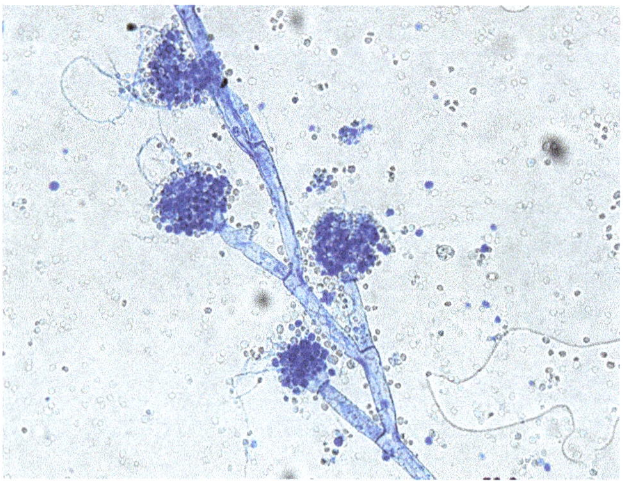

Fig. 6.4.22 *Mucor racemosus*.

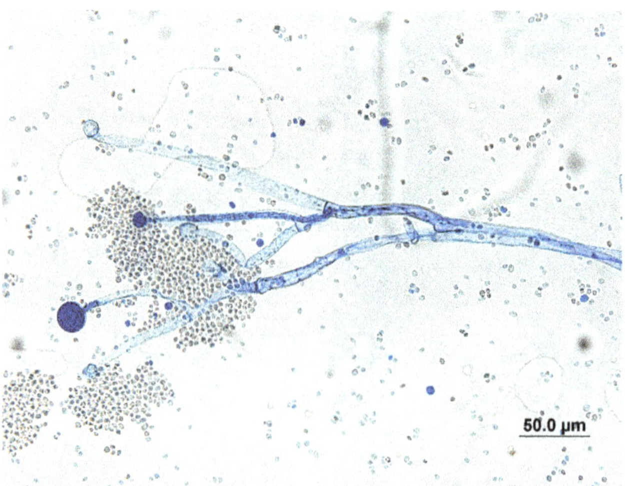

Fig. 6.4.23 *Mucor racemosus*.

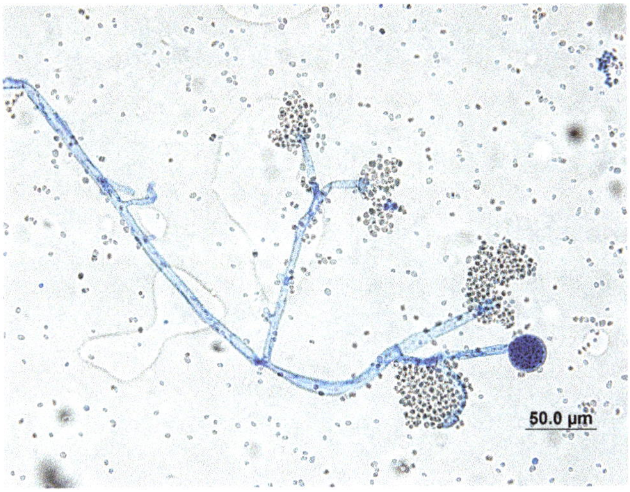

Fig. 6.4.24 *Mucor racemosus*.

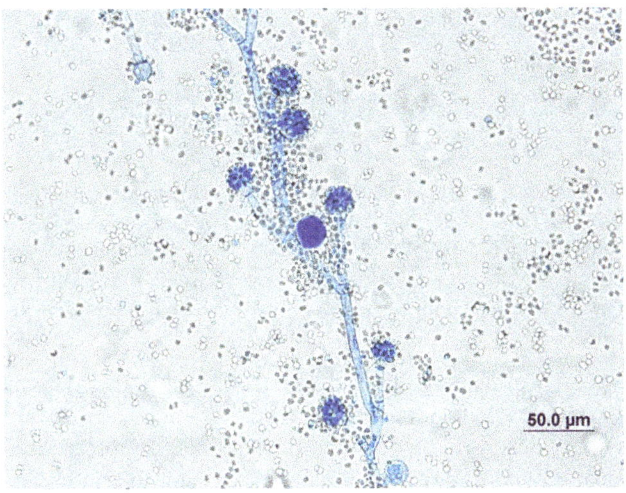

Fig. 6.4.25 *Mucor racemosus*.

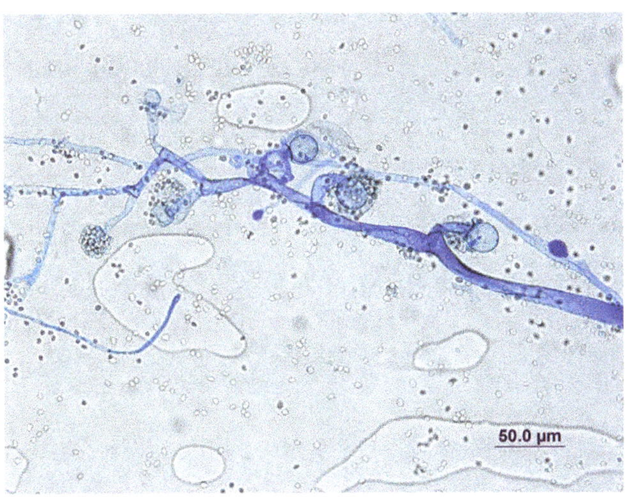

Fig. 6.4.26 *M. racemosus*, round columnella with flanged collarette.

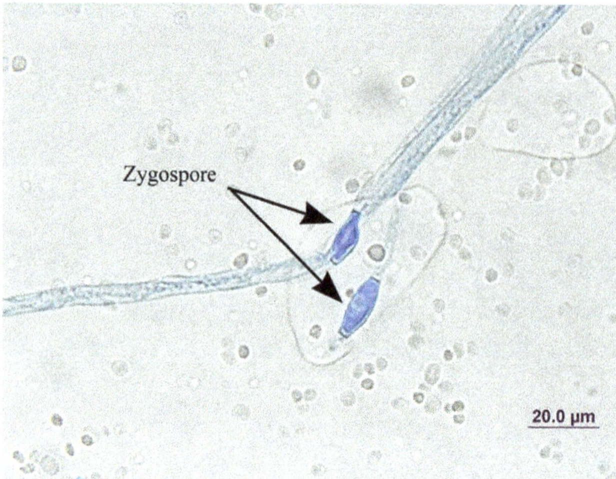

Fig. 6.4.27 *M. racemosus*, intercalary zygospore.

6.5 *Rhizomucor* spp

Colonial morphology: colonies mature rapidly and are thermophilic (optimal growth at 54 °C). They fill the culture plate and have a fluffy appearance; brown-black on the surface with a white reverse. The cultures darken with age and the reverse may become dark gray to black. Growth is inhibited by cycloheximide.

Microscopic morphology: hyphae are hyaline, flat and ribbon-like and non-septate. Sporangia are round (60–100 µm) with a protruding oval or pear-shaped columella but no apophysis. Sporangia always have a septum below the sporangium and sometimes a collarette is visible. Spores are round and dark brown. Occasionally, sterile, spade-like structures that resemble chili peppers may be seen on sporangiophores. Chlamydoconidia are absent. Rarely, zygospores are produced and are rough walled, reddish brown to black, 45–65 µm in diameter. Zygospores are mixed into hyphal mats and are often seen in young cultures where mating is frequent.

Organisms resemble both *Rhizopus* and *Mucor* but can be distinguished from each by the following characteristics.

- Rudimentary rhizoids and stolons are present but are rare. Rhizoids are less developed than those of *Rhizopus*.
- Rhizoids may occur along the stolons, between sporangiophores rather than at the base. This feature easily distinguishes *Rhizomucor* from *Rhizopus*.
- Sporangiophores are sympodially branched (most often with 2–3 branches) and may occur in umbelliform (umbrella-like) arrangements, unlike *Mucor* that lacks branching.

Rhizomucor

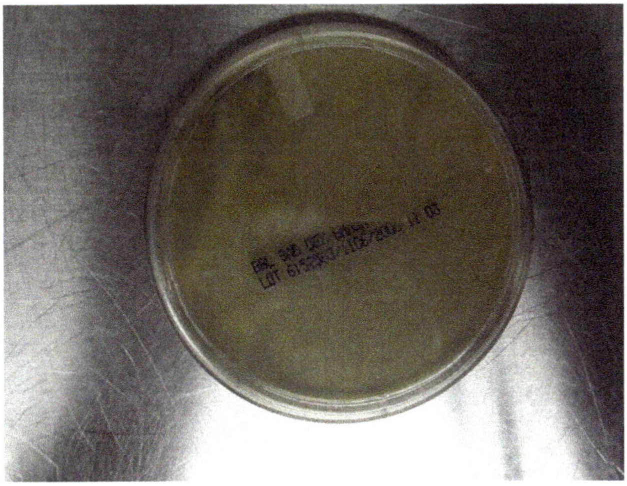

Fig. 6.5.1 *Rhizomucor* on SABDX (bottom).

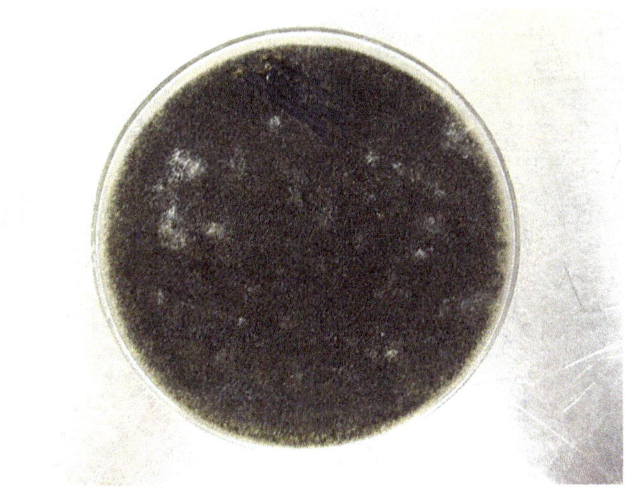

Fig. 6.5.2 *Rhizomucor*, SABDX (top).

Fig. 6.5.3 *Rhizomucor*, SABDX (side).

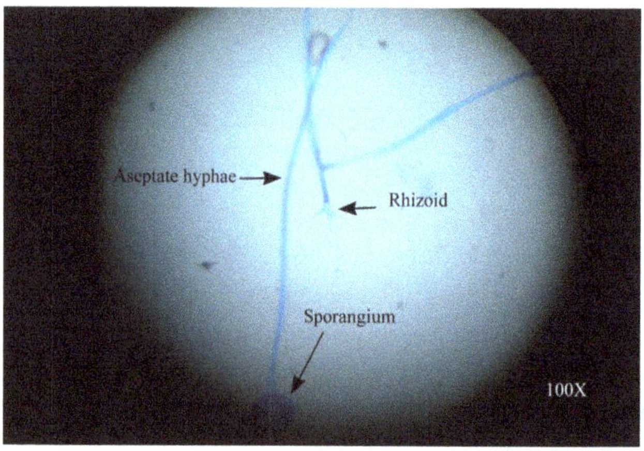

Fig. 6.5.4 *Rhizomucor*, complete structure.

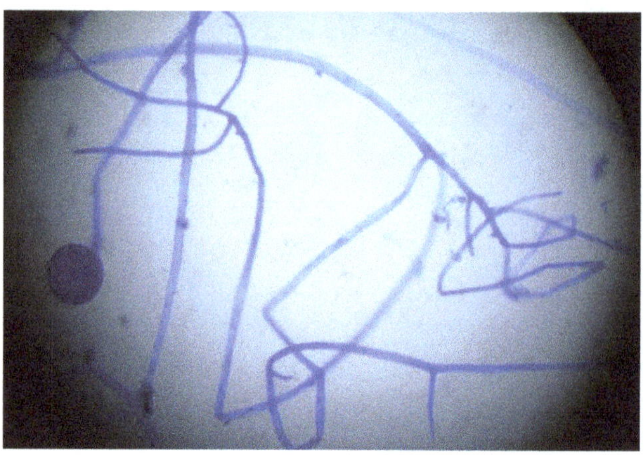

Fig. 6.5.5 Rhizomucor, complete structure; 100×.

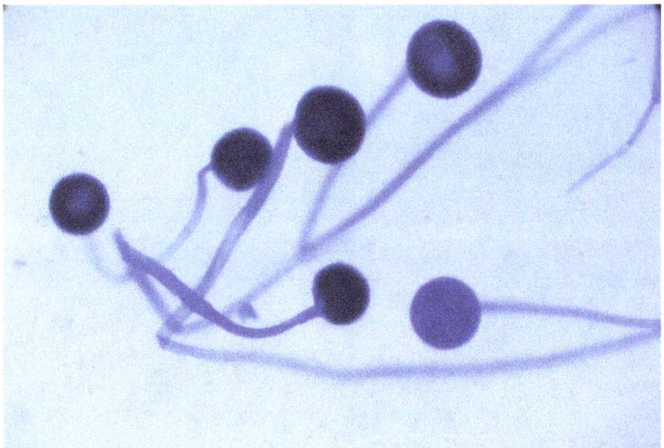

Fig. 6.5.6 Branching.

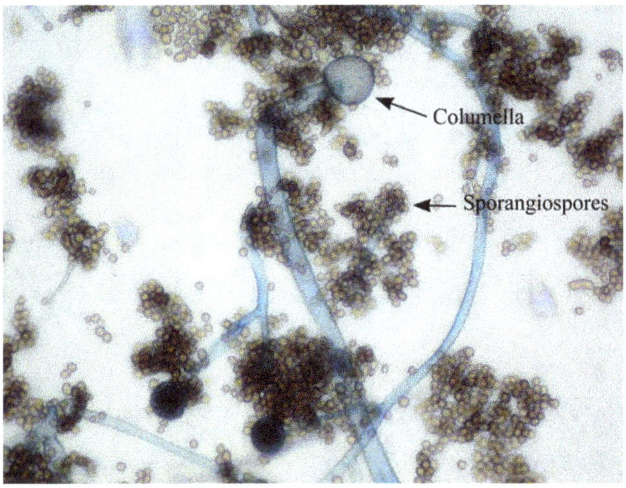

Fig. 6.5.7 Ruptured sporangium.

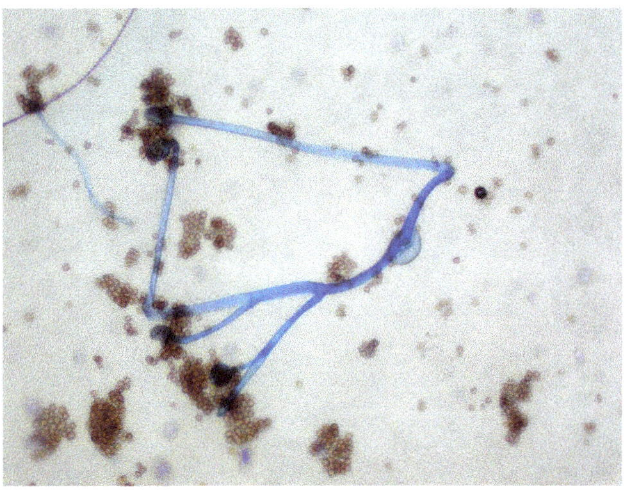

Fig. 6.5.8 Branched sporangiophores.

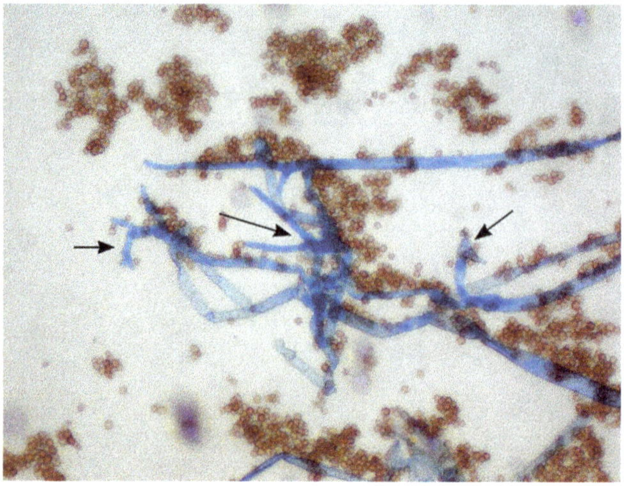

Fig. 6.5.9 Primitive rhizoids.

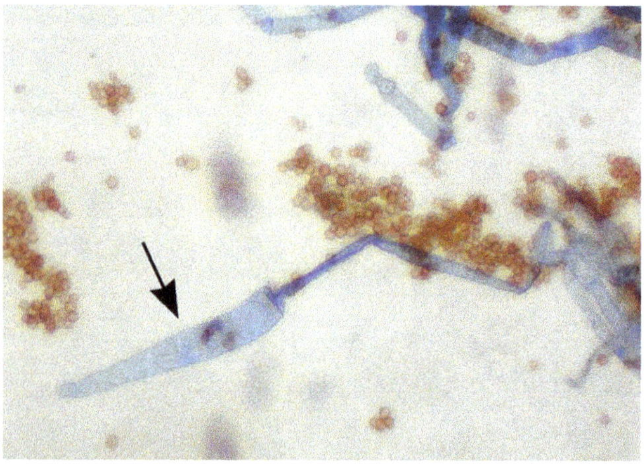

Fig. 6.5.10 Conical structures present.

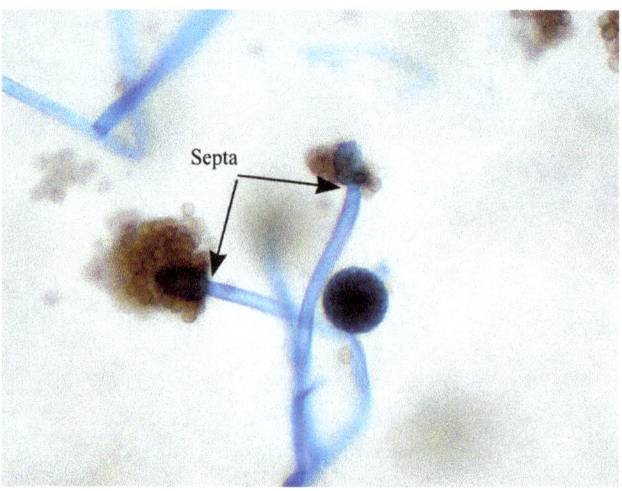

Fig. 6.5.11 Shows septa at base of sporangia.

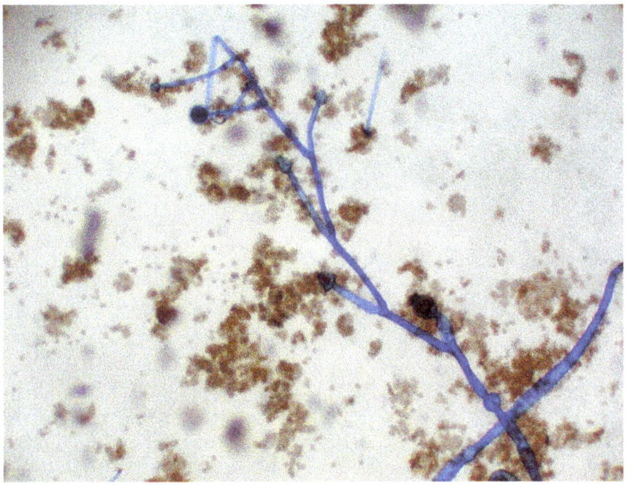

Fig. 6.5.12 Branching is common.

6.6 *Rhizopus* spp

Colonial morphology: colonies grow rapidly, filling the plate within 4 days. Growth is fluffy, white, becoming gray- or yellow-brown. The reverse is white, and turns light gray-brown with age. Optimal growth is at 37 °C and is inhibited by cycloheximide. It does not grow at 45 °C.

Microscopic morphology: hyphae are hyaline, ribbon-like, varying in size, and non-septate. Sporangiophores are dark brown, unbranched, and at the point where the sporangiophore meets the stolon, rhizoids are produced. Foot cells are often seen at the end opposite the sporangium, and rhizoids are abundantly produced from the foot cells. Rhizoids occasionally radiate along the sporangiophore, not the stolon. Stolons connect sporangiophores, where multiple sporangiophores arise from a single point along the stolon. Sporangiophores are up to 1500 µm in length and 18 µm in width. Sporangia are round, dark brown to grayish black, with a flattened base, up to 175 µm in diameter and contain many spores. The columella is small with an apophysis (rare). The collarette disintegrates after the sporangium ruptures and sporangiospores are released. Sporangiospores are oval to round, brown, and in some species, striations occur.

Rhizopus

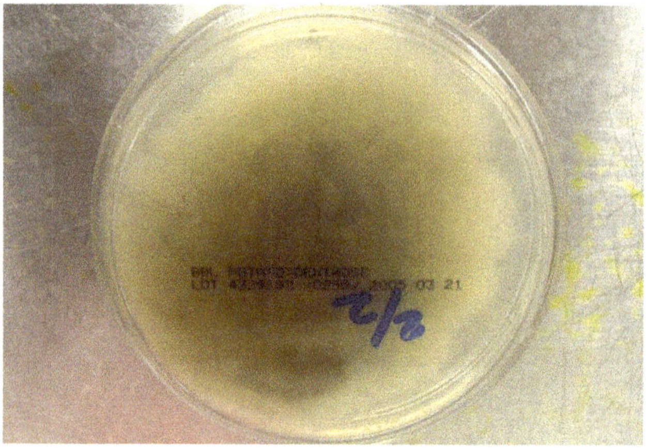

Fig. 6.6.1 *Rhizopus* (bottom).

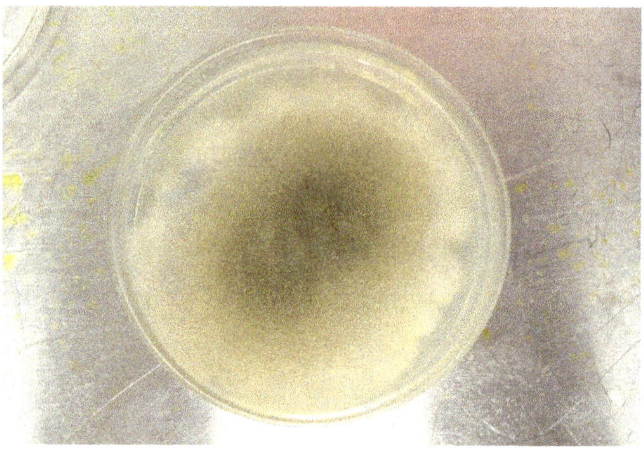

Fig. 6.6.2 *Rhizopus* (top).

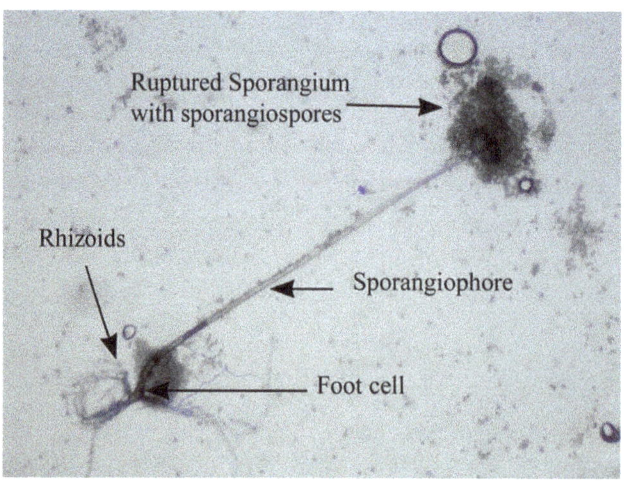

Fig. 6.6.3 Complete sporangiophore with rhizoids and sporangium.

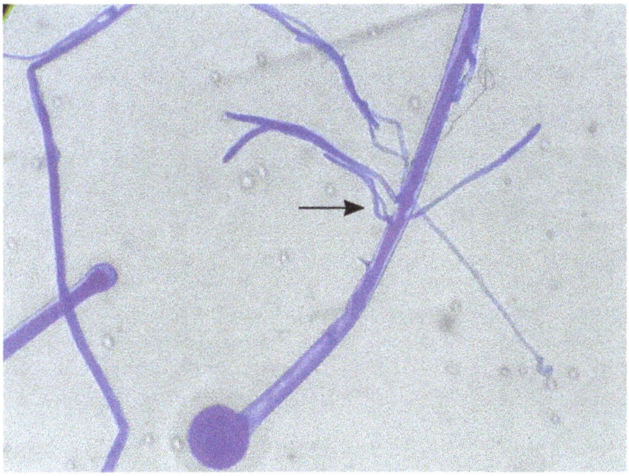

Fig. 6.6.4 Stolons along sporangiophore.

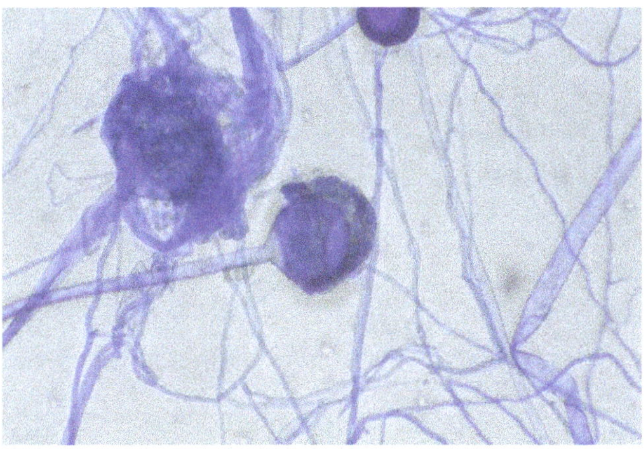

Fig. 6.6.5 Collapsed sporangium showing columella.

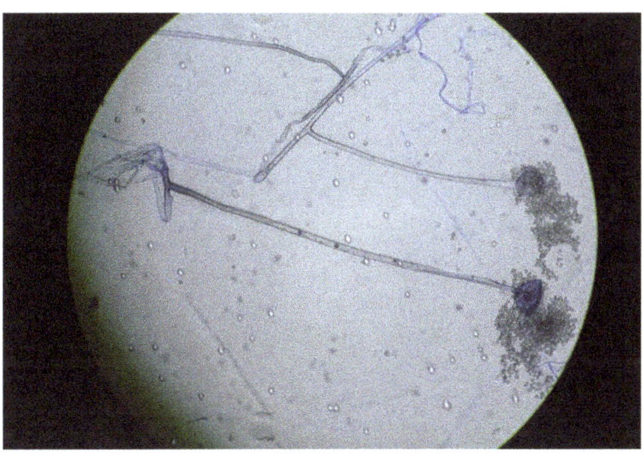

Fig. 6.6.6 100× magnification.

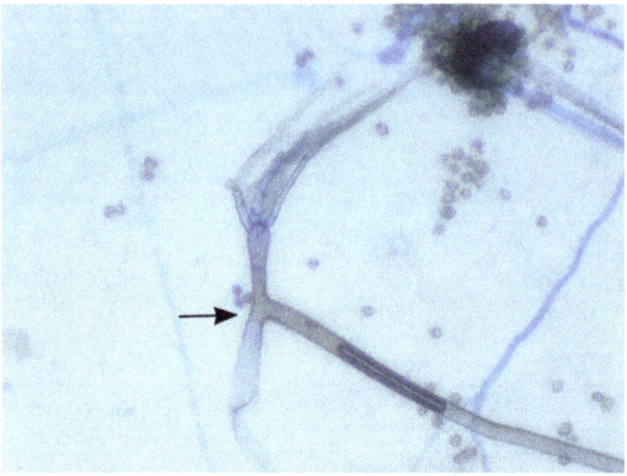

Fig. 6.6.7 Foot cell.

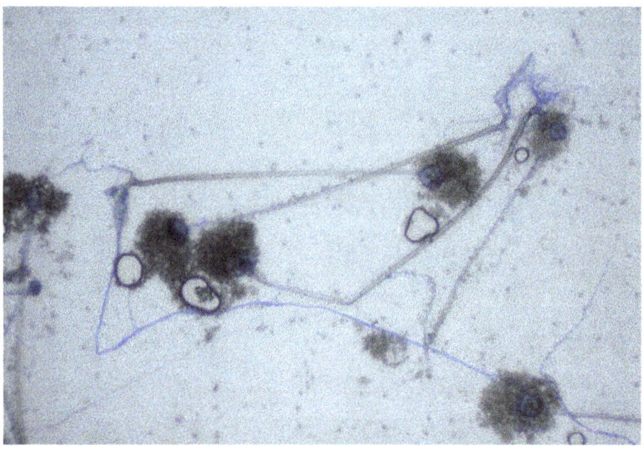

Fig. 6.6.8 Complete structures showing rhizoids and sporangia.

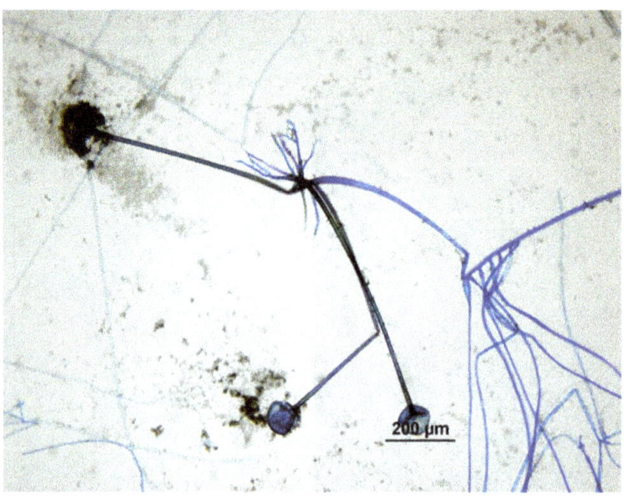

Fig. 6.6.9 Rhizoids at stolon (middle) and radiating along the sporangiophore (right).

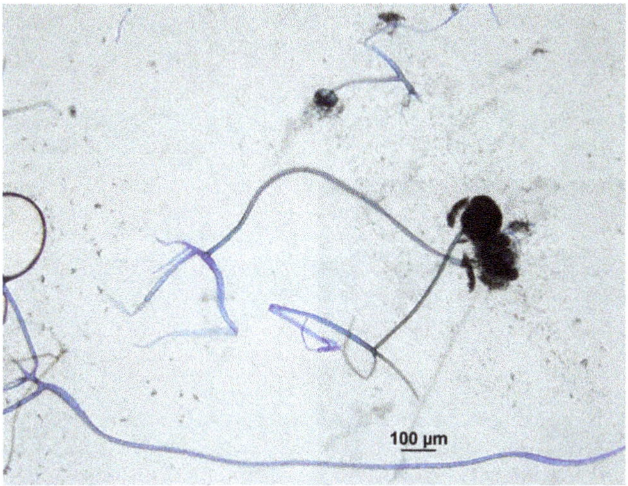

Fig. 6.6.10 Complete structure, 100×.

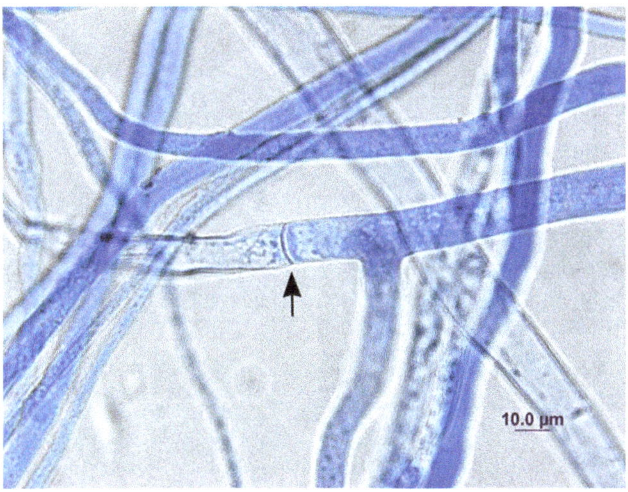

Fig. 6.6.11 Typical septum formation.

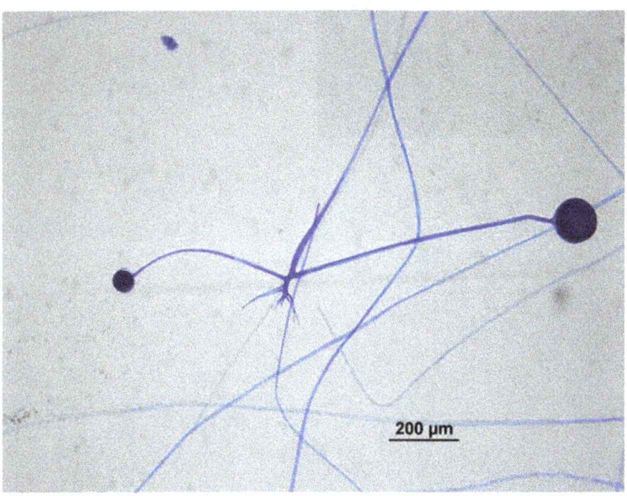

Fig. 6.6.12 Intact sporangia.

Rhizopus oryzae

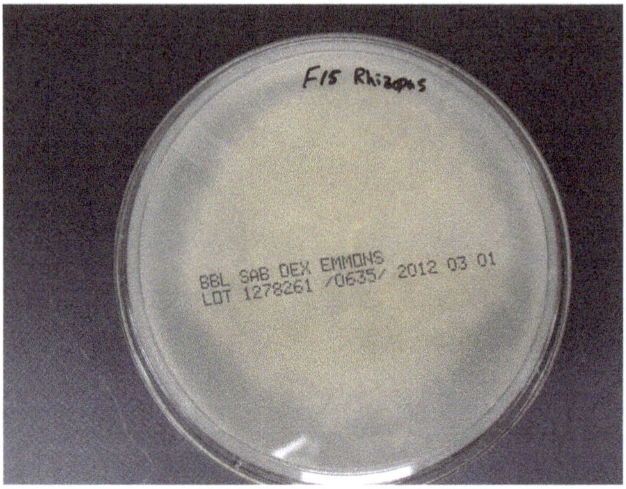

Fig. 6.6.13 *Rhizopus oryzae* (bottom).

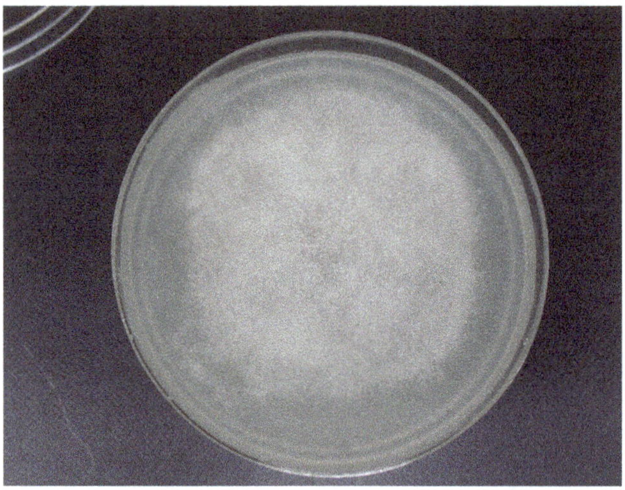

Fig. 6.6.14 *Rhizopus oryzae* (top).

Fig. 6.6.15 *Rhizopus oryzae* (side).

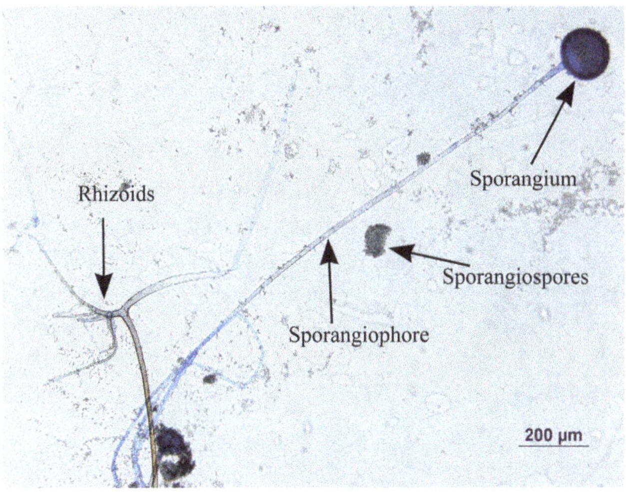

Fig. 6.6.16 *R. oryzae*, basic structure with intact sporangium.

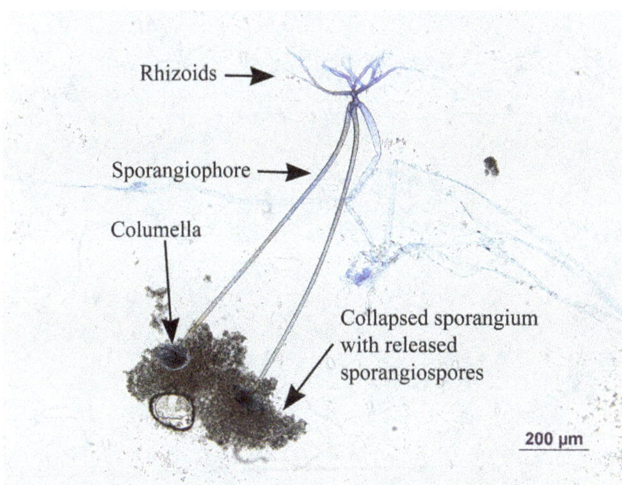

Fig. 6.6.17 *R. oryzae*, basic structure with collapsed sporangia.

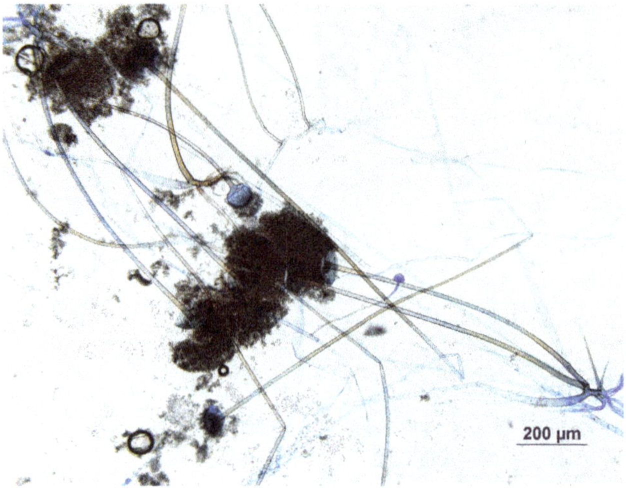

Fig. 6.6.18 *R. oryzae*, basic structure with collapsed sporangia.

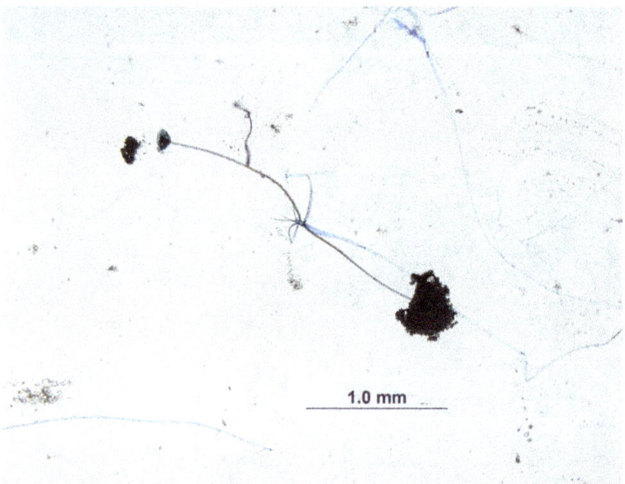

Fig. 6.6.19 *R. oryzae*, basic structure with collapsed sporangia.

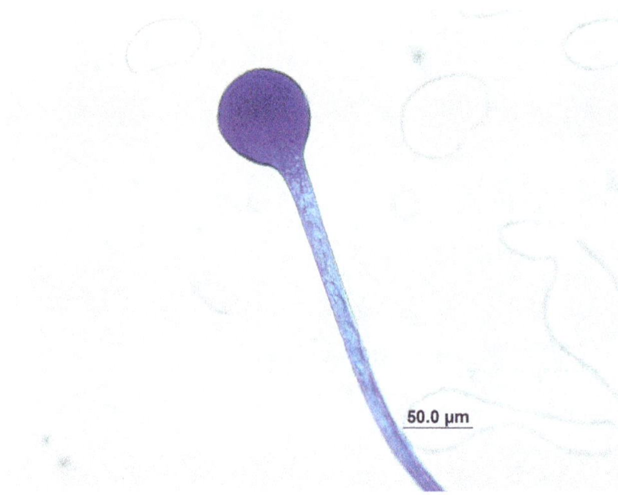

Fig. 6.6.20 *R. oryzae*, showing immature sporangium.

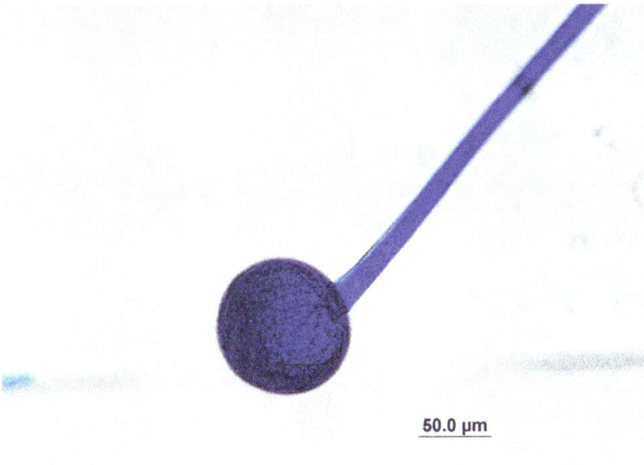

Fig. 6.6.21 *R. oryzae*, mature sporangium.

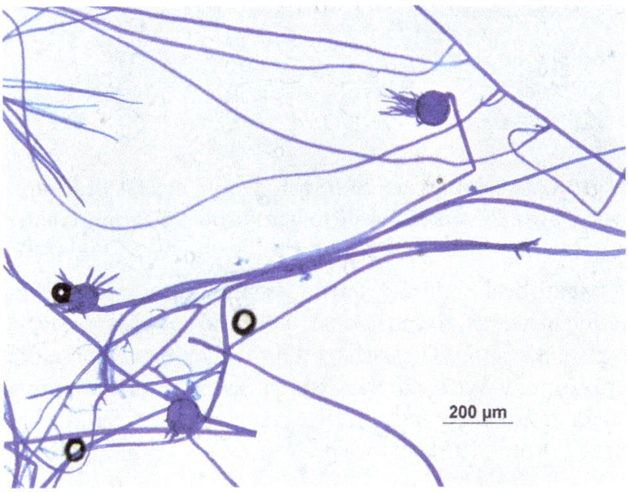

Fig. 6.6.22 Rarely seen, stolons produced from sporangia; 100×.

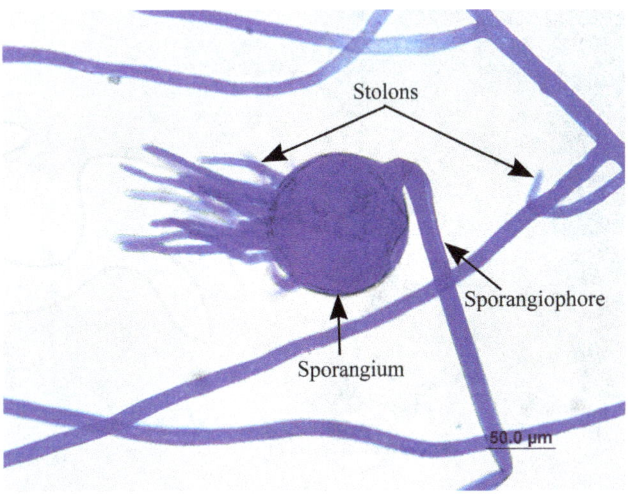

Fig. 6.6.23 Abundant stolons, 400×.

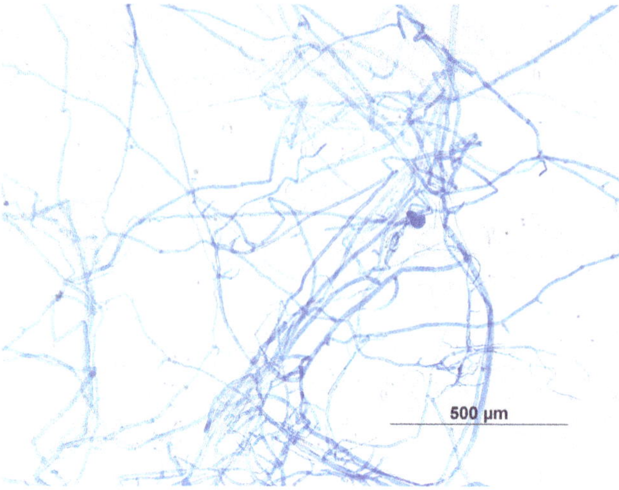

Fig. 6.6.24 Aseptate irregular hyphae, 100×.

6.7 *Syncephalastrum racemosum*

Colonial morphology: rapidly growing colonies are cottony to wooly, at first white and later developing concentric rings of gray. Reverse is white to light brown. A single colony will fill the petri dish in 5 days.

Microscopic morphology: hyphae are hyaline, broad, and non-septate. Sporangiophores are single or branched, terminating in a large, round vesicle. A conical apophysis may be seen to attach the vesicle to the sporangiophore. Cigar-shaped sporangia (merosporangia) (4–6 × 9–60 µm) radiate outward from the vesicle in the shape of a sunflower. Sporangia produce many sporangiospores in consecutive fashion that chain together. Spores are round or sometimes rectangular (5–8 µm). Rhizoids are present.

Syncephalastrum

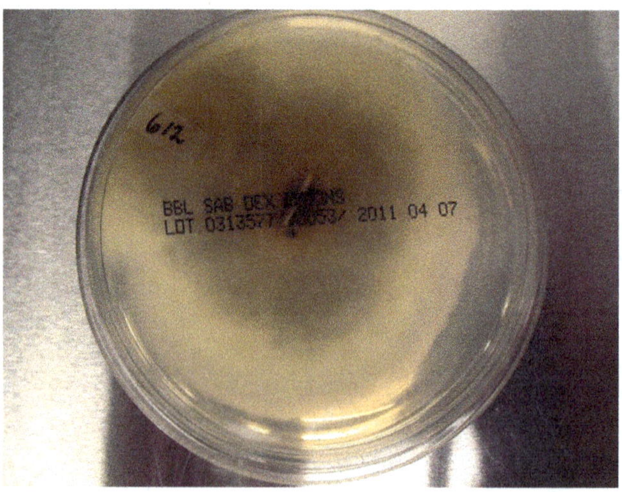

Fig. 6.7.1 *Syncephalastrum* (bottom).

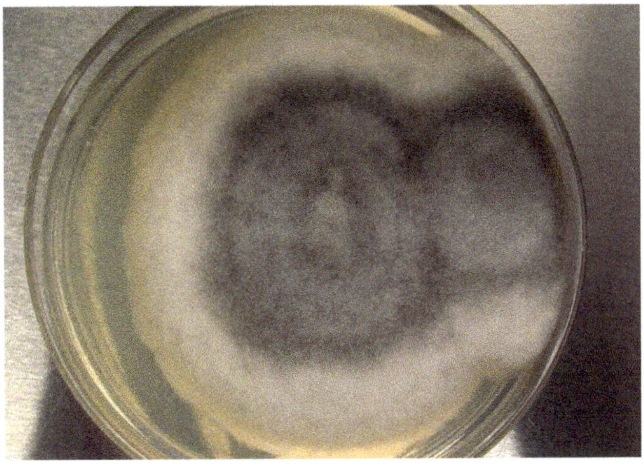

Fig. 6.7.2 *Syncephalstrum* (young, top).

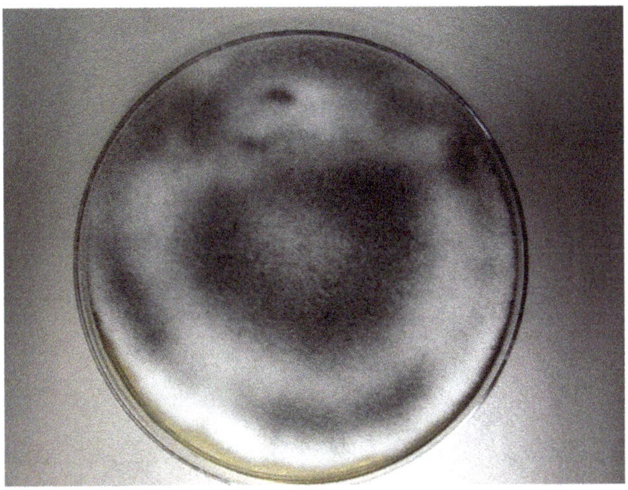

Fig. 6.7.3 *Syncephalastrum* (mature, top).

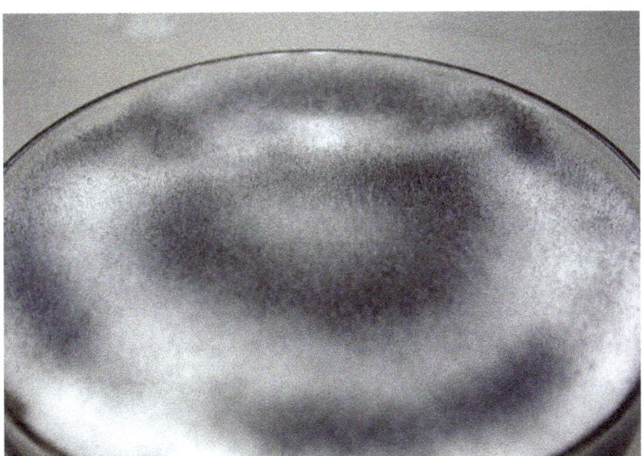

Fig. 6.7.4 *Syncephalastrum* (mature, side).

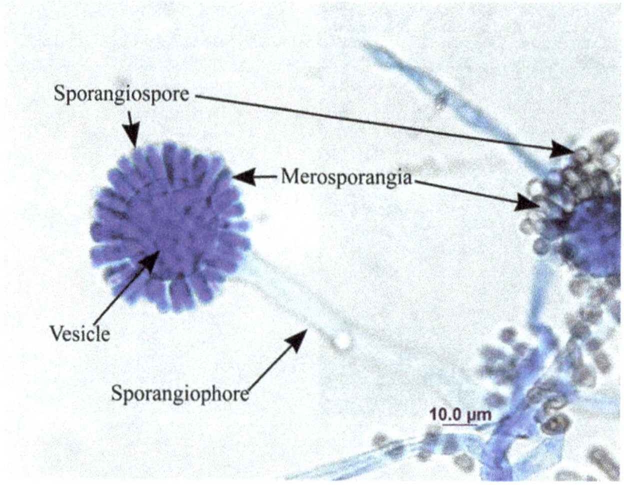

Fig. 6.7.5 *S. racemosum*, showing basic structure.

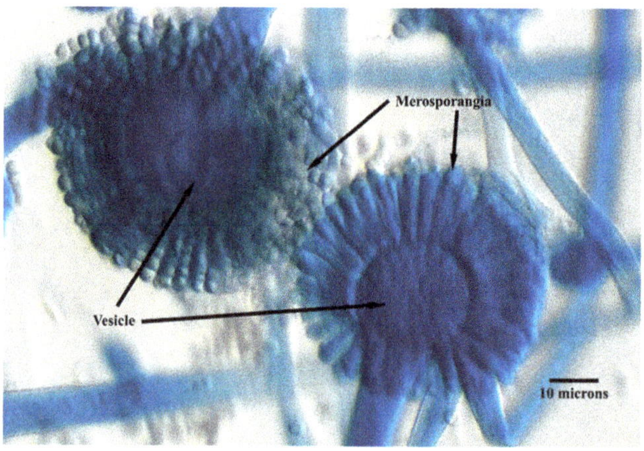

Fig. 6.7.6 Showing vesicle and merosporangia.

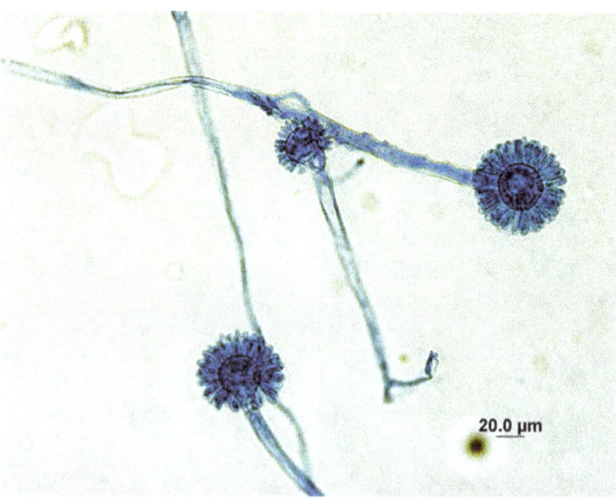

Fig. 6.7.7 Early LCPB preparation (no spores).

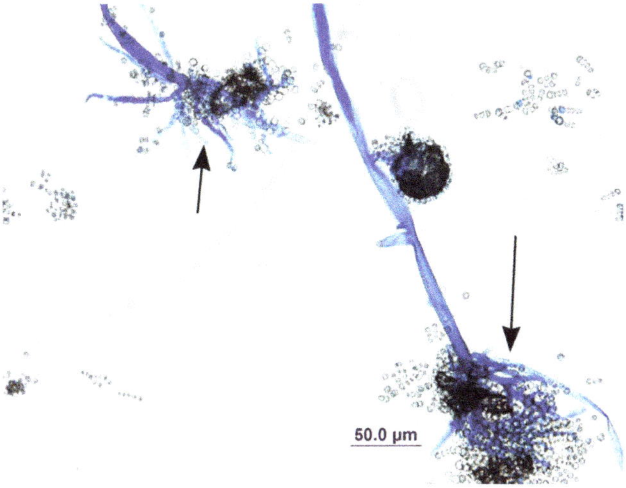

Fig. 6.7.8 Rhizoids present.

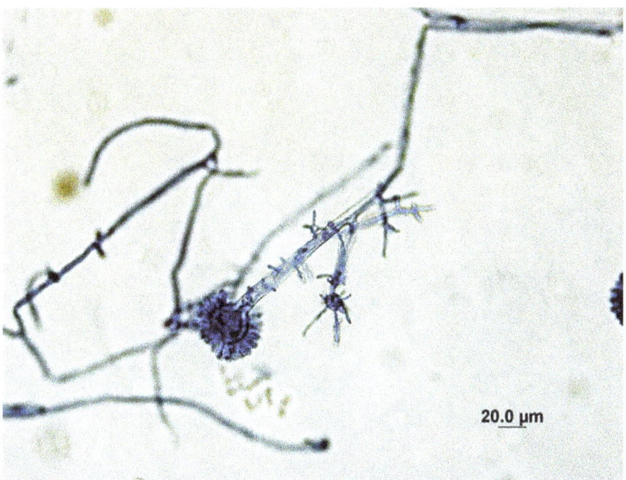

Fig. 6.7.9 Rhizoids along sporangiophore.

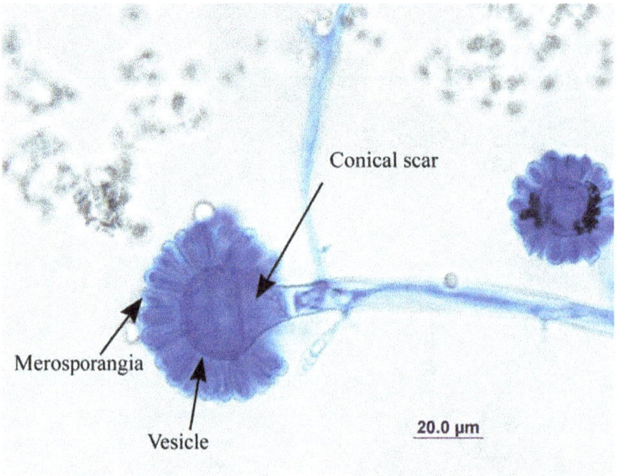

Fig. 6.7.10 Vesicle with conical scar.

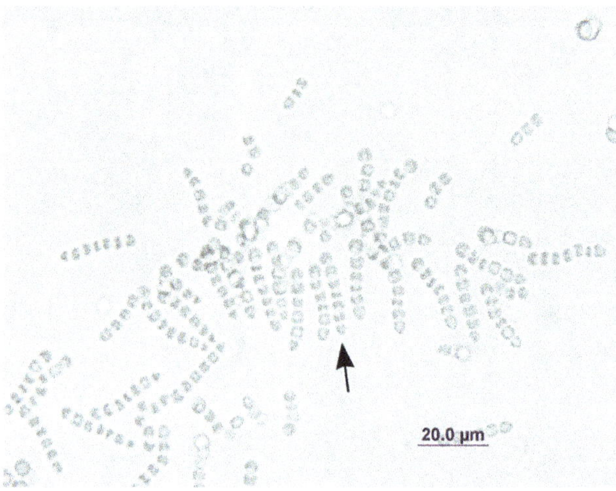

Fig. 6.7.11 Sporangiospores in chains.

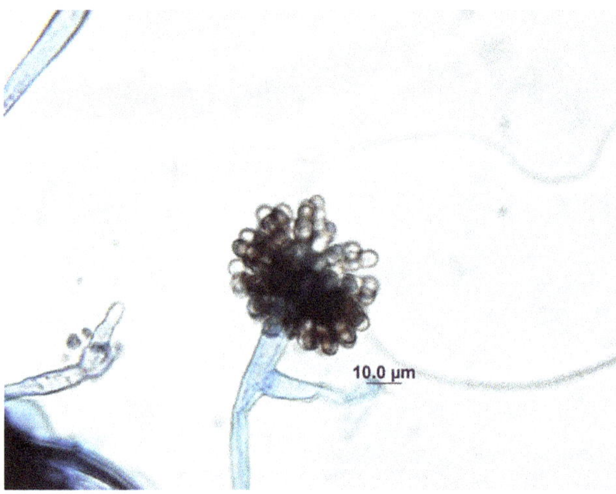

Fig. 6.7.12 Mature sporangia with sporangiospores.

7

Dermatophytes

Dermatophytes are those fungi associated with invasion of the skin, nails, hair, and scalp. They include *Trichophyton, Microsporum,* and *Epidermophyton.* Keratin is their main source of nutrients. Culture and morphological growth characteristics are insufficient to adequately speciate these fungi. This text does not discuss the many culture and morphological characteristics of these fungi.

Only a few characteristic photos of these fungi are shown. Culture media for the identification and differentiation of these fungi are comprehensive and include several selective and differential media. More recent advances in microbiology for identification of these microorganisms include matrix-assisted laser desorption ionization time-of-flight mass spectroscopy, RNA analysis by PCR or PFGE, and oligonucleotide array analysis. Therefore, each laboratory should establish its own protocol and technique for the identification of these microorganisms.

7.1 *Microsporum* spp

Colonial morphology: colonies have a moderate growth rate. Colonies are white to tan, with the center of the colony of some species becoming light yellow or pink. Colonies often fill the surface of the petri dish and are powdery or rough. They are flat and not heaped and seldom form depressions or frills. The reverse color varies with species but is mostly yellow, with some species displaying patches of orange, red, brown or violet.

Microscopic morphology: septate hyphae display irregular sides and are fine to coarse. Species morphology varies, so identification is mostly based upon the characteristic developmental structure of macro- and microconidia. All species except *M. ferrugineum* produce both macroconidia and microconidia. *M. ferrugineum* characteristically displays favic chandeliers with few other morphological characteristics. Macroconidia are fusiform and segmented into 2–15 cells. *M. nanum* characteristically has balloon-shaped macroconidia with only 2–3 cells. Typically, the number of cells/macroconidia and the outer cell wall morphology, which is either echinulate or smooth, are used to speciate isolates. Microconidia may be rare or numerous. Chlamydospores and hyphal racquet cells are frequently seen.

Atlas of Clinically Important Fungi, First Edition. Carmen V. Sciortino.

Speciation of *Microsporum* is primarily accomplished by comparing morphology with biochemical and physiological (hair perforation) tests. Therefore, due to the variation in morphology, the reader is referred to more specialized texts that describe morphology in detail for each species.

Microsporum spp

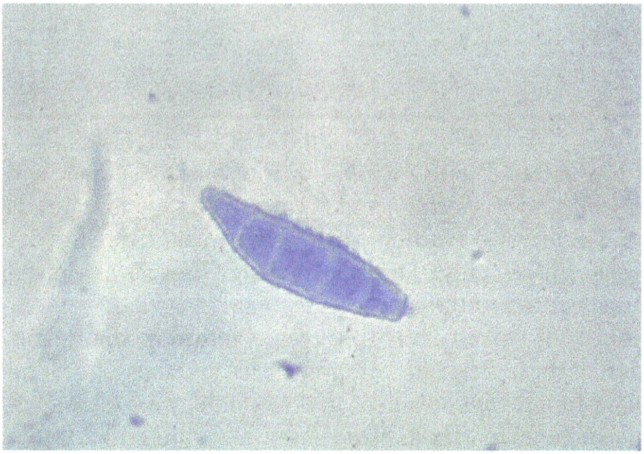

Fig. 7.1.1 *Microsporum audouinii* conidium.

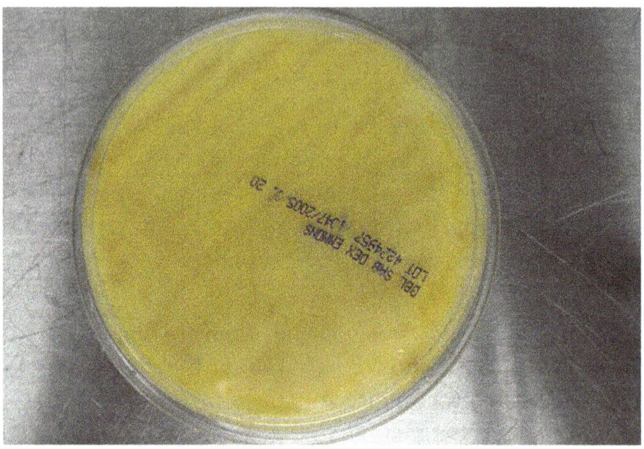

Fig. 7.1.2 *Microsporum canis* (bottom).

Fig. 7.1.3 *Microsporum canis* (top).

Fig. 7.1.4 *Microsporum canis* (side).

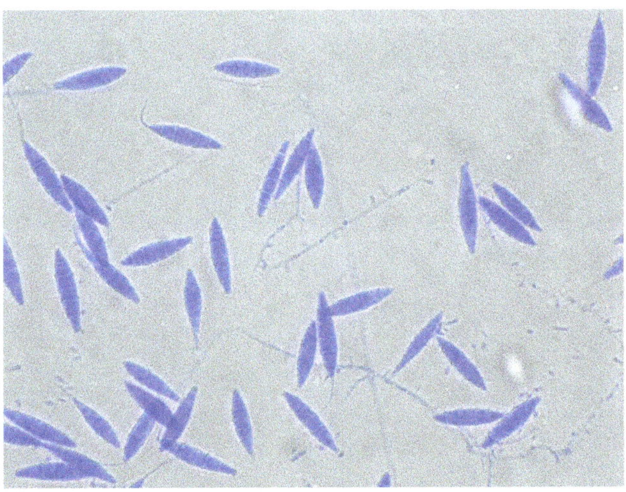

Fig. 7.1.5 *M. canis*.

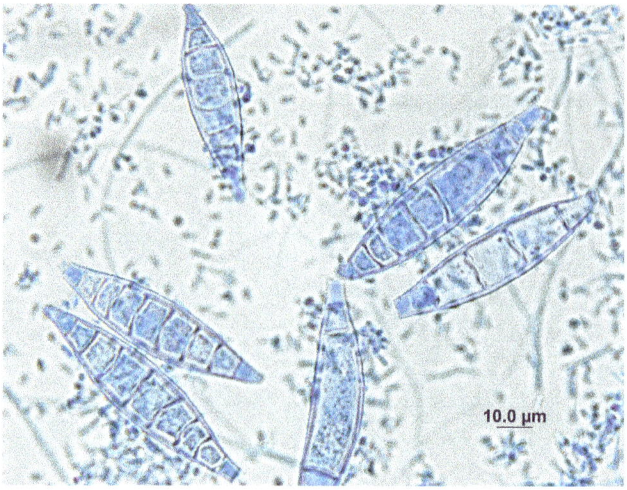

Fig. 7.1.6 *M. canis*.

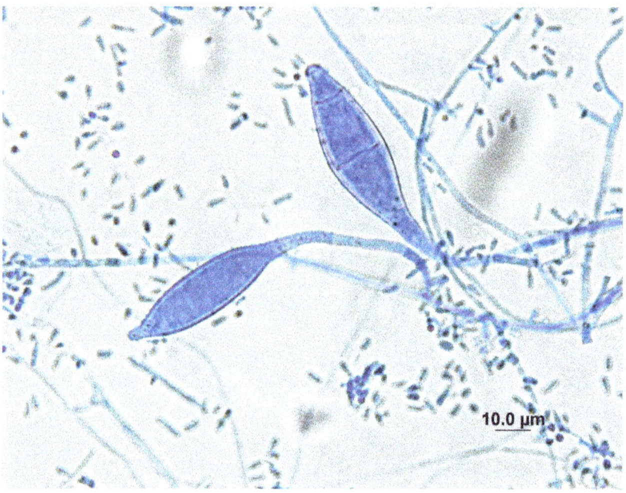

Fig. 7.1.7 *M. canis*.

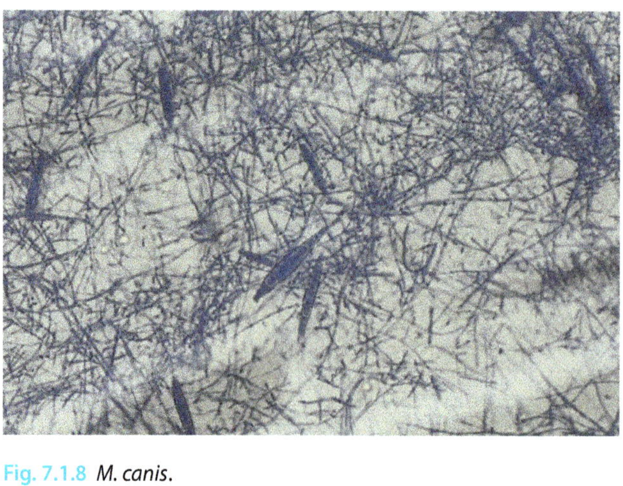

Fig. 7.1.8 *M. canis*.

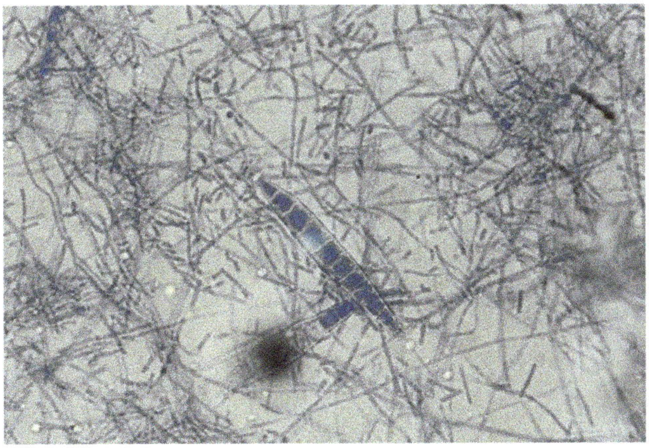

Fig. 7.1.9 *M. canis*.

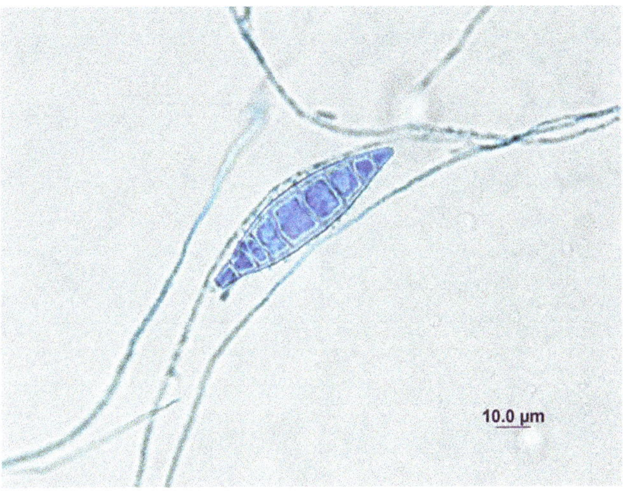

Fig. 7.1.10 *M. canis*.

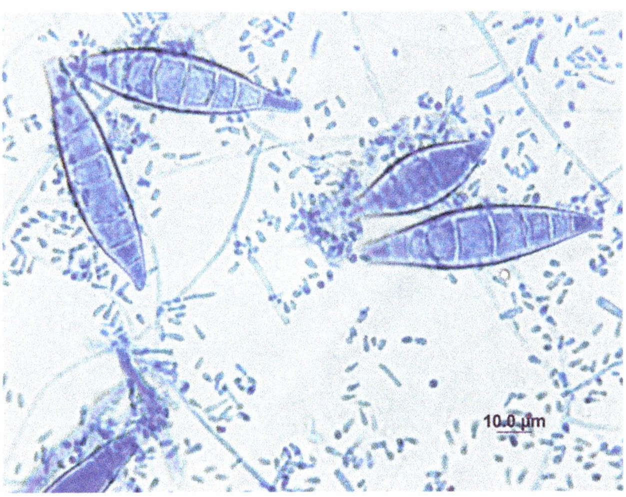

Fig. 7.1.11 *M. canis*.

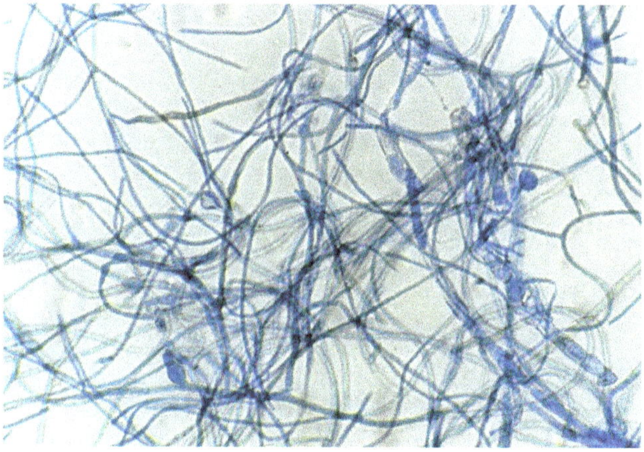

Fig. 7.1.12 *Microsporum ferrugineum*.

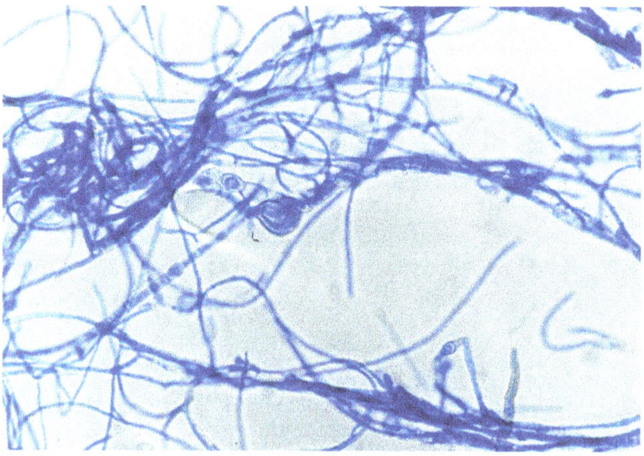

Fig. 7.1.13 *Microsporum ferrugineum*.

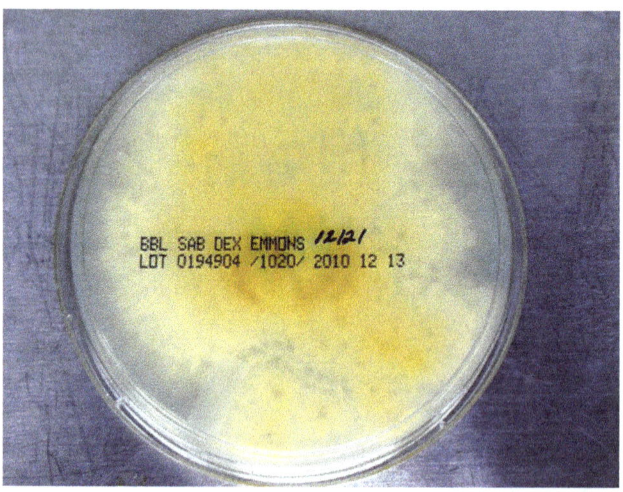

Fig. 7.1.14 *Microsporum gypseum* (bottom).

Fig. 7.1.15 *Microsporum gypseum* (top).

Fig. 7.1.16 *Microsporum gypseum* (side).

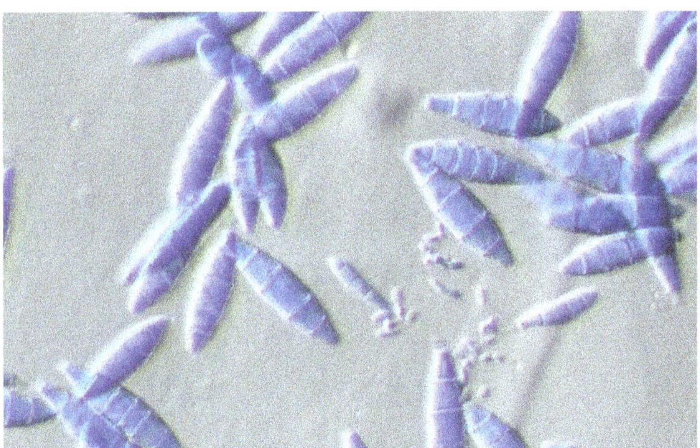

Fig. 7.1.17 *M. gypseum*.

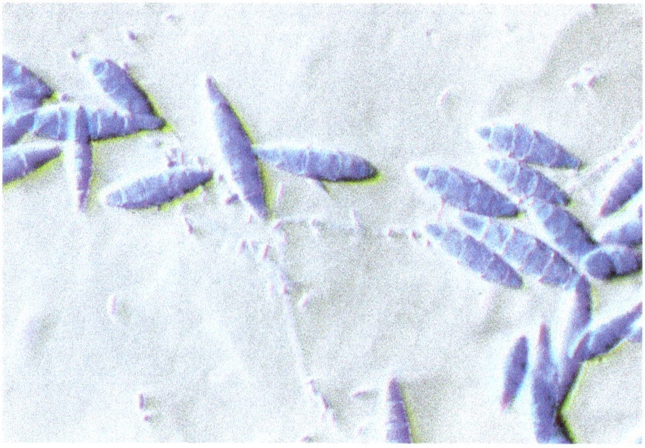

Fig. 7.1.18 *M. gypseum*.

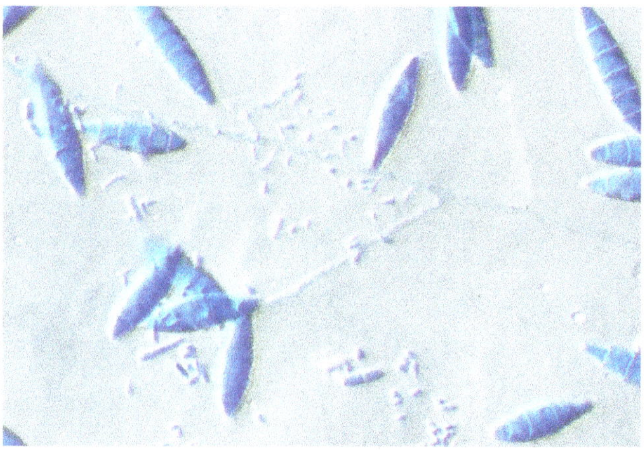

Fig. 7.1.19 *M. gypseum*.

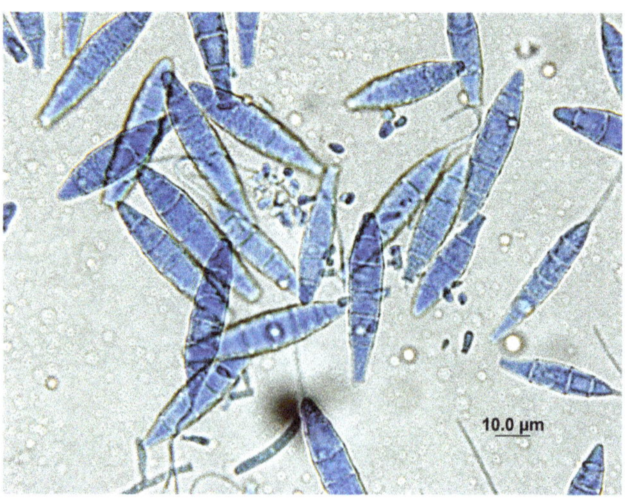

Fig. 7.1.20 *M. gypseum*.

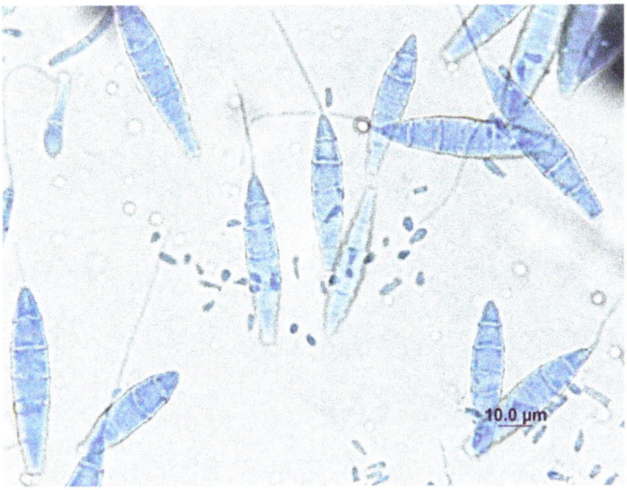

Fig. 7.1.21 *M. gypseum*.

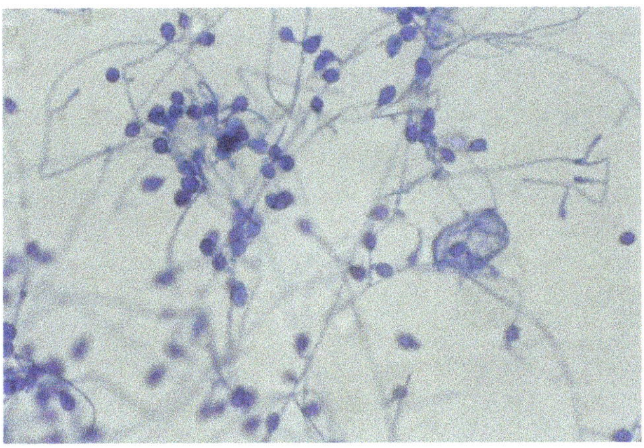

Fig. 7.1.22 *Microsporum nanum*.

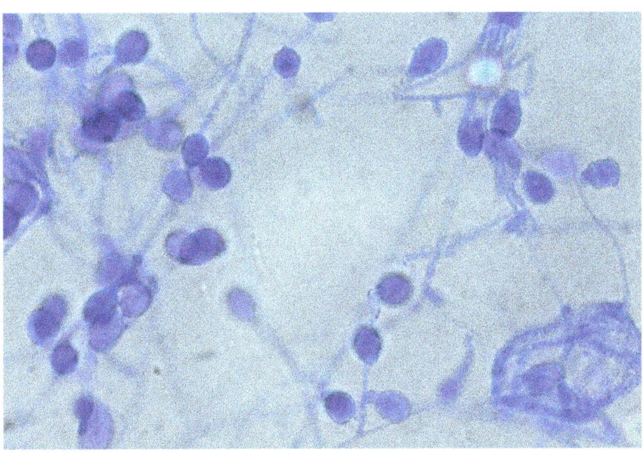

Fig. 7.1.23 *M. nanum*.

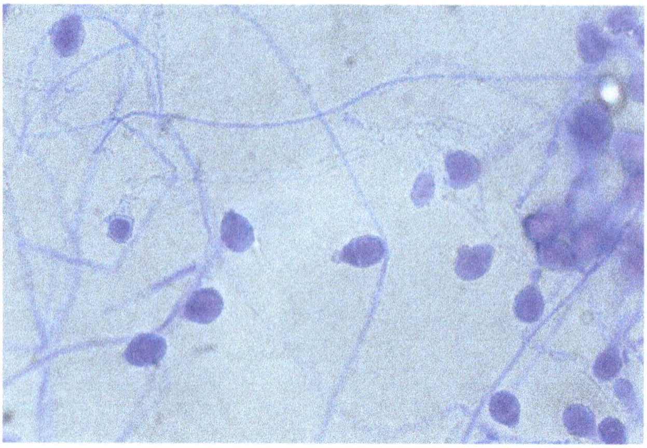

Fig. 7.1.24 *M. nanum.*

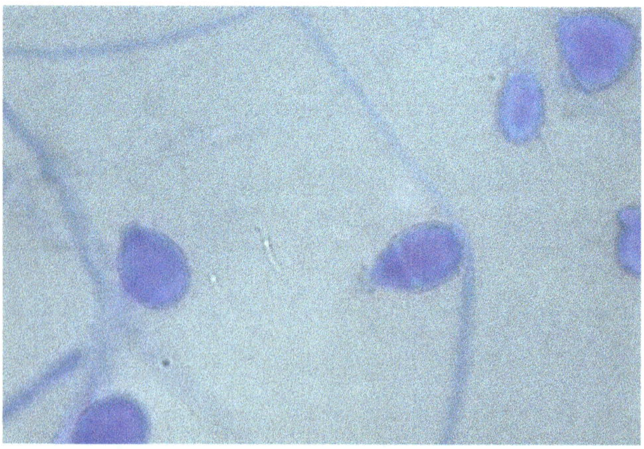

Fig. 7.1.25 *M. nanum.*

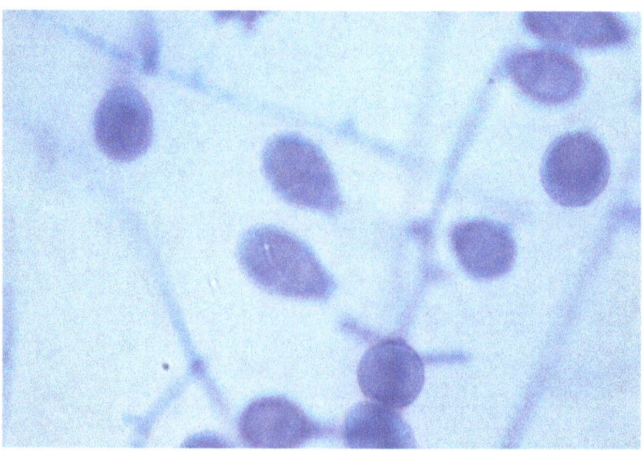

Fig. 7.1.26 *M. nanum.*

7.2 *Trichophyton* spp

Colonial morphology: colonies grow slowly and are white to buff, at first smooth, becoming fluffy. Color varies with species, from white, yellow, beige, red or violet. The reverse is pale, yellow, brown or red-brown.

Microscopic morphology: hyphae are hyaline and septate. The morphology of the various species varies, depending upon the development of smooth-walled macro- and microconidia. Microconidia form directly on the hyphae or on short conidiophores not different from hyphae. Microconidia are single cell, round, tear shaped or club shaped and may form singly or in clusters. The presence of microconidia separates this genus from *Epidermophyton*. Macroconidia are multicellular, cigar or club shaped. Species producing microconidia may not produce macroconidia. Some species do not produce conidia; rather, they produce chlamydoconidia or arthroconidia. Some species do not produce fruiting structures, but form unusual structures such as favic chandeliers (*T. schoenleinii*). *Trichophyton* is distinguished from *Microsporum* by its smooth-walled macroconidia.

Due to the variation in morphology, the reader is referred to more specialized texts that describe morphology in detail for each species. Also, growth on *Trichophyton* agars is most often used to speciate this group. However, due to the diversity of this genus, no single test can be reliably used for speciation.

Trichophyton mentagrophytes

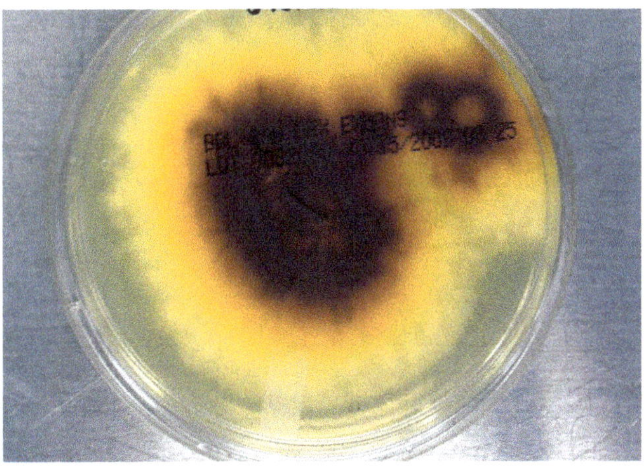

Fig. 7.2.1 *Trichophyton mentagrophytes* (bottom).

Fig. 7.2.2 *Trichophyton mentagrophytes* (top).

Fig. 7.2.3 *Trichophyton mentagrophytes* (side).

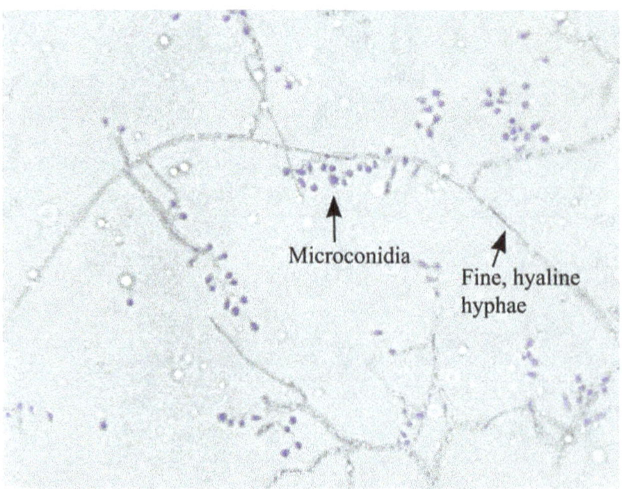

Fig. 7.2.4 *T. mentagrophytes*.

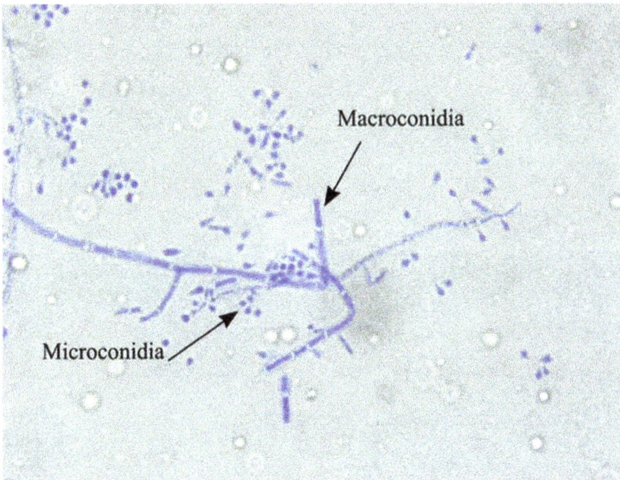

Fig. 7.2.5 *T. mentagrophytes*.

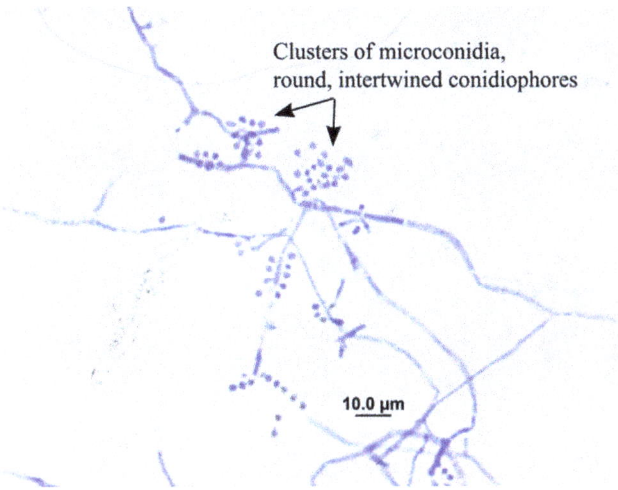

Fig. 7.2.6 *T. mentagrophytes*.

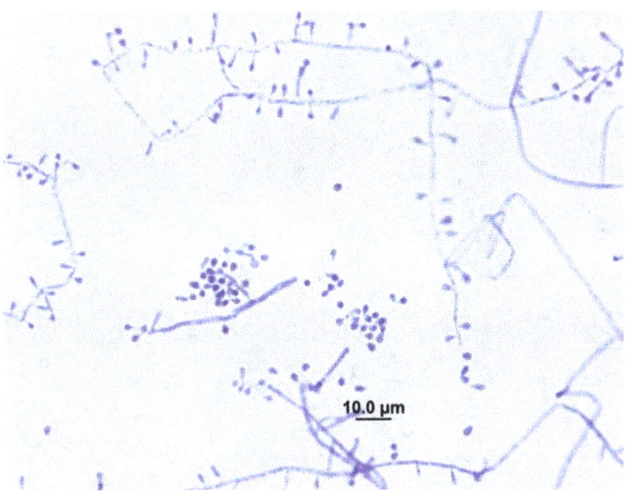

Fig. 7.2.7 *T. mentagrophytes*.

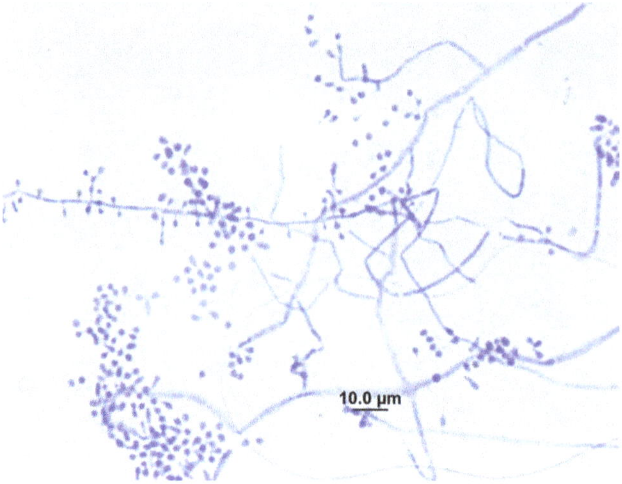

Fig. 7.2.8 *T. mentagrophytes*.

Trichophyton rubrum

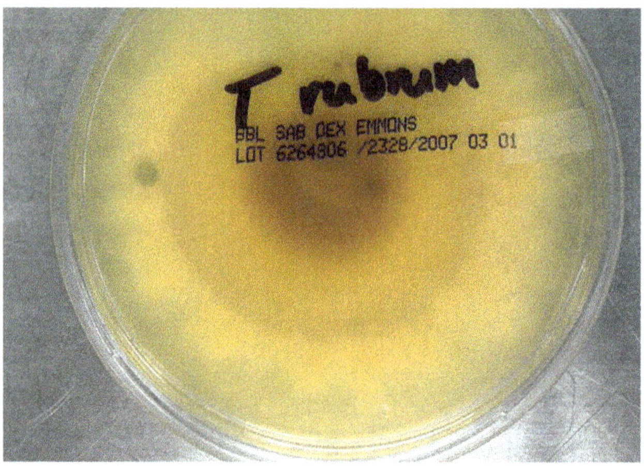

Fig. 7.2.9 *Trichophyton rubrum* (bottom).

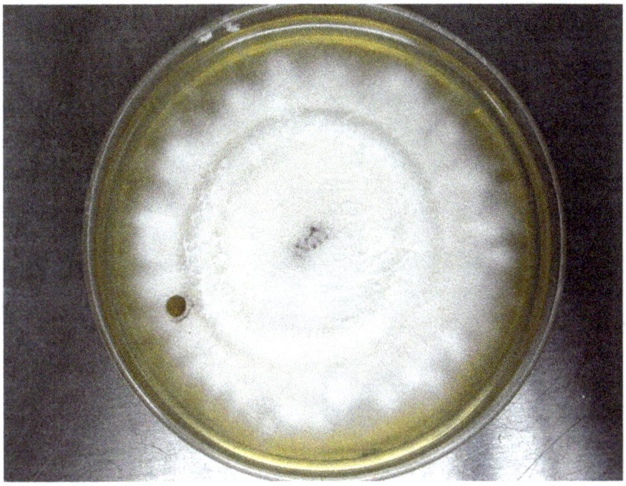

Fig. 7.2.10 *Trichophyton rubrum* (top).

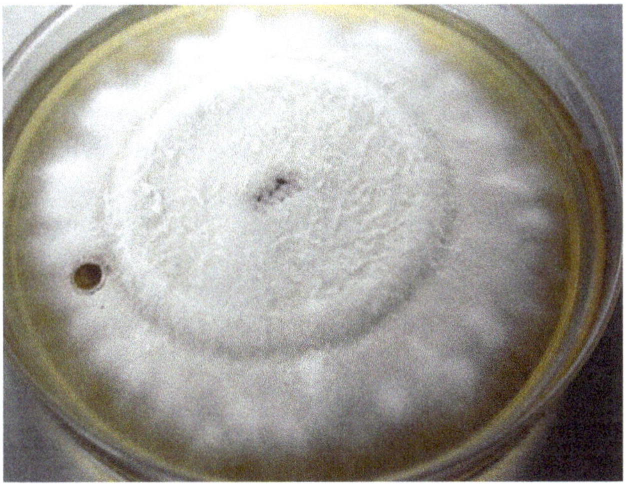

Fig. 7.2.11 Trichophyton rubrum (side).

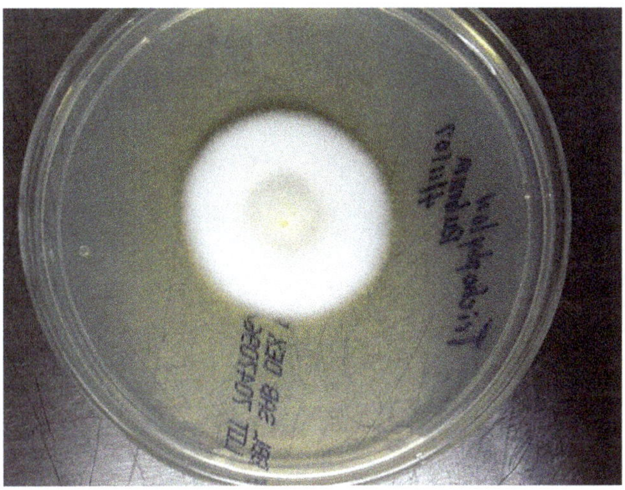

Fig. 7.2.12 *Trichophyton rubrum* (top).

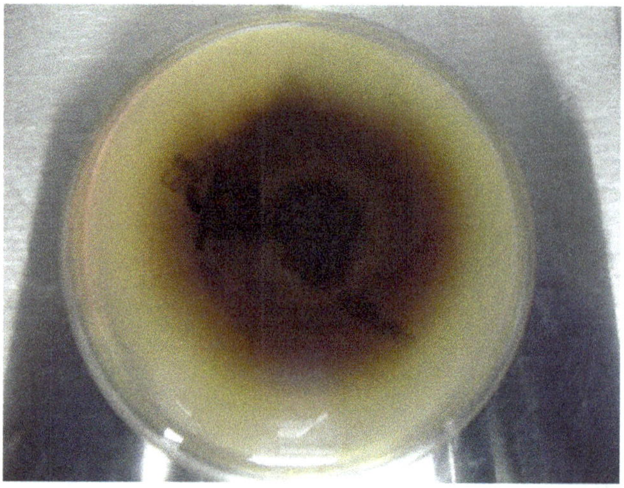

Fig. 7.2.13 *Trichophyton rubrum* (bottom).

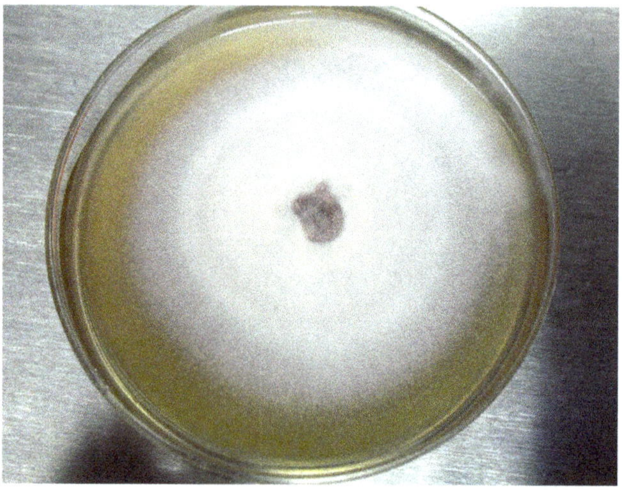

Fig. 7.2.14 *Trichophyton rubrum* (top).

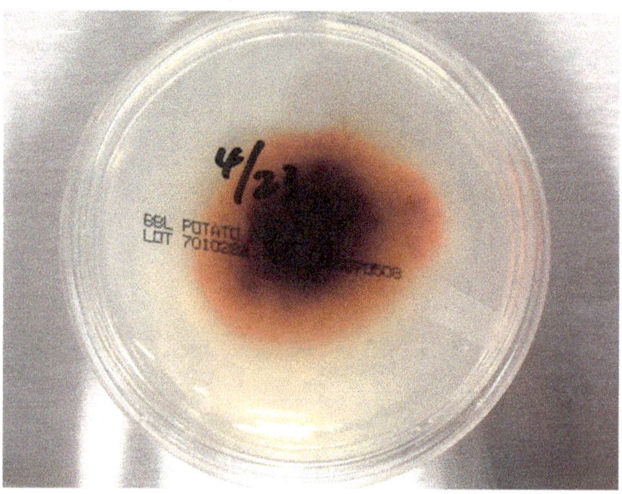

Fig. 7.2.15 *Trichophyton rubrum* (bottom).

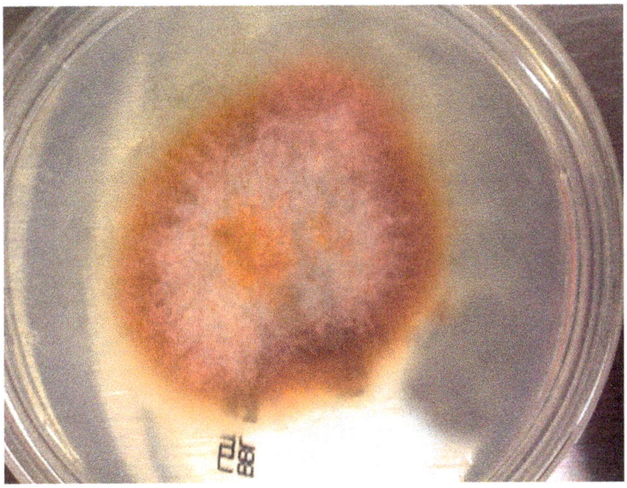

Fig. 7.2.16 *Trichophyton rubrum* (top).

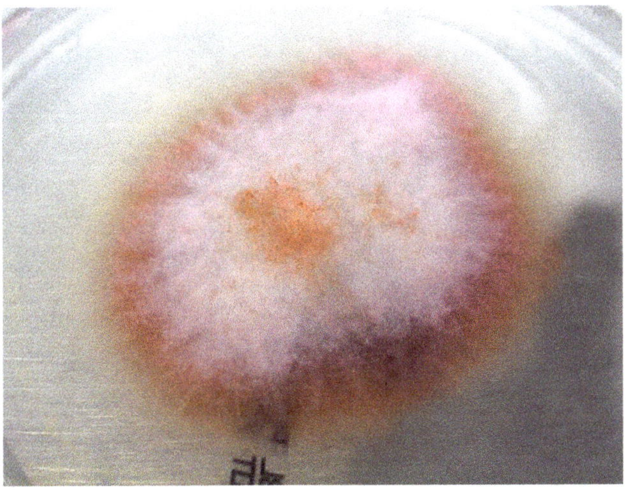

Fig. 7.2.17 *Trichophyton rubrum* (side).

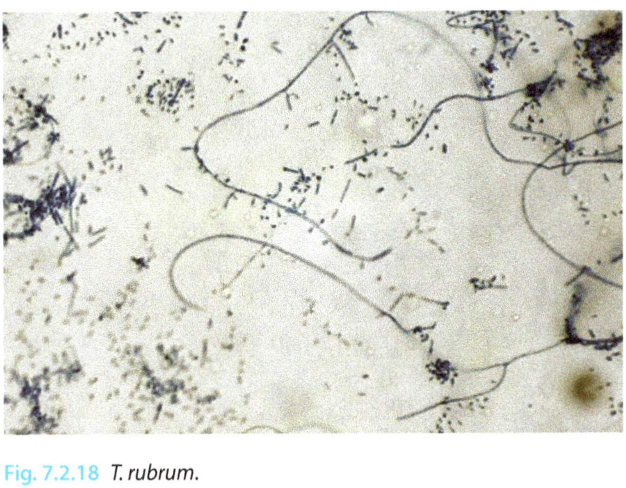

Fig. 7.2.18 *T. rubrum*.

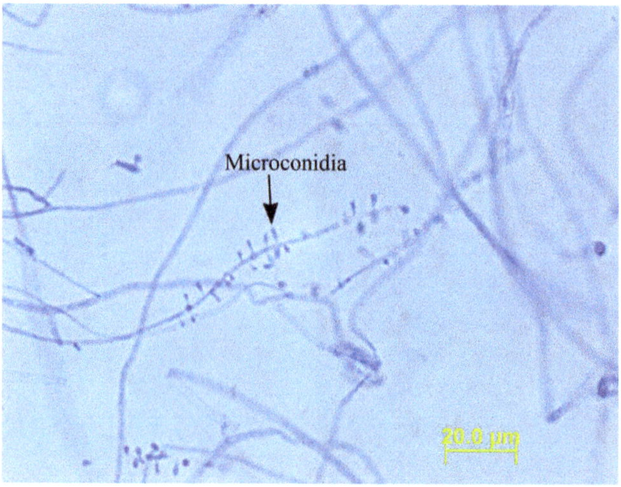

Fig. 7.2.19 *T. rubrum*.

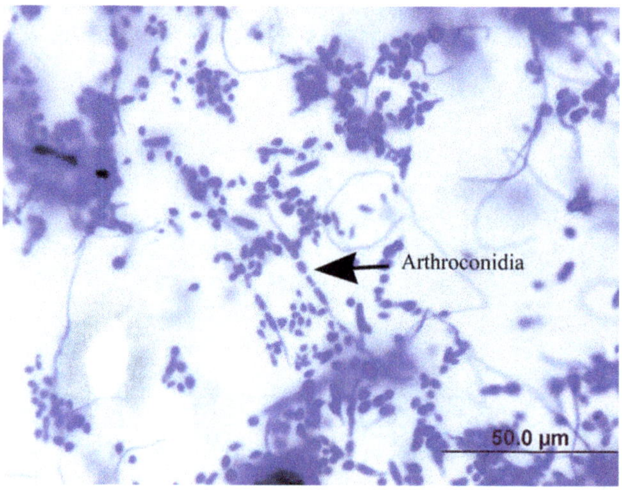

Fig. 7.2.20 *T. rubrum*.

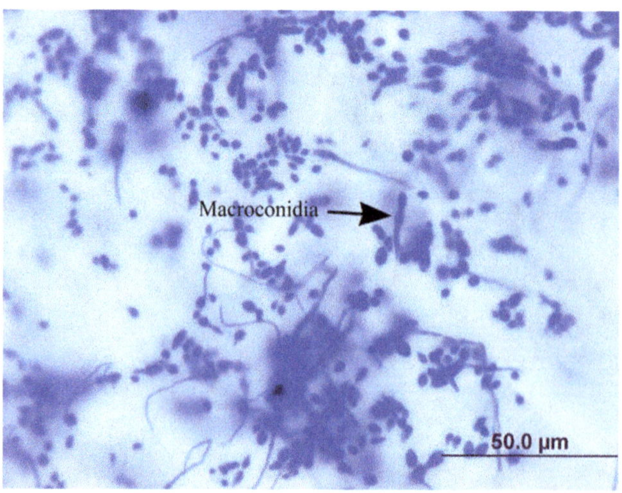

Fig. 7.2.21 *T. rubrum*.

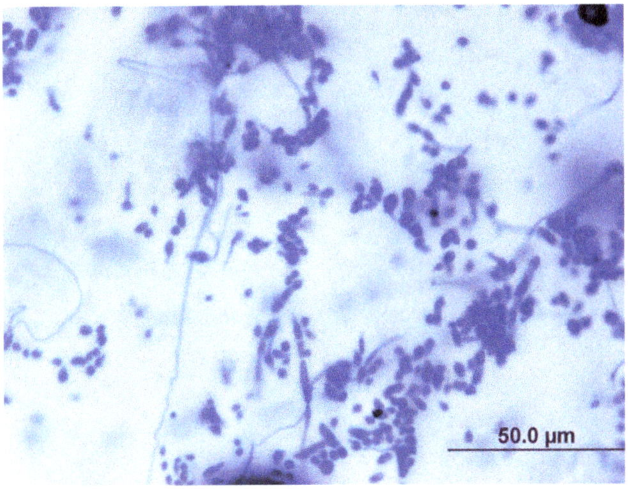

Fig. 7.2.22 *T. rubrum*.

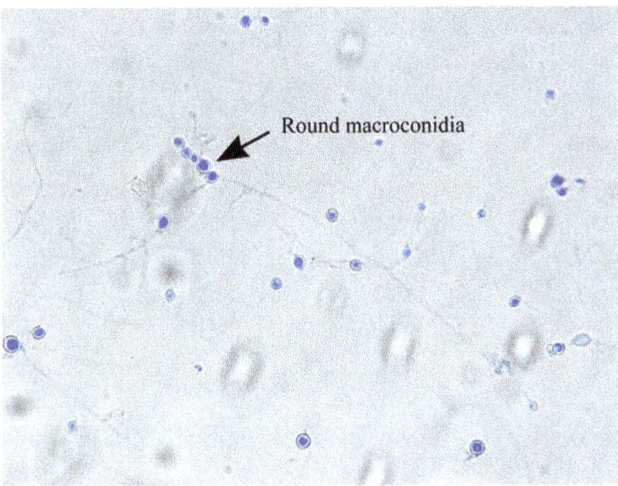

Fig. 7.2.23 *T. rubrum*.

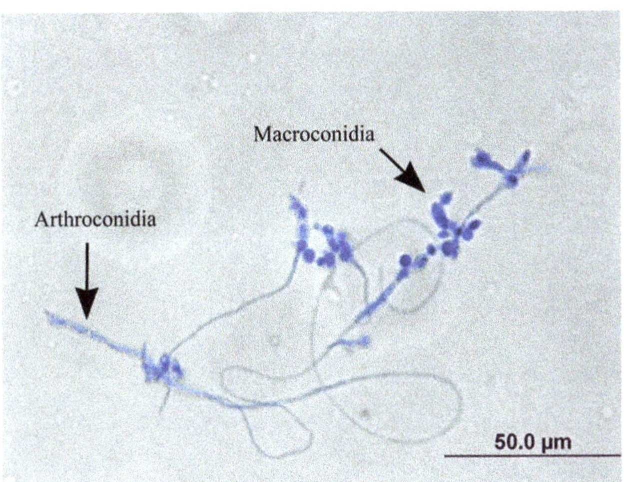

Fig. 7.2.24 *T. rubrum*.

Trichophyton schoenleinii

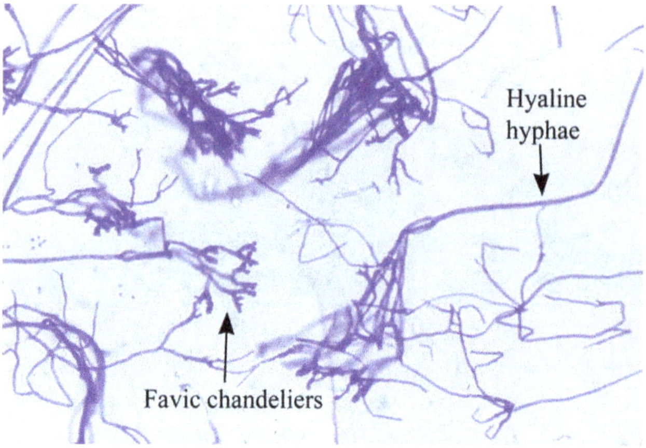

Fig. 7.2.25 *Trichophyton schoenleinii.*

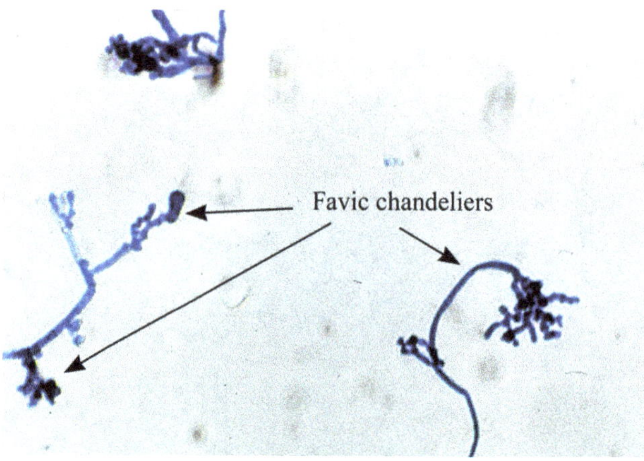

Fig. 7.2.26 *Trichophyton schoenleinii.*

Trichophyton tonsurans

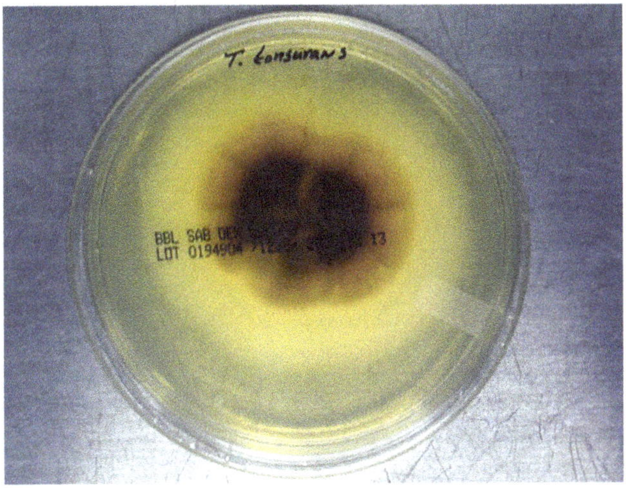

Fig. 7.2.27 *Trichophyton tonsurans* (bottom).

Fig. 7.2.28 *Trichophyton tonsurans* (top).

Fig. 7.2.29 *Trichophyton tonsurans* (side).

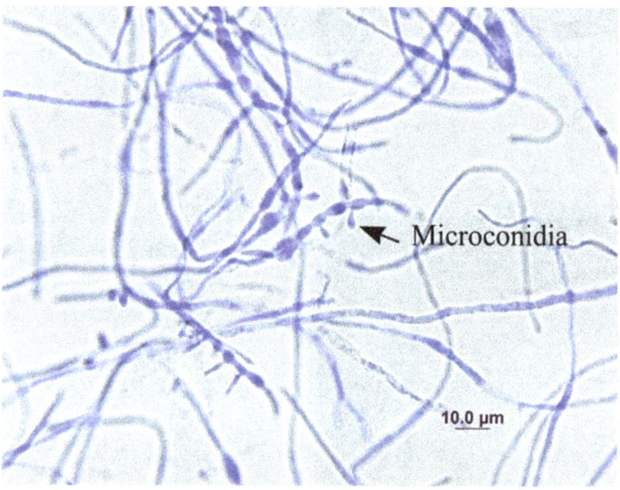

Fig. 7.2.30 *T. tonsurans*.

Fig. 7.2.31 *T. tonsurans*.

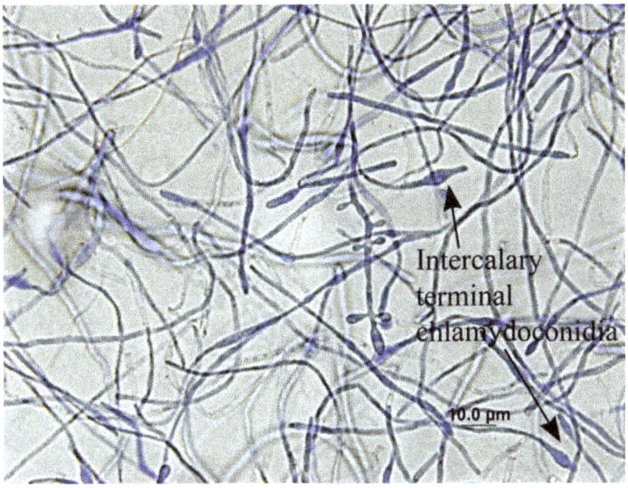

Fig. 7.2.32 *T. tonsurans.*

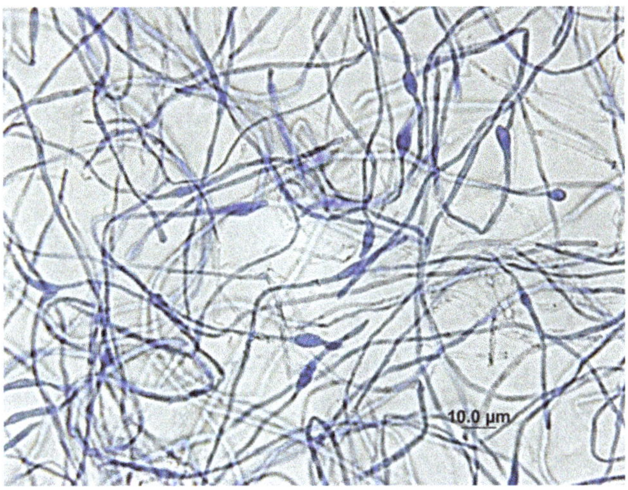

Fig. 7.3.33 *T. tonsurans.*

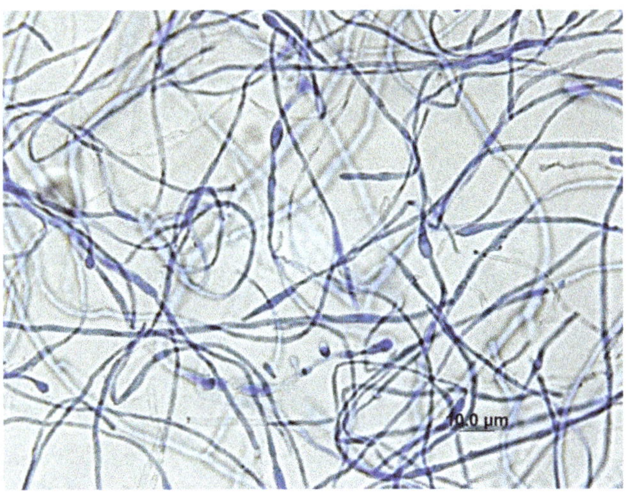

Fig. 7.2.34 *T. tonsurans.*

Trichophyton verrucosum

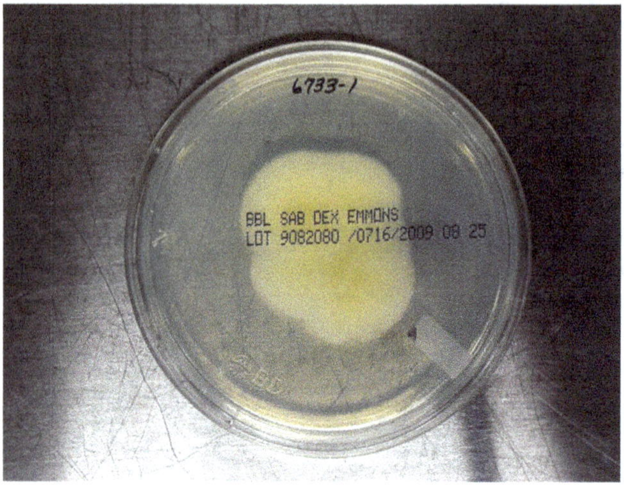

Fig. 7.2.35 *Trichophyton verrucosum* (bottom).

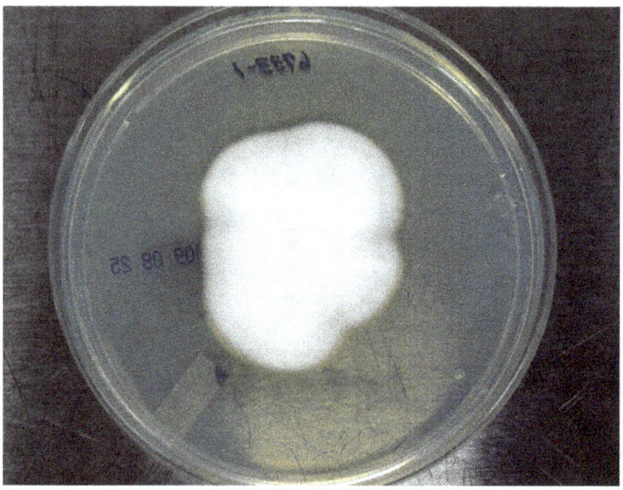

Fig. 7.2.36 *Trichophyton verrucosum* (top).

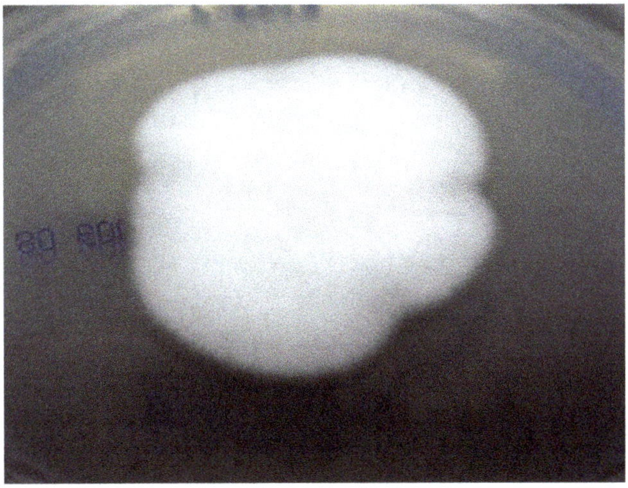

Fig. 7.2.37 *Trichophyton verrucosum* (side).

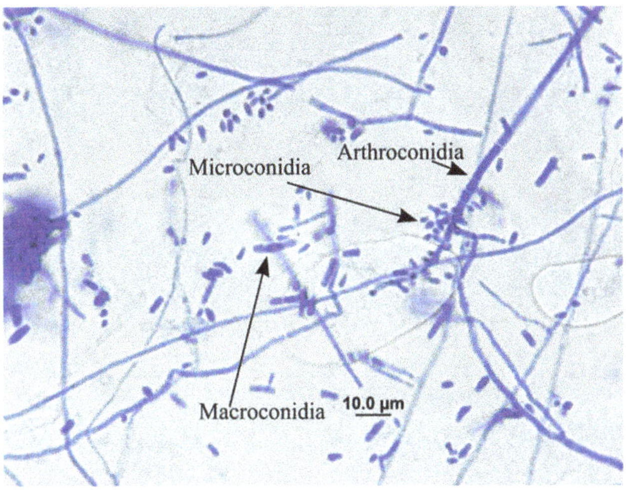

Fig. 7.2.38 *T. verrucosum*.

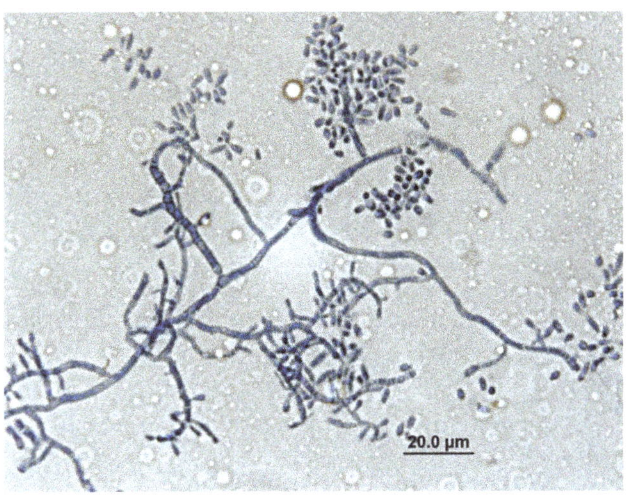

Fig. 7.2.39 *T. verrucosum*.

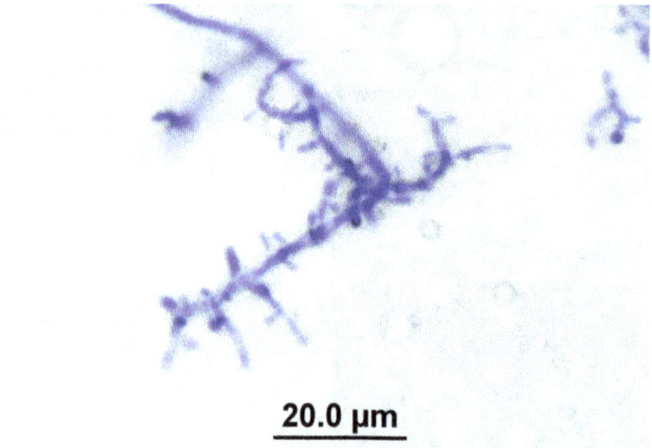

Fig. 7.2.40 *T. verrucosum*.

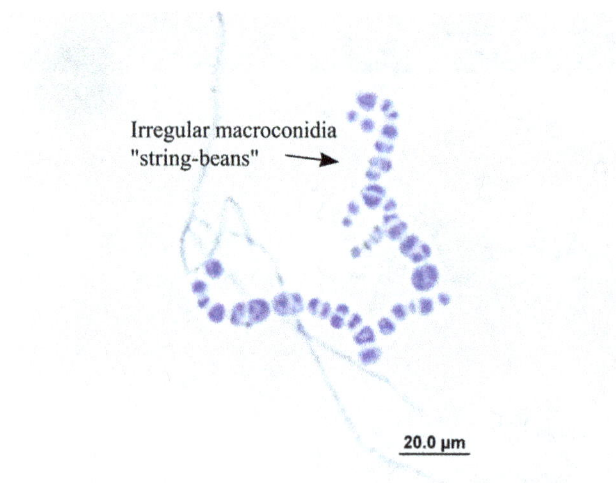

Fig. 7.2.41 *T. verrucosum*.

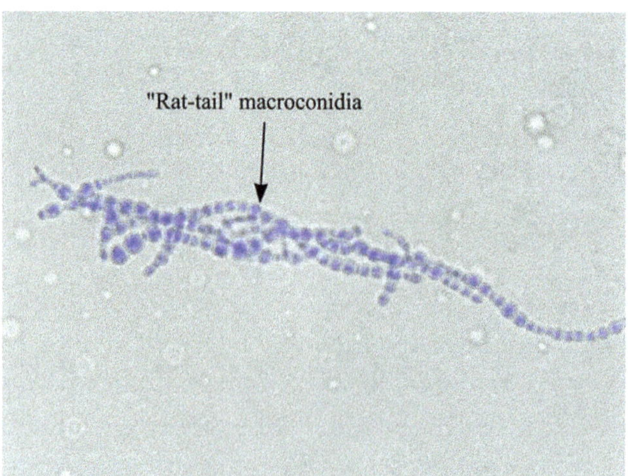

Fig. 7.2.42 *T. verrucosum*.

7.3 *Epidermophyton floccosum*

Colonial morphology: colonies are slow growing on Sabouraud dextrose agar, usually taking 10 days to mature. Colonies begin flat, white or khaki colored, and have a suede-like surface that later becomes raised and folded in the center. Color may darken to yellow or brown with age and may develop white radiating tufts of mycelia on the periphery. The reverse shows a deep yellowish-brown pigment. On potato dextrose agar and blood agar, colonies are white. Blood agar-grown cultures show colonies that appear as tiny white doughnuts on the surface of the agar. Cultures often become sterile after several weeks of incubation due to sterile aerial hyphae being produced on the surface. Sometimes, these sterile hyphae form white patches within the original colony, giving the appearance of a contaminant. To maintain viability, it is important to subculture colonies on Sabouraud dextrose agar with 3–5% sodium chloride, at least weekly.

Microscopic morphology: hyphae are septate, and may produce numerous chlamydoconidia. No microconidia are produced. Thin-walled macroconidia (7–12 × 20–40 μm) arise directly from the hyphae, singly or in clusters. These are smooth walled, club shaped, multiseptate (4–5 septations), and may develop rapidly (overnight) in some cultures. Sterile aerial hyphae are seen in cultures older than 10 days. Failure to perform weekly subcultures may cause a loss of viability and the inability to identify isolates due to the disintegration of macroconidia and replacement by sterile hyphae. Hence, the timing of microscopic observations of cultures is important to the overall identification of isolates.

Epidermophyton floccosum

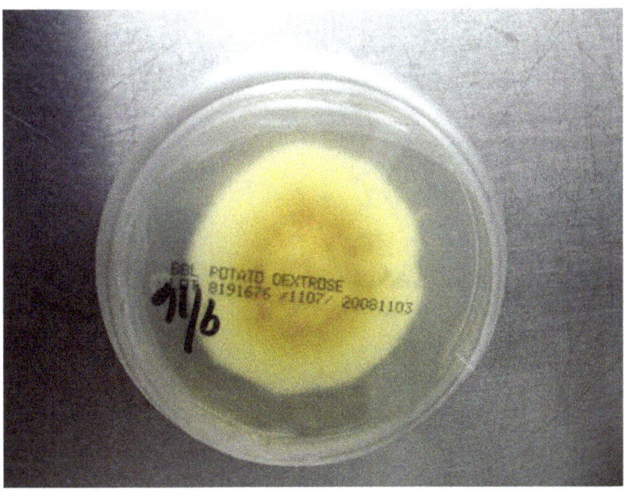

Fig. 7.3.1 *Epidermophyton* (bottom).

Fig. 7.3.2 *Epidermophyton*.

Fig. 7.3.3 *Epidermophyton* (side).

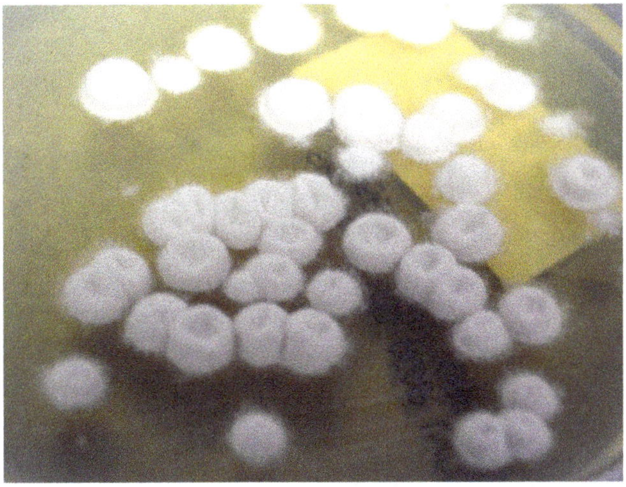

Fig. 7.3.4 *Epidermophyton*, SABHI agar.

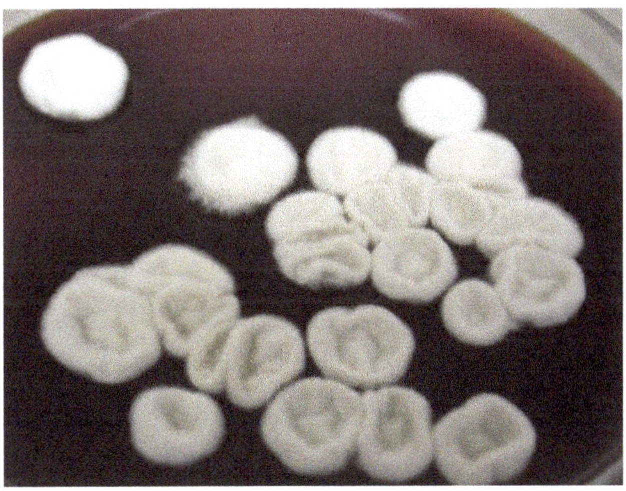

Fig. 7.3.5 *Epidermophyton*, BHI blood agar.

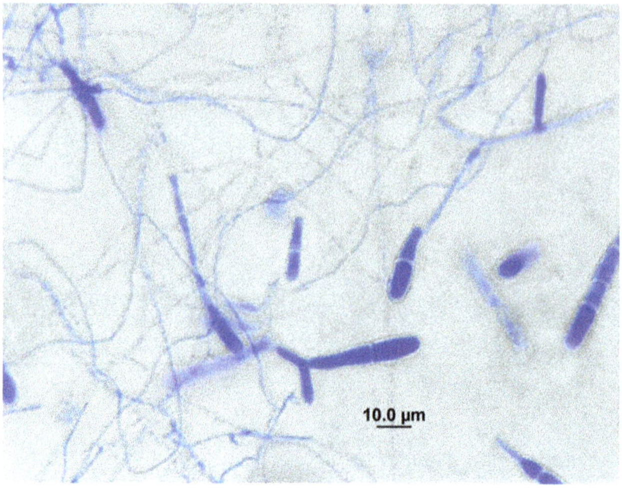

Fig. 7.3.6 *Epidermophyton*.

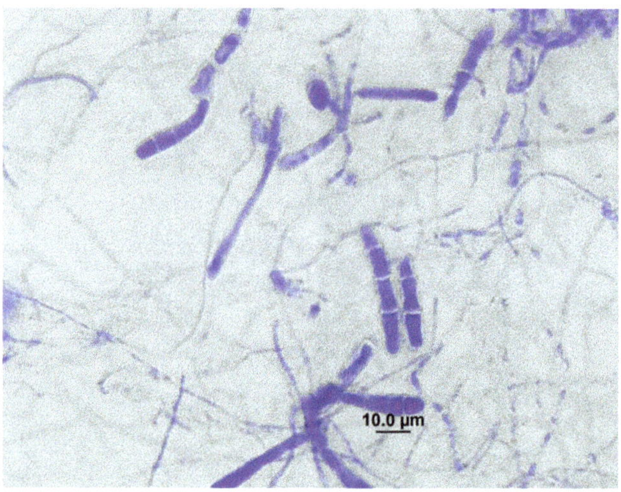

Fig. 7.3.7 *Epidermophyton*.

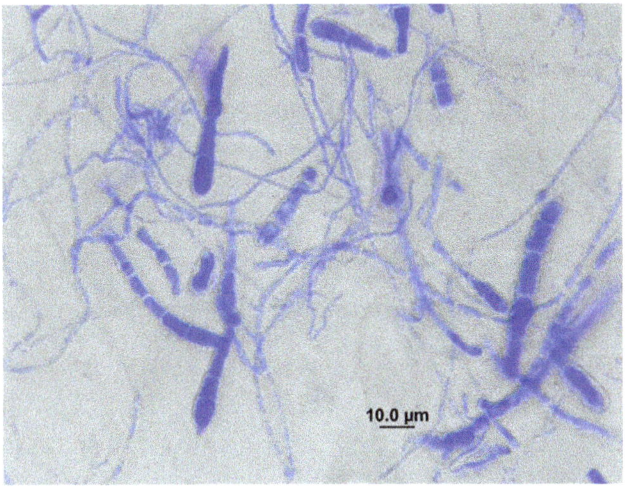

Fig. 7.3.8 *Epidermophyton*.

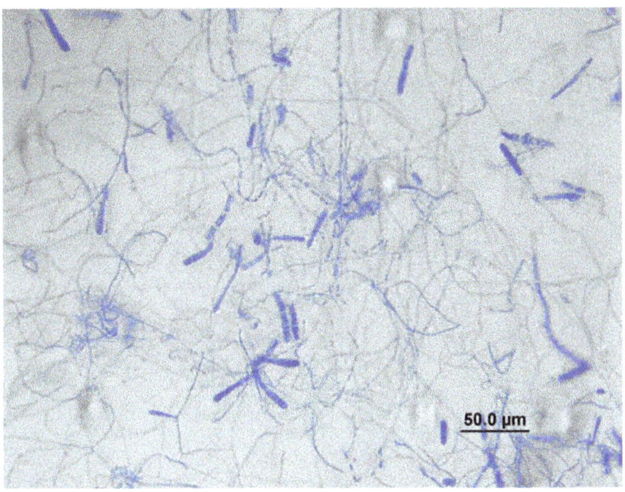

Fig. 7.3.9 *Epidermophyton*.

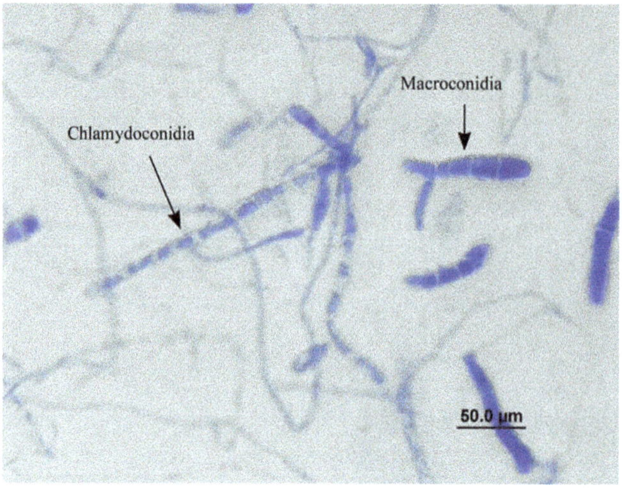

Fig. 7.3.10 *Epidermophyton*.

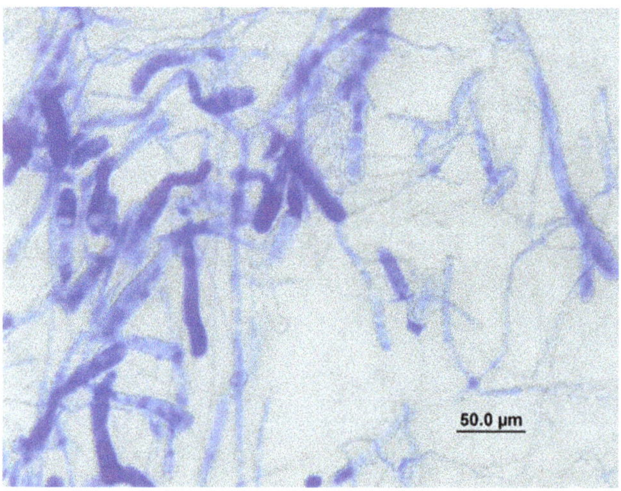

Fig. 7.3.11 *Epidermophyton*.

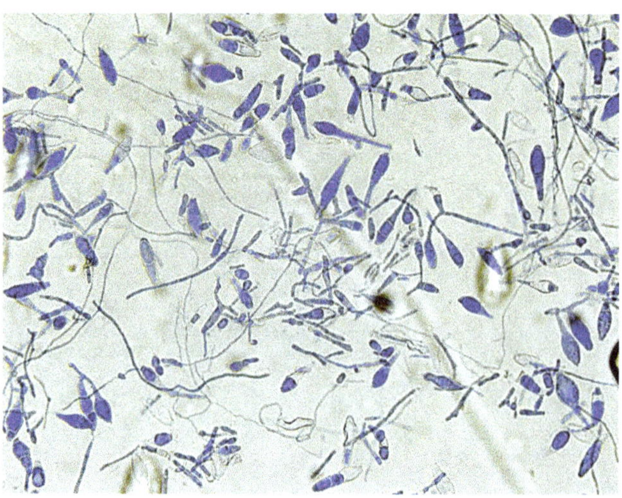

Fig. 7.3.12 *Epidermophyton*.

8

Hyaline Molds (Septate)

8.1 *Aspergillus* spp

Colonial morphology: the variations of colony morphology in *Aspergillus* are many, depending upon the species. Most species begin as white colonies, but rapidly develop colors of green, yellow, orange, black or brown. Colonies are fluffy to velvet-like and are mature in 3–5 days. Some form concentric circles of color while others are uniform in color.

Microscopic morphology: hyphae are septate and in tissue samples, branching at 45° angles. The long conidiophores arise from hyphae via a foot cell, ending in a swollen vesicle. Some environmental species form hyphae that are spiral or zig-zag in appearance. All form a vesicle, that may be round, semi-circular or club shaped. The vesicle becomes covered with phialides, occurring singly (seriate) or in pairs (biseriate) connected end to end. Some species such as *A. flavus* are both uniseriate and biseriate. From the phialides emerge the conidia that form chains. The conidia are round with a smooth or rough surface.

Morphologically, the arrangement of the phialides and their orientation on the vesicle can be used to help identify the various species. Some species such as *A. nidulans*, in their sexual stage (*Emericella nidulans*), form cleistothecia that contain ascospores. Cleistothecia appear as large round structures in which there are many ascospores, often obscuring the presence of the cleistothecia wall. When seen on LPCB preparations, gently pressing on the cleistothecia will express the ascospores. The sexual stage of *A. nidulans* also forms thickened large sterile cells that have a lumen, called Hülle cells. The basic morphological appearance and growth studies at various temperatures are used for the primary identification of the various clinically significant species. For instance, *A. fumigatus* characteristically grows at 45 °C whereas most others do not. The teleomorph stage of some *Aspergillus nidulans*, *Emericella nidulans* and *Emericella quadrilineata*, will grow at 48 °C.

Atlas of Clinically Important Fungi, First Edition. Carmen V. Sciortino.

Aspergillus spp

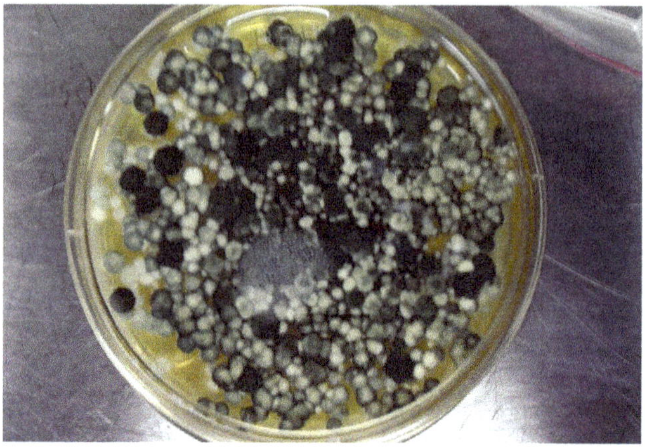

Fig. 8.1.1 *Aspergillus fumigatus* on SabHI.

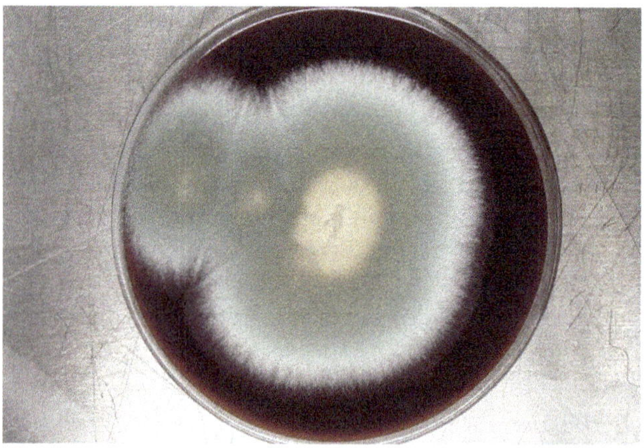

Fig. 8.1.2 *A. fumigatus* BA.

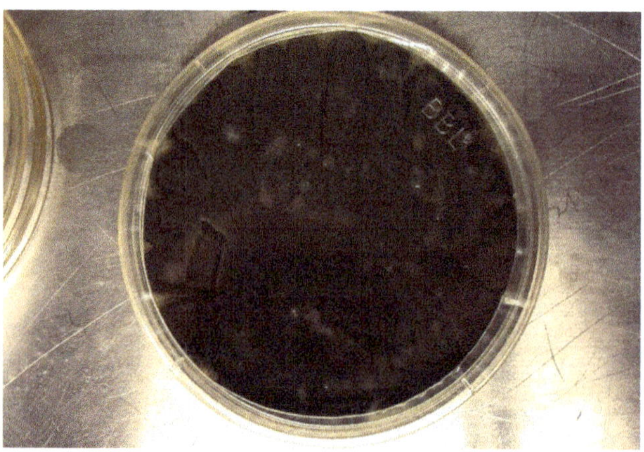

Fig. 8.1.3 *A. fumigatus* on SDX.

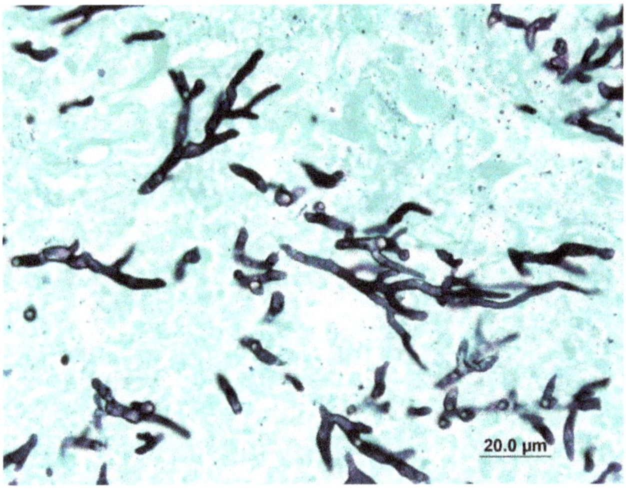

Fig. 8.1.4 *Aspergillus* in tissue.

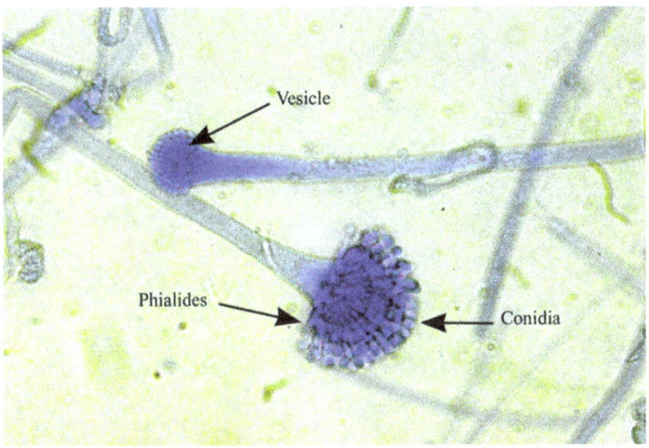

Fig. 8.1.5 *A. fumigatus*.

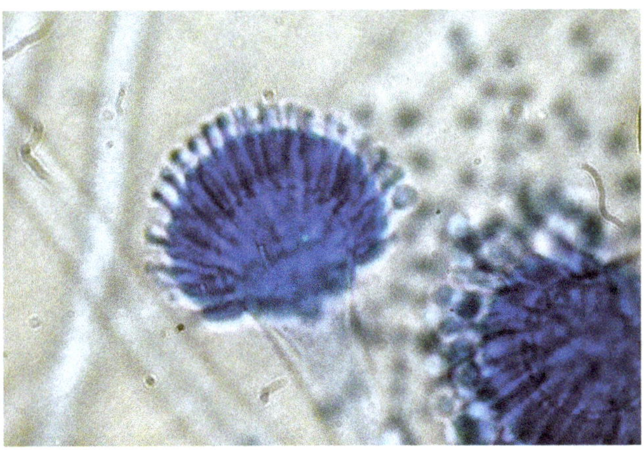

Fig. 8.1.6 *A. fumigatus* uniseriate.

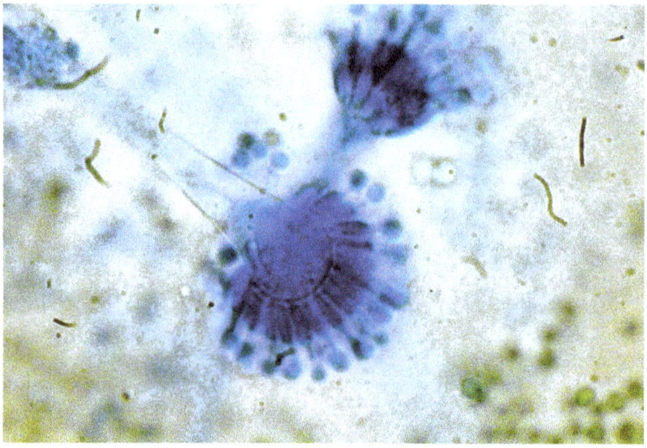

Fig. 8.1.7 *A. fumigatus*.

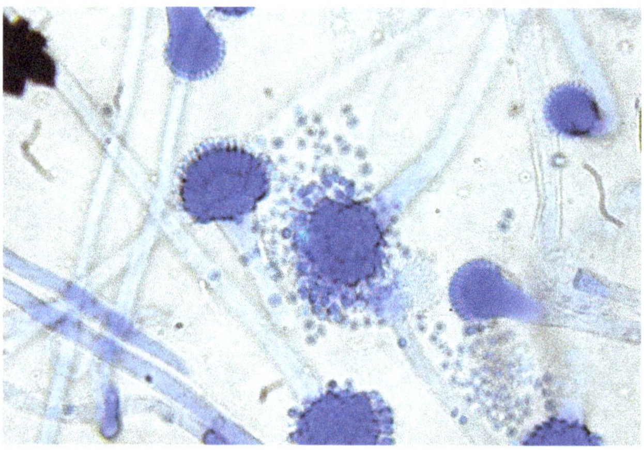

Fig. 8.1.8 *A. fumigatus*.

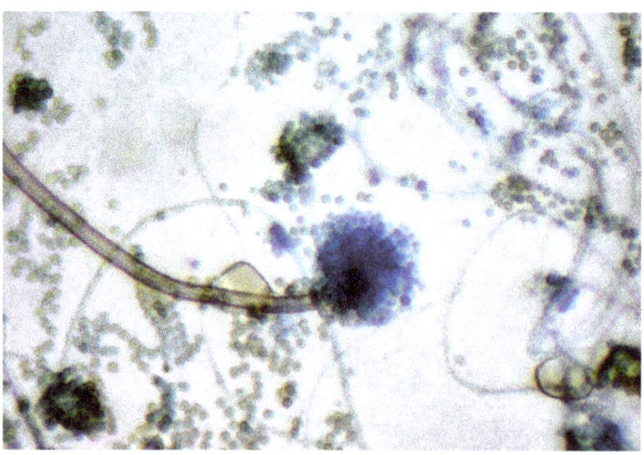

Fig. 8.1.9 *A. flavus*.

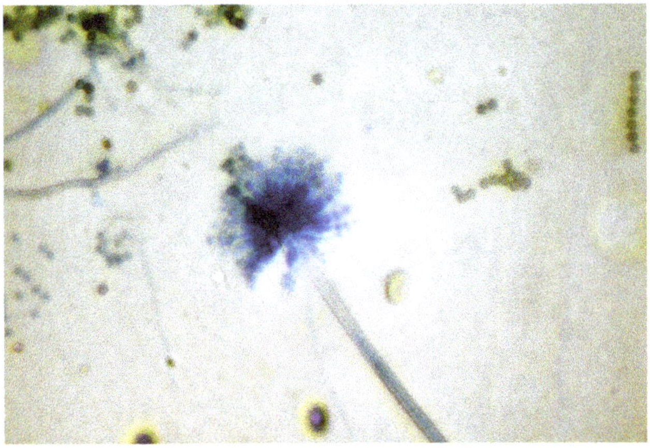

Fig. 8.1.10 *A. flavus*.

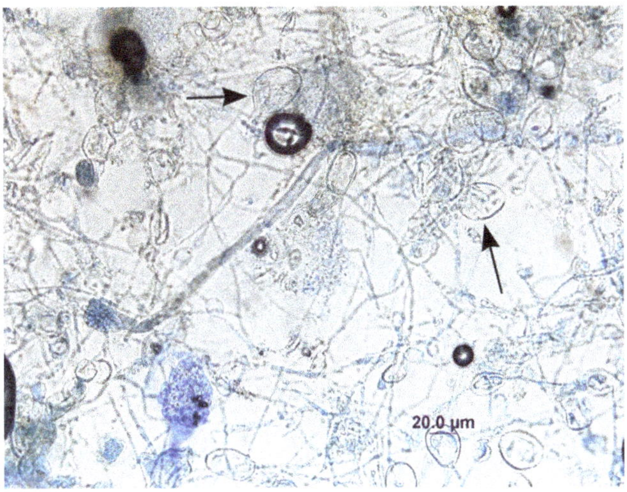

Fig. 8.1.11 *A. ustus* with Hülle cells.

Fig. 8.1.12 *A. terreus*.

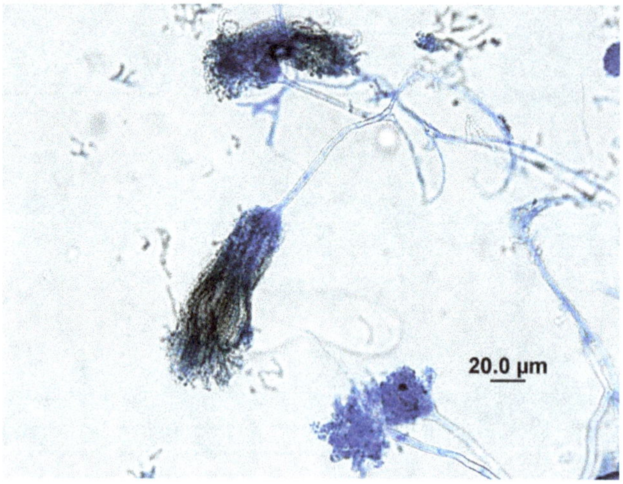

Fig. 8.1.13 *A. terreus*.

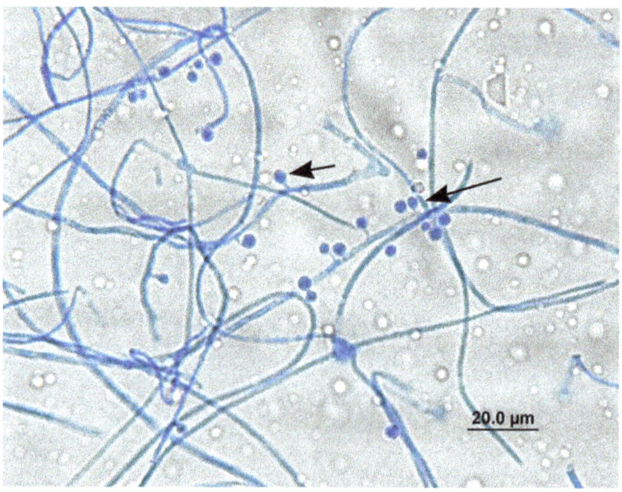

Fig. 8.1.14 *A. terreus* aleuriospores.

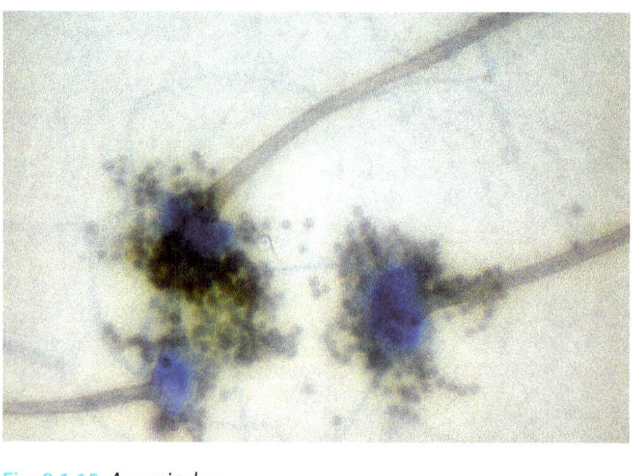

Fig. 8.1.15 *A. versicolor*.

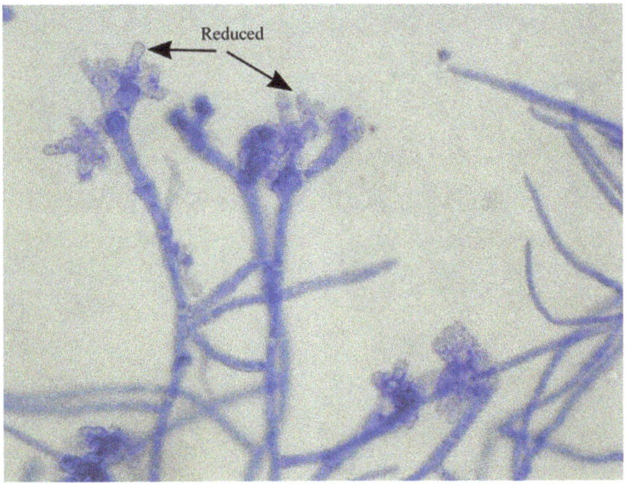

Fig. 8.1.16 *A. versicolor* with reduced structures.

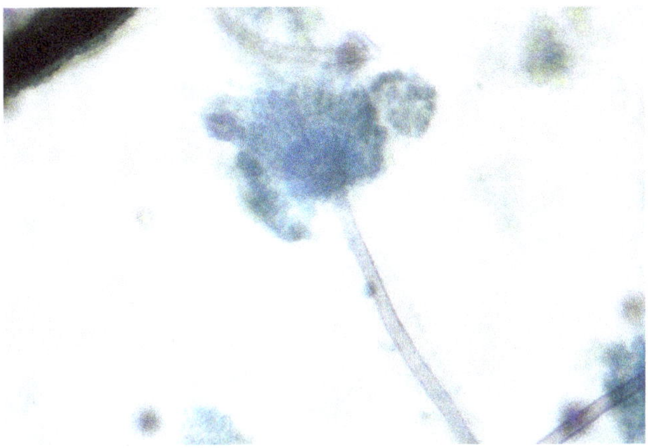

Fig. 8.1.17 *A. versicolor*.

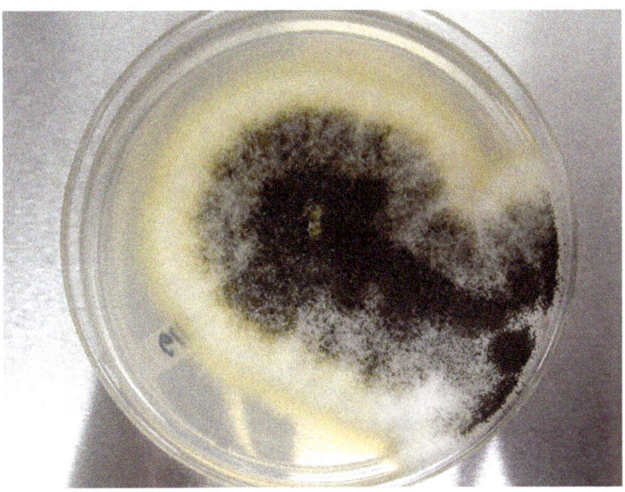

Fig. 8.1.18 *A. niger* (top).

Fig. 8.1.19 *A. niger* (side).

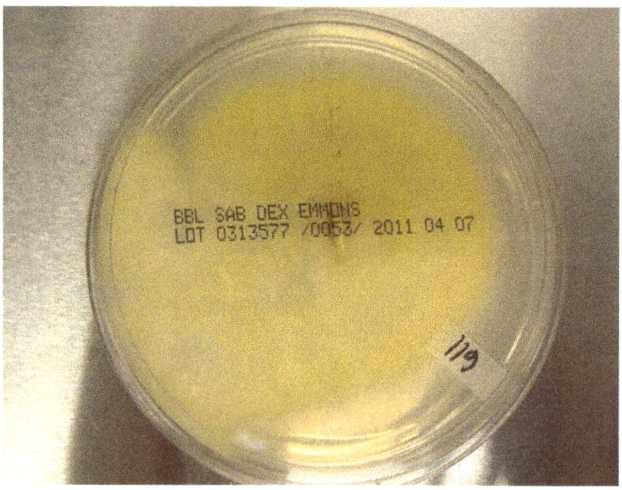

Fig. 8.1.20 *A. niger* (bottom).

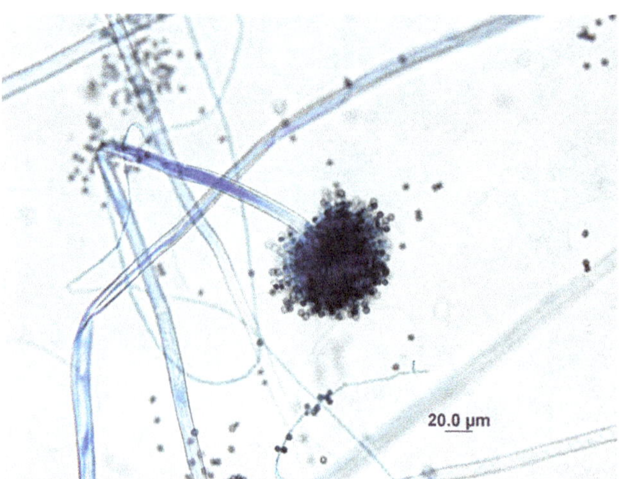

Fig. 8.1.21 *A. niger*.

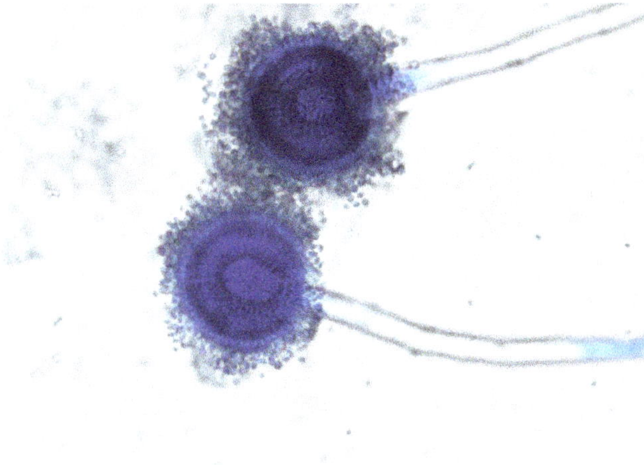

Fig. 8.1.22 *A. niger*.

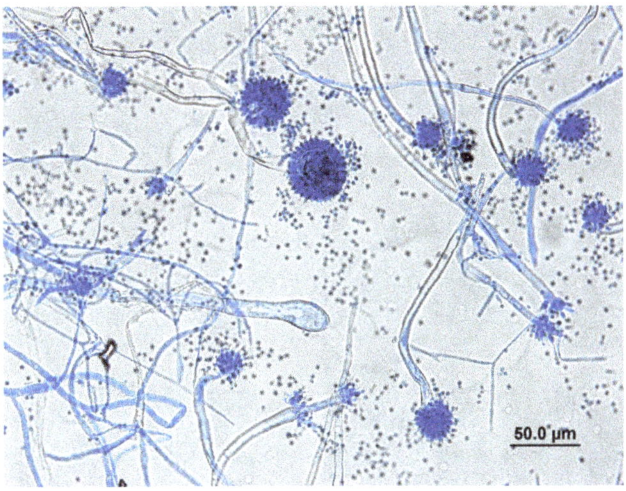

Fig. 8.1.23 *A. niger*.

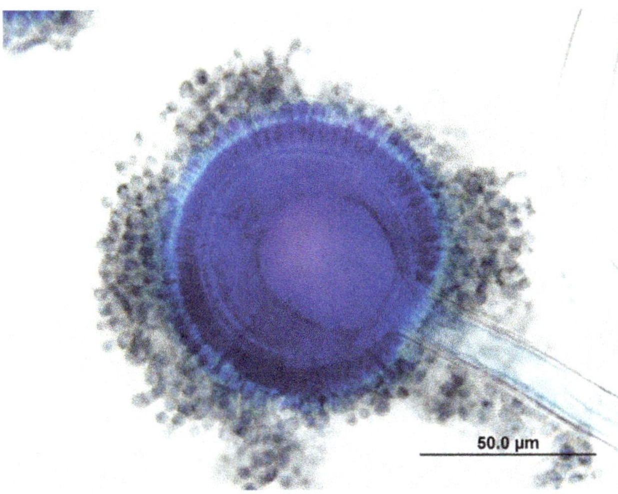

Fig. 8.1.24 *A. niger*.

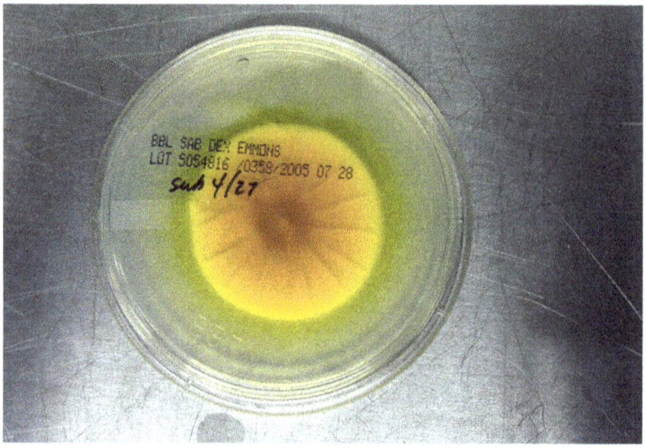

Fig. 8.1.25 *A. glaucus* (bottom).

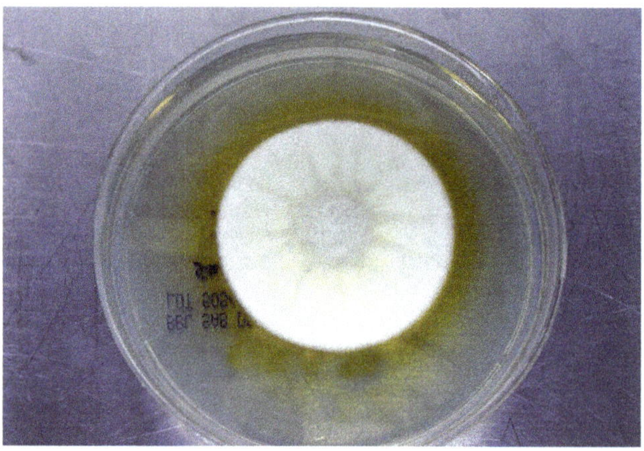

Fig. 8.1.26 *A. glaucus* (top).

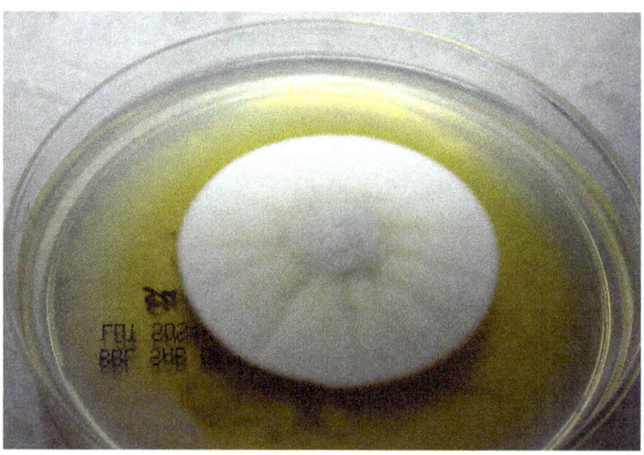

Fig. 8.1.27 *A. glaucus* (side).

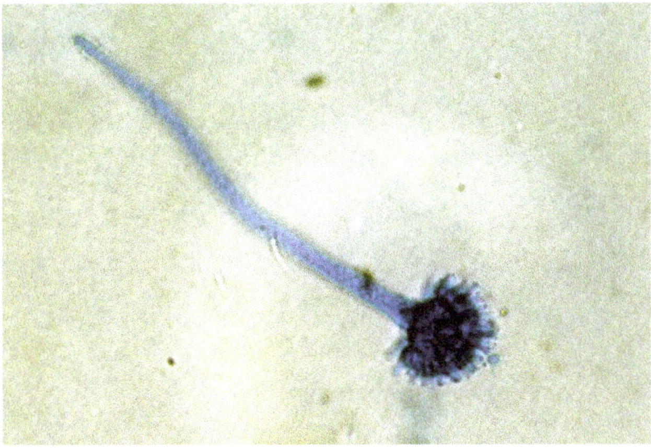

Fig. 8.1.28 *A. glaucus*.

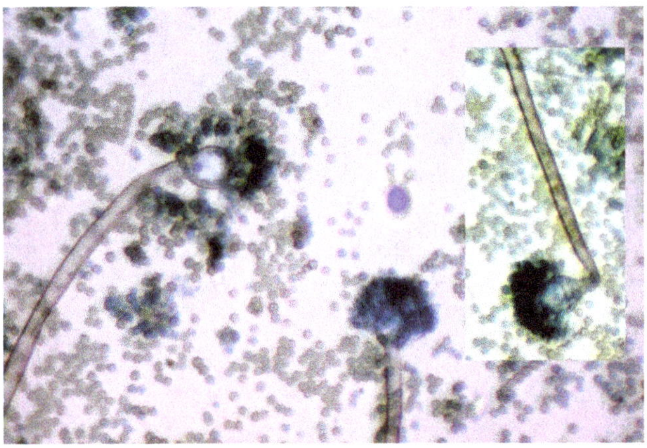

Fig. 8.1.29 *A. glaucus*.

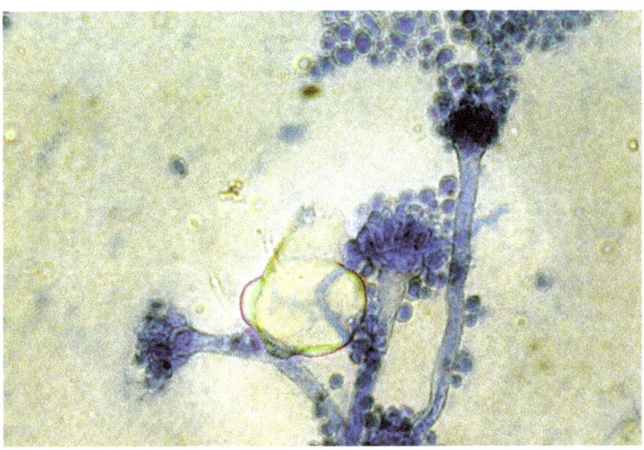

Fig. 8.1.30 *A. glaucus* BHI agar.

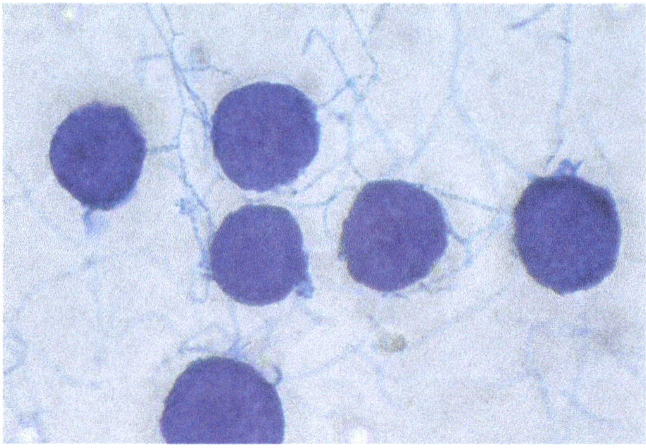

Fig. 8.1.31 *A. glaucus* with cleistothecium on PDA.

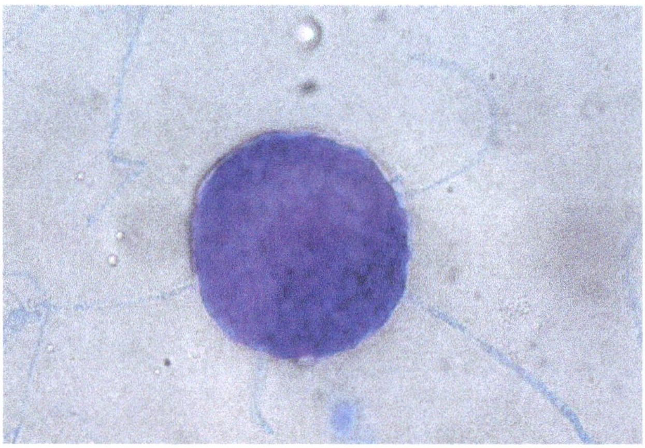

Fig. 8.1.32 *A. glaucus* with cleistothecium on PDA.

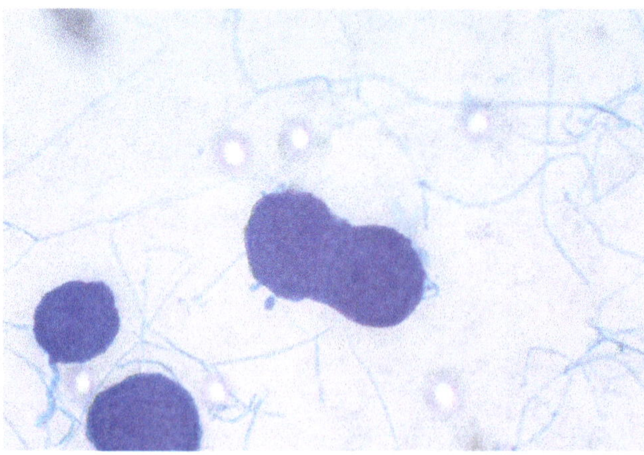

Fig. 8.1.33 *A. glaucus* PDA.

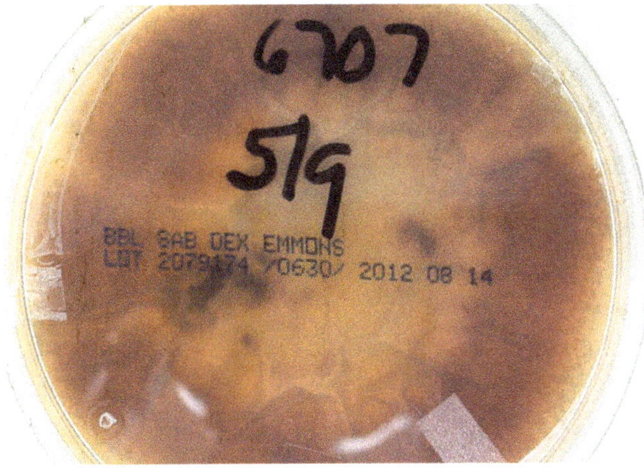

Fig. 8.1.34 *A. nidulans* (bottom).

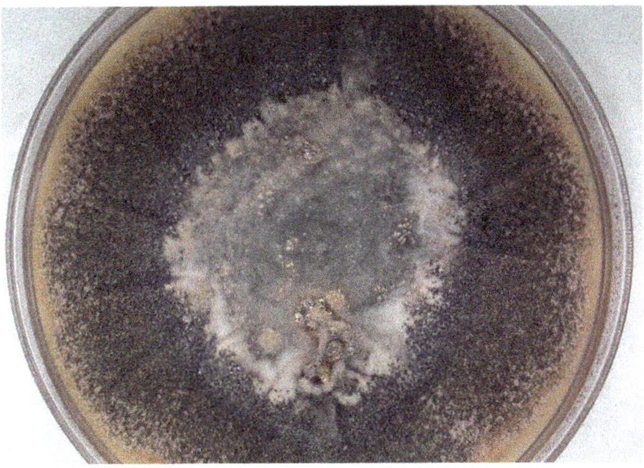

Fig. 8.1.35 *A. nidulans* (top).

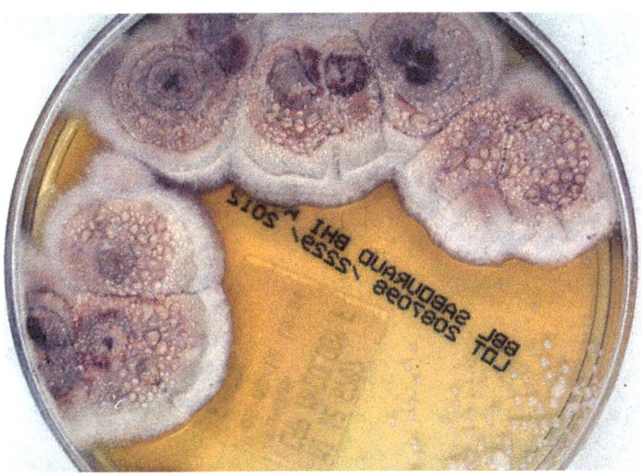

Fig. 8.1.36 *A. nidulans* (top).

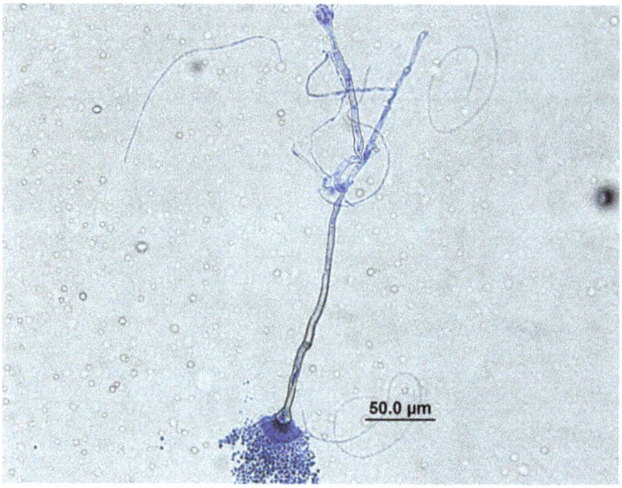

Fig. 8.1.37 *A. nidulans*.

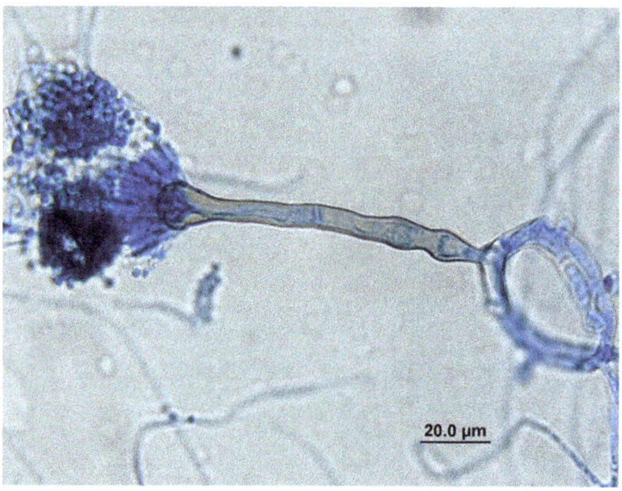

Fig. 8.1.38 *A. nidulans*.

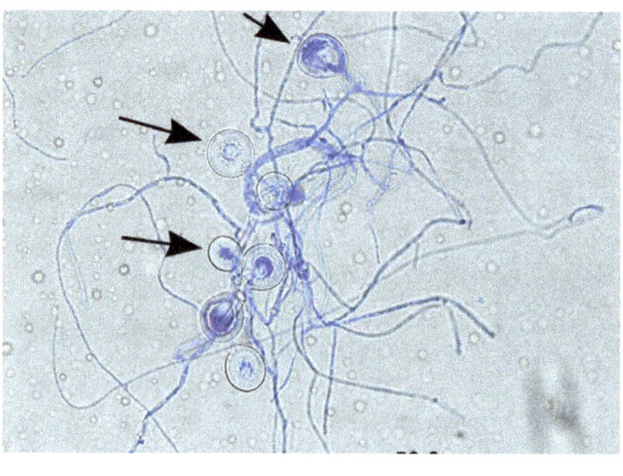

Fig. 8.1.39 *A. nidulans* Hülle cells.

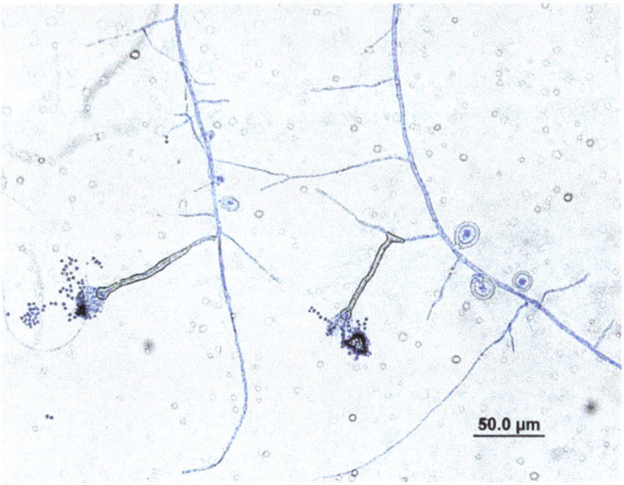

Fig. 8.1.40 *A. nidulans*.

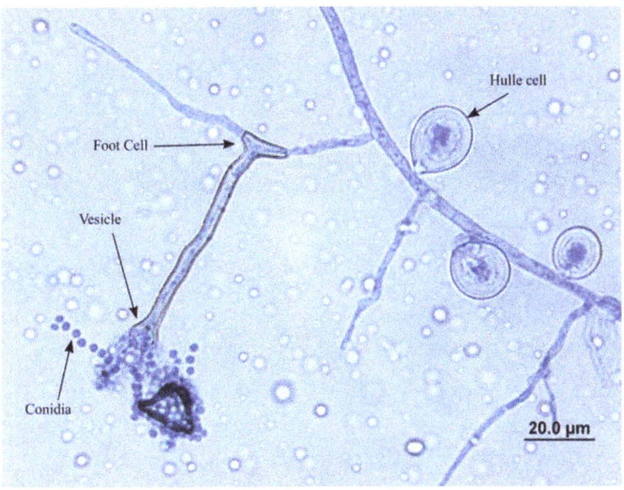

Fig. 8.1.41 *A. nidulans*.

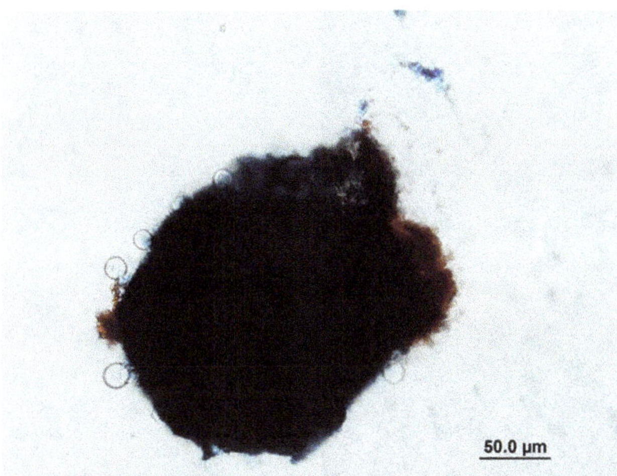

Fig. 8.1.42 *A. nidulans* with cleistothecium.

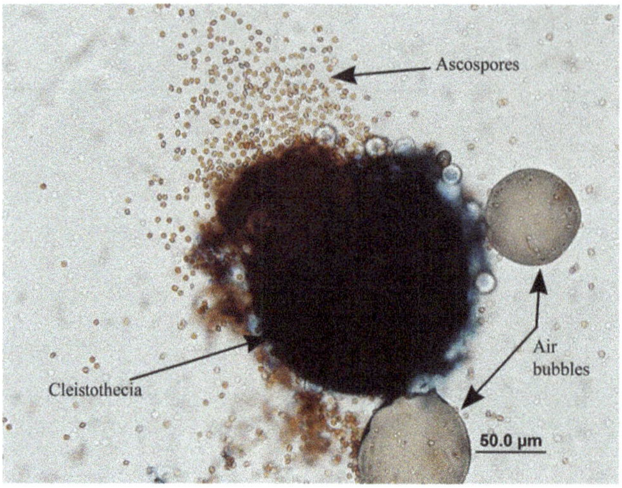

Fig. 8.1.43 *A. nidulans* with cleistothecium with ascospores.

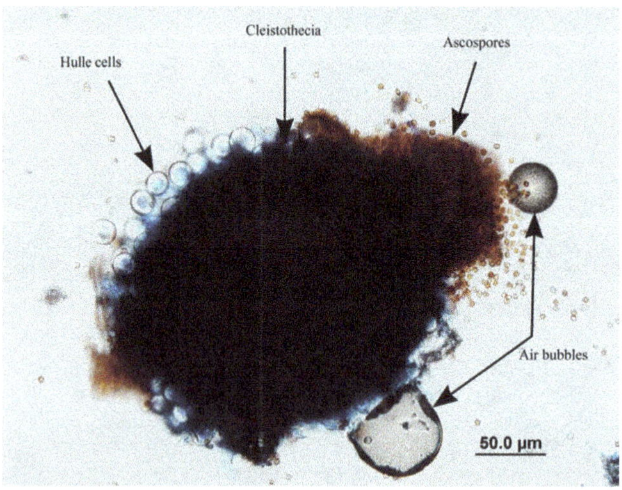

Fig. 8.1.44 *A. nidulans*.

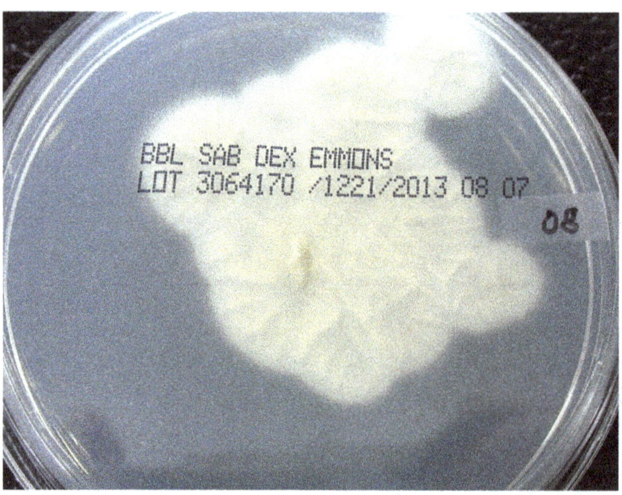

Fig. 8.1.45 *A. clavatus* (bottom).

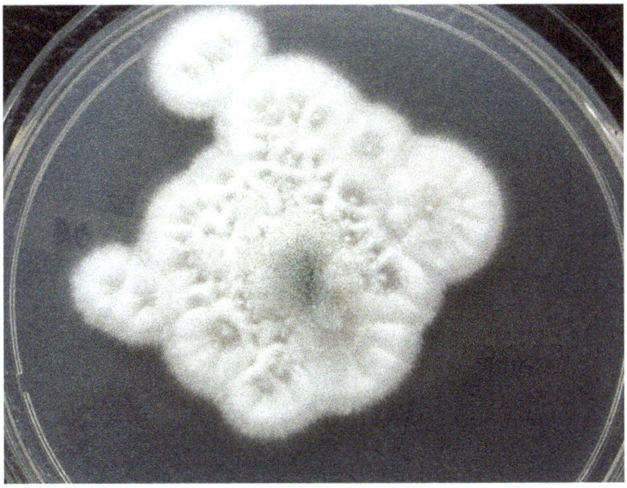

Fig. 8.1.46 *A. clavatus* (top).

Fig. 8.1.47 *A. clavatus* (side).

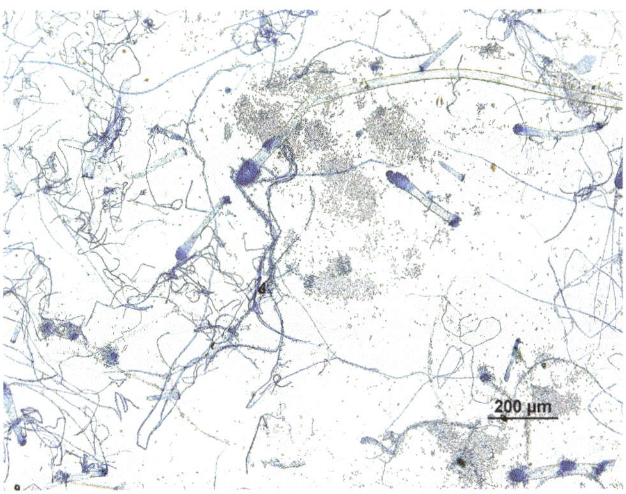

Fig. 8.1.48 *A. clavatus*.

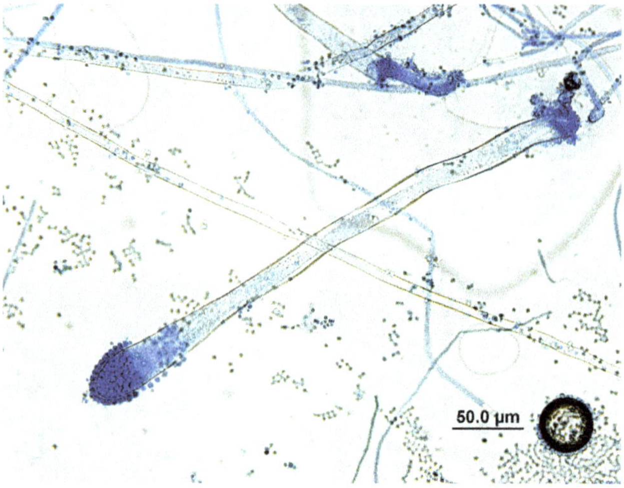

Fig. 8.1.49 *A. clavatus*.

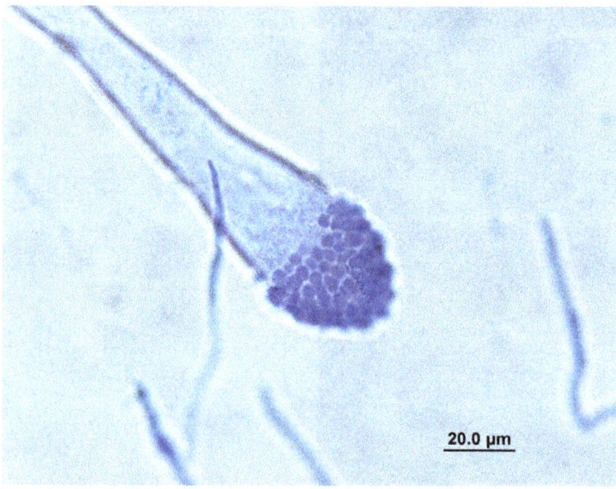

Fig. 8.1.50 *A. clavatus*.

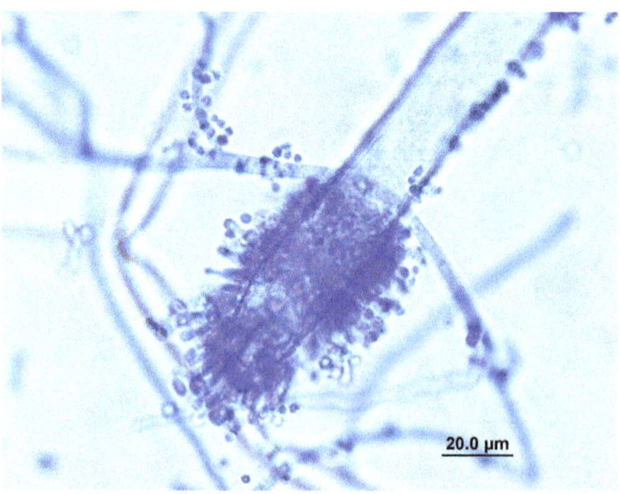

Fig. 8.1.51 *A. clavatus*.

8.2 *Beauveria bassiana*

Colonial morphology: colonies are first white then become yellow to pink. The reverse may show more pink or peach color. A pink pigment may be secreted into the agar, confusing this with *P. marneffei*. Colonies are fluffy to cottony on most media, but colonies grown on Sabouraud brain heart infusion agar have a unique feathery appearance. Resistant to cycloheximide.

Microscopic morphology: hyphae are septate, hyaline, delicate, and tiny (1–3 µm in width). Some hyphae may appear flat or ribbon-like. Conidiophores arise sympodially (zig-zag) along hyphae and form flask-shaped cells. These cells develop thin, rachiform projections resulting in terminal conidia production. Clusters of hyphae and conidiophores are often aggregated into sporodochia or synnema. Conidia are hyaline, round or tear shaped (1.5 × 2–3 µm). May be easily misidentified as *Sporothrix*, *Trichoderma*, *P. marneffei* or *Exophiala*.

Beauveria

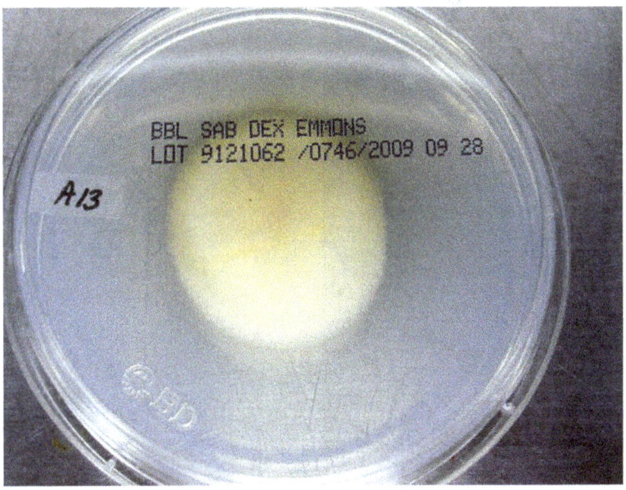

Fig. 8.2.1 *Beauveria* Sab Dex Emmons (bottom).

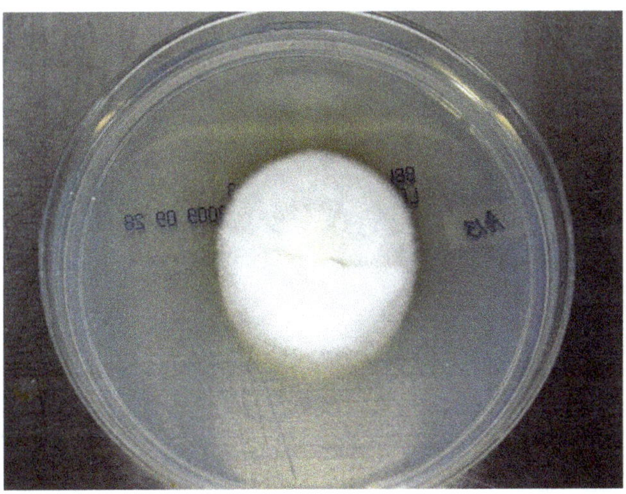

Fig. 8.2.2 *Beauveria* Sab Dex Emmons.

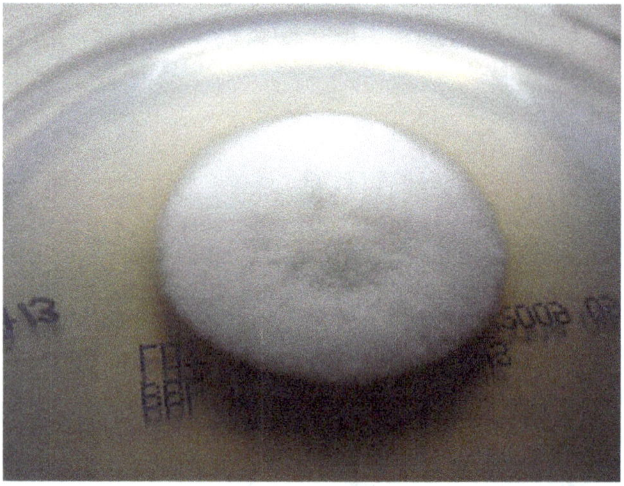

Fig. 8.2.3 *Beauveria* Sab Dex Emmons (side).

Fig. 8.2.4 *Beauveria* Sab Dex Emmons.

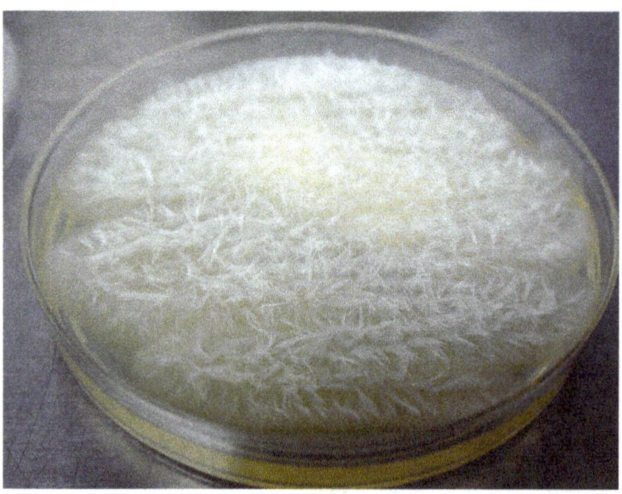

Fig. 8.2.5 *Beauveria* Sabouraud BHI agar (side).

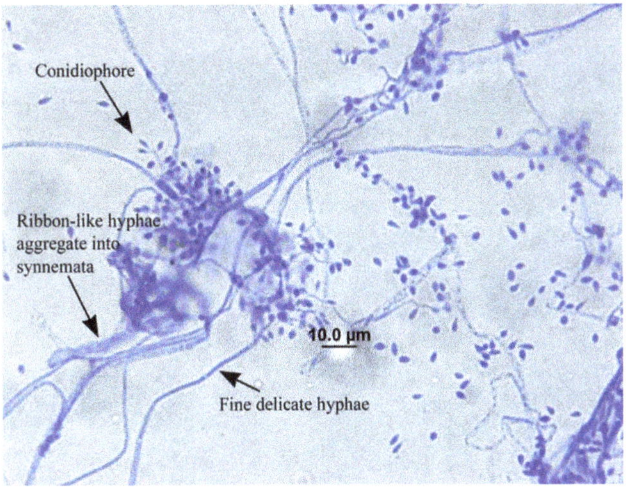

Fig. 8.2.6 *Beauveria*.

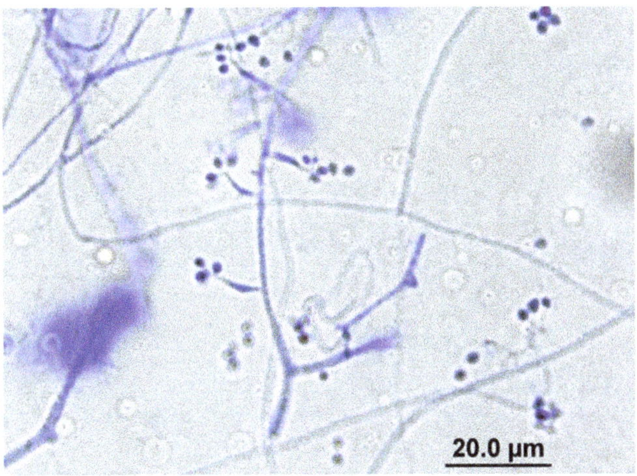

Fig. 8.2.7 *Beauveria*, flask-shaped basal cells.

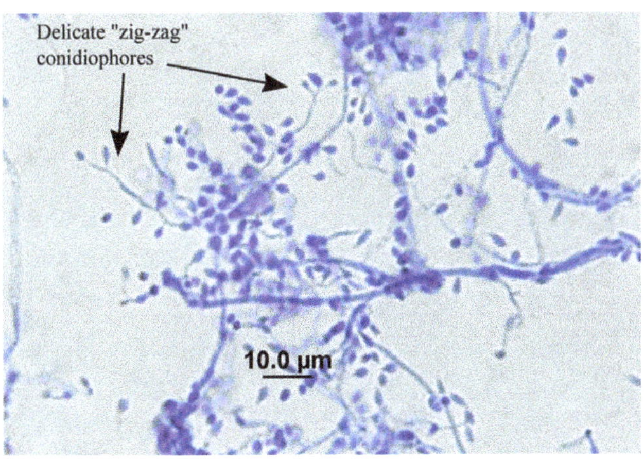

Fig. 8.2.8 *Beauveria*.

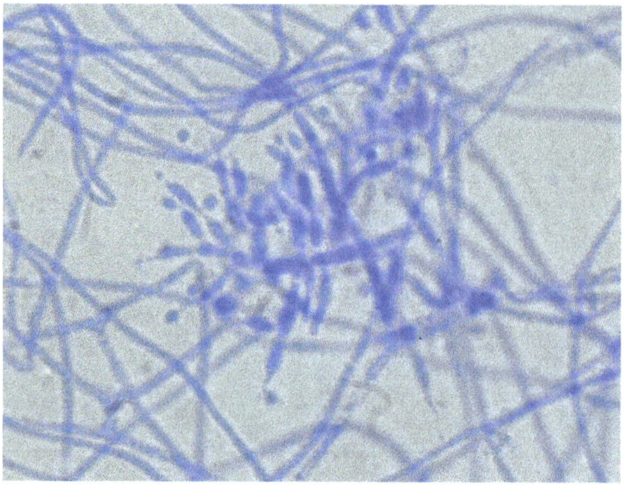

Fig. 8.2.9 Early sporulation resembles *Trichoderma*.

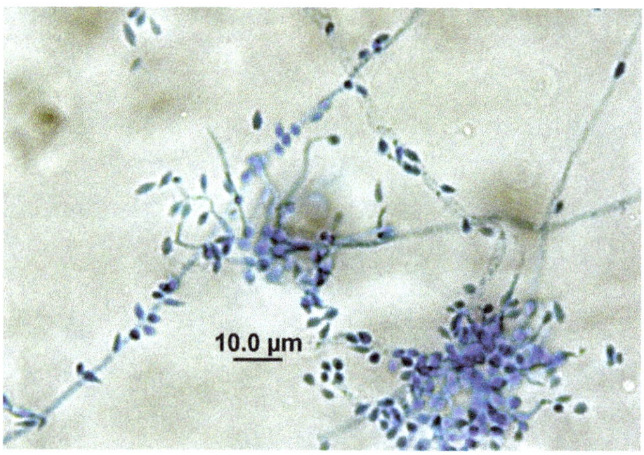

Fig. 8.2.10 *Beauveria*, with fine delicate denticles, 1000×.

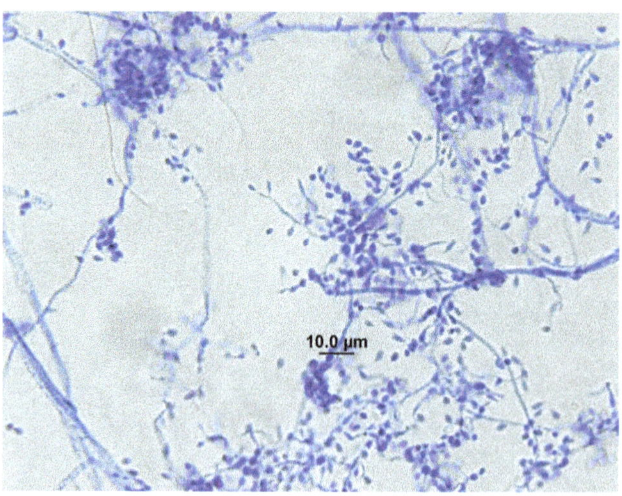

Fig. 8.2.11 *Beauveria*, 400×.

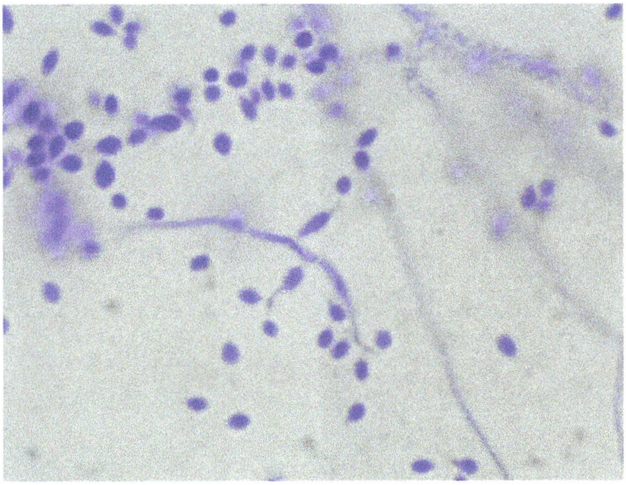

Fig. 8.2.12 *Beauveria*, 1000×.

8.3 *Emmonsia* spp

Colonial morphology: the colony surface is white with tan edges and the reverse is tan to brown. Colonies appear velvety at first and then become fluffy. Black to brown "pepper-like" spots develop in the center of the colonies with age.

Microscopic morphology: hyphae are septate, of two types: (1) fine, narrow, and hyaline or (2) broad, bamboo-like with darkening at septa. Conidia are produced on fine conidiophores that branch perpendicular to the hyphae. Some conidiophores show branching but most look like "lollipops" attached to the hyphae, similar to that seen with *Blastomyces*. Conidia are round (2–5 µm in diameter), hyaline, unicellular, and may become rough with age. Some conidia may form chains of 3–4 in length. When incubated at >35 °C, conidia enlarge to form round, thick-walled adiaspores. Adiaspores vary in size and for *E. crescens* can reach 20–140 µm in diameter whereas *E. parva* adiaspores rarely exceed 25 µm. The presence of adiaspores is unmistakable. Microscopic examination of pepper-like droplets formed on the surface of the colony shows needle-like crystals in a yellow to brown liquid (see Fig. 8.3.2, arrow). Crystals polarize on polarized microscopy. Unlike *Blastomyces*, there is no yeast phase. *Emmonsia* can be mistaken for *Sporotrichum* but *Emmonsia* does not produce chlamydospores.

Emmonsia

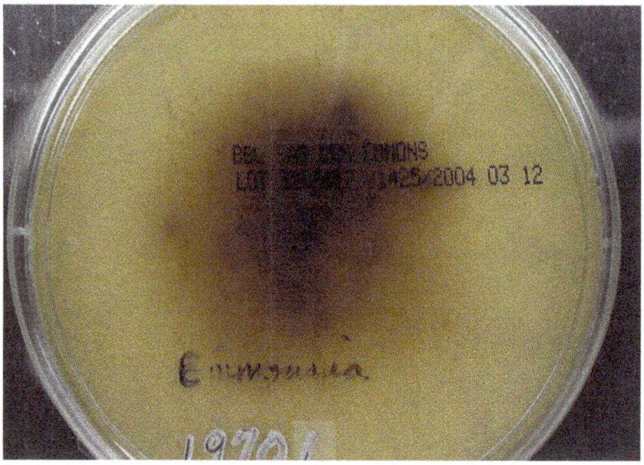

Fig. 8.3.1 *Emmonsia* (bottom).

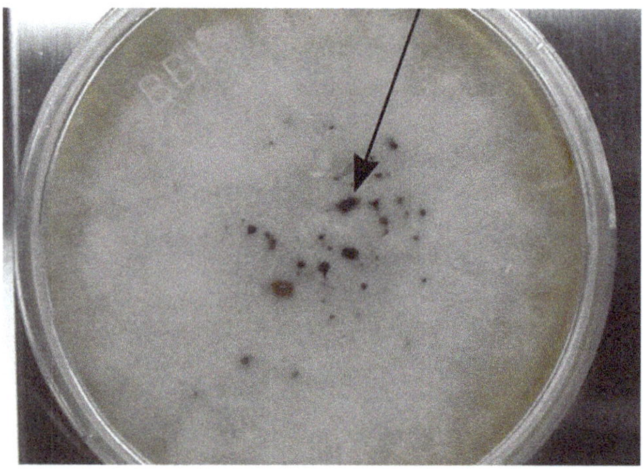

Fig. 8.3.2 *Emmonsia*.

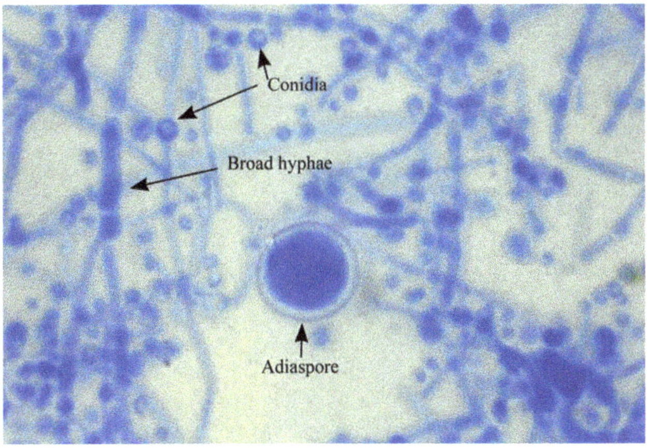

Fig. 8.3.3 *Emmonsia*, morphology.

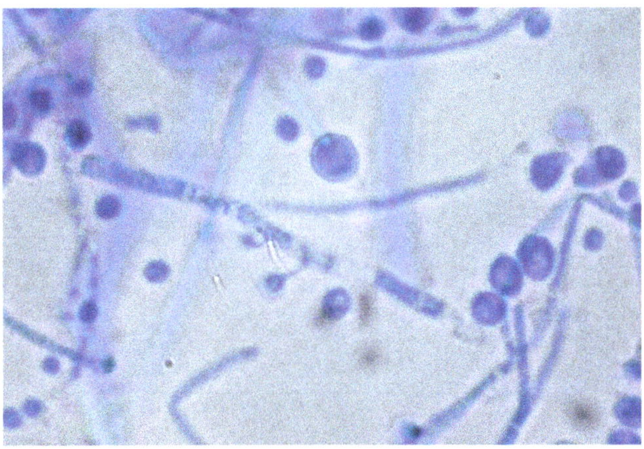

Fig. 8.3.4 Combined morphology.

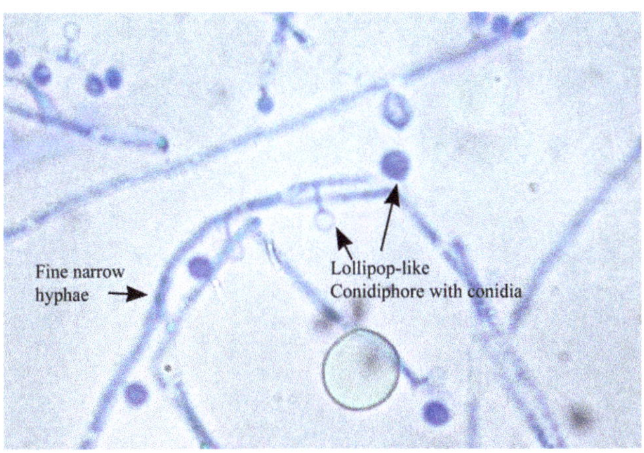

Fig. 8.3.5 *Emmonsia*, morphology.

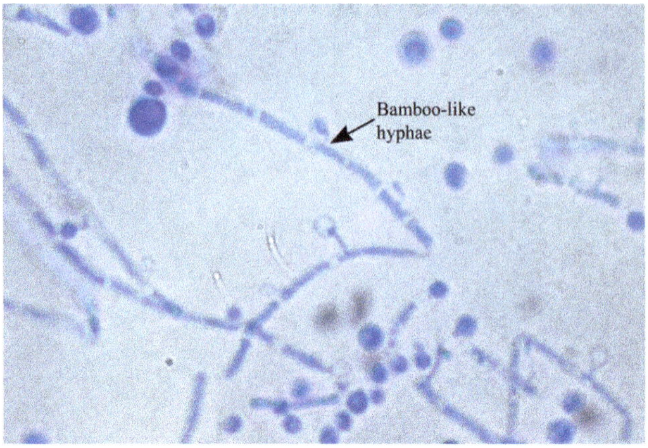

Fig. 8.3.6 *Emmonsia*, with bamboo-like hyphae.

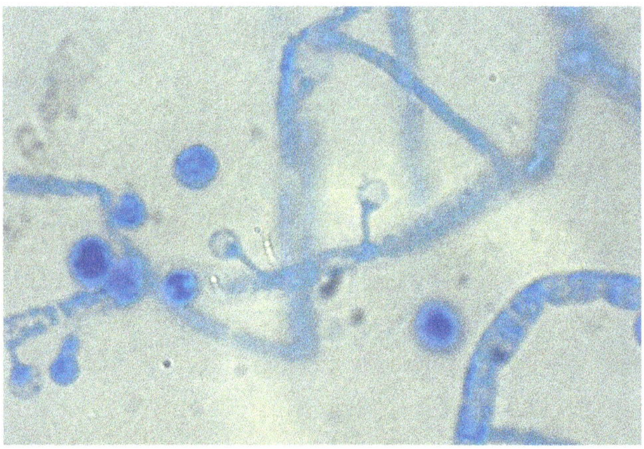

Fig. 8.3.7 *Emmonsia*, 1000×.

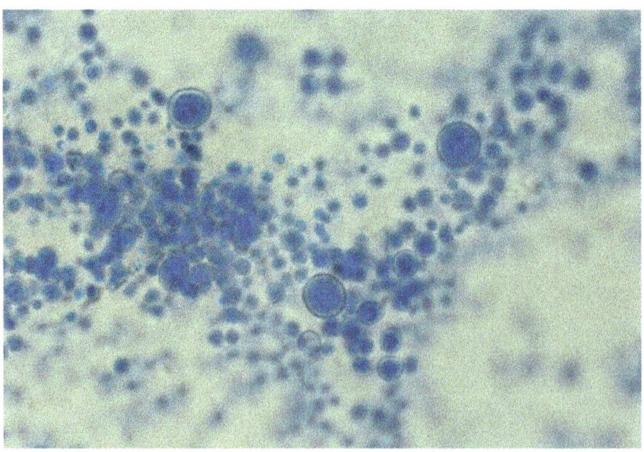

Fig. 8.3.8 Varying sizes and abundance of adiaspores.

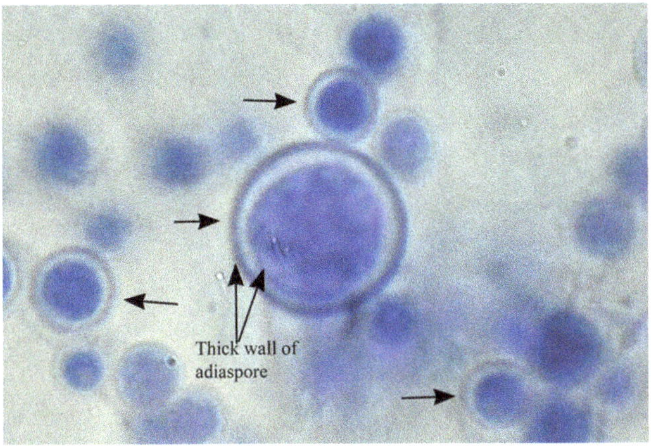

Fig. 8.3.9 Varying sizes of adiaspores.

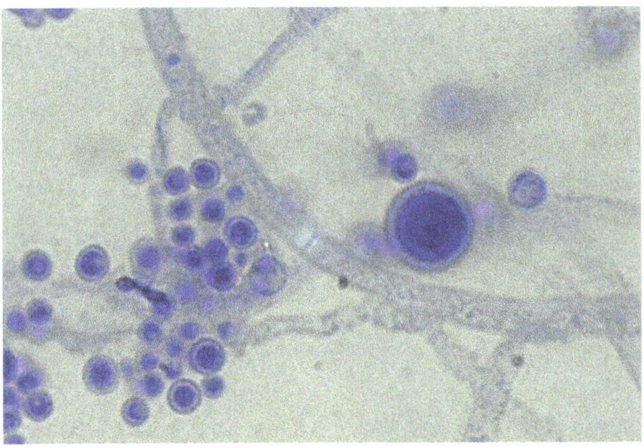

Fig. 8.3.10 *Emmonsia*.

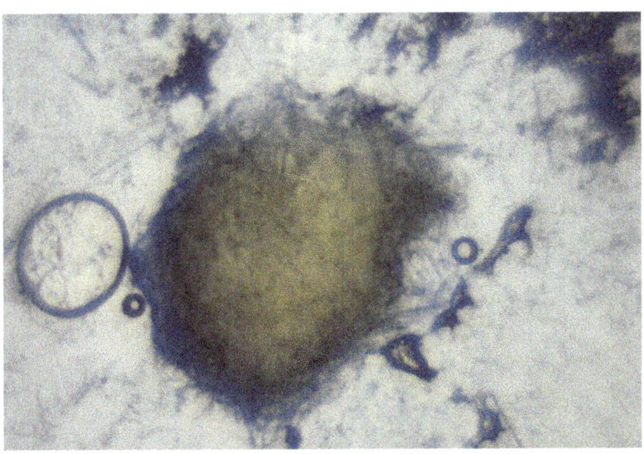

Fig. 8.3.11 Crystaline deposit on colony.

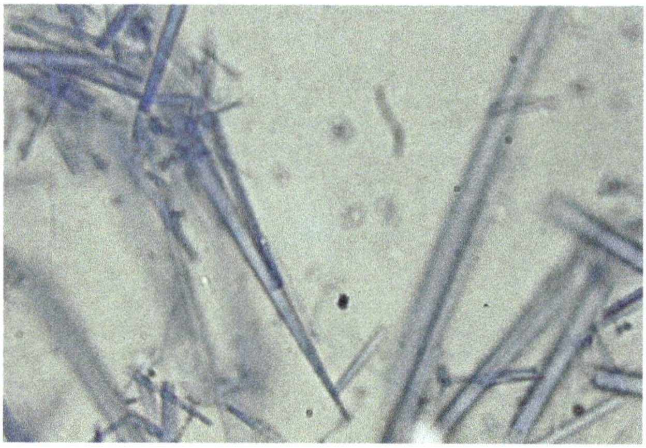

Fig. 8.3.12 Crystalline deposit, 100×.

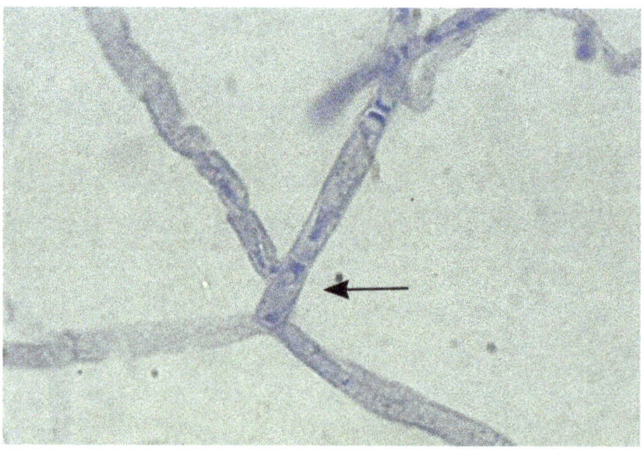

Fig. 8.3.13 Hyphal branching at right angles.

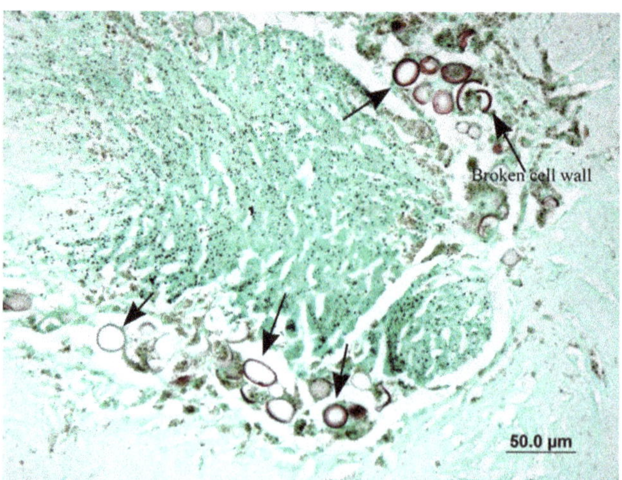

Fig. 8.3.14 Adiaspores in human lung, silver stain.

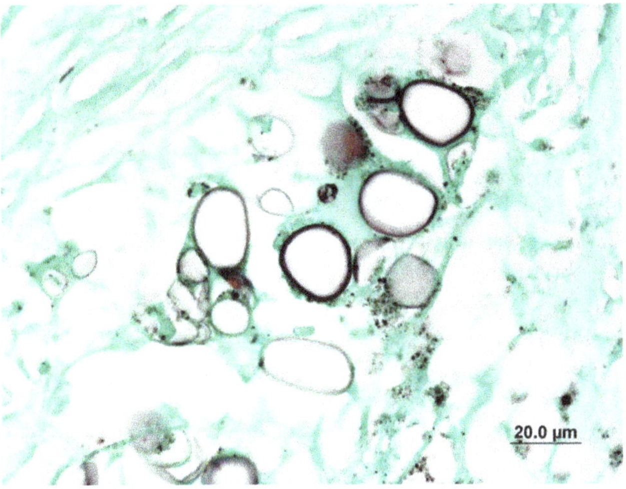

Fig. 8.3.15 Adiaspores in human lung, silver stain.

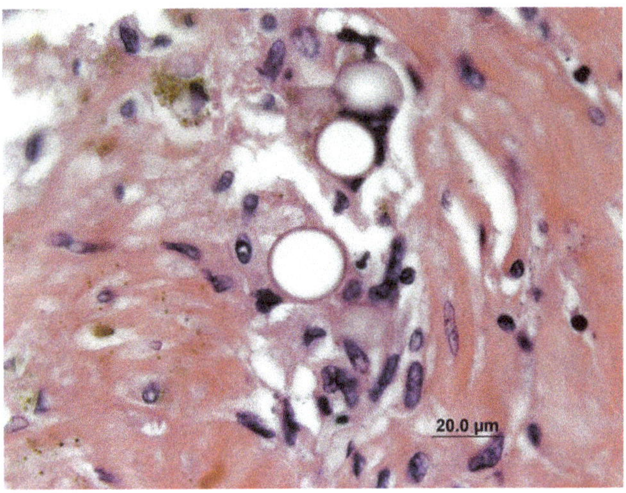

Fig. 8.3.16 Adiaspores in human lung, H&E stain.

8.4 *Fusarium* spp

Colonial morphology: colonies usually grow rapidly and are mature in 4 days. However, some may show only sterile hyphae for weeks before germination of fruiting structures. Some species will not germinate on nutrient media, so specialized media such as carnation leaf agar or sterile water agar may be required for germination. Colonies are first white and cottony. Later, beginning in the center, various colors may form such as yellow, orange, pink, blue-green, violet or pale brown. The border often stays white and sometimes the culture remains all white. Growth usually fills the culture plate and aerial mycelia touch the lid. The reverse color varies with species.

Microscopic morphology: hyphae are septate and hyaline, usually very fine. Conidiation occurs in the form of both macroconidia and microconidia. Macroconidia are produced on sporodochia which may be surrounded by slime. Macroconidia may also be produced on monophialides (single pore opening) or polyphialides (multiple pores). Microconidia are produced on aerial hyphae (conidiophores) but not on sporodochia. Chlamydoconidia are also produced by some species. Chlamydoconidia appear as singles, pairs, clumps or chains and have a rough or smooth outer wall. Phialides are formed along hyphae and may be long or short, cylindrical, simple or branched with a fine collarette at the apex. Microconidia (2–4 × 4–8 µm) are abundantly produced, usually on short conidiophores. They are oval, 1–2 celled, occurring in clusters at the ends of conidiophores, giving them the appearance of *Acremonium* spp. *Fusarium* can be distinguished from *Acremonium* if macroconidia are produced, and by colony morphology. Macroconidia of *Fusarium* are geniculate whereas *Acremonium* and *Cylindrocarpon* are not. Macroconidia vary in size but are usually 2–6 × 14–80 µm, banana shaped, with 1–5 septa and geniculate at the base. Chlamydoconidia of *F. solani* or *F. equiseti* are hyaline, round, smooth or rough walled (look like popcorn balls), borne singly or in pairs on short parallel hyphal branches or they are intercalary. *Fusarium moniliforme* has recently been renamed as *Fusarium verticilliodes*.

Fusarium avenaceum

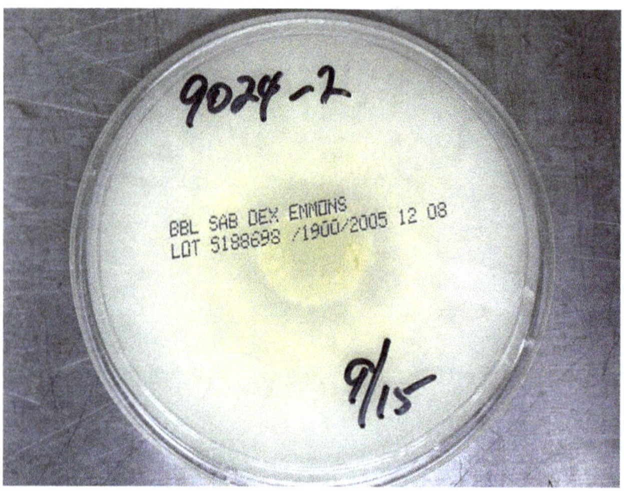

Fig. 8.4.1 *Fusarium avenaceum* (bottom).

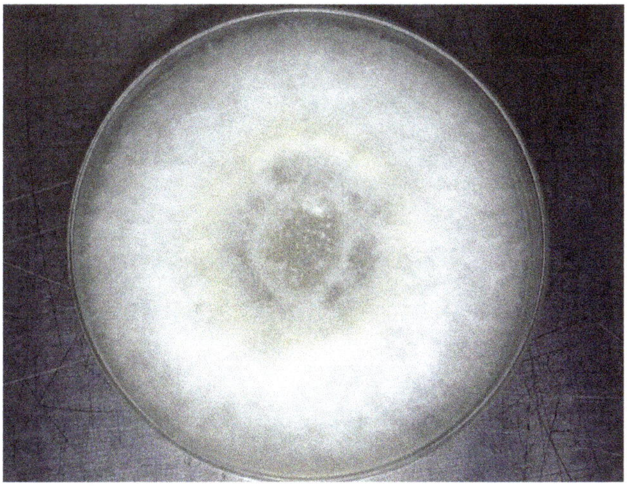

Fig. 8.4.2 *Fusarium avenaceum* (top).

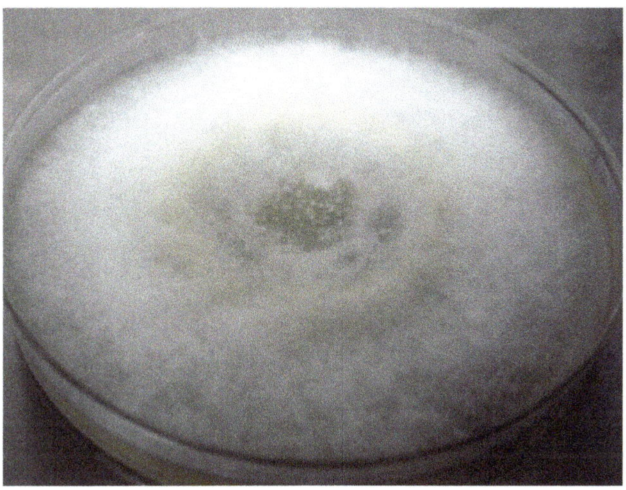

Fig. 8.4.3 *Fusarium avenaceum* (side).

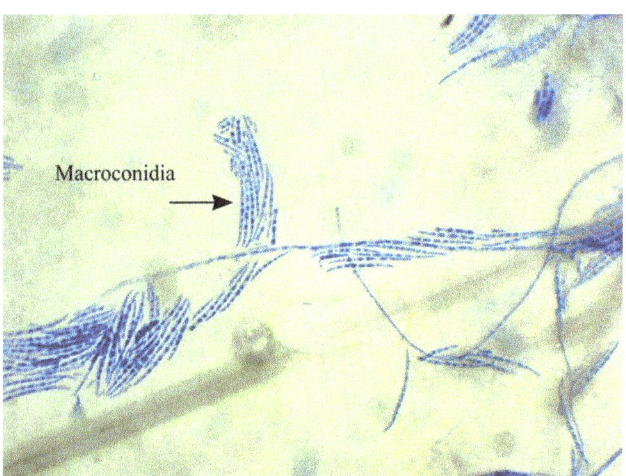

Fig. 8.4.4 *F. avenaceum*.

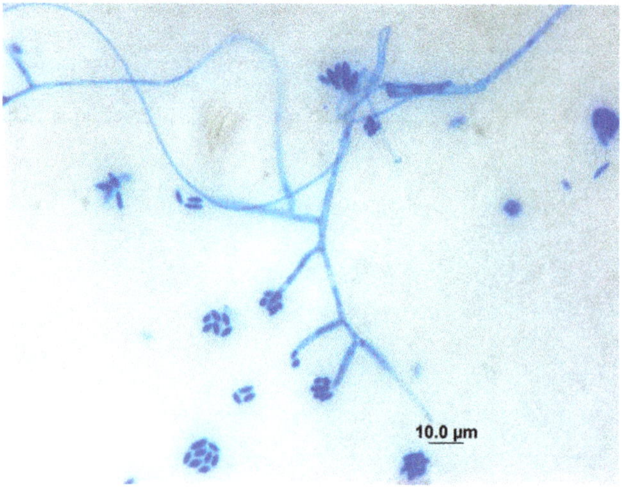

Fig. 8.4.5 *F. avenaceum.*

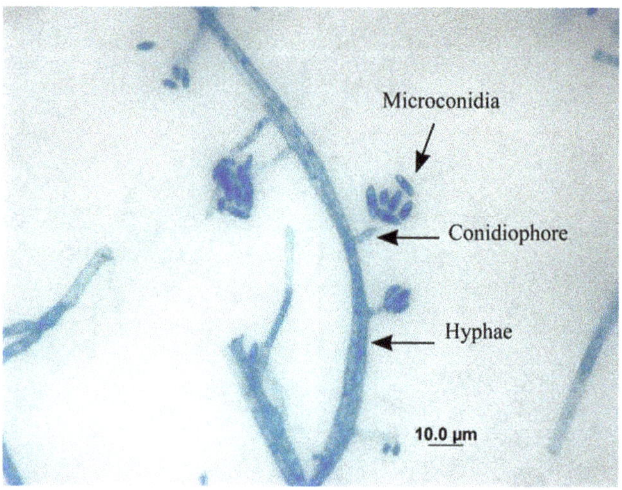

Fig. 8.4.6 *F. avenaceum.*

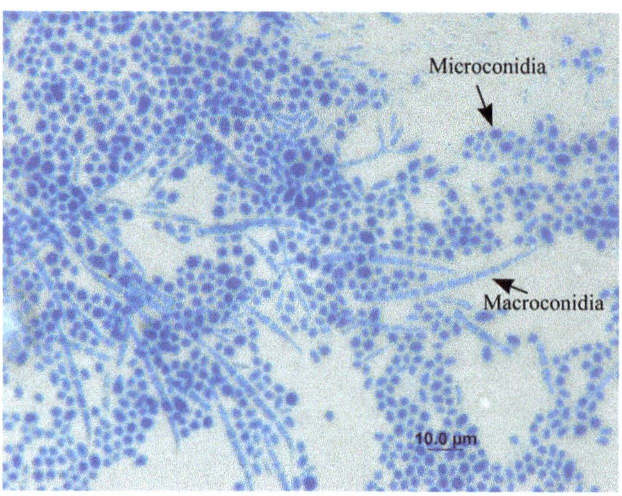

Fig. 8.4.7 *F. avenaceum.*

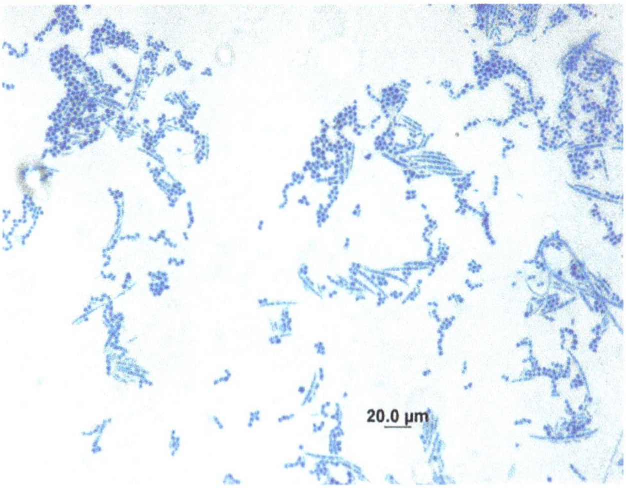

Fig. 8.4.8 *F. avenaceum*.

Fusarium equiseti

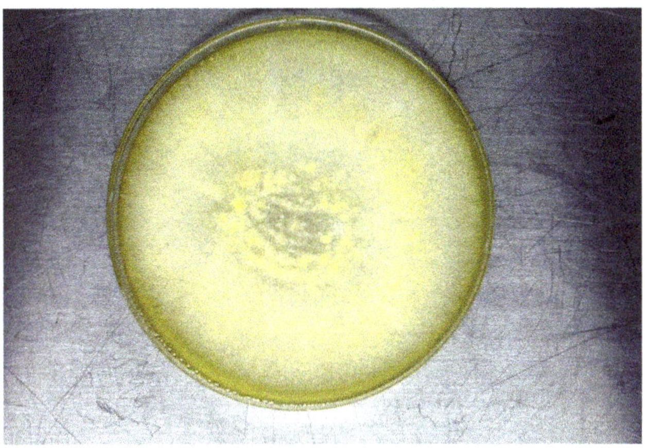

Fig. 8.4.9 *Fusarium equiseti*.

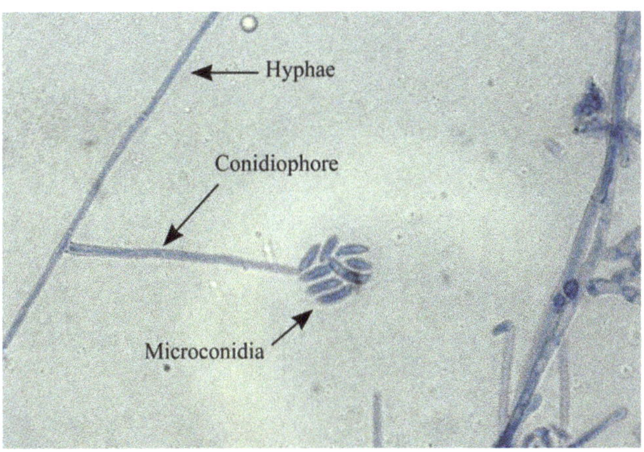

Fig. 8.4.10 *F. equiseti* microconidia.

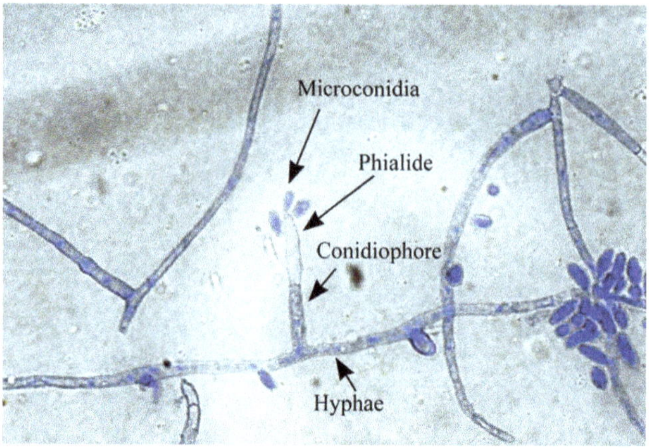

Fig. 8.4.11 *F. equiseti* microconidia.

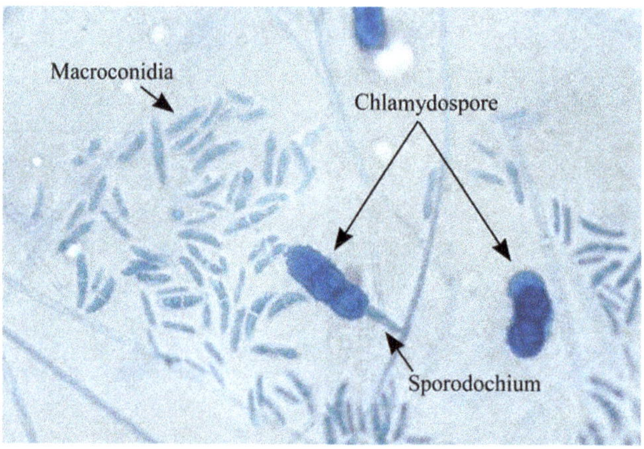

Fig. 8.4.12 *F. equiseti* chlamydospore on sporodochium.

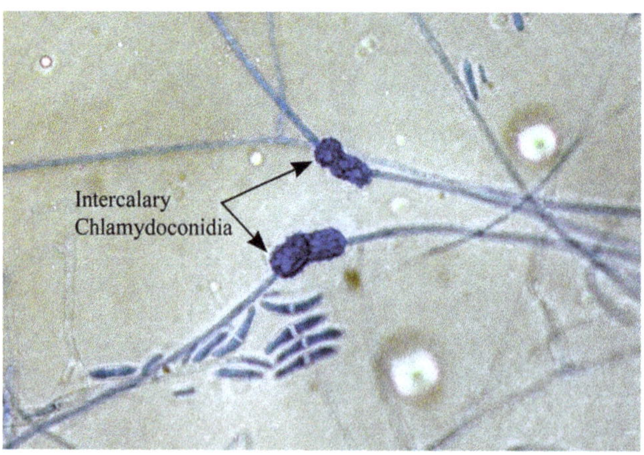

Fig. 8.4.13 *F. equiseti* chlamydospores intercalary.

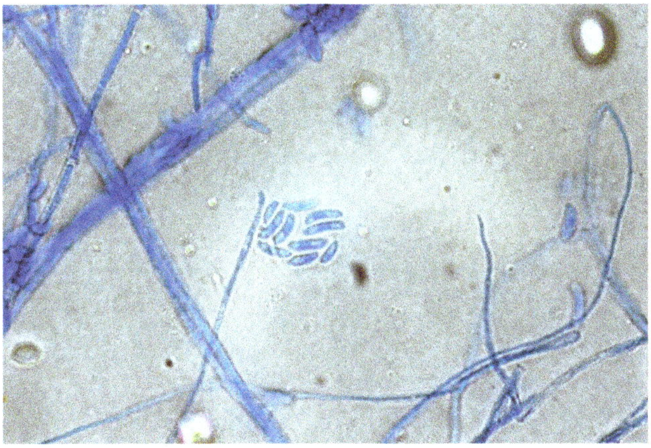

Fig. 8.4.14 *F. equiseti*.

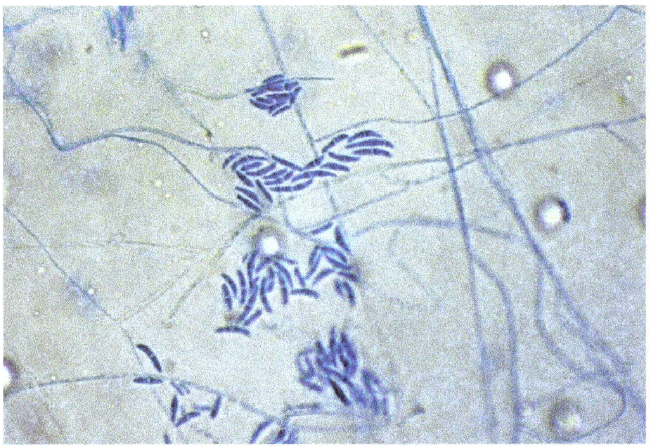

Fig. 8.4.15 *F. equiseti* macroconidia.

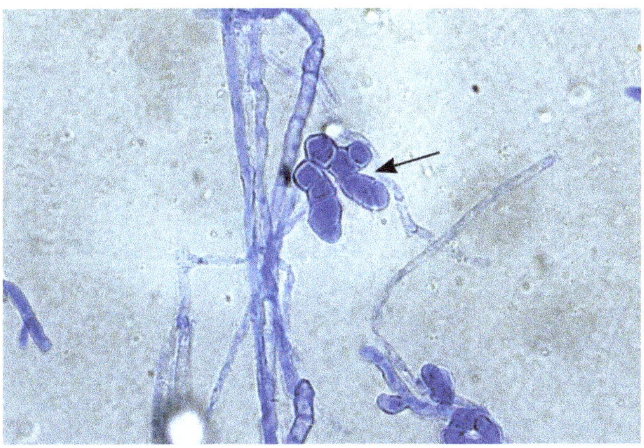

Fig. 8.4.16 *F. equiseti* smooth-walled chlamydospore.

Fusarium moniliforme

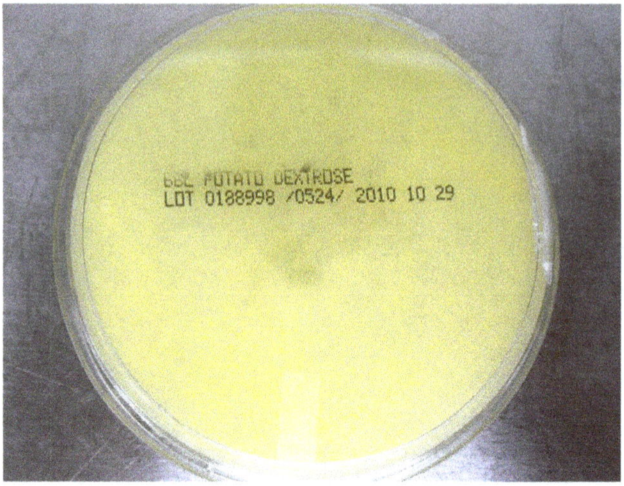

Fig. 8.4.17 *Fusarium moniliforme*, PDA (bottom).

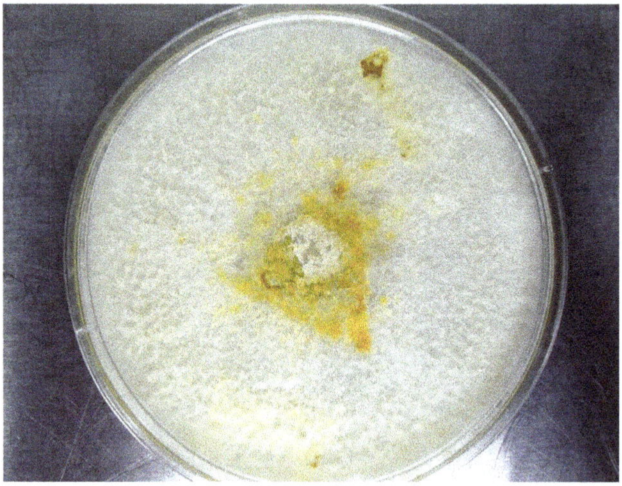

Fig. 8.4.18 *Fusarium moniliforme*, PDA (top).

Fig. 8.4.19 *Fusarium moniliforme*, PDA, (side).

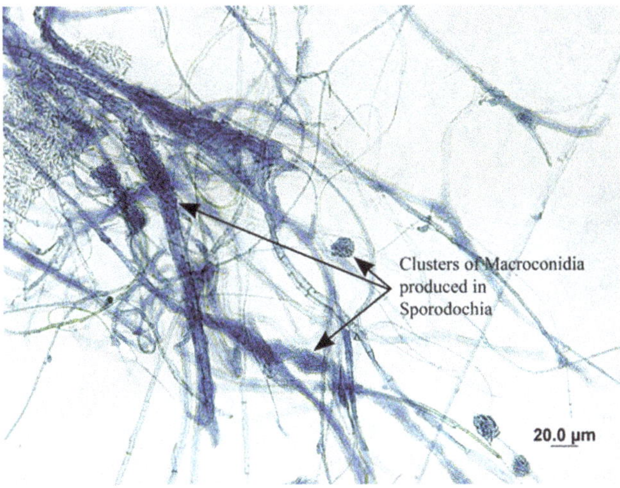

Fig. 8.4.20 *F. moniliforme*.

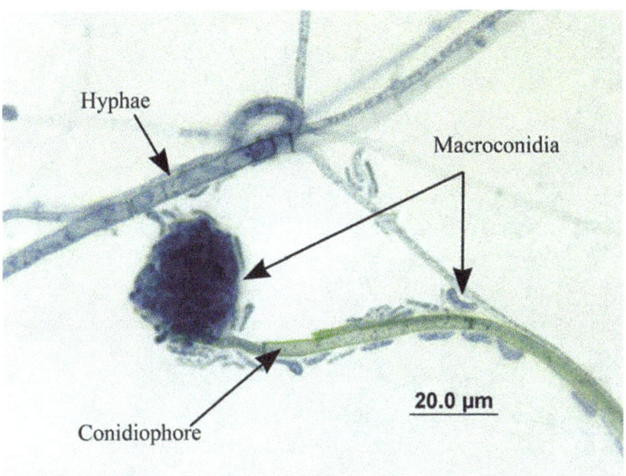

Fig. 8.4.21 *F. moniliforme*.

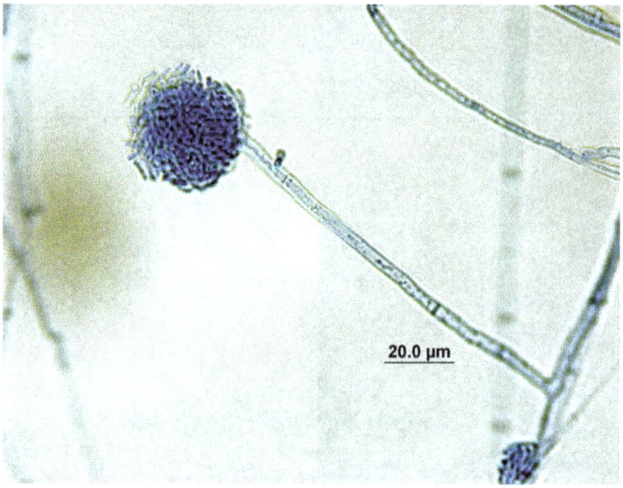

Fig. 8.4.22 *F. moniliforme.*

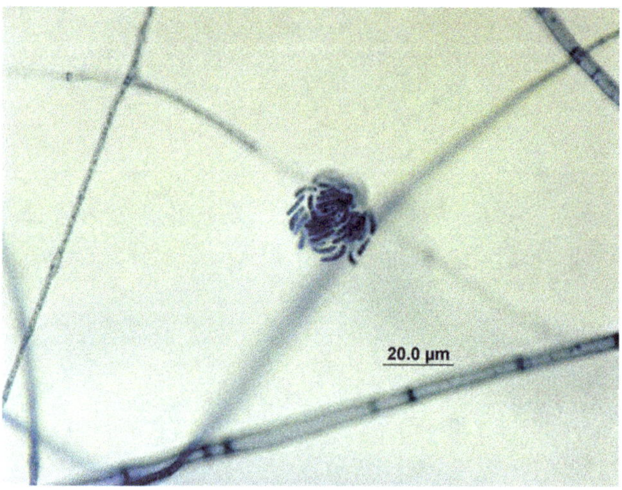

Fig. 8.4.23 *F. moniliforme.*

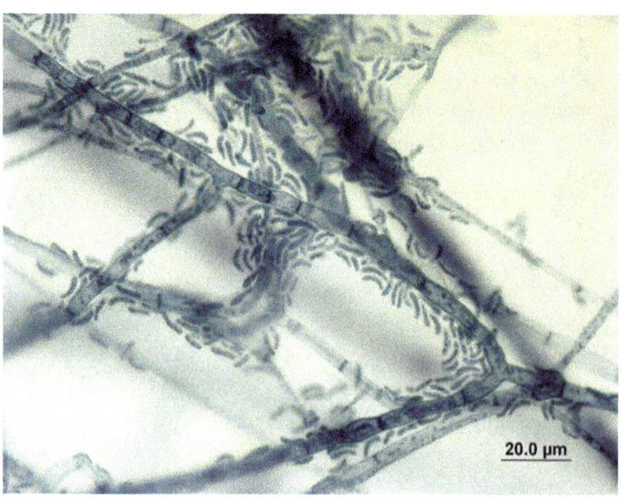

Fig. 8.4.24 *F. moniliforme.*

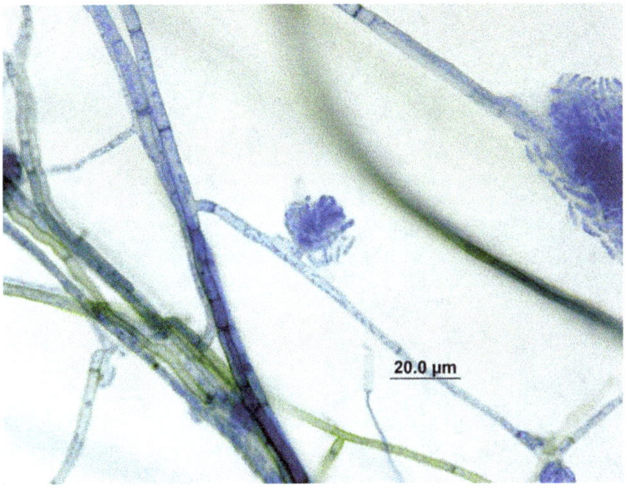

Fig. 8.4.25 *F. moniliforme*.

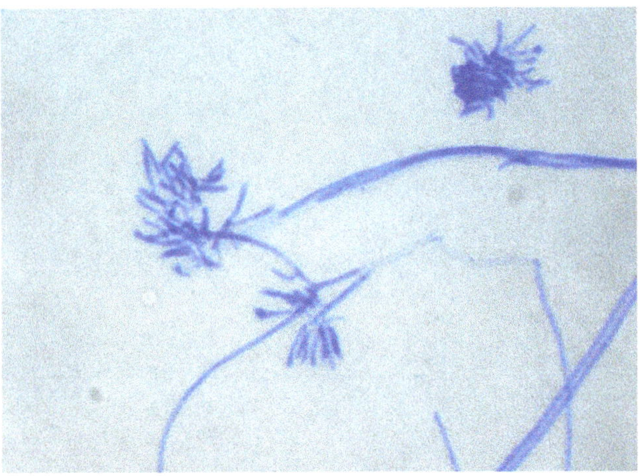

Fig. 8.4.26 *F. moniliforme*.

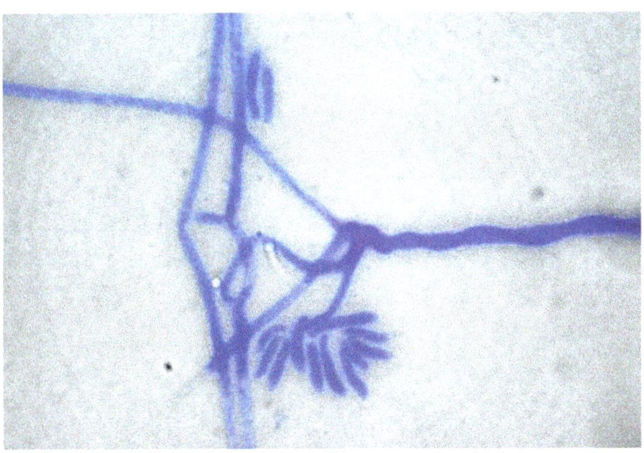

Fig. 8.4.27 *F. moniliforme*.

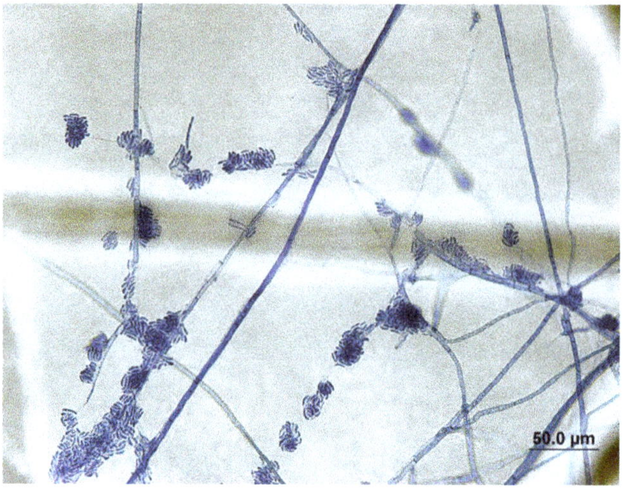

Fig. 8.4.28 *F. moniliforme*.

Fusarium oxysporum

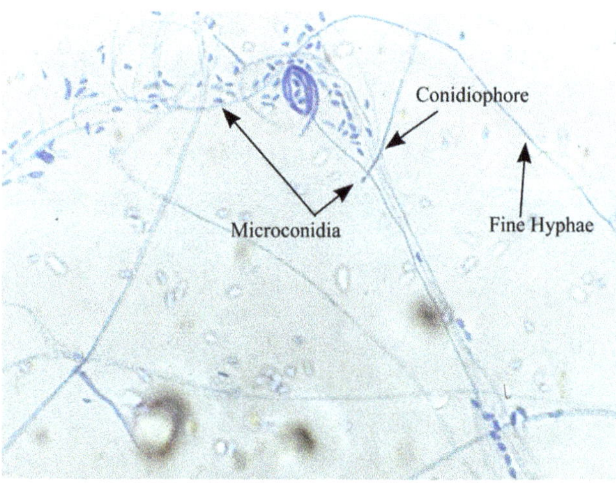

Fig. 8.4.29 *Fusarium oxysporum*.

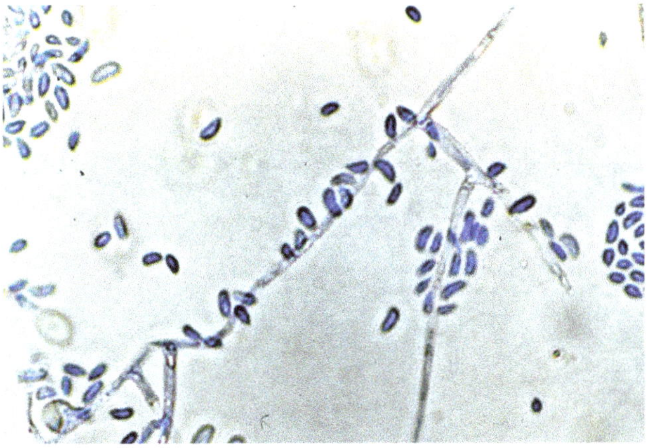

Fig. 8.4.30 *Fusarium oxysporum* microconidia.

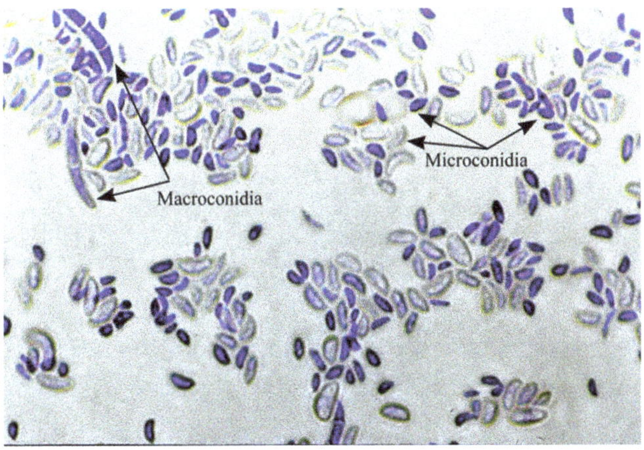

Fig. 8.4.31 *Fusarium oxysporum*.

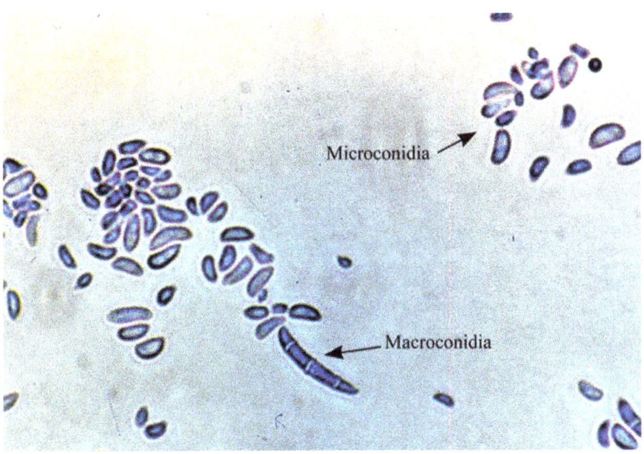

Fig. 8.4.32 *Fusarium oxysporum*.

Fusarium roseum

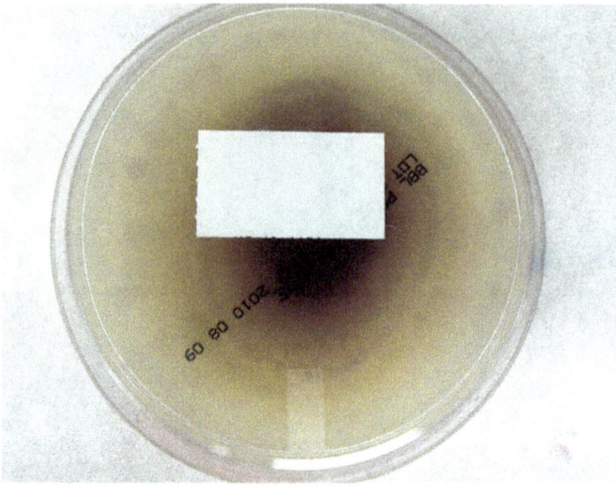

Fig. 8.4.33 *Fusarium roseum* (bottom).

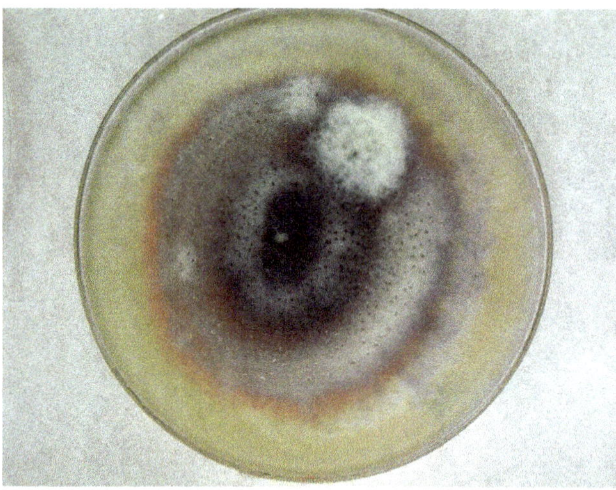

Fig. 8.4.34 *Fusarium roseum* (top).

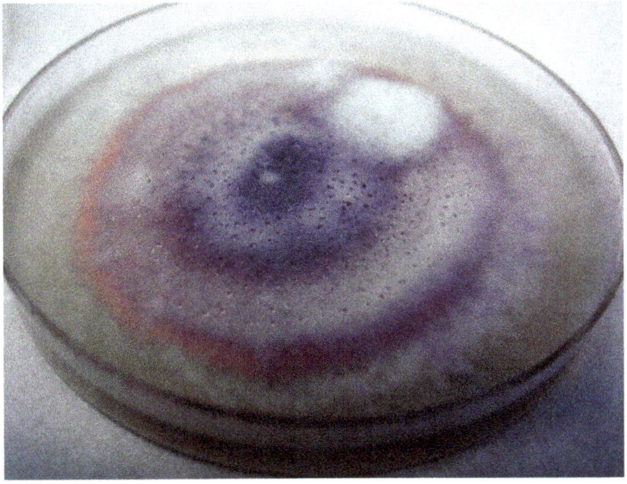

Fig. 8.4.35 *Fusarium roseum* (side).

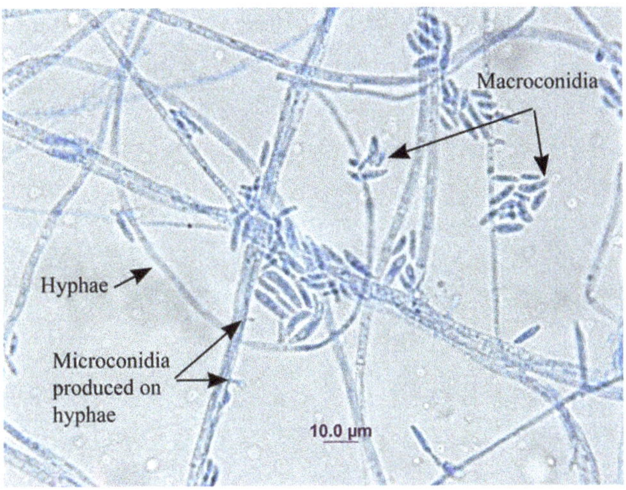

Fig. 8.4.36 *F. roseum*.

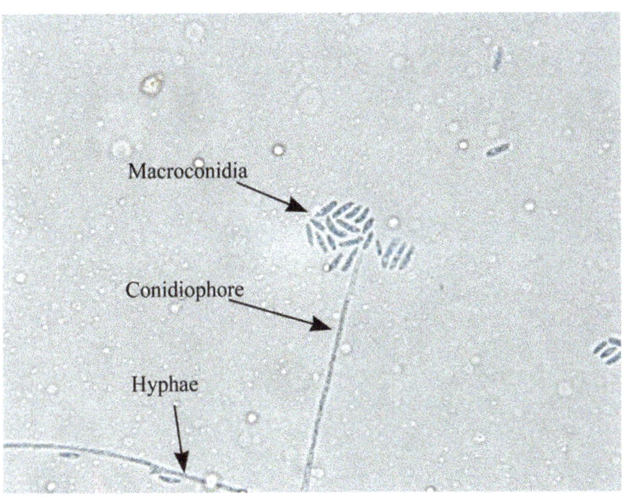

Fig. 8.4.37 *F. roseum*.

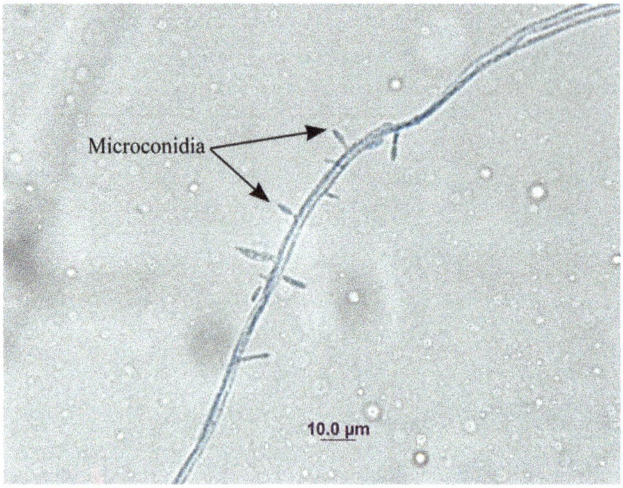

Fig. 8.4.38 *F. roseum*.

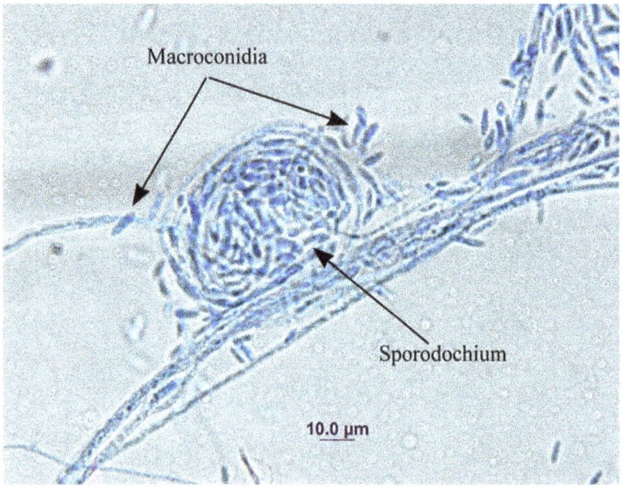

Fig. 8.4.39 *F. roseum*.

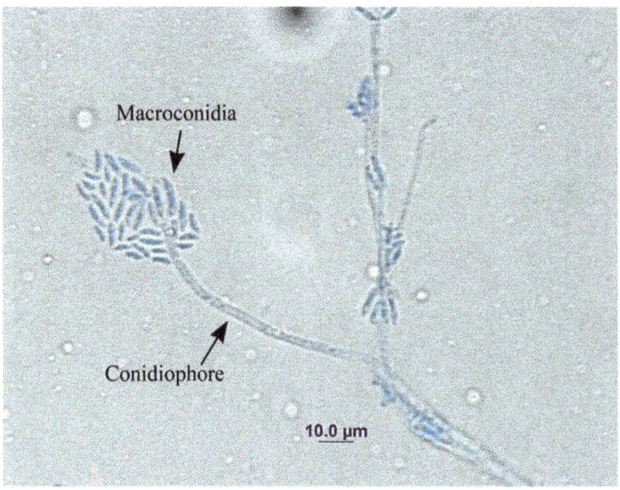

Fig. 8.4.40 *F. roseum*.

Fusarium scripi

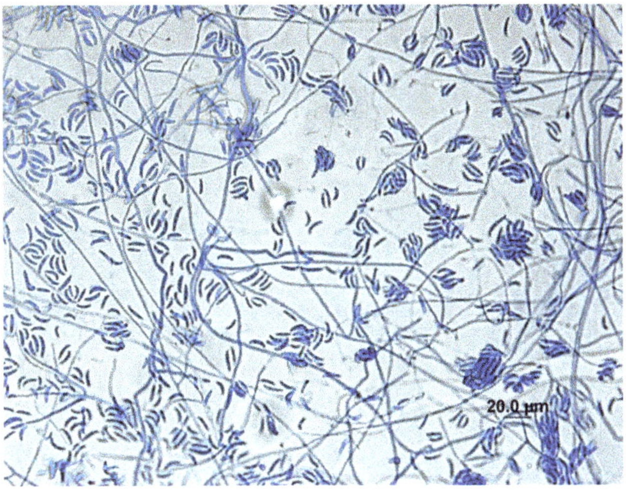

Fig. 8.4.41 *Fusarium scripi* macroconidia.

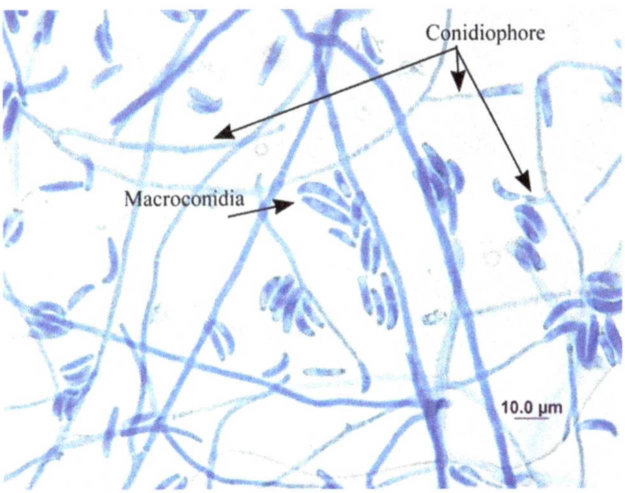

Fig. 8.4.42 *Fusarium scripi*.

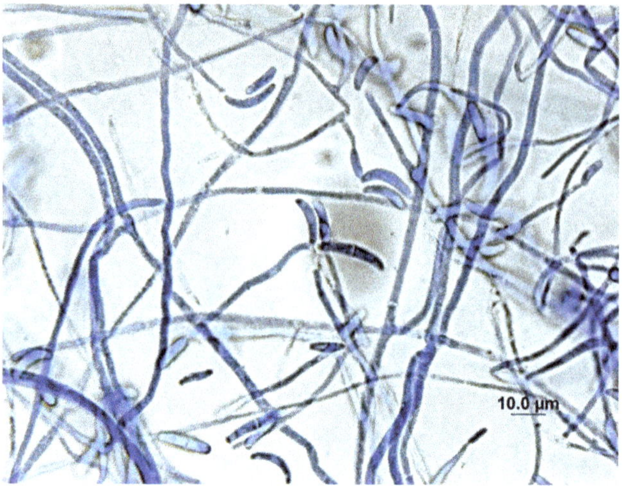

Fig. 8.4.43 *Fusarium scripi.*

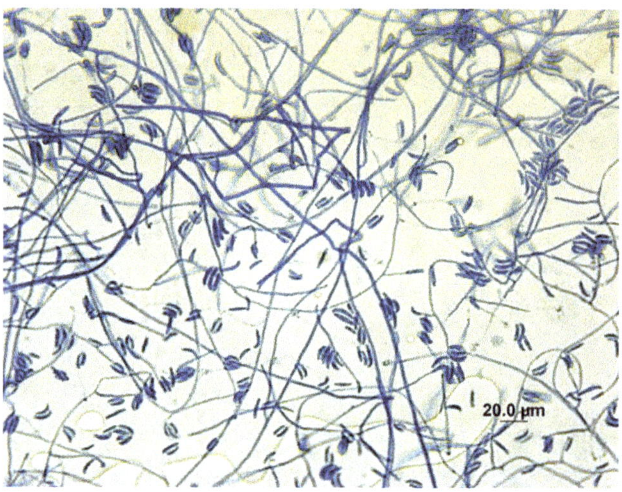

Fig. 8.4.44 *Fusarium scripi.*

Fusarium solani

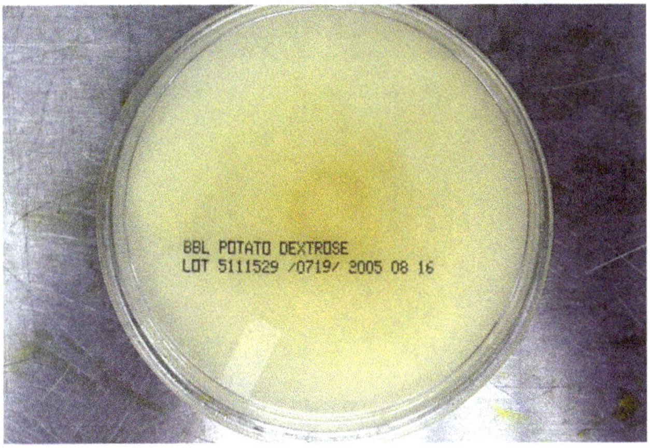

Fig. 8.4.45 *Fusarium solani*, PDA (bottom).

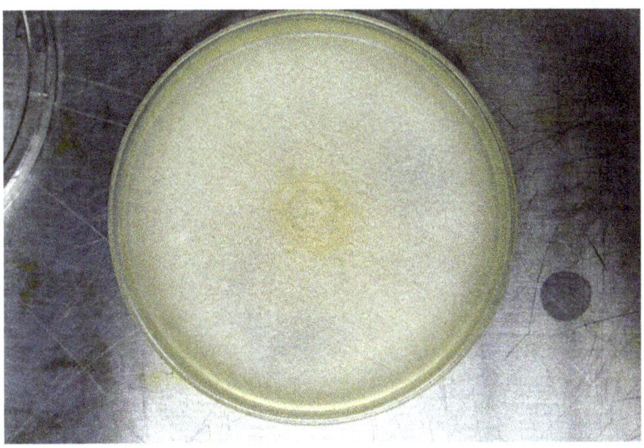

Fig. 8.4.46 *Fusarium solani*, PDA (top).

Fig. 8.4.47 *Fusarium solani*, PDA (side).

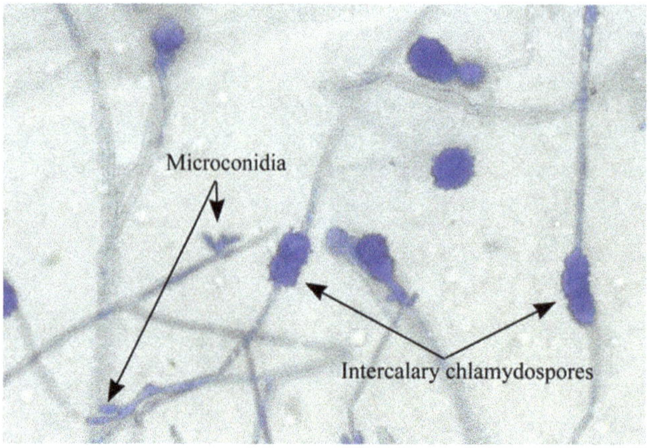

Fig. 8.4.48 *F. solani* rough chlamydospores in singles and pairs, intercalary.

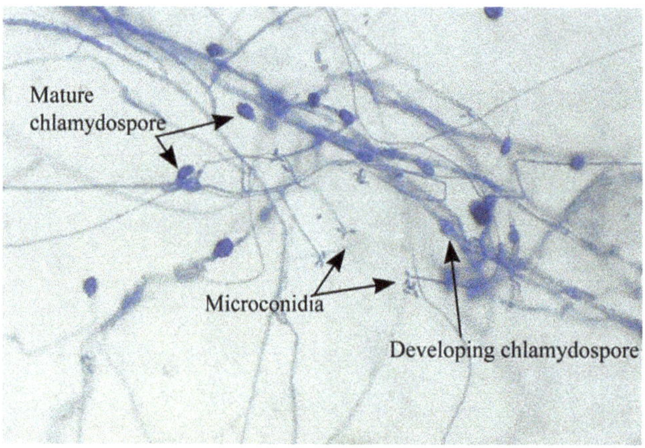

Fig. 8.4.49 *F. solani* chlamydospores and microconidia.

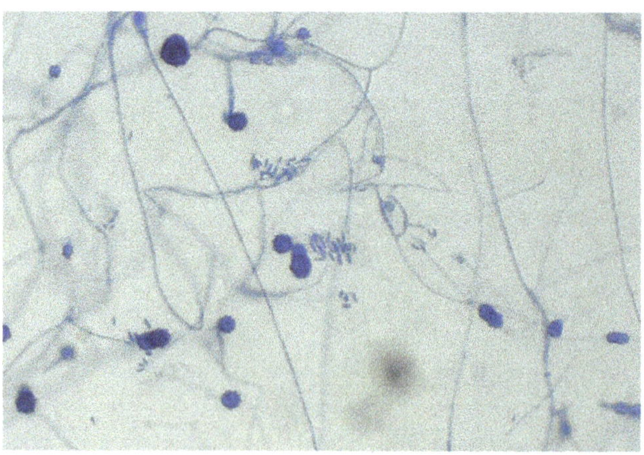

Fig. 8.4.50 *F. solani*.

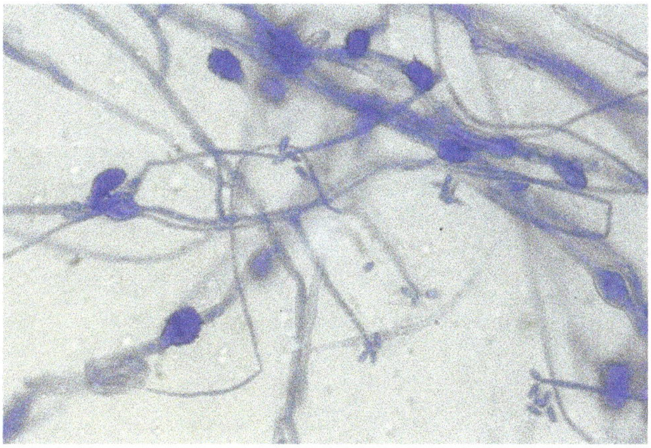

Fig. 8.4.51 *F. solani*.

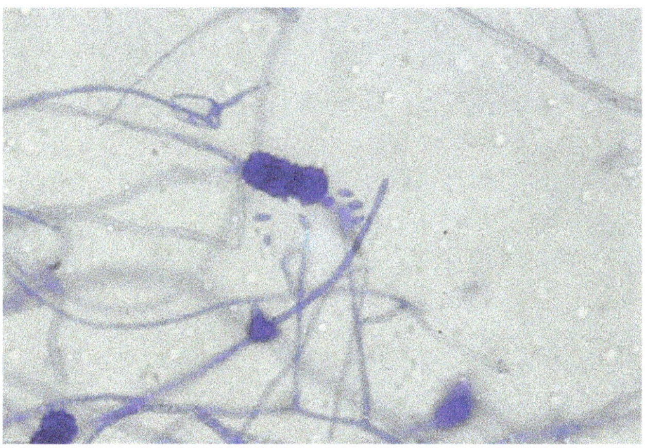

Fig. 8.4.52 *F. solani*.

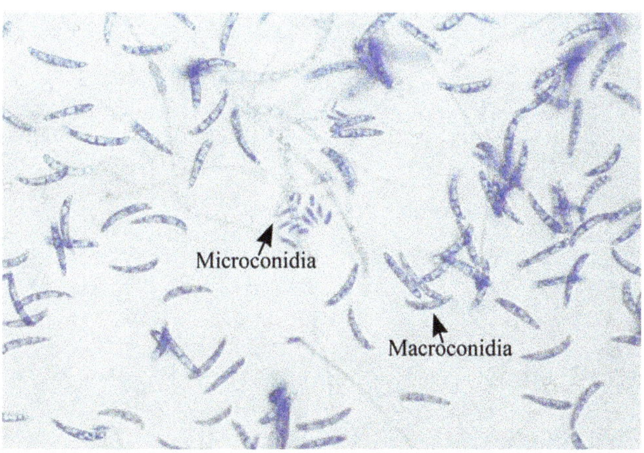

Fig. 8.4.53 *F. solani* macroconidia.

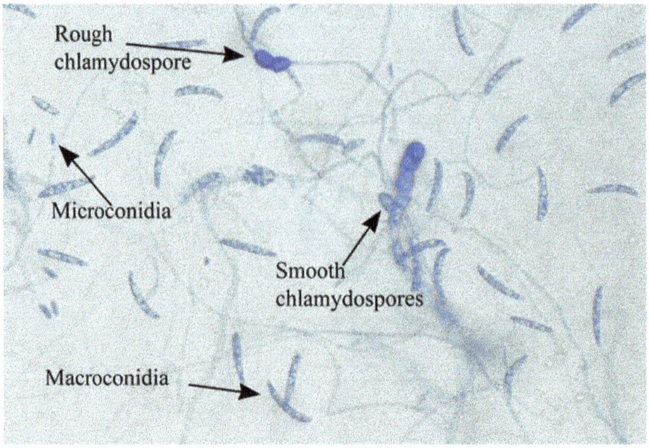

Fig. 8.4.54 *F. solani* macroconidia and chlamydospores.

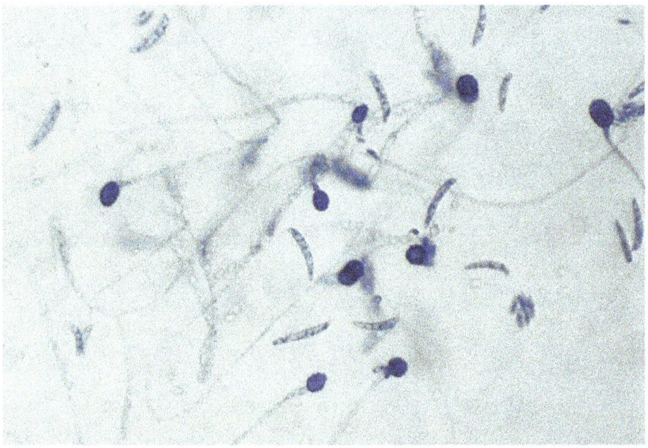

Fig. 8.4.55 *F. solani*.

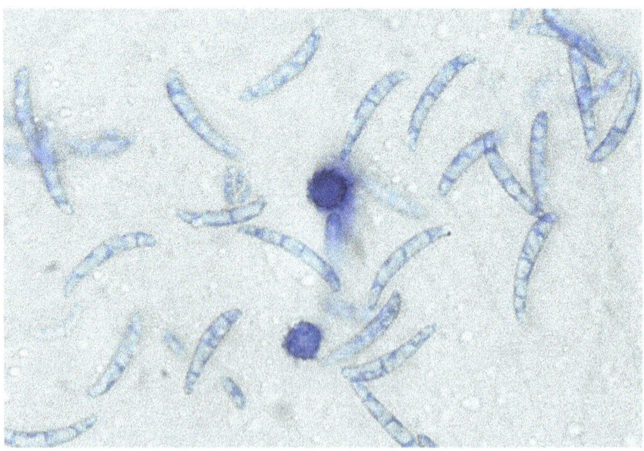

Fig. 8.4.56 *F. solani*.

8.5 *Gliocladium*

Colonial morphology: Colonies grow rapidly and develop a cottony, fluffy morphology, filling the plate. Colonies are first white then change to pink or green, with some species having spokes of white radiating out like the spokes of a wagon wheel.

Microscopic morphology: hyphae are septate and hyaline. Conidiophores are penicillinate, consisting of erect conidiophores producing primary and secondary phialides. Conidia are produced at the apices, abundantly, and cluster into a ball rather than forming chains like *Penicillium*. Conidia are unicellular, slimy, smooth walled, and green or yellow. In addition to penicillinate fruiting heads, verticillate production of conidia may be present. Thus, *Gliocladium* is sometimes confused with *Verticillium* or *Trichoderma*. Chlamydospore production is rarely seen or described, but Figures 8.5.10 and 8.5.11 demonstrate clustering and production of chlamydospores. *G. roseum* has recently been described as producing diesel fuel.

Gliocladium

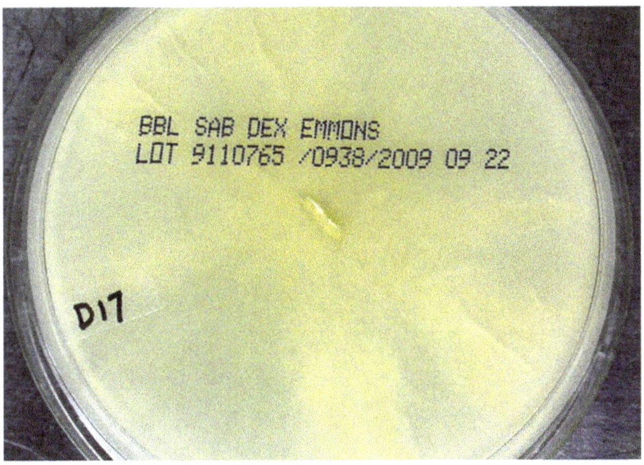

Fig. 8.5.1 *Gliocladium* (bottom).

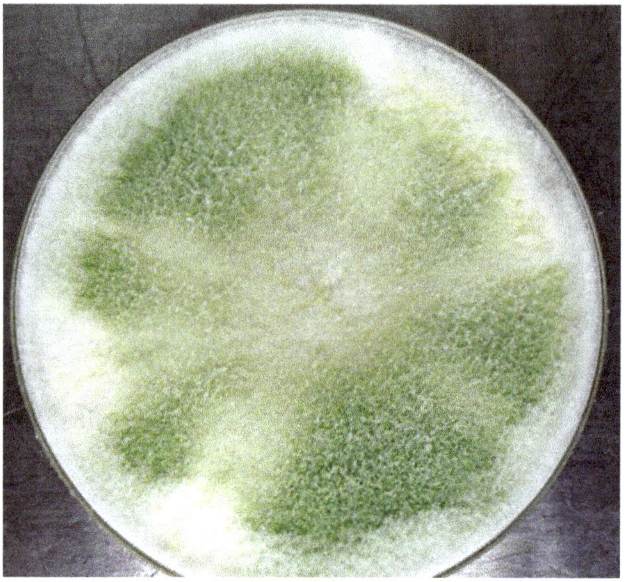

Fig. 8.5.2 *Gliocladium* (top).

Fig. 8.5.3 *Gliocladium* (side).

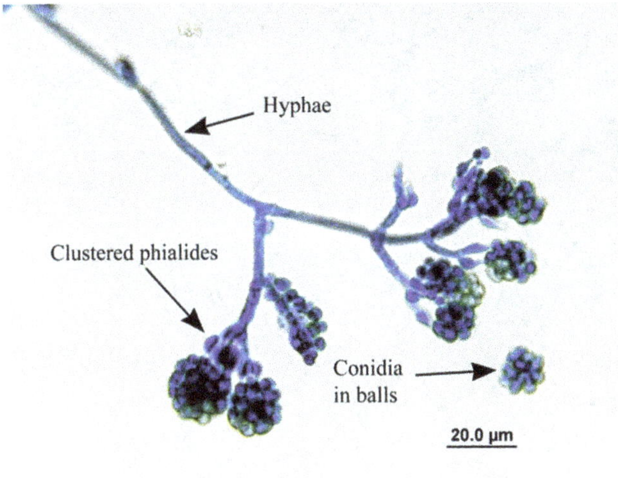

Fig. 8.5.4 *Gliocladium*, morphology.

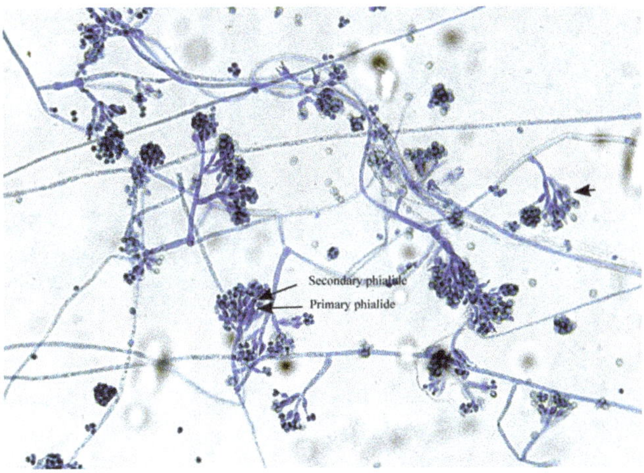

Fig. 8.5.5 Showing primary and secondary phialides.

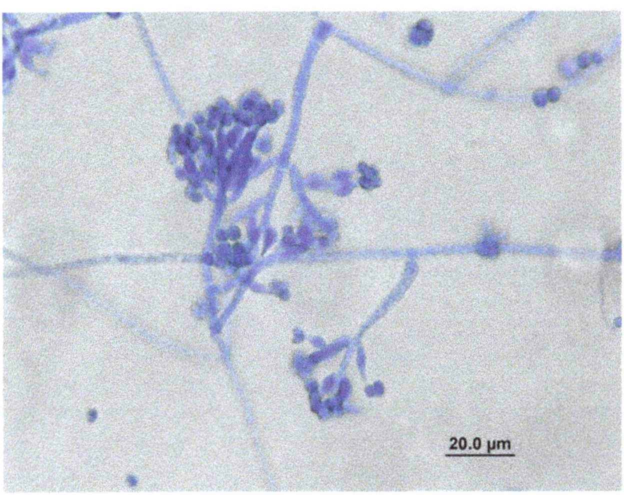

Fig. 8.5.6 *Gliocladium*.

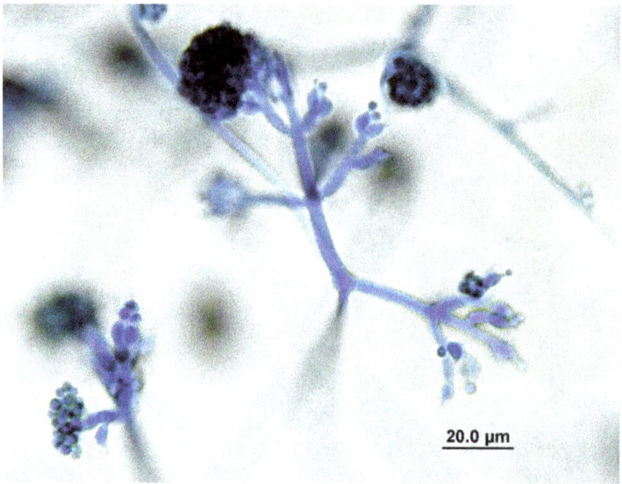

Fig. 8.5.7 *Gliocladium*.

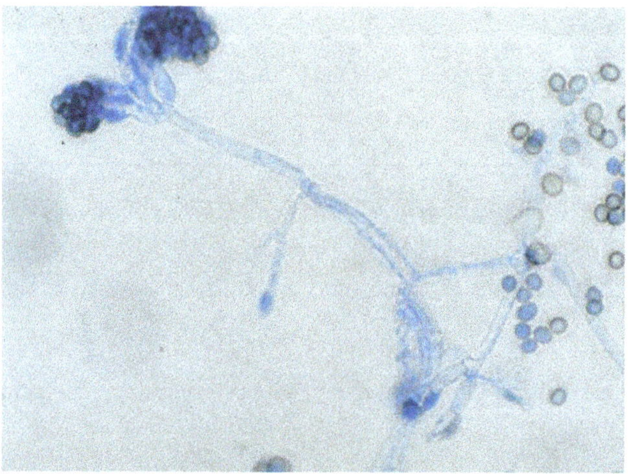

Fig. 8.5.8 *Gliocladium*.

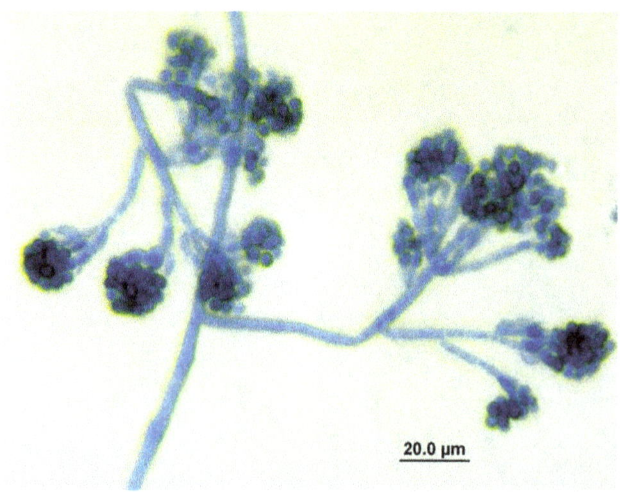

Fig. 8.5.9 *Gliocladium*.

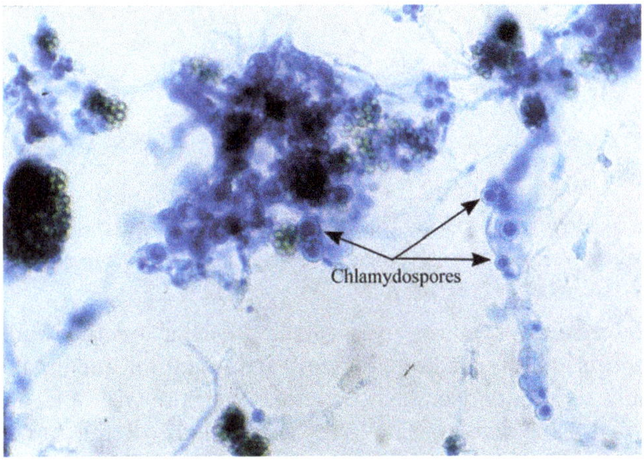

Fig. 8.5.10 Showing chlamydospores.

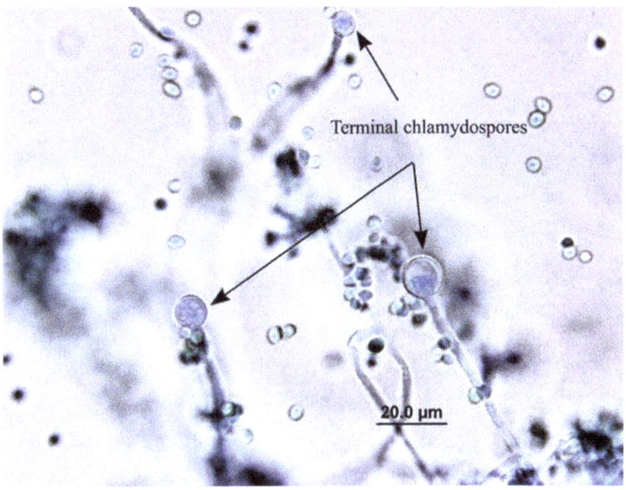

Fig. 8.5.11 Chlamydospore production.

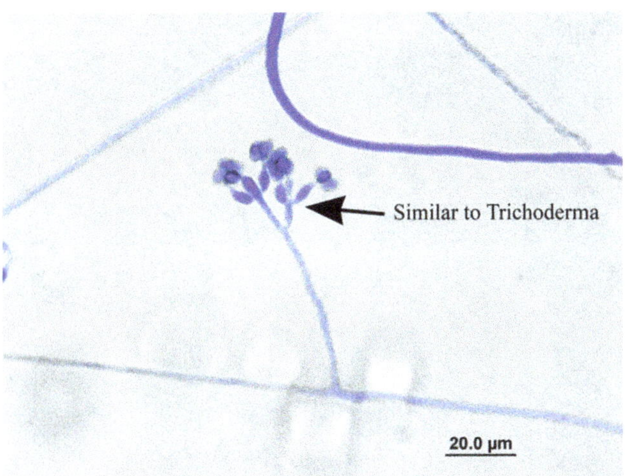

Fig. 8.5.12 Verticillate production of conidia.

8.6 *Malbranchea* spp

Colonial morphology: colonies are white, orange, beige, pink or sometimes brown and have a cottony, fluffy appearance. Reverse is yellow to orange. On potato dextrose agar, colonies are more wrinkled, develop more intense color, and are suede-like rather than cottony.

Microscopic morphology: hyphae are hyaline and septate. Mycelia are fine and resemble those of *Coccidioides immitis*. Hyphae separate into arthroconidia and break up to release free arthroconidia in abundance. Arthroconidia alternate with a clear space between them, much like *Coccidioides*. The arthroconidia are of the same diameter as mycelia. An annular frill occurs at the ends of arthroconidia when they break apart. Because of the similarity to *Coccidioides*, isolates should be transferred to a BSL-3 facility for work-up until a final identification is made. Conidiophores are absent and no other fruiting structures are present.

Malbranchea

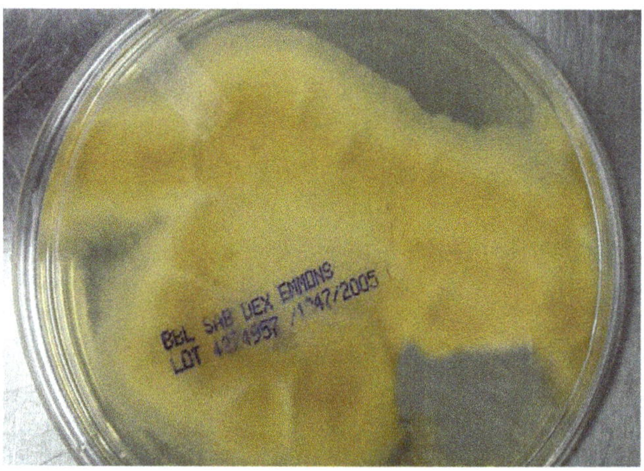

Fig. 8.6.1 *Malbranchea*, SABDX (bottom).

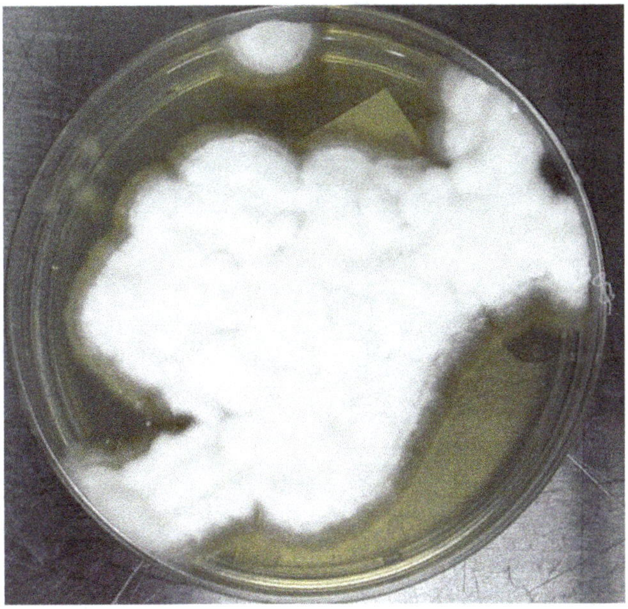

Fig. 8.6.2 *Malbranchea*, SABDX (top).

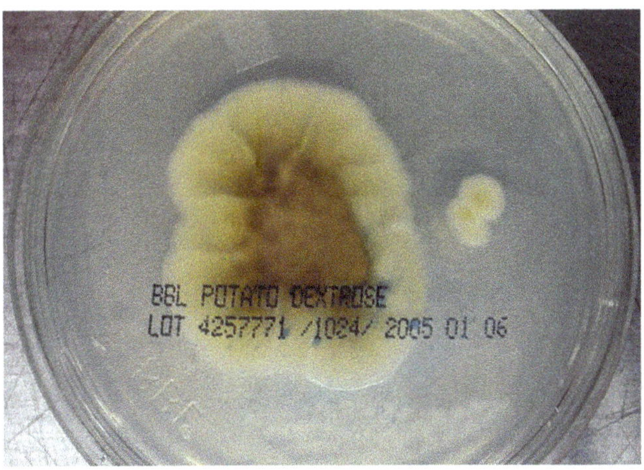

Fig. 8.6.3 *Malbranchea*, PDA (bottom).

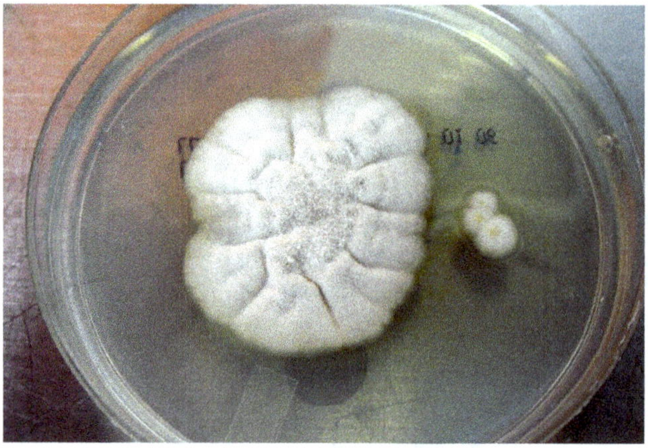

Fig. 8.6.4 *Malbranchea*, PDA (top).

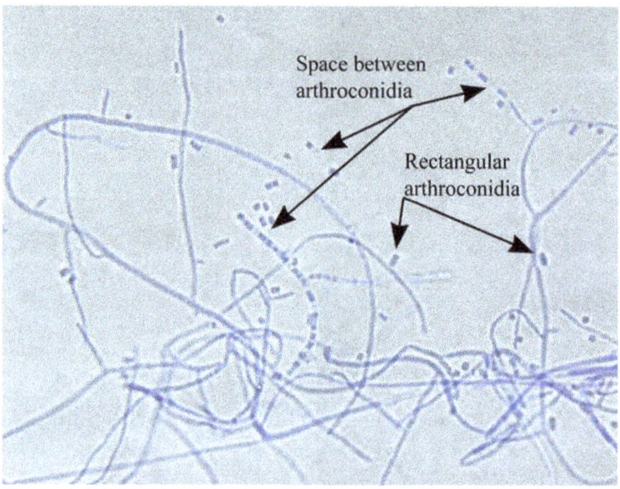

Fig. 8.6.5 Alternating arthroconidia.

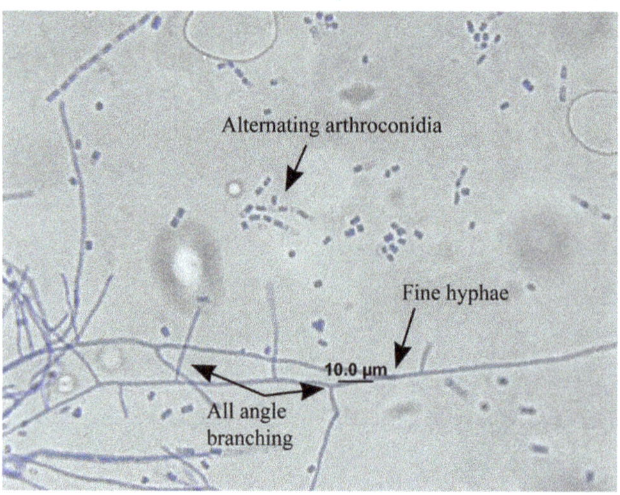

Fig. 8.6.6 Alternating arthoconidia with parallel sides.

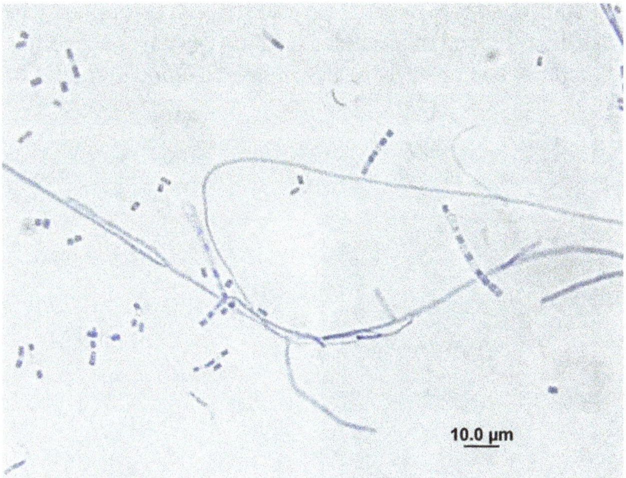

Fig. 8.6.7 *Malbranchea.*

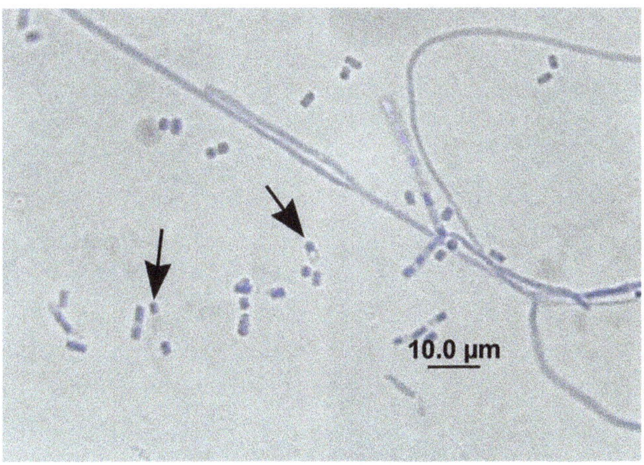

Fig. 8.6.8 Arthroconidia with annular frills.

8.7 *Paecilomyces* spp

Colonial morphology: colonies develop rapidly and colors may be yellow, tan, rust, gold, lilac, pink, light brown or white. Colonies are never bright green or blue-green, differentiating them from *Penicillium.* Reverse is cream, pink, yellow or light brown. The surface of the colony is powdery or wooly-velvety.

Microscopic morphology: hyphae are hyaline and septate. Phialides are formed either directly from the hyphae or from short branched primary conidiophores. Phialides are swollen at their bases, gradually descending into a long and slender neck, and occur solitarily, in pairs, verticillate or in penicillate heads. Conidia are produced from phialides in long slender chains and are dry, hyaline to dark, smooth or rough, elongate to spindle shaped. Phialides are distantly spaced, and V or W shaped as opposed to *Penicillium* which produces brush-like clusters of phialides. Chlamydospores are sometimes produced. *Paecilomyces lilacinum* has recently been renamed as *Purpureocillium lilacinum*.

Paecilomyces

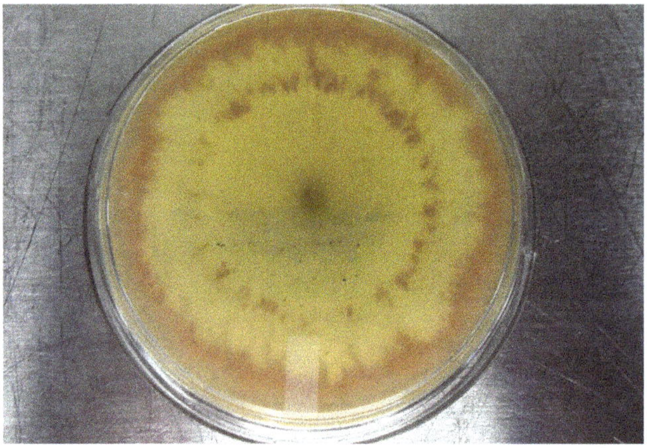

Fig. 8.7.1 *Paecilomyces* on SABDX (bottom).

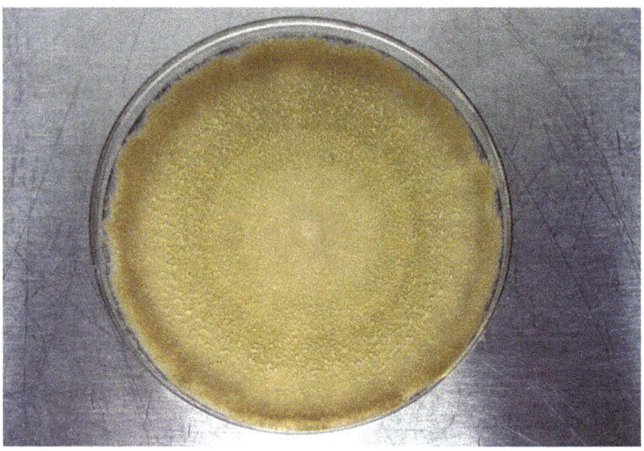

Fig. 8.7.2 *Paecilomyces* on SABDX (top).

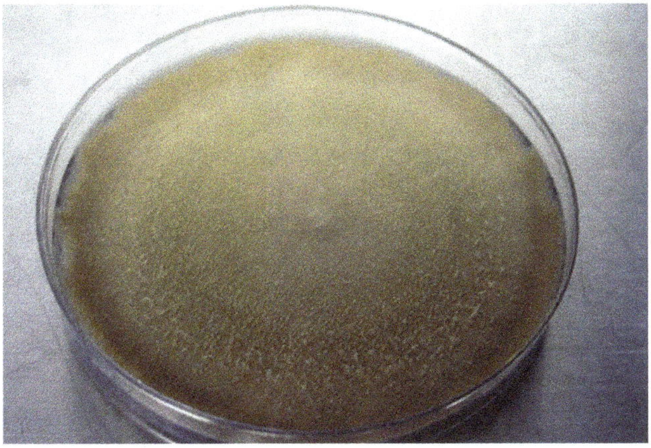

Fig. 8.7.3 *Paecilomyces* on SABDX (side).

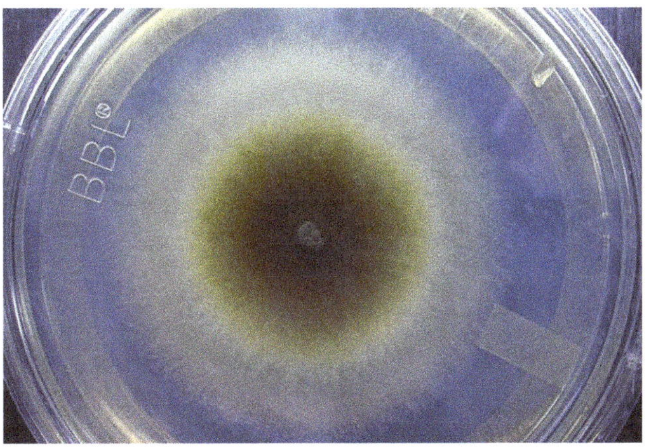

Fig. 8.7.4 *Paecilomyces* on CMA (top).

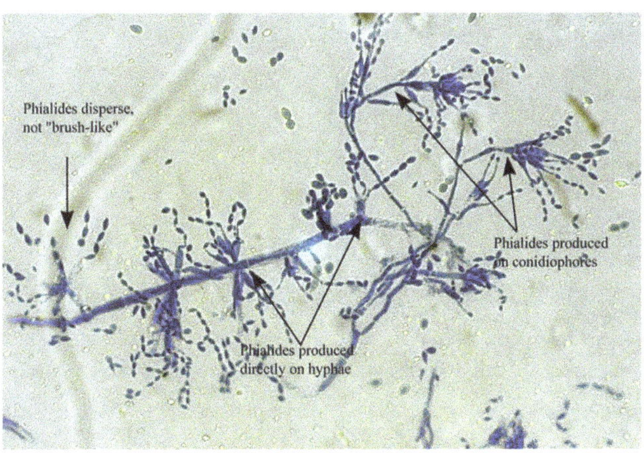

Fig. 8.7.5 *Paecilomyces*, basic morphology.

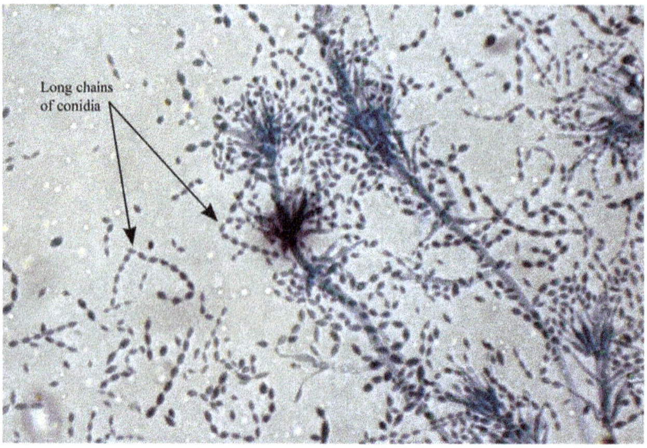

Fig. 8.7.6 *Paecilomyces*, with long chains of conidia.

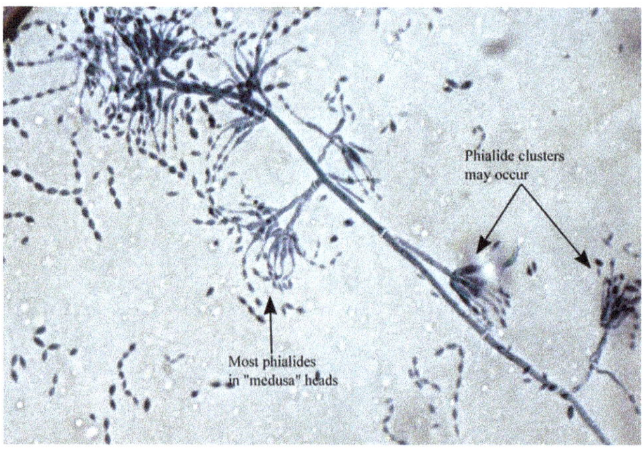

Fig. 8.7.7 *Paecilomyces*, with "medusa head" phialides.

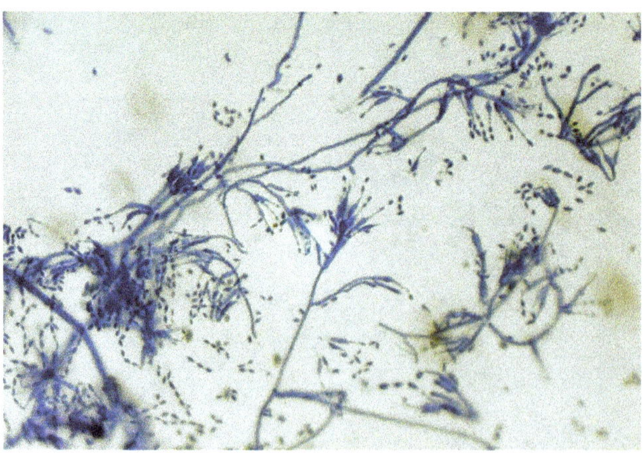

Fig. 8.7.8 *Paecilomyces*.

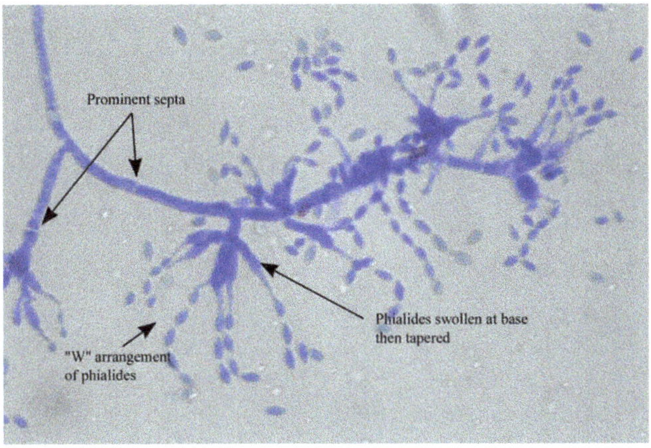

Fig. 8.7.9 *Paecilomyces*, with "W" arrangement of phialides.

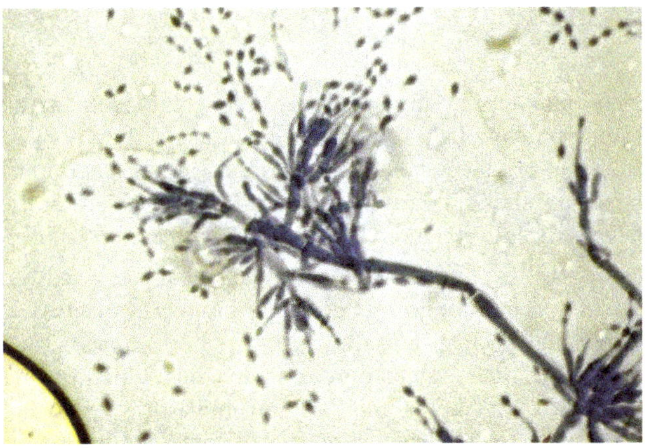

Fig. 8.7.10 *Paecilomyces*.

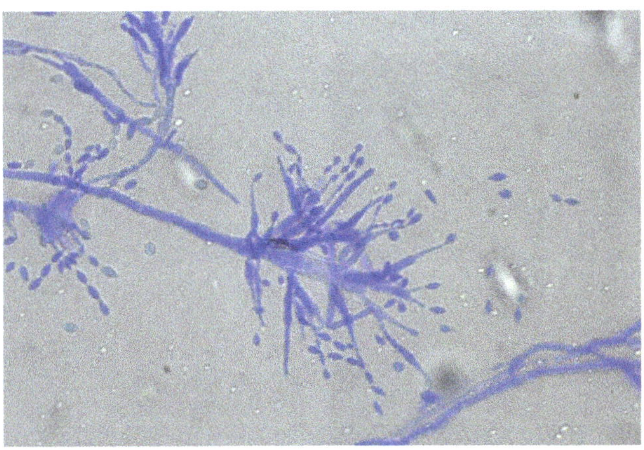

Fig. 8.7.11 *Paecilomyces*.

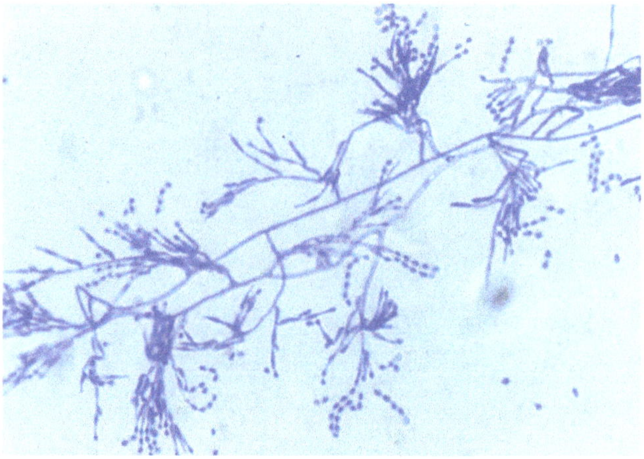

Fig. 8.7.12 *P. lilacinus.*

8.8 *Penicillium* spp

Colonial morphology: colonies develop rapidly and mostly are green, blue-green, gray-green, white, yellow or pink. Reverse is yellow, light brown or red. Surface of the colony is powdery or wooly-velvety. Isolates that produce a red pigment in the agar are suspicious for *Penicillium marneffei*. Such isolates should be tested for thermal dimorphism or by PCR for *P. marneffei*. *P. marneffei* has recently been renamed as *Talaromyces marneffei*.

Microscopic morphology: hyphae are hyaline and septate. Phialides are formed on short branched or unbranched primary conidiophores. Secondary branching occurs via metula. Phialides may be produced singly or in groups. Branching may occur via metula formation. This gives a brush-like appearance referred to as a penicillus. Phialides are swollen at their bases descending into a neck. Conidia (ameroconidia) are produced from phialides in chains and are smooth, ellipsoidal, cylindrical or spindled, hyaline or greenish, smooth or rough walled. Sclerotia may be produced by some species.

Penicillim spp

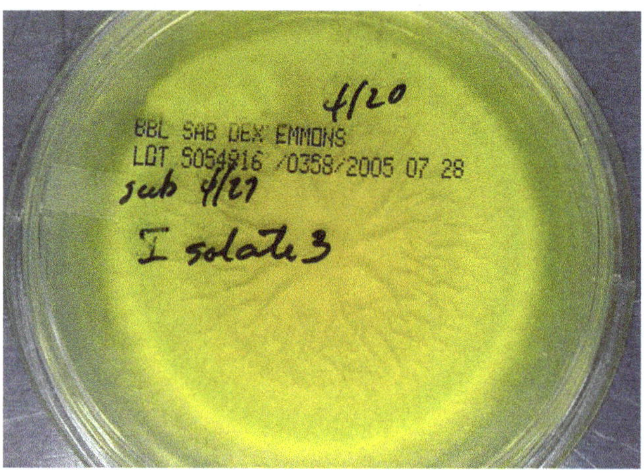

Fig. 8.8.1 *Penicillium*, (bottom).

Fig. 8.8.2 *Penicillium* (top).

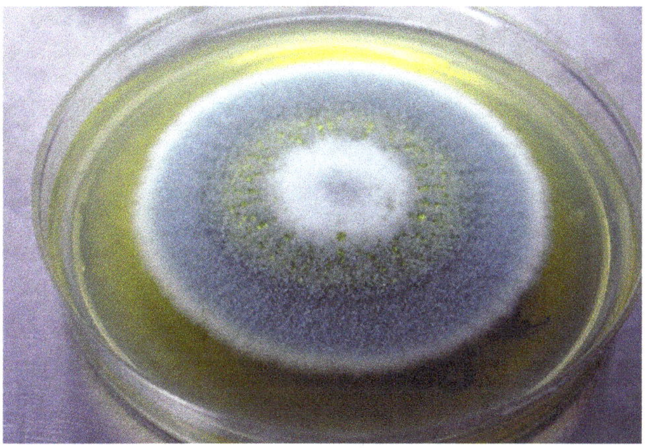

Fig. 8.8.3 *Penicillium* (side).

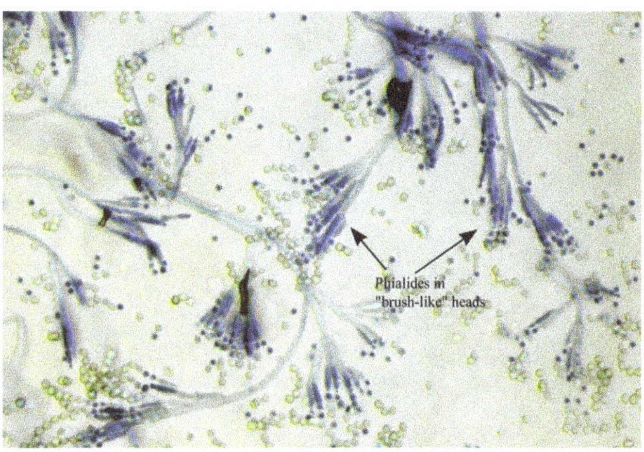

Fig. 8.8.4 *Penicillium*, phialides in "brush-like" heads.

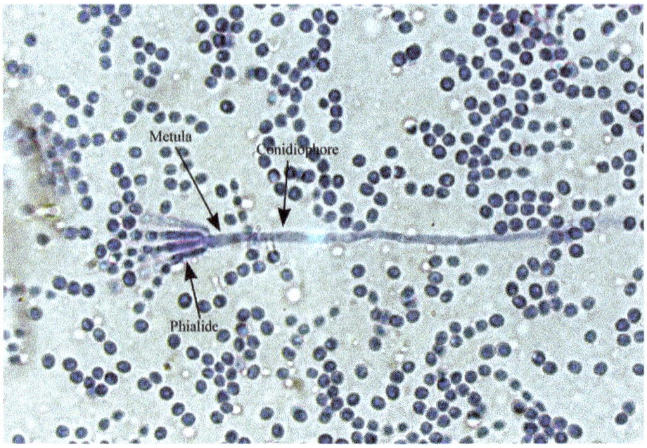

Fig. 8.8.5 *Penicillium*, showing conidiophore, metula, phialide and conidia.

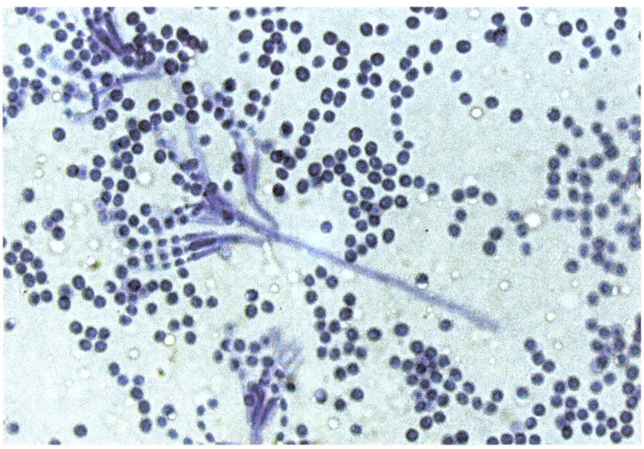

Fig. 8.8.6 *Penicillium*.

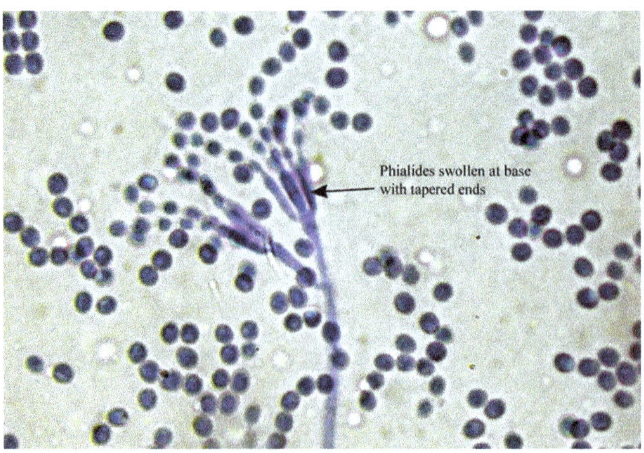

Fig. 8.8.7 *Penicillium*.

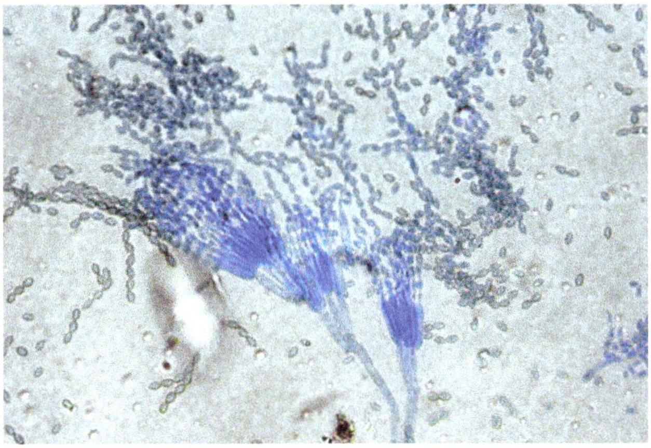

Fig. 8.8.8 *Penicillium*.

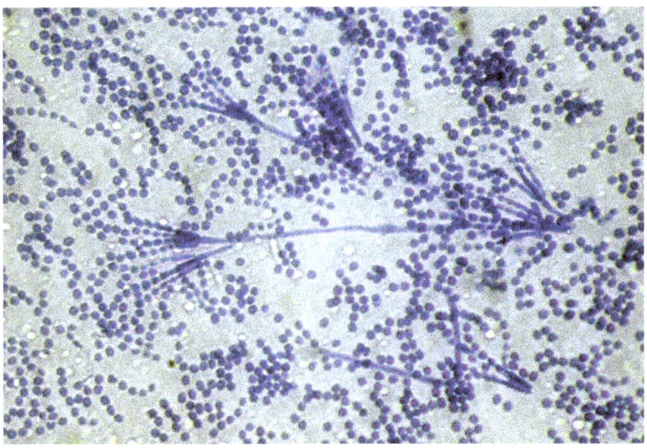

Fig. 8.8.9 *Penicillium*.

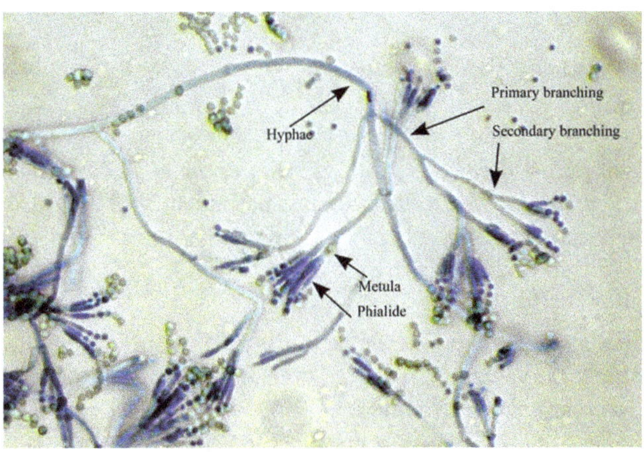

Fig. 8.8.10 *Penicillium*, branching characteristics.

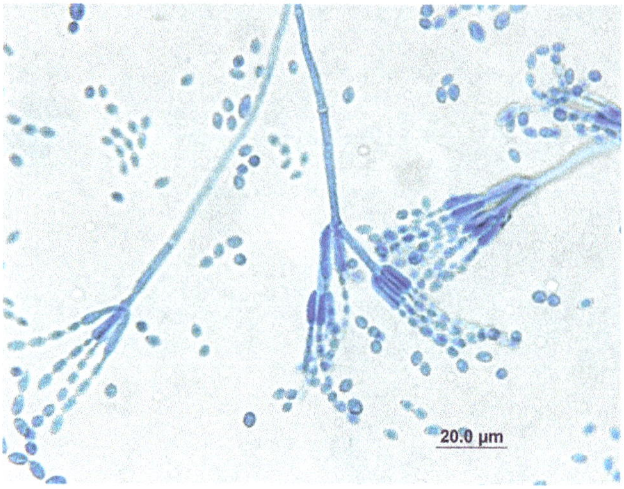

Fig. 8.8.11 *Penicillium.*

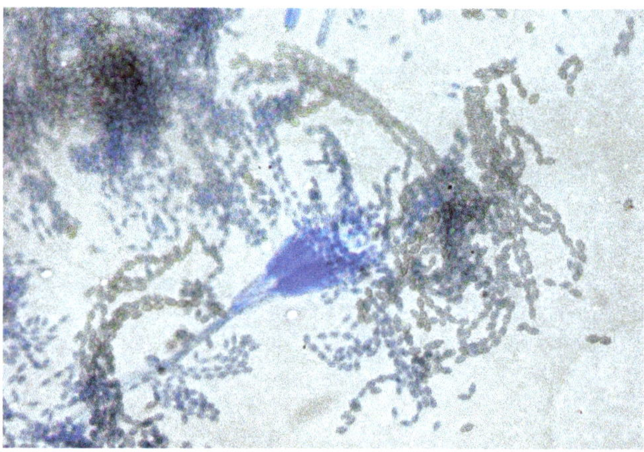

Fig. 8.8.12 Long chains of conidia.

8.9 *Ramichloridium* spp

The serious neurotrophic human pathogen of this group is *R. mackenziei* whereas two other human pathogens, *R. schulzeri* and *R. basitonum*, are less pathogenic but may also cause cerebral pheohyphomycosis. *R. mackenziei* was recently reclassified as *Rhinocladiella mackenziei*. *R. schulzeri* is also known as *Myrmecridium schulzeri*. Other species are common soil saprobes.

Colonial morphology: colonies appear yeast-like or leathery on the surface of the plate. With age, the surface may become velvety. *R. schulzeri* is yellow-orange pigmented, whereas other species are tan to cream or olive to black.

Microscopic morphology: hyphae are hyaline and septate. Mycelia are submerged with aerial hyphae produced in older cultures. Conidiophores are straight, not branched, thick walled, forming a descending rachis toward the tip. Some species have brown conidiophores. A septum is regularly observed at the base of the conidiophores where they attach to the hyphae. Often, conidiophores are swollen at the base, appearing "spade-like." Some species show scarring (or pimples) on the rachis where conidia are produced. Conidia (2–5 µm) are produced singly and may have 1–2 septa in some species but septa in pathogenic species are rare. Conidia are elongate and tear shaped. Foot cells are produced.

The key feature of this fungus is that both sympodial (rachiform) and opposing (non-sympodial, raduliform) arrangements of conidia occur. Therefore, *Ramichloridium* must be differentiated from *Rhinocladiella* using discrete criteria for conidia formation along with colonial morphology.

Ramichloridium shulzeri

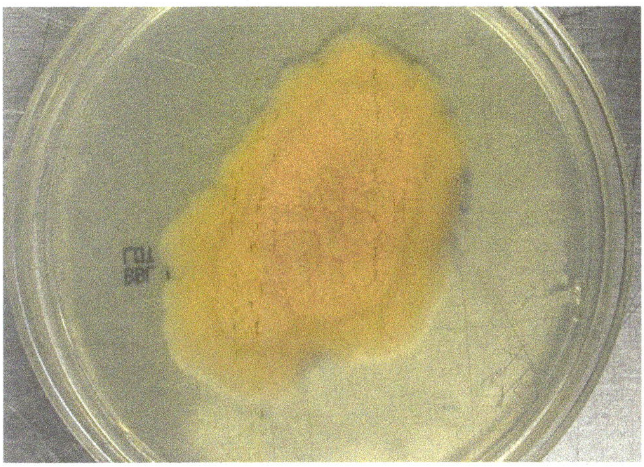

Fig. 8.9.1 *Ramichloridium schulzeri* (top).

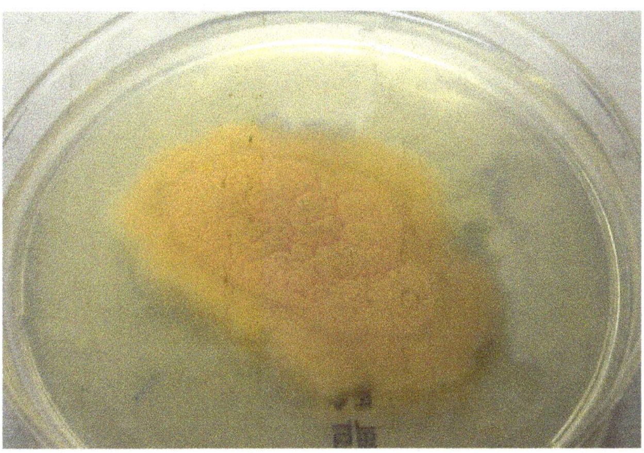

Fig. 8.9.2 *R. schulzeri*, at 2 weeks (side).

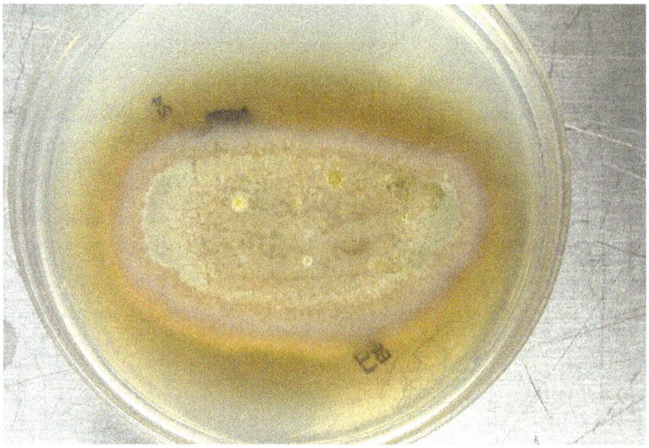

Fig. 8.9.3 *R. schulzeri*, at 3 weeks.

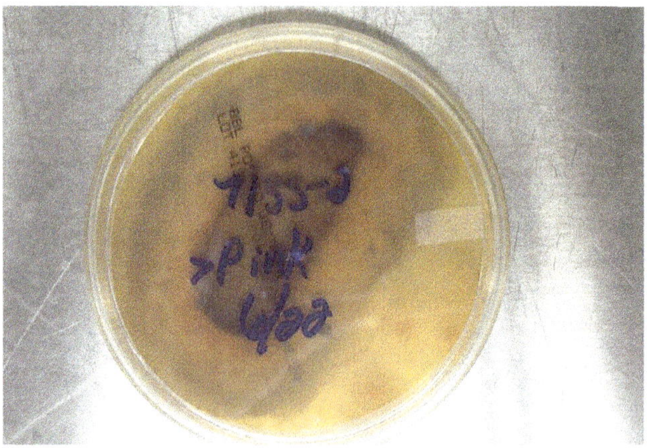

Fig. 8.9.4 *R. schulzeri*, at 5 weeks (bottom).

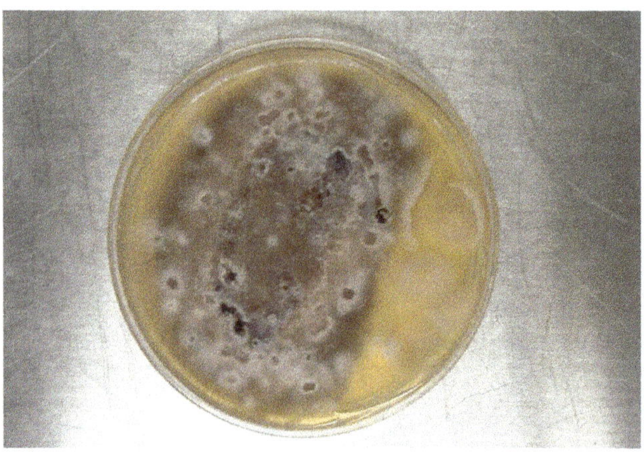

Fig. 8.9.5 *R. schulzeri*, at 5 weeks (top).

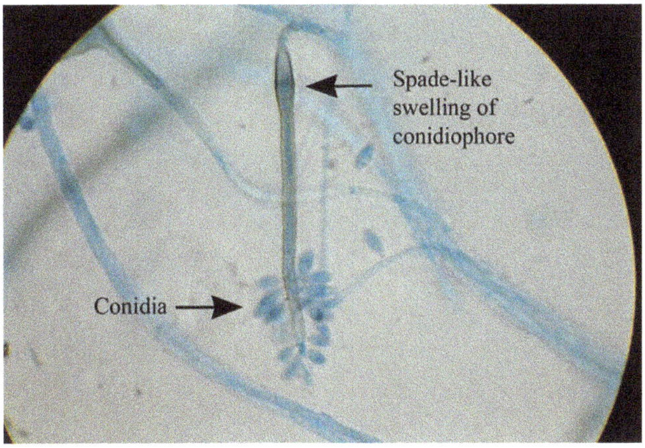

Fig. 8.9.6 *Rhinocladiella* type conidiation.

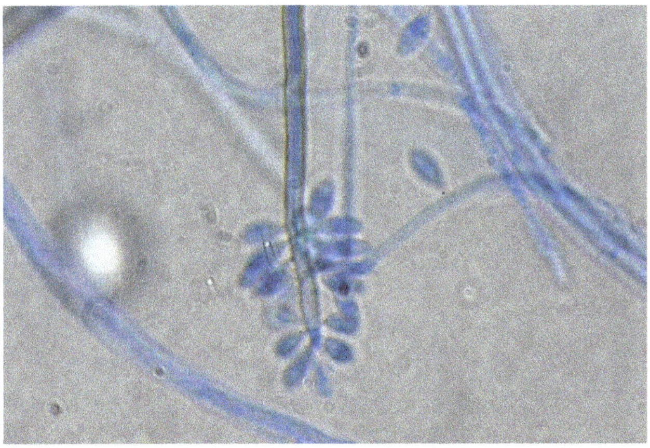

Fig. 8.9.7 *Rhinocladiella*, 2000×.

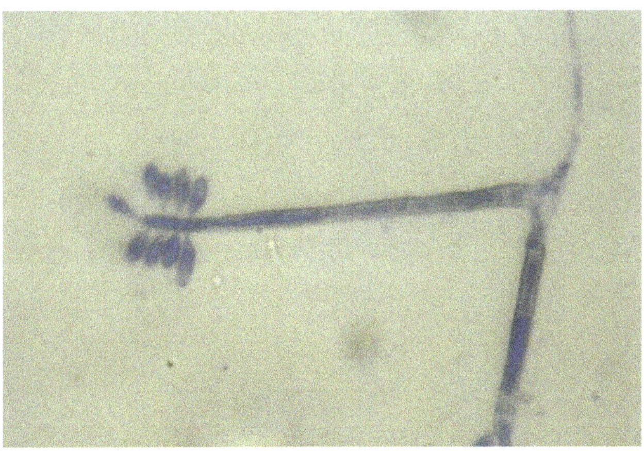

Fig. 8.9.8 Non-sympodial but opposing conidial formation.

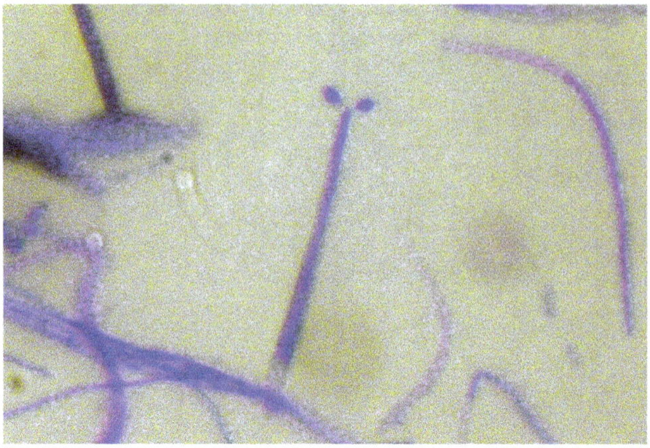

Fig. 8.9.9 Mouse ears formation at tip of conidiopore.

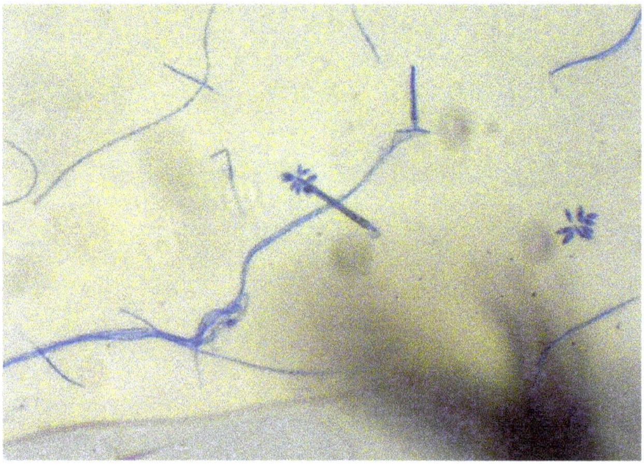

Fig. 8.9.10 *Rhinocladiella*, 400×.

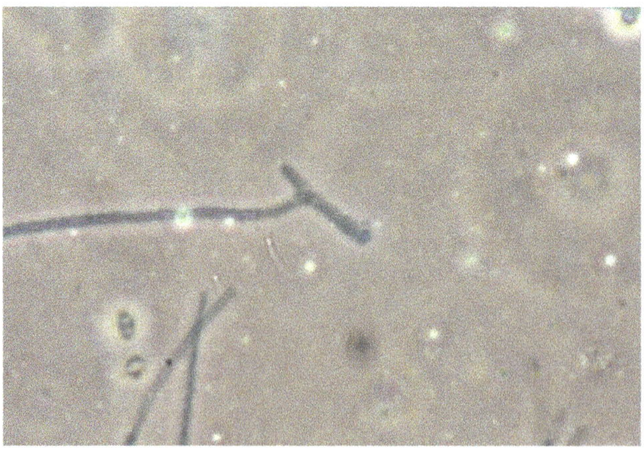

Fig. 8.9.11 Foot cell.

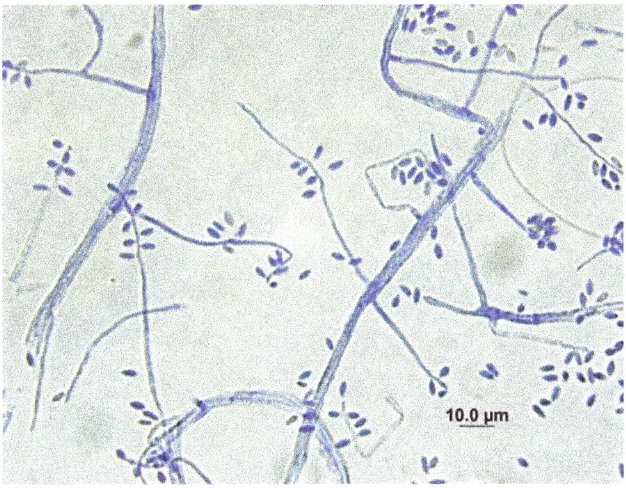

Fig. 8.9.12 *Rhinocladiella*, 400×.

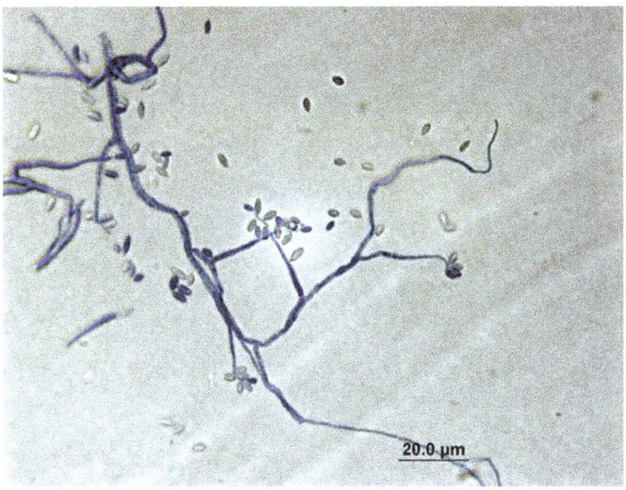

Fig. 8.9.13 *Rhinocladiella*.

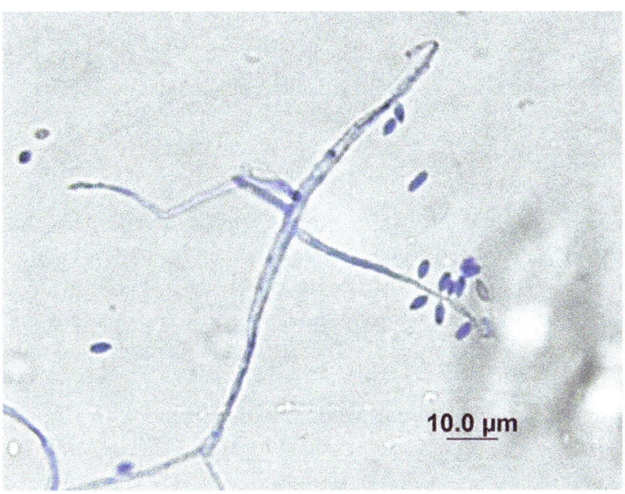

Fig. 8.9.14 *Rhinocladiella*, 1000×.

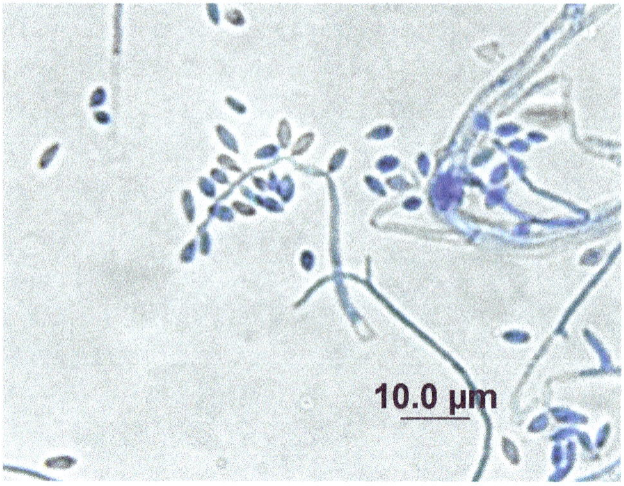

Fig. 8.9.15 *Rhinocladiella*, 1500×.

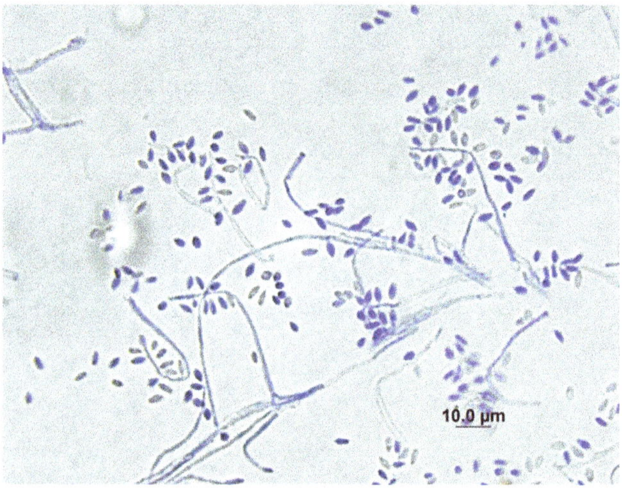

Fig. 8.9.16 *Rhinocladiella*.

8.10 *Schizophyllum commune*

Colonial morphology: colonies begin as white fluffy molds. Within a few days, the mold turns dark brown and spreads over the surface of the plate, becoming rough and leathery when grown in the dark at 22 °C. When cultures are grown in the light at 22 °C, gilled fruiting bodies may develop or the culture may develop thick mushroom-like structures on the surface. After 5–7 weeks, colonies begin to lift the culture plate lid and grow along the sides of the culture plate. This thick, spongy mat of growth (basidiocarp) is consistent with the *Basidiomycetes* (mushrooms). Deposits of golden-yellow liquid may form on the surface of the colony. The reverse is dark brown to orange. The cultures shown were sealed with parafilm, yet the basidiocarp continued to force the parafilm away from the sides of the plate.

Microscopic morphology: hyphae are hyaline and sparsely septate, yet bumpy and not parallel. They sometimes become pigmented with a golden, brown or yellow pigment (grown in light). Clamp connections develop along the hyphae in some but not all cultures. Diagnostic spicules are often numerous along the hyphae of the dikaryotic phase as well as cell fusions and the development of basidiospores. As shown in Fig. 8.10.5, this particular species formed unique rectangular spring-like projections along the hyphae. Monokaryotic cultures may remain sterile, but rare golden-brown basidiospores may be produced when cultures are grown in the dark at 30 °C (Fig. 8.10.6).

Schizophyllum commune

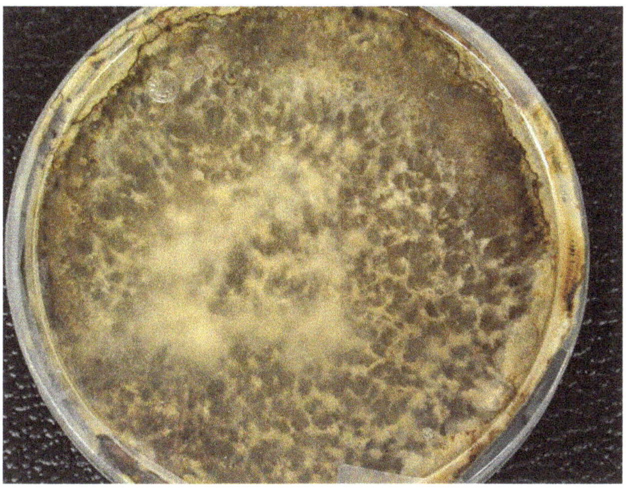

Fig. 8.10.1 *Schizophyllum commune*, dark at 30 °C (top).

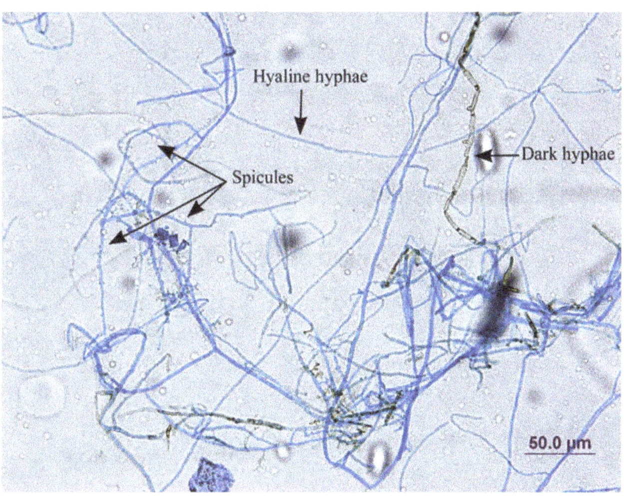

Fig. 8.10.2 *S. commune*, dark at 30 °C.

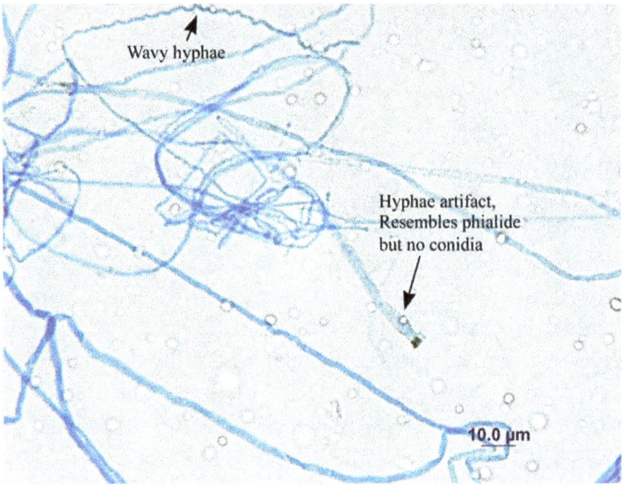

Fig. 8.10.3 *S. commune*, dark at 30 °C.

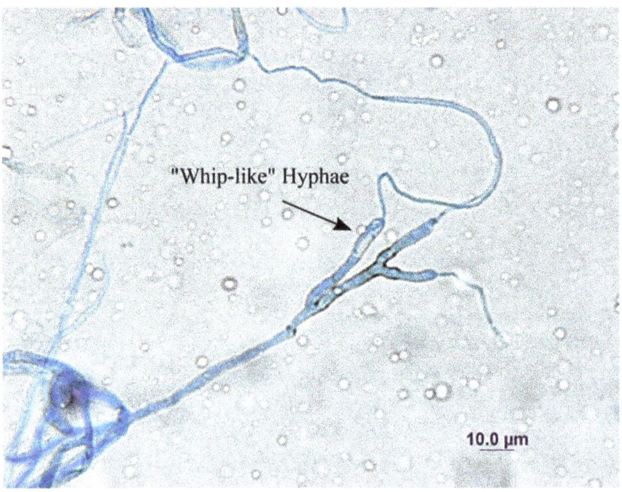

Fig. 8.10.4 *S. commune*, dark at 30 °C.

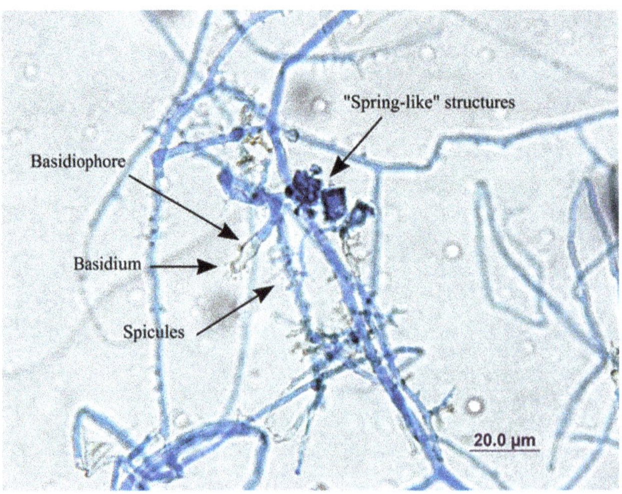

Fig. 8.10.5 *S. commune*, dark at 30 °C.

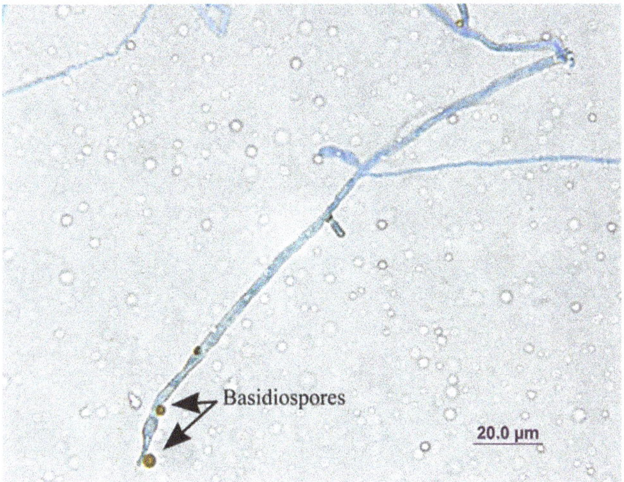

Fig. 8.10.6 *S. commune*, dark at 30 °C.

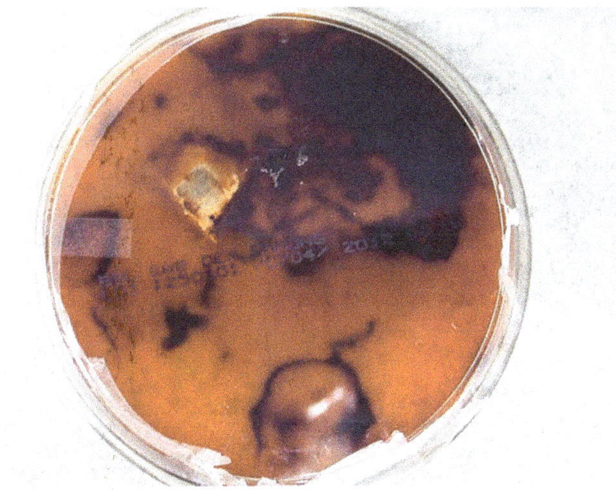

Fig. 8.10.7 *S. commune* (bottom), light at 22 °C (bottom).

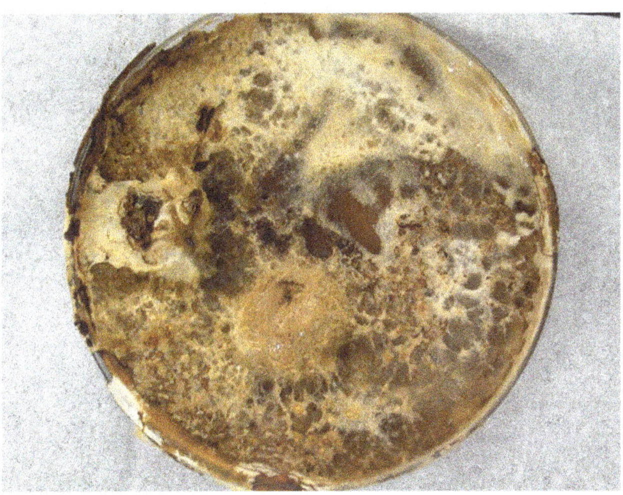

Fig. 8.10.8 *S. commune*, light at 22 °C (top).

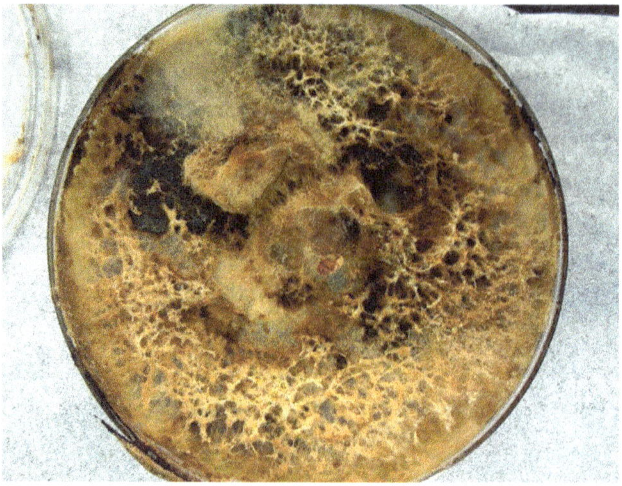

Fig. 8.10.9 *S. commune*, light at 22 °C (top).

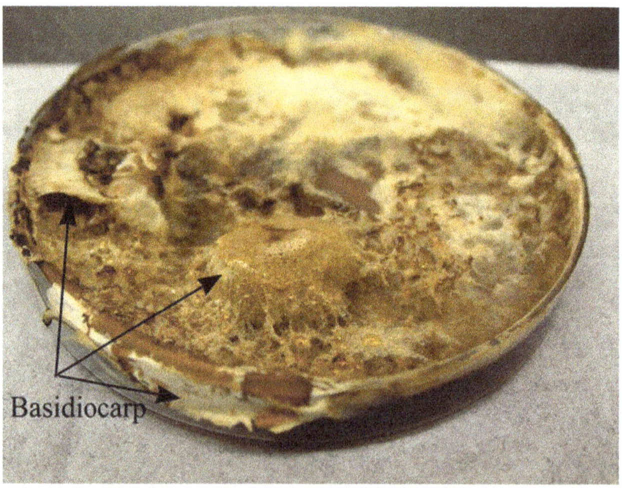

Fig. 8.10.10 *S. commune*, light at 22 °C (side).

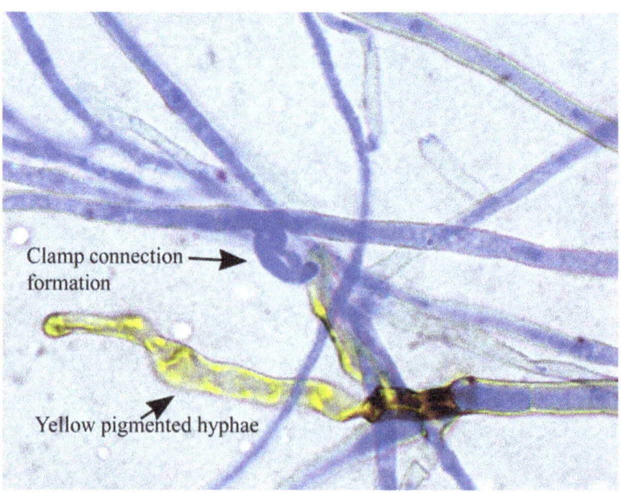

Fig. 8.10.11 *S. commune*, light at 22 °C.

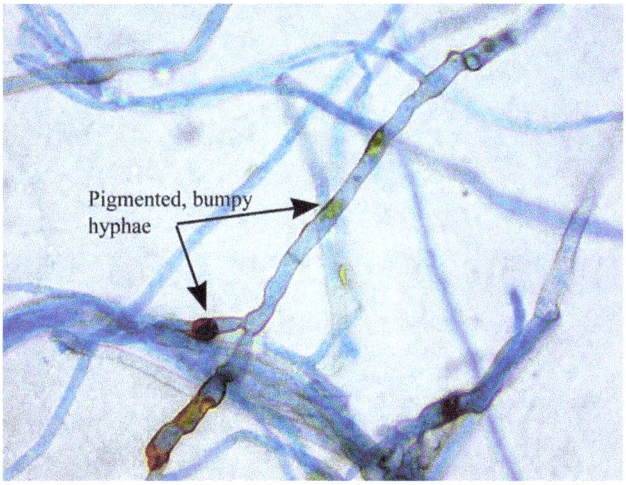

Fig. 8.10.12 *S. commune*, light at 22 °C.

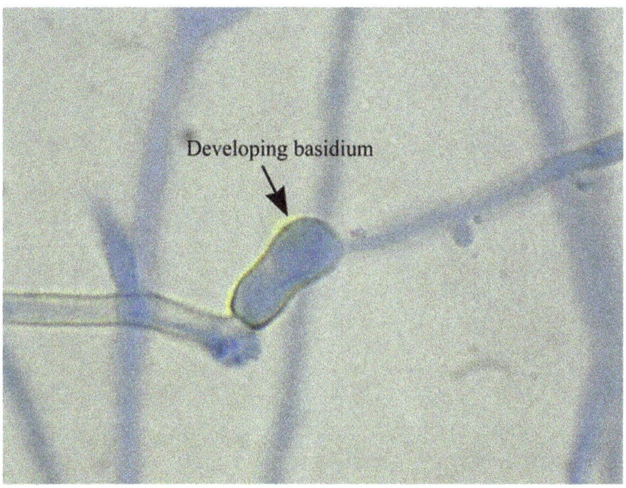

Fig. 8.10.13 *S. commune*, light at 22 °C.

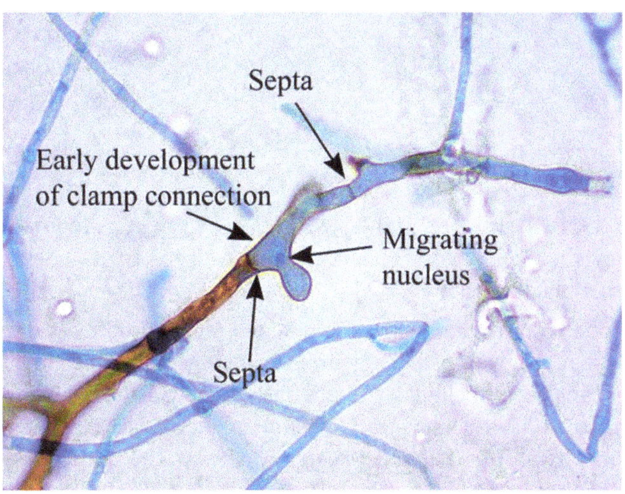

Fig. 8.10.14 *S. commune*, light at 22 °C.

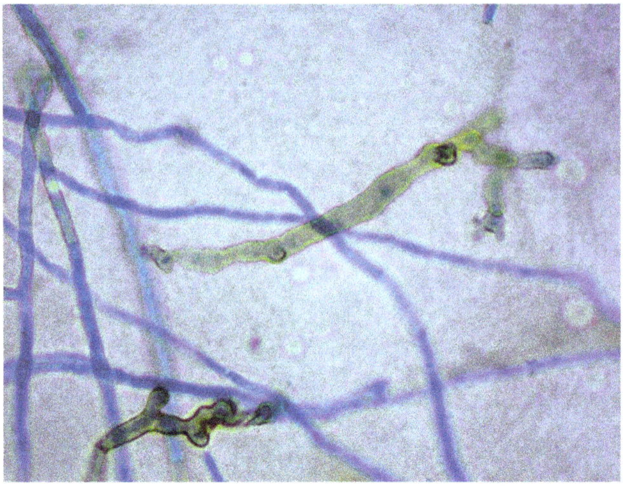

Fig. 8.10.15 *S. commune*, light at 22 °C.

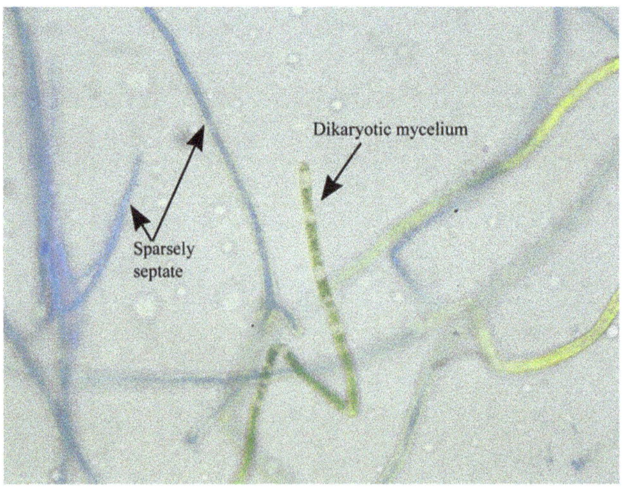

Fig. 8.10.16 Yellow mycelia.

8.11 *Scopulariopsis brevicaulis*

Colonial morphology: colonies are first white, and later develop a gray or cinnamon center, then developing a grayish black color. The surface is velvety to powdery. The reverse is white or tan with a black center.

Microscopic morphology: hyphae are hyaline or dark and septate. Conidiophores are produced laterally along the hyphae. They are single or branched with medium to long (9–25 μm in length) secondary branches (annellides) that are flask shaped. The annellides form a brush-like penicillus similar to those seen with *Penicillium*. Conidia (single-celled ameroconidia) are produced in chains in basipetal progression from the annellides.

The youngest conidium is at the basal end of the chain. Chains of conidia are 3–15 in length. Conidia (5–7 μm in diameter) are round with a "neck" at the base, then cut off; flat where attached to the annellide or other conidia in the chain. Conidia are spiny and rough, hyaline or light brown, and sometimes pointed at the top.

Scopulariopsis brevicaulis

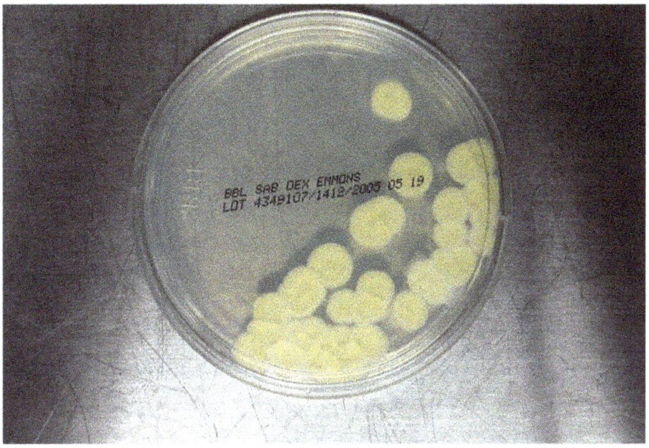

Fig. 8.11.1 *Scopulariopsis brevicaulis* (bottom).

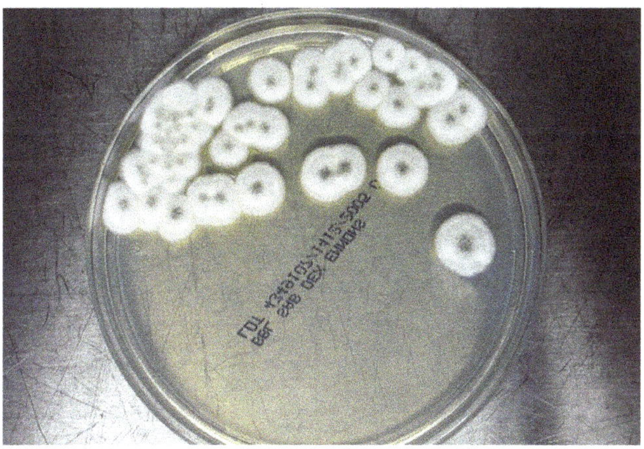

Fig. 8.11.2 *Scopulariopsis brevicaulis* (top).

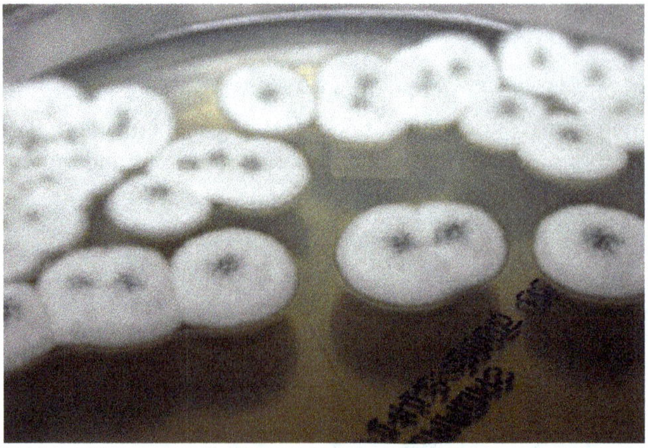

Fig. 8.11.3 *Scopulariopsis brevicaulis* (side).

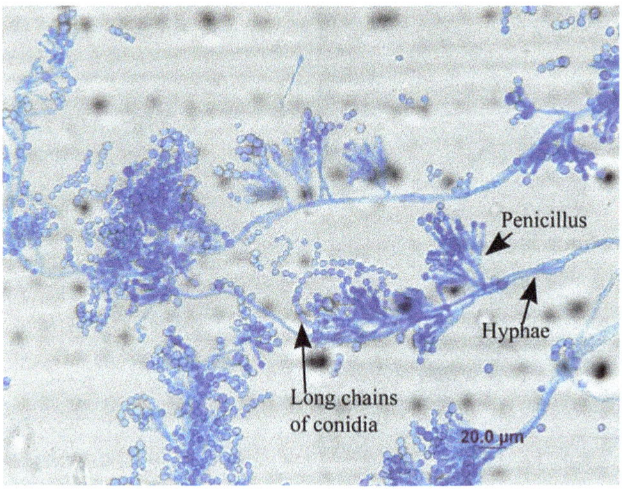

Fig. 8.11.4 *Scopulariopsis*, basic morphology.

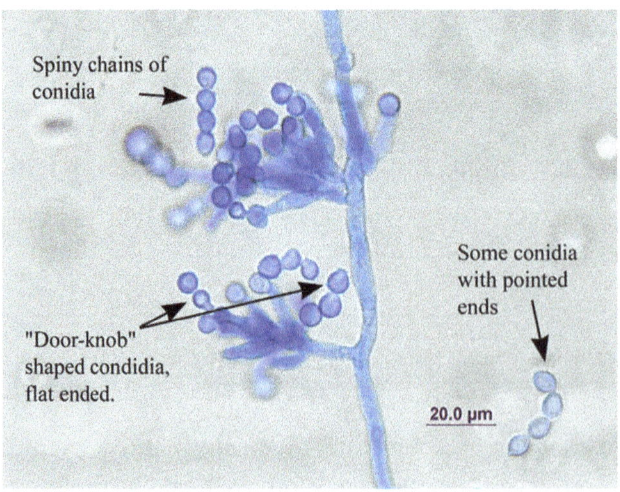

Fig. 8.11.5 *Scopulariopsis*, basic morphology.

Fig. 8.11.6 *S. brevicaulis*, 400×.

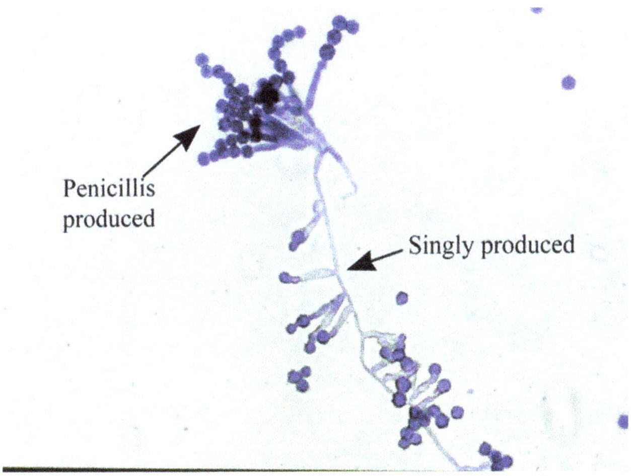

Fig. 8.11.7 Annellide arrangements.

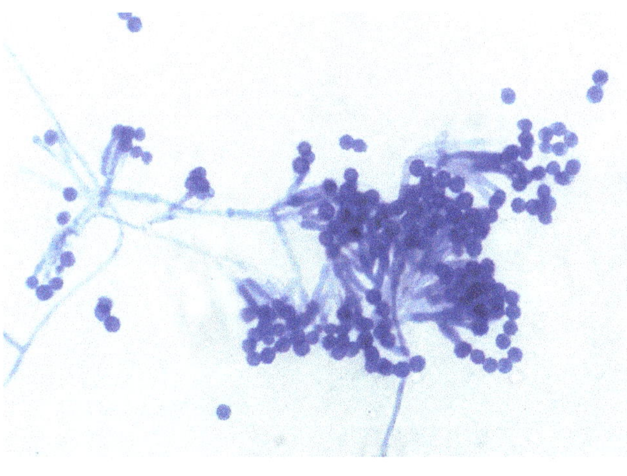

Fig. 8.11.8 *S. brevicaulis*, 400×.

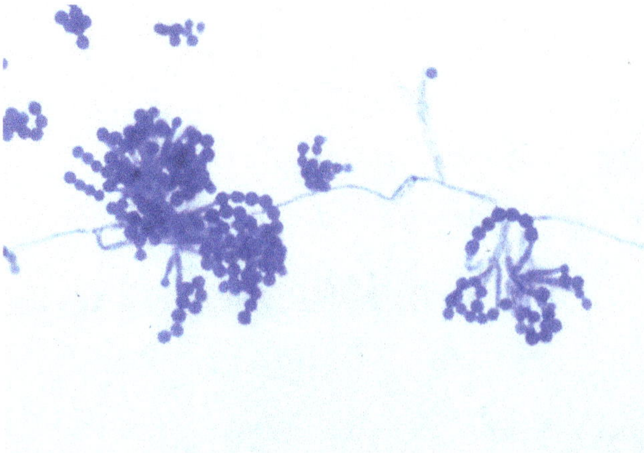

Fig. 8.11.9 *S. brevicaulis*, 400×.

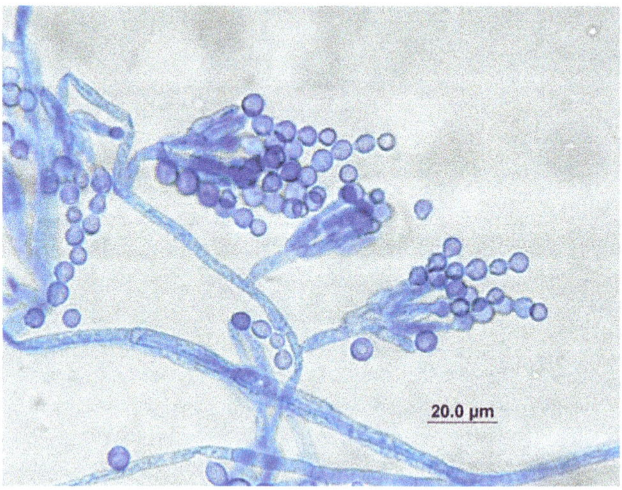

Fig. 8.11.10 *S. brevicaulis*, 1000×.

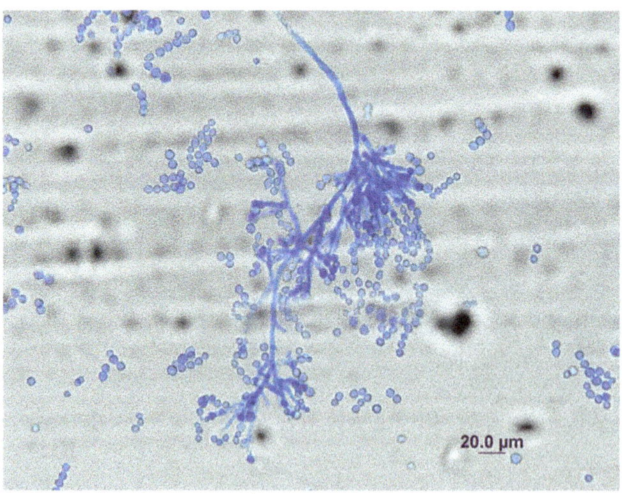

Fig. 8.11.11 *S. brevicaulis*, 400×.

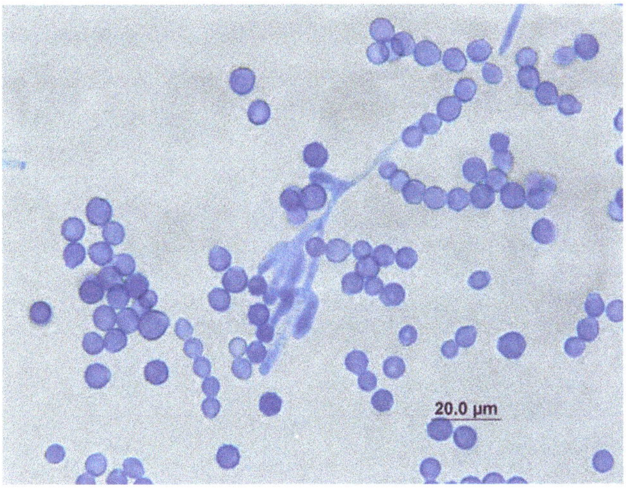

Fig. 8.11.12 Conidia morphology.

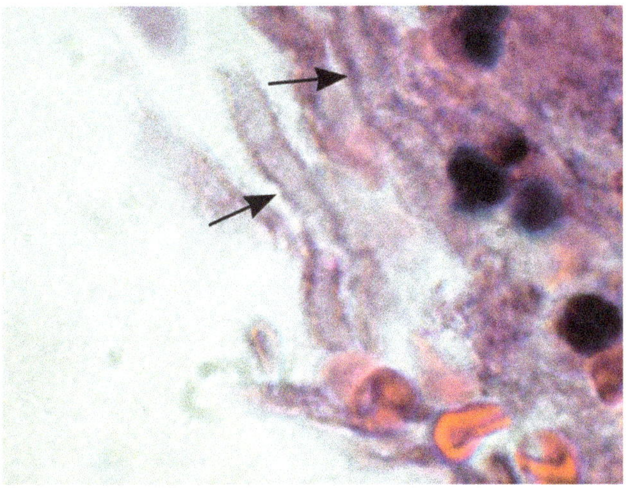

Fig. 8.11.13 Scopularia hyphae in sinus by H&E stain.

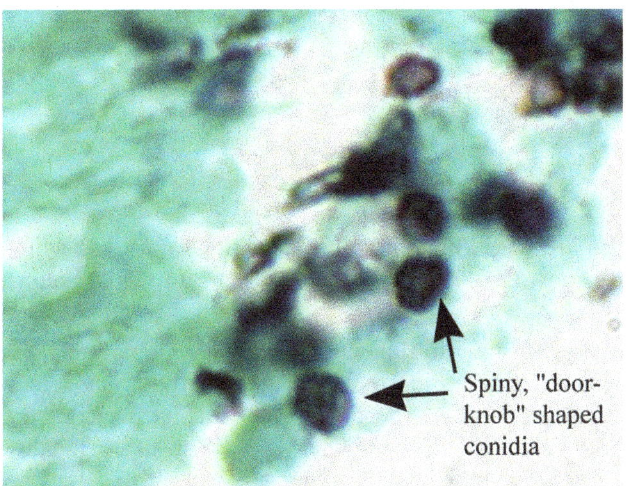

Fig. 8.11.14 Scopularia in sinus tissue biopsy, GMS stain.

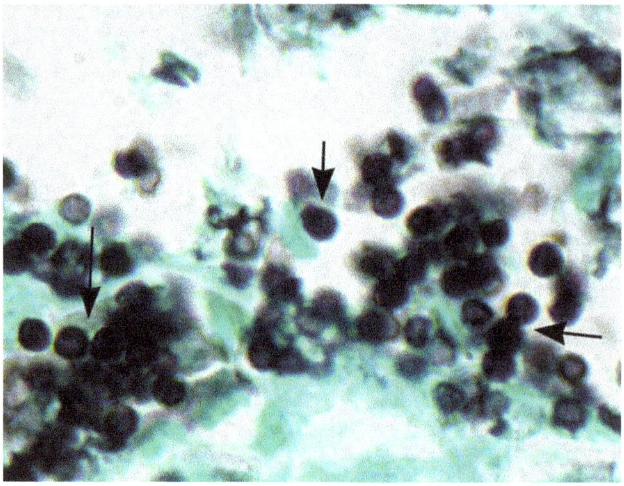

Fig. 8.11.15 Truncated spores and chains of spores in sinus biopsy.

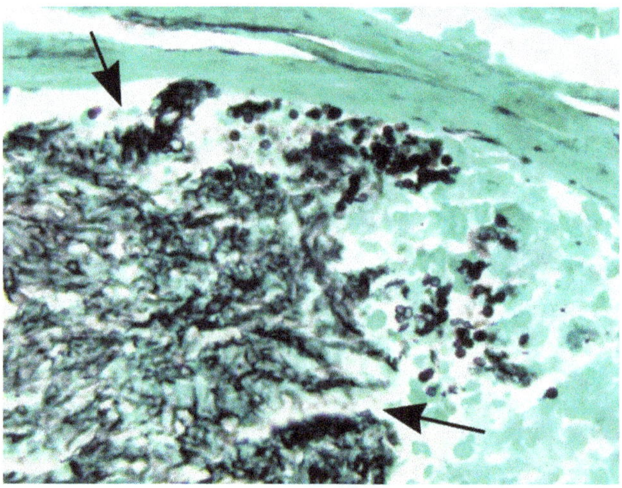

Fig. 8.11.16 Scopularia fungal ball in human sinus, GMS stain.

8.12 *Scopulariopsis brumptii*

Colonial morphology: colonies are first white, and later become gray with a white border. The surface is velvety to powdery. The reverse is white or tan with a yellow center.

Microscopic morphology: hyphae are fine, hyaline, and septate. Conidiophores are produced laterally along the hyphae. They are mostly single and tapered or short and swollen, but can be branched with medium to short (5–10 μm in length) secondary branches (annellides) that are flask shaped. The annellides form brush-like penicillus like those seen with *Penicillium* but fewer annellides per group are seen with *S. brumptii* than *S. brevicaulis* or not at all. Conidia (single-celled ameroconidia) are produced in chains in basipetal progression from the annellides. The youngest conidium is at the basal end of the chain. Chains of conidia are 1–8 in length. Conidia (5–7 μm in diameter) are round. Conidia are spiny and rough, hyaline, and sometimes flat at the base. The microscopic features of *S. brevicaulis* are distinct whereas those of *S. brumptii* are subtle. Sporodochia (mats of twisted hyphae) are present.

On certain media, where secondary annellides are rare, *S. brumptii* can be misidentified as *Acremonium* or *Sporothrix*.

Scopulariopsis brumptii

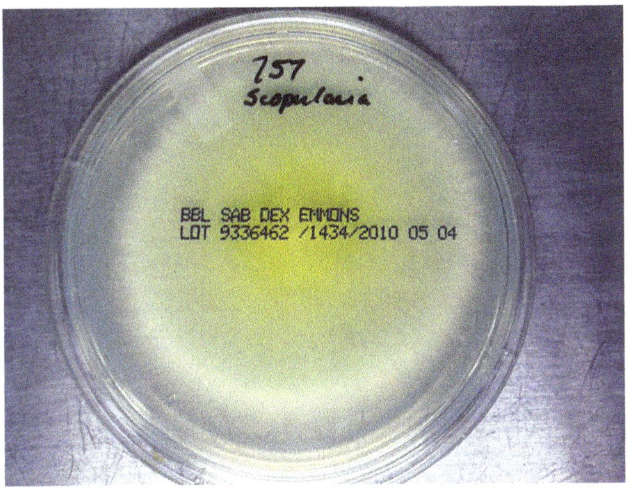

Fig. 8.12.1 *Scopulariopsis brumptii* (bottom).

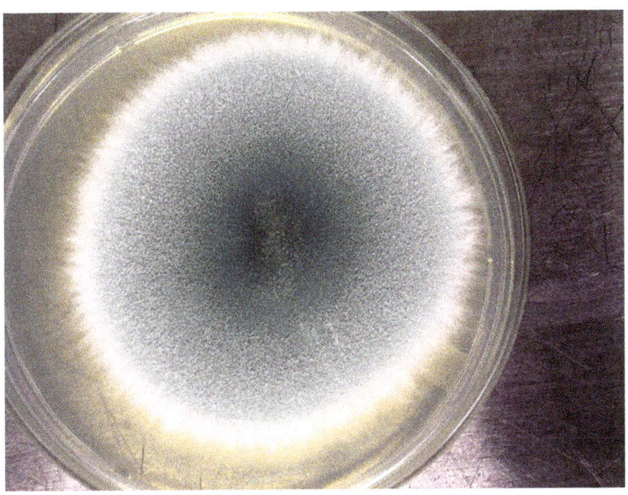

Fig. 8.12.2 *Scopulariopsis brumptii* (top).

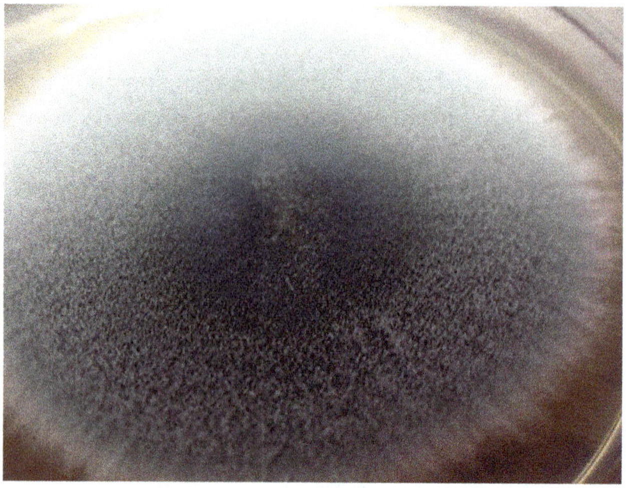

Fig. 8.12.3 *Scopulariopsis brumptii* (side).

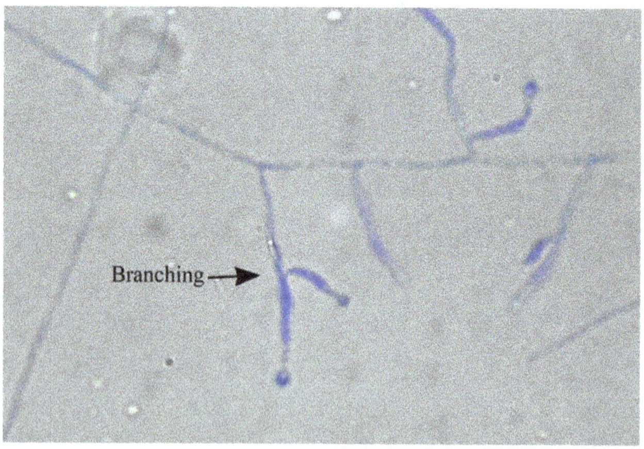

Fig. 8.12.4 *S. brumptii*, rare branching annellides.

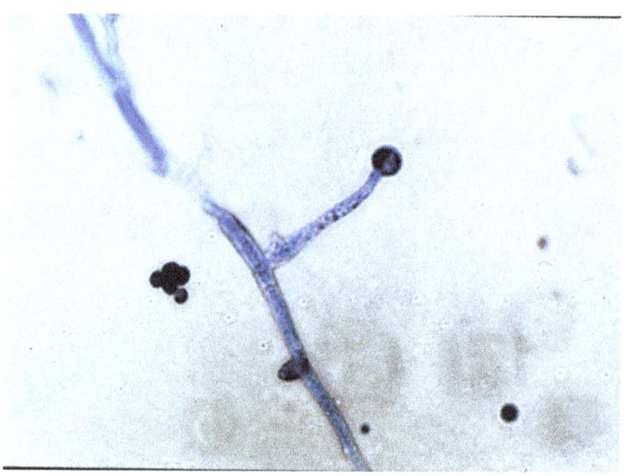

Fig. 8.12.5 Single annellides.

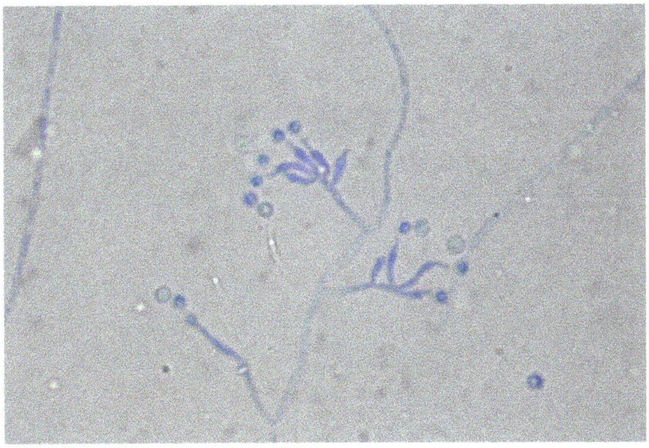

Fig. 8.12.6 Multiple annellides.

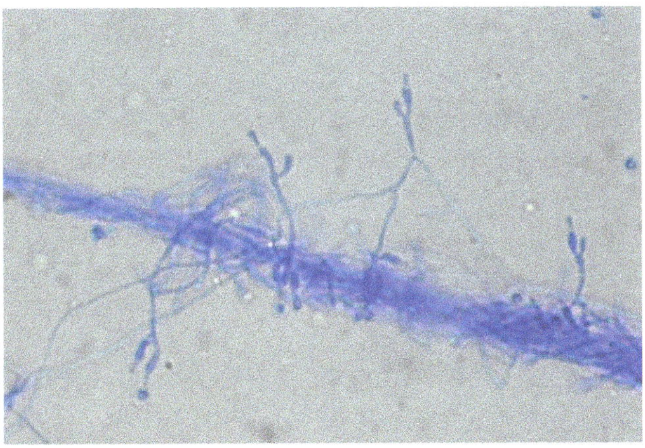

Fig. 8.12.7 Sporodochia.

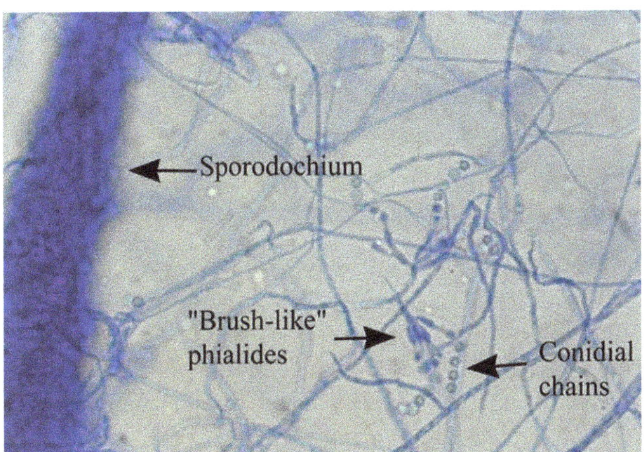

Fig. 8.12.8 Brush-like annellides with sporodochium present (left).

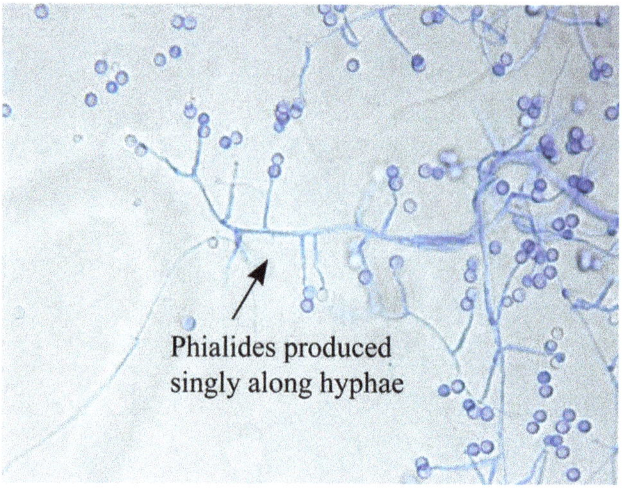

Fig. 8.12.9 *S. brumptii.* Annellides (phialides) long and thin; conidia in chains.

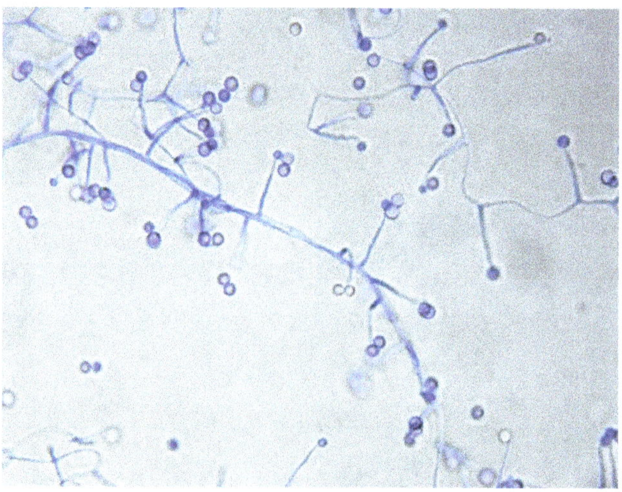

Fig. 8.12.10 *S. brumptii.* Round conidia are flat on one side.

8.13 *Sepedonium* spp

Colonial morphology: colonies are first white and smooth, then become wooly. Colonies tend to wrinkle in a radiating pattern from the center that becomes heaped. With age, colonies may turn cream to golden yellow with a pale reverse.

Microscopic morphology: hyphae are hyaline and septate. Conidiophores are non-specialized and resemble vegetative hyphae. Tuberculate macroconidia are produced that resemble macroconidia of *Histoplasma capsulatum.* Unlike *H. capsulatum,* no microconidia are produced. Large egg-shaped conidia with smooth walls may be produced in some cultures (Fig. 8.13.8). These should not be confused with microconidia formation. Often, tuberculate macroconidia develop a golden yellow color (if colonies are yellow), but in some cultures they remain hyaline. *Sepedonium* will not convert to a yeast phase at 37 °C whereas *H. capsulatum* will. It is recommended that cultures yielding *Sepedonium* be tested by gene-probe technology to rule out *H. capsulatum*.

Sepedonium spp

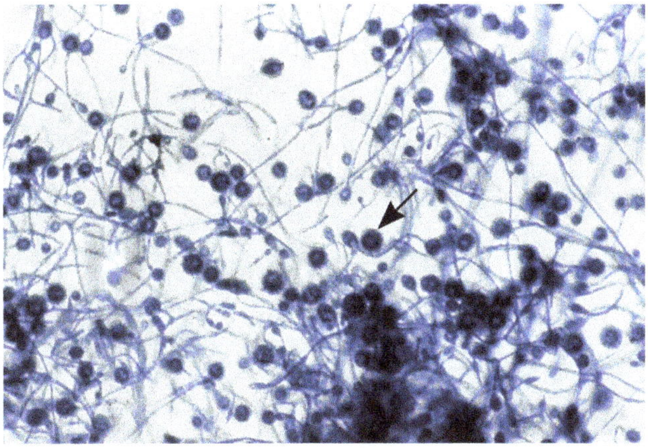

Fig. 8.13.1 *Sepedonium*, echinulate conidia.

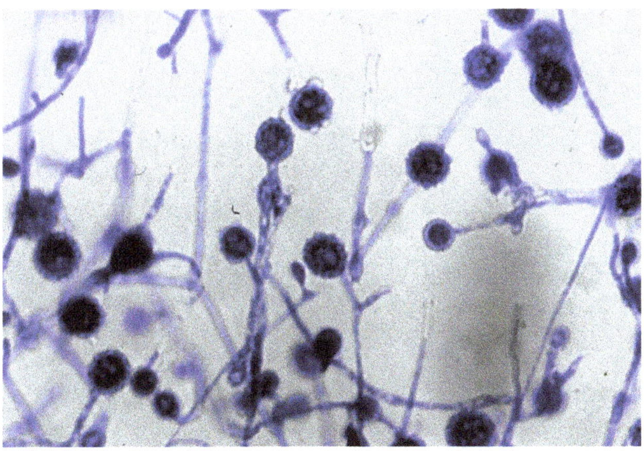

Fig. 8.13.2 *Sepedonium*.

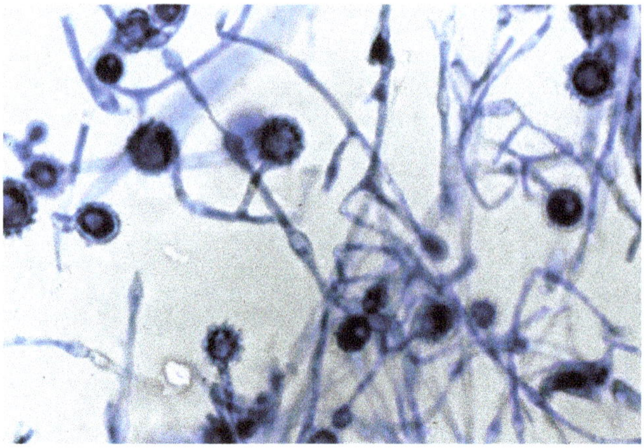

Fig. 8.13.3 *Sepedonium.*

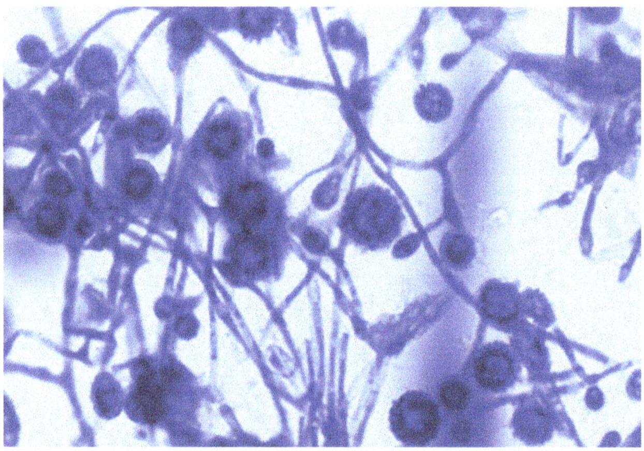

Fig. 8.13.4 *Sepedonium.*

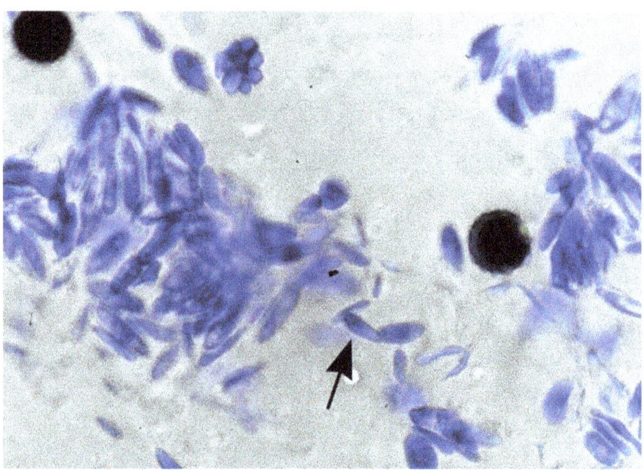

Fig. 8.13.5 Ellipsoidal macroconidia.

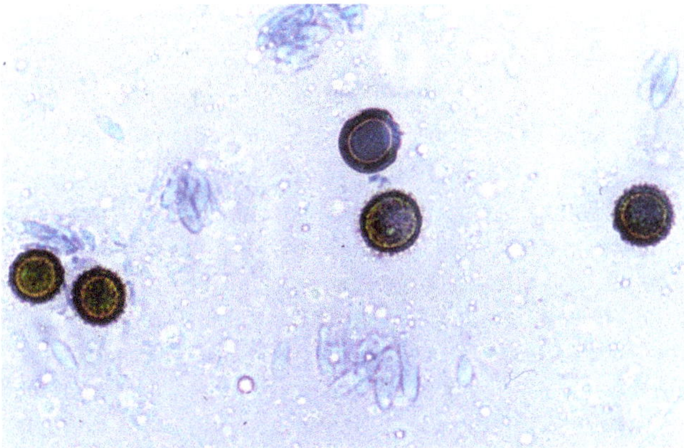

Fig. 8.13.6 Yellow pigmented echinulate conidia.

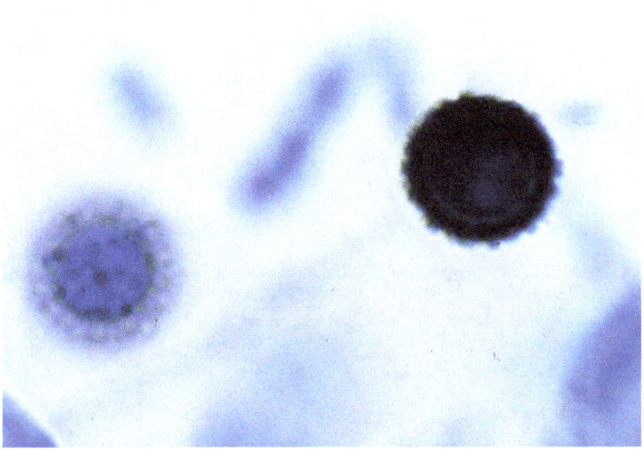

Fig. 8.13.7 Tuberculate macroconidia.

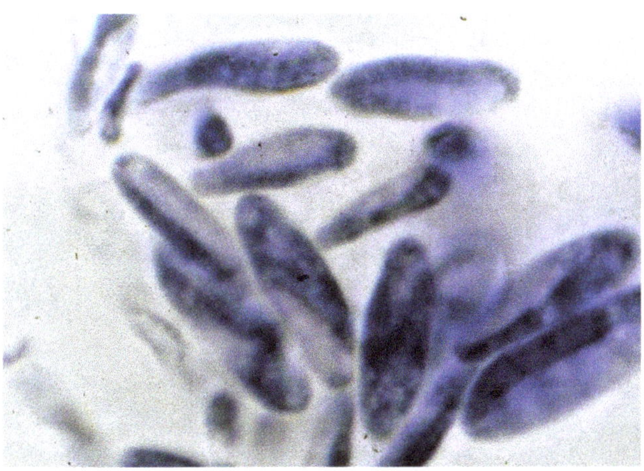

Fig. 8.13.8 Smooth-walled macroconidia.

8.14 *Sporotrichum pruinosum*

Colonial morphology: colonies are fluffy, initially white with a black, gray, yellow, rose, orange or tan center developing with age. Reverse color is tan. Cultures are inhibited by cycloheximide but grow at 25–45 °C.

Microscopic morphology: hyphae are hyaline to dark, and septate. They are of broad and narrow widths. Clamp connections are rare. Conidiophores branch at 90° angles from hyphae but bend to 45° angles. Both opposite and sympodial branching occur. Single conidia are produced at the ends of conidiophores, with conidiophores bending toward the central one, giving the appearance of a candelabrum. Conidia are one-celled, oval or avocado shaped with a flat base. A thick ring-like septum or scar forms at the point where the conidia attach to the conidiophores. The scar sometimes remains attached to the conidia as they are released. The conidia are thick walled, and sometimes yellow-brown or dark. Alternating arthroconidia with spaces between are abundant at 25 °C and they break up readily. Chlamydospores (65 µm) are always formed at 25 °C and sometimes at 37 °C. This helps differentiate *Sporotrichum* from *Chrysosporium*. *Chrysosporium* produces alternating arthroconidia that remain chained together rather than breaking up. *Chrysosporium* does not produce chlamydospores at 37 °C, does not grow well above 37 °C, and is cycloheximide resistant. Temperature growth studies (at 25 °C, 37 °C and 40–45 °C) are beneficial in the identification of this fungus.

Sporotrichum pruinosum

Fig. 8.14.1 *Sporotrichum pruinosum*, young SABDX (bottom).

Fig. 8.14.2 *S. pruinosum*, young, SABDX (top).

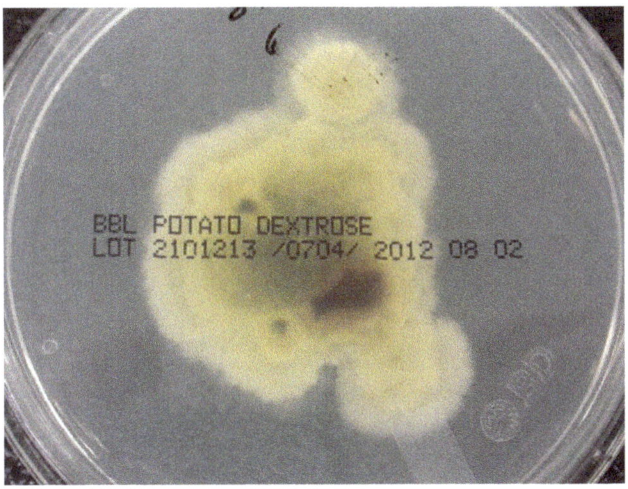

Fig. 8.14.3 *S. pruinosum*, older, PDA (bottom).

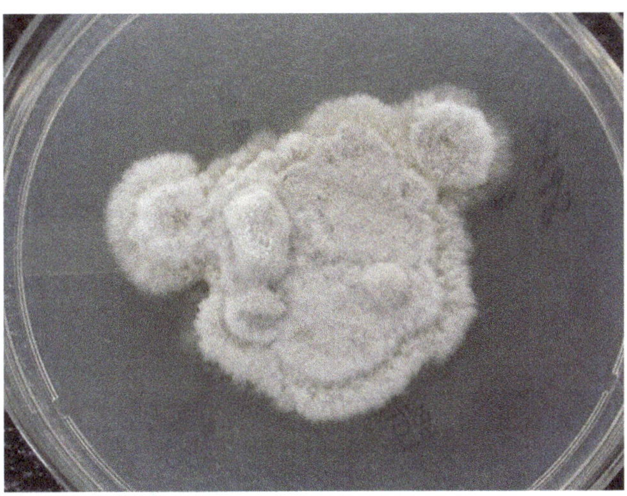

Fig. 8.14.4 *S. pruinosum*, older, PDA (top).

Fig. 8.14.5 *S. pruinosum* (side), older.

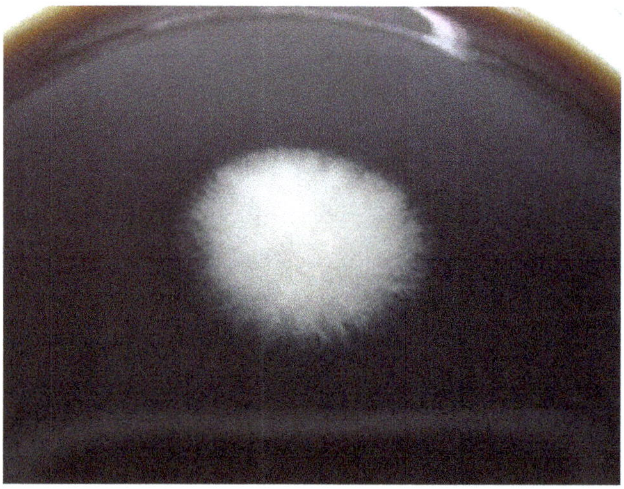
Fig. 8.14.6 *S. pruinosum* on blood agar.

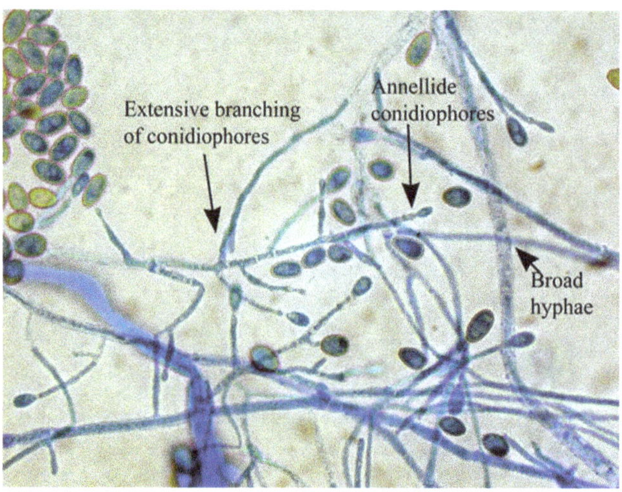

Fig. 8.14.7 Broad hyphae (*right*) may form arthrospores.

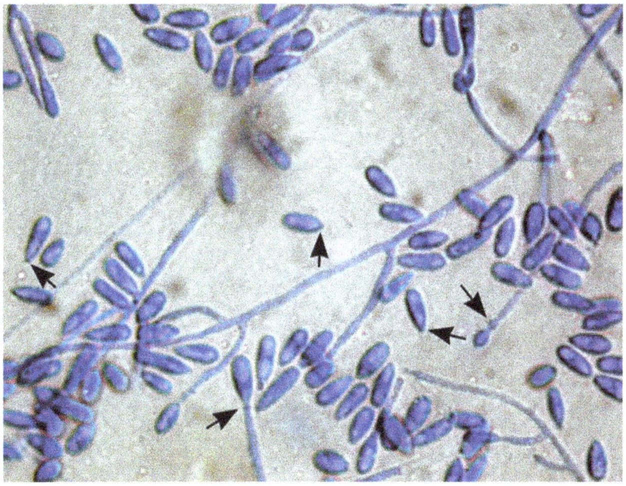

Fig. 8.14.8 Conidia showing septal scars at attachment.

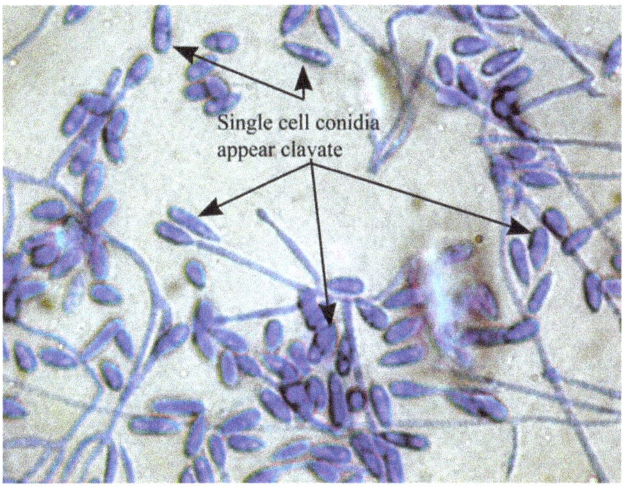

Fig. 8.14.9 Single cell conidia.

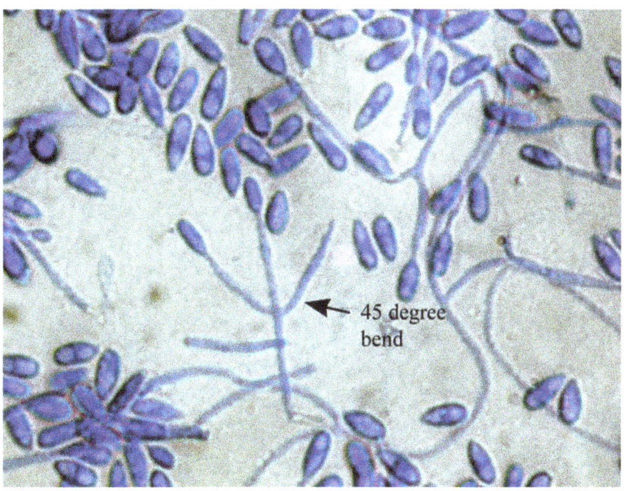

Fig. 8.14.10 Conidiophores branch at 45 degree angles.

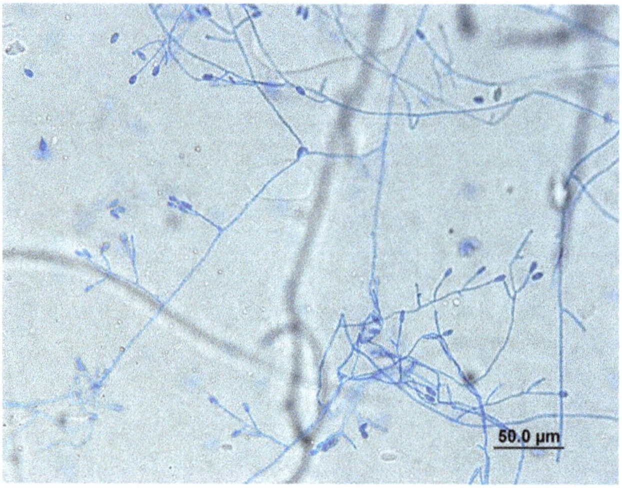

Fig. 8.14.11 Typical conidiophore morphology.

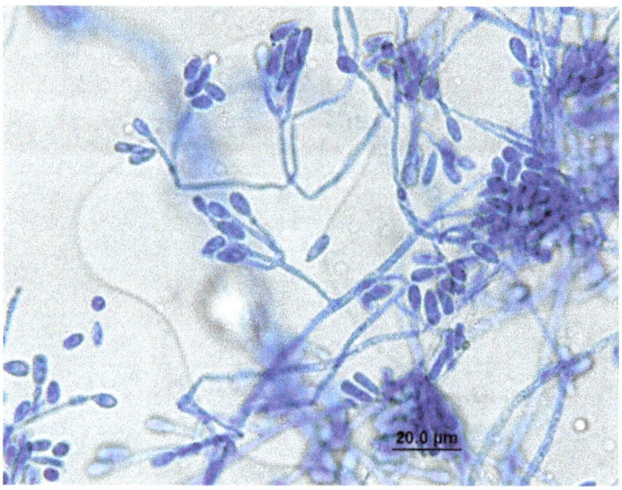

Fig. 8.14.12 Branching conidiophores.

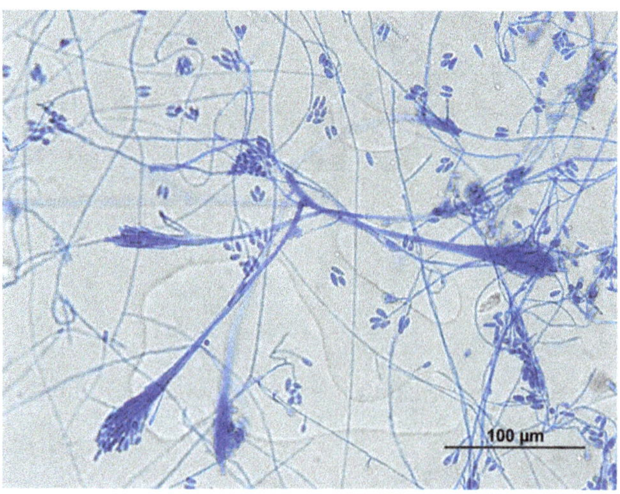

Fig. 8.14.13 *Graphium* type hyphae.

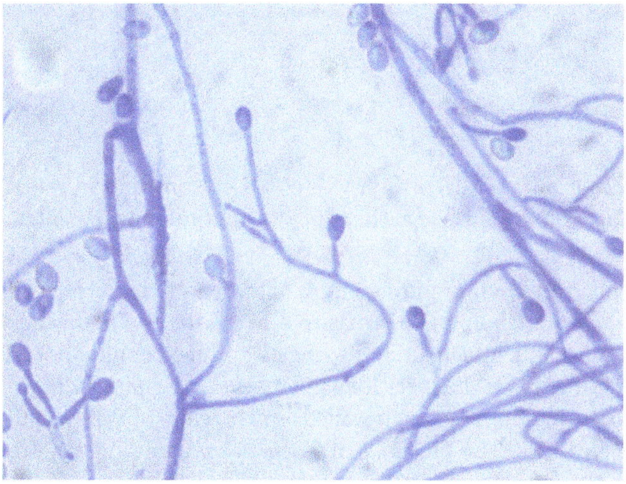

Fig. 8.14.14 Single, non-branching conidiophores.

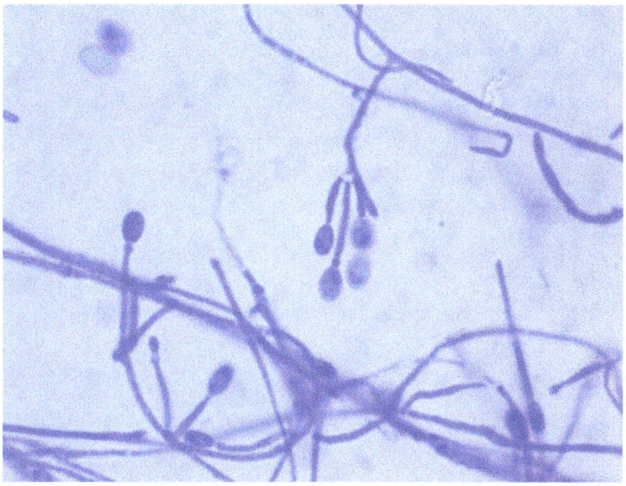

Fig. 8.14.15 Candelabrum arrangement of conidiophores.

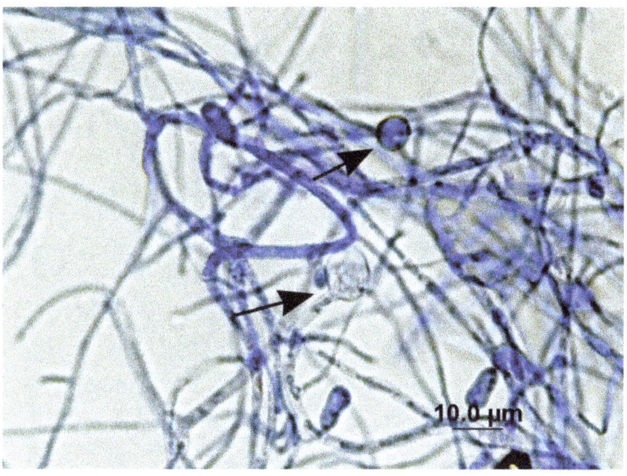

Fig. 8.14.16 Chlamydospores present.

8.15 *Trichoderma* spp

Colonial morphology: colonies grow rapidly, at first white and cottony and in a few days, yellow to green patches form on the surface at the center, sometimes in concentric rings. The reverse is white to yellow.

Microscopic morphology: hyphae are hyaline and septate. Conidiophores are produced perpendicular to the hyphae, then branch in a Christmas tree-like fashion. The hyaline phialides are swollen at the base and attach at right angles to the conidiophores. Conidiophores may be short in some species but very long in others (*T. longibrachiatum*). Conidia (2–5 µm) are oval or round, smooth or rough walled, green, white or yellow, and form clusters at the tips of the phialides. The phialides of *Trichoderma* are singular, not clustered like in *Gliocladium*. Chlamydospores (5–10 µm) may be produced.

Trichoderma spp

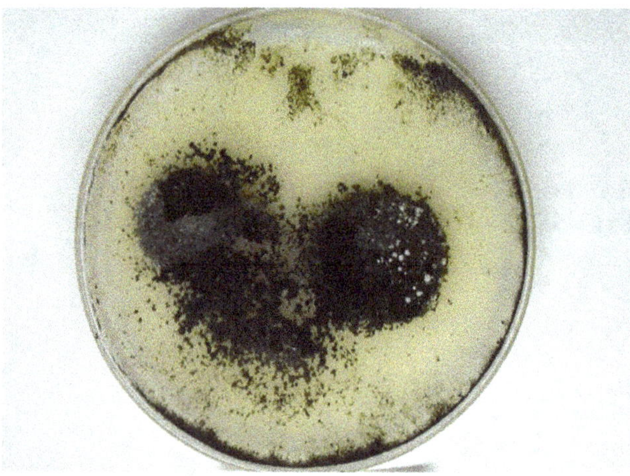

Fig. 8.15.1 *Trichoderma* (top).

Fig. 8.15.2 *Trichoderma* (top).

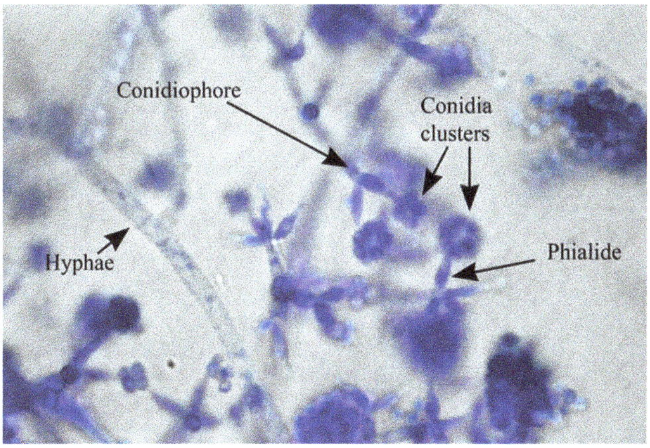

Fig. 8.15.3 Basic morphology.

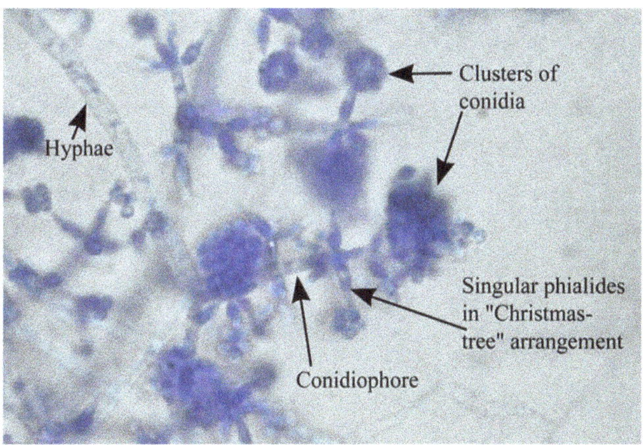

Fig. 8.15.4 Basic morphology.

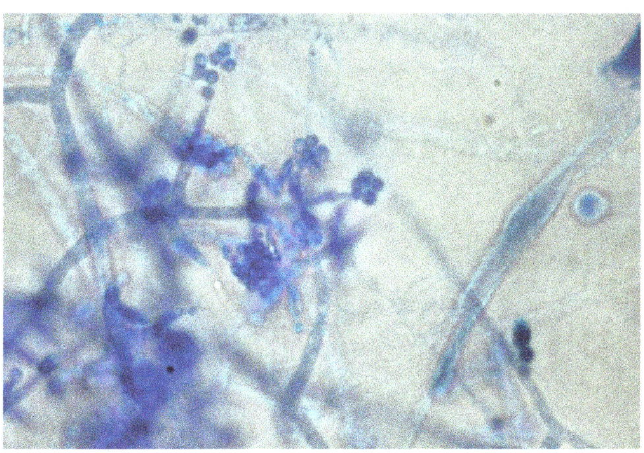

Fig. 8.15.5 Typical *Trichoderma* branching.

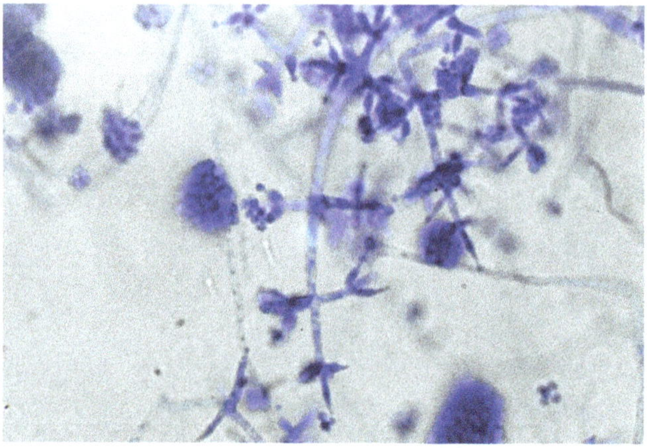

Fig. 8.15.6 "Christmas tree-like" branching.

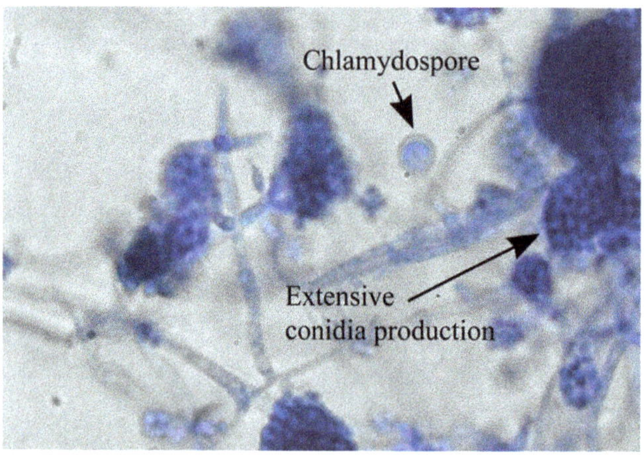

Fig. 8.15.7 Many conidia and rare chlamydospores.

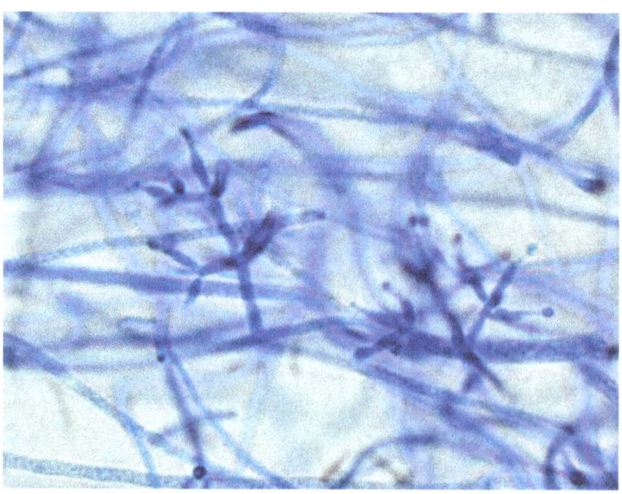

Fig. 8.15.8 Phialides are swollen at base and constricted at the tip.

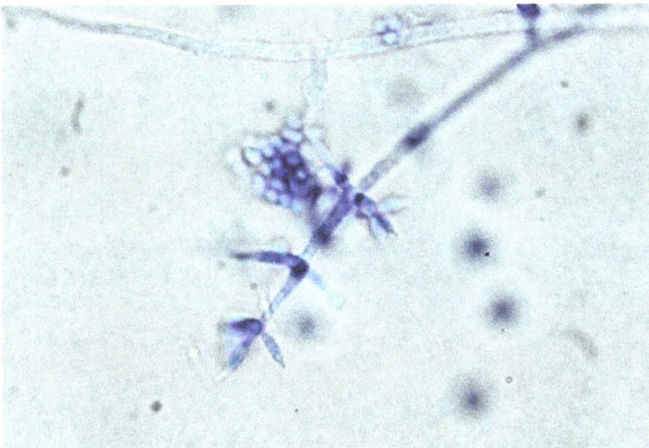

Fig. 8.15.9 *Trichoderma*.

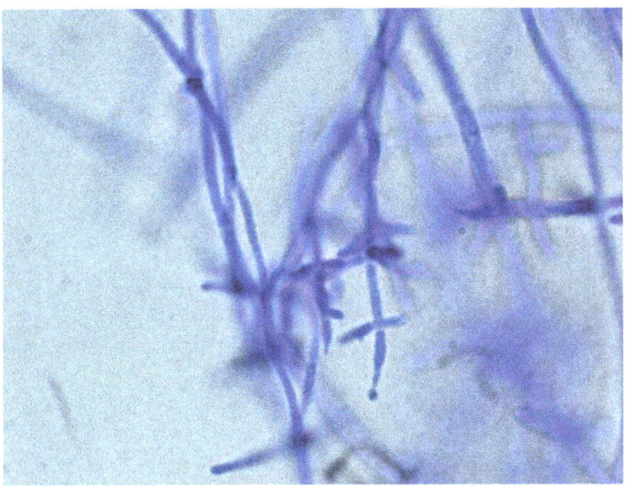

Fig. 8.15.10 Ninety degree branching of phialides.

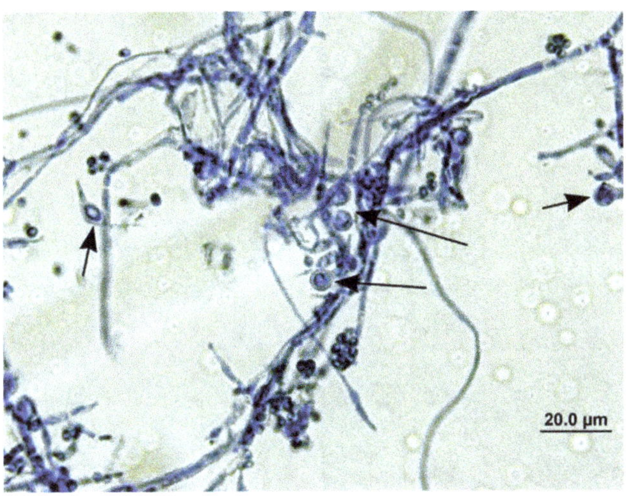

Fig. 8.15.11 Chlamydospores.

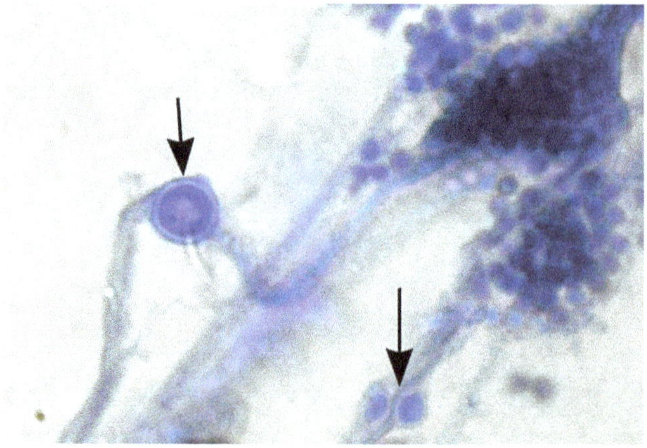

Fig. 8.15.12 Chlamydospore.

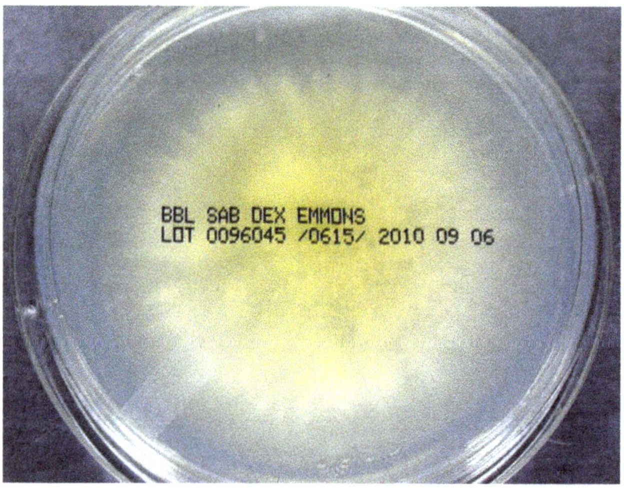

Fig. 8.15.13 *Trichoderma longibrachiatum* (bottom).

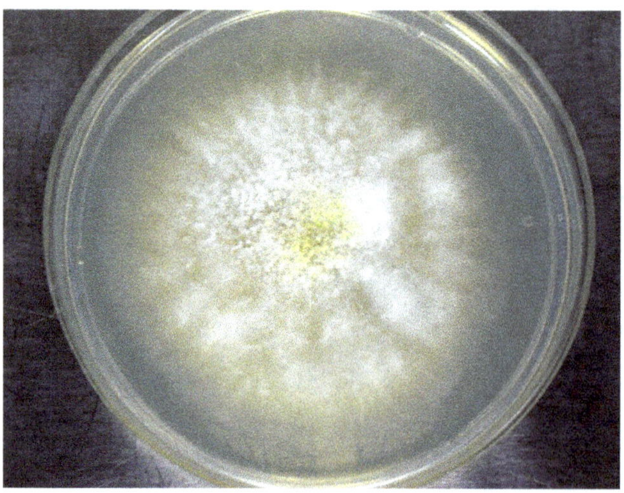

Fig. 8.15.14 *T. longibrachiatum* (top).

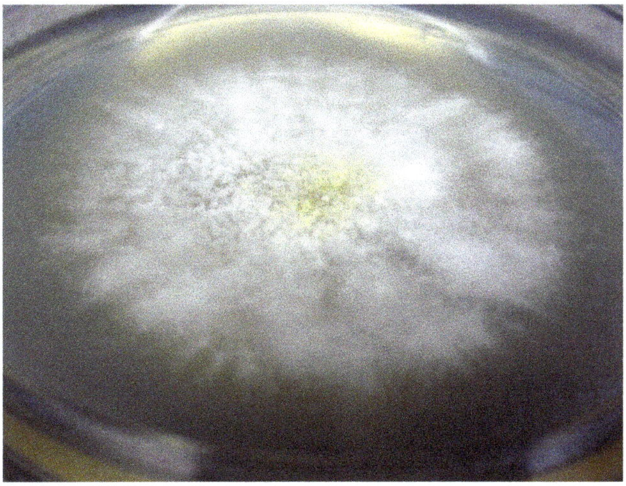

Fig. 8.15.15 *T. longibrachiatum* (side).

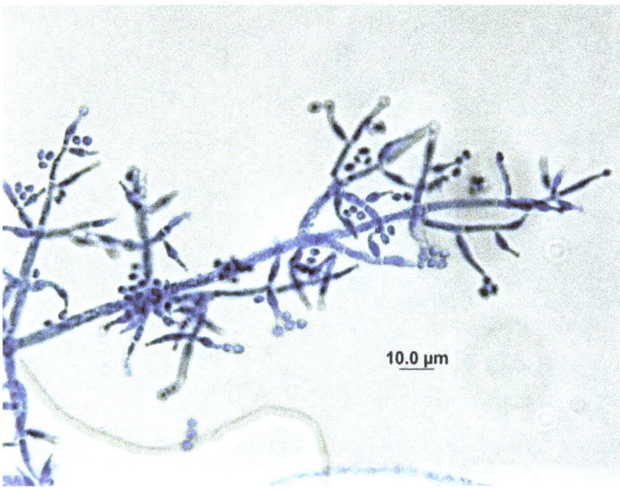

Fig. 8.15.16 *T. longibrachiatum*.

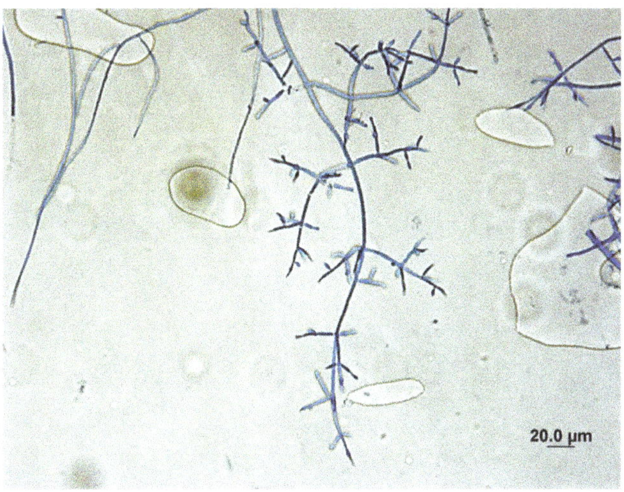

Fig. 8.15.17 *T. longibrachiatum*.

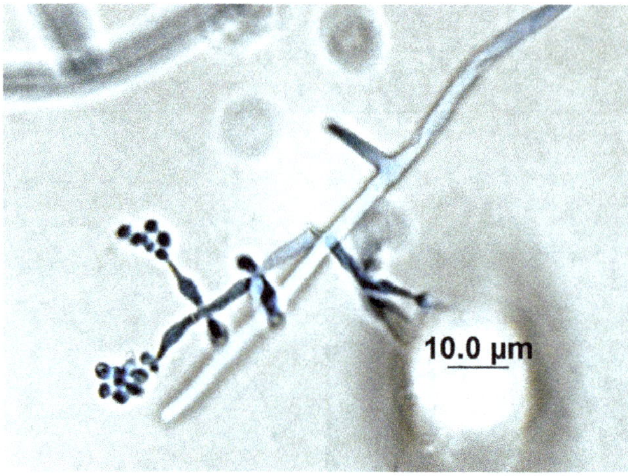

Fig. 8.15.18 *T. longibrachiatum*.

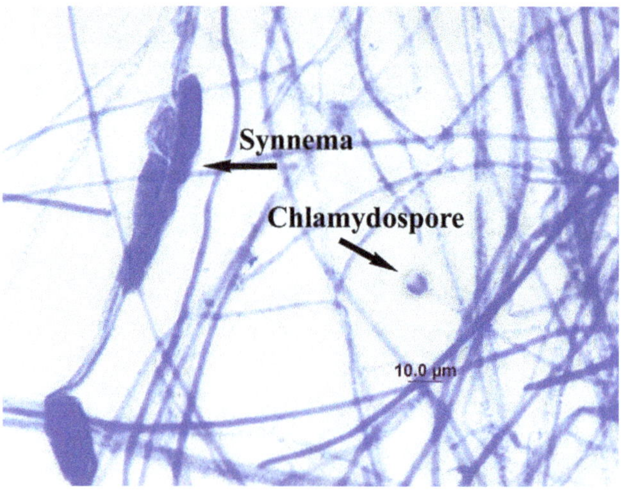

Fig. 8.15.19 *T. longibrachiatum*.

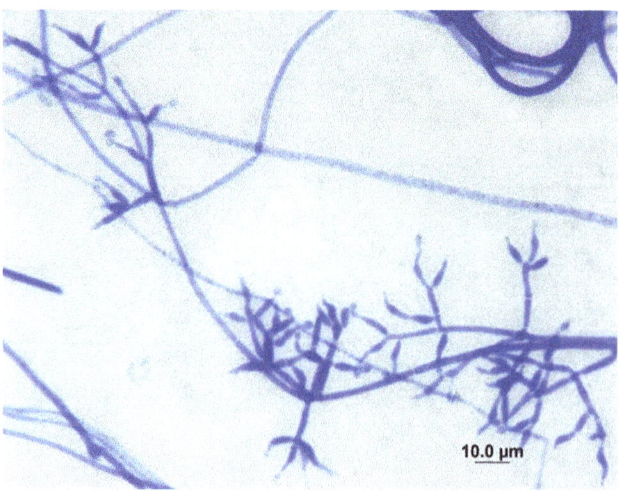

Fig. 8.15.20 T. longibrachiatum

8.16 *Trichothecium roseum*

Colonial morphology: colonies on Sabouraud dextrose agar are at first white and powdery, then become peach-orange and rough, like pig skin. The reverse is white. Colonies on potato dextrose agar have white fluffy concentric rings, followed by the development of a rosy, peach-orange overlay color.

Microscopic morphology: hyphae are hyaline and septate. Conidiophores appear hyphae like – unbranched. As conidia are formed, the conidiophores differentiate, becoming rough walled, septate at the base, and bearing club-shaped conidia along the conidiophores in a zig-zag arrangement. The conidiophore shortens as conidia develop, undergoing retrogressive conidial development. Conidia are two-celled (biclavate, didymoconidia), with a curved end bearing a flattened attachment point, often called a "foot." The conidia (8–10 × 12–18 µm) are oval, hyaline, smooth, and thick walled. Conidia seem to crowd each other, forming a side-to-side arrangement that sticks together even after detachment from the conidiophore. Care must be taken not to confuse *Trichothecium* with *Microsporum nanum*. Colonies are similar, and conidia share the same appearance, with the exception that *Microsporum* conidia walls are thin and verrucose, and formed singly on conidiophores, lacking the coherence of *Trichothecium*.

Trichothecium roseum

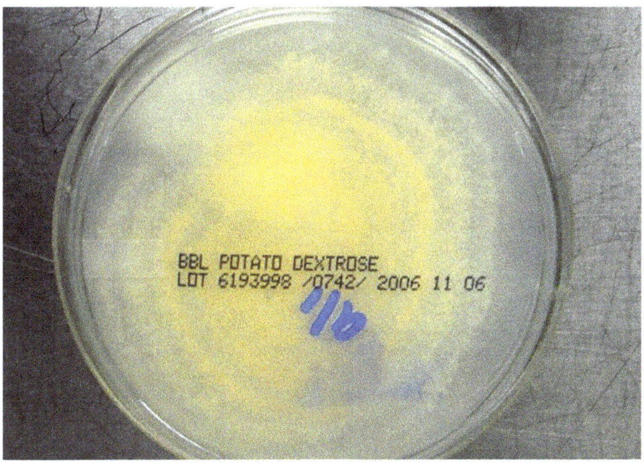

Fig. 8.16.1 *Trichothecium roseum*, PDA (bottom).

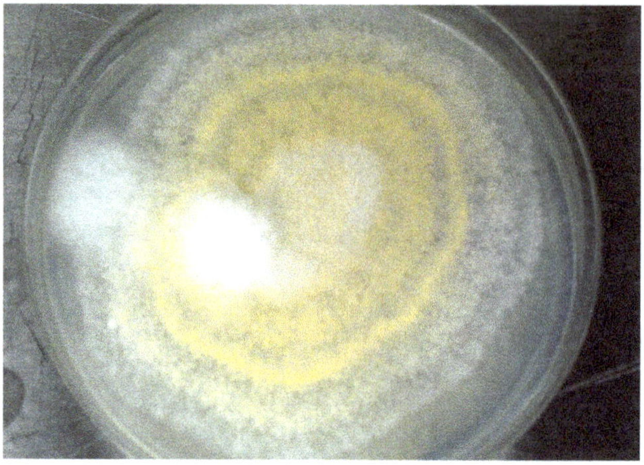

Fig. 8.16.2 *Trichothecium roseum* (top).

Fig. 8.16.3 *Trichothecium roseum* (side).

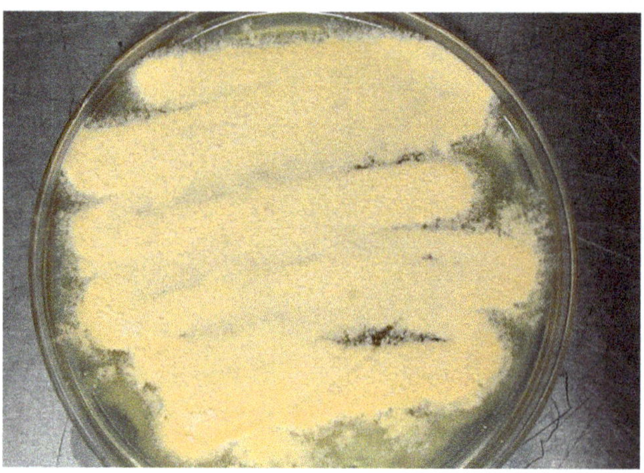

Fig. 8.16.4 *Trichothecium roseum*, SABDX (top).

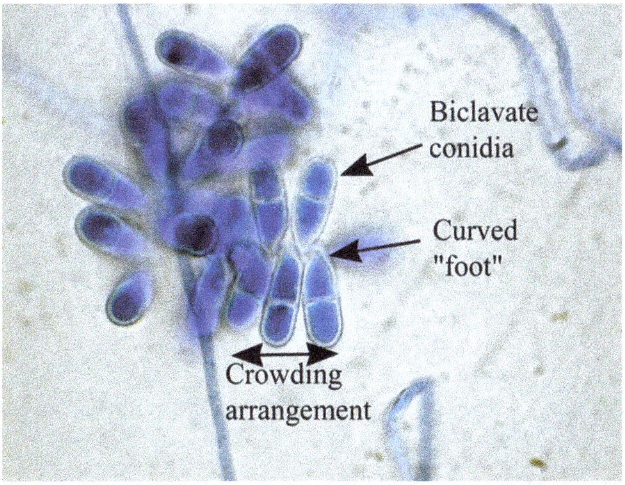

Fig. 8.16.5 Basic morphology.

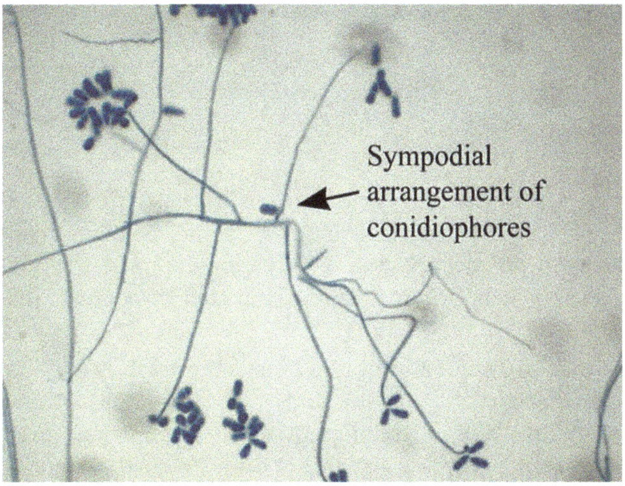

Fig. 8.16.6 Conidiophore arrangement.

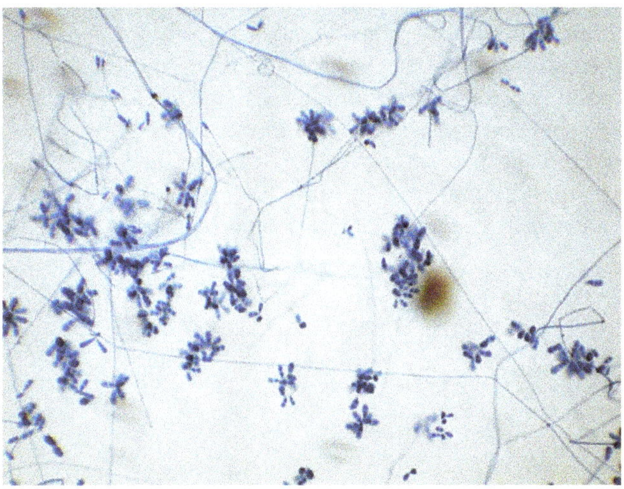

Fig. 8.16.7 Daisy-like appearance at 100×.

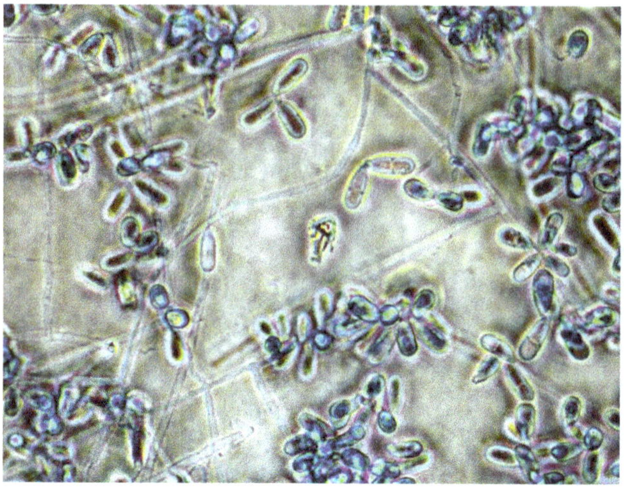

Fig. 8.16.8 Phase contrast.

8.17 *Chrysosporium tropicum*

Colonial morphology: colonies are white, fluffy and cottony (especially on blood agar), and mature within about 6–7 days. The surface of older colonies on Sabouraud dextrose agar may become grainy or powdery, and develop pigmentation ranging in color from yellow, tan or orange to peach. The reverse is pigmented throughout the entire colony.

Microscopic morphology: hyphae are generally hyaline, but large hyphae may be present with the appearance of arthroconidia-like septations. The alternating arthroconidia may bulge on one side, and give rise to conidiophores. The large hyphae may have unusual shapes with thick walls. The hyaline hyphae are branched with multiple, short or long conidiophores that are produced perpendicular to the hyphae. They look much like *Trichophyton* in appearance. At the point where conidiophores branch, septation is common. Often a terminal conidiopore is observed and double septation is seen at the base of the conidium. Conidia are one-celled, with a rounded apex and broad flattened base. Part of the conidiophore may be seen to adhere to the flattened base of the conidium as it separates from the conidiophore. Some conidia become thick walled with age. Terminal chlamydospores are seen, and when present, are almost round, with a slight bump or projection at the apex. *Chrysosporium* is often misidentified as *Trichophyton* (on *Trichophyton* media), *Emmonsia*, *Blastomyces dermatitidis*, *Scedosporium*, and *Sporotrichum*. *Chrysosporium* is not thermally dimorphic.

Chrysosporium tropicum

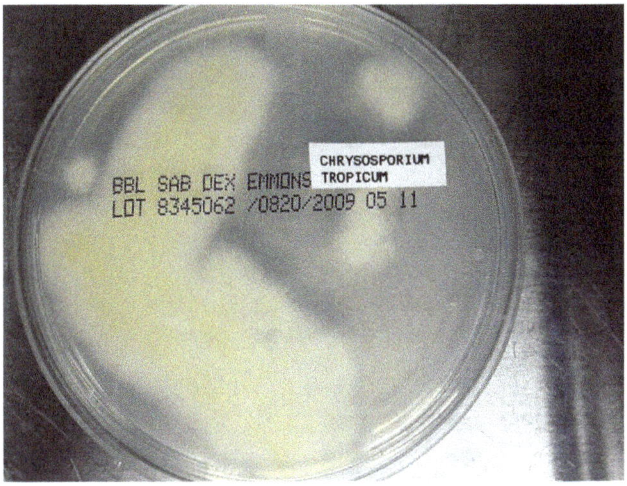

Fig. 8.17.1 *Chrysosporium*, Sab DX (bottom).

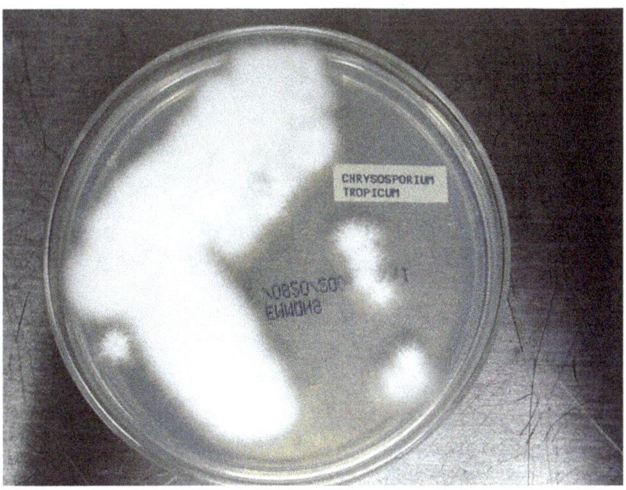

Fig. 8.17.2 *Chrysosporium*, Sab DX.

Fig. 8.17.3 *Chrysosporium*, Sab DX (side).

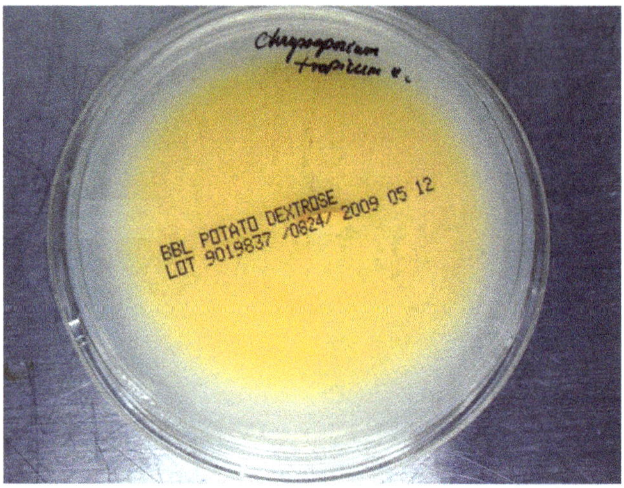

Fig. 8.17.4 *Chrysosporium*, PDA (bottom).

Fig. 8.17.5 *Chrysosporium*, PDA (side).

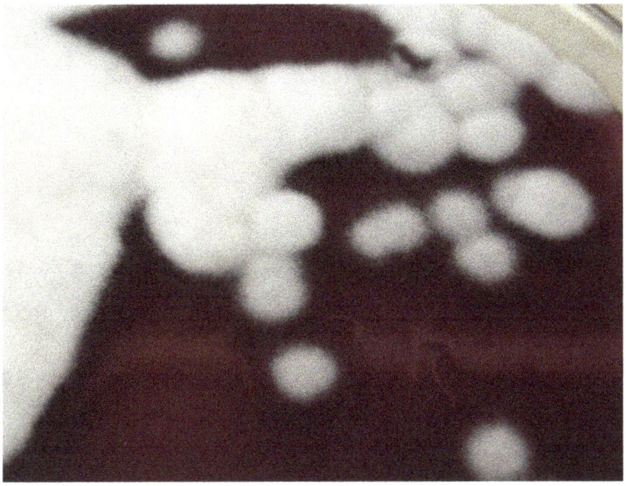

Fig. 8.17.6 *Chrysosporium*, blood agar (side).

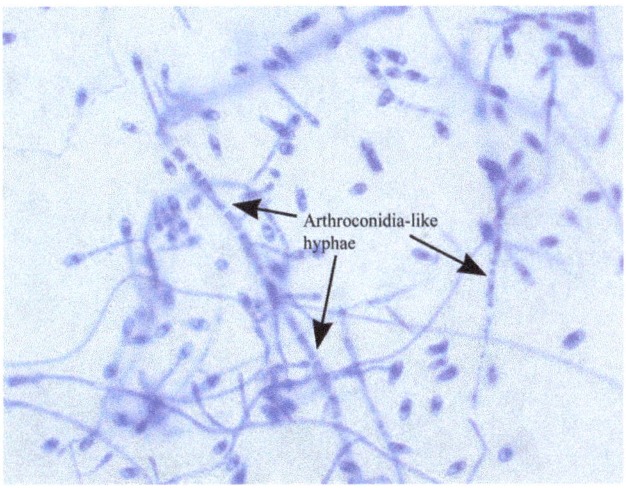

Fig. 8.17.7 Hyphae vary in morphology.

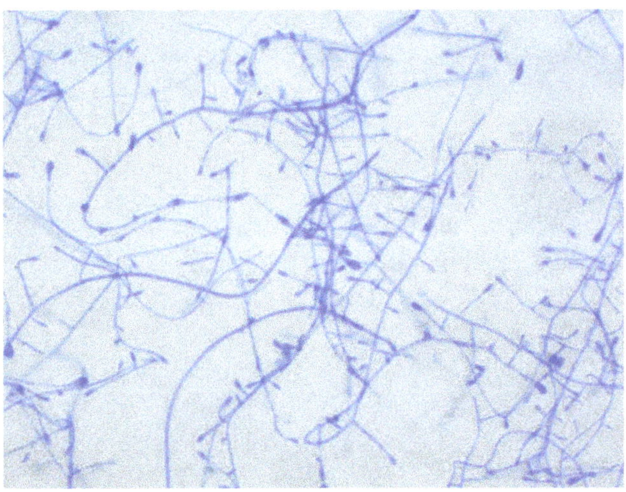

Fig. 8.17.8 *Chrysosporium*, 100×.

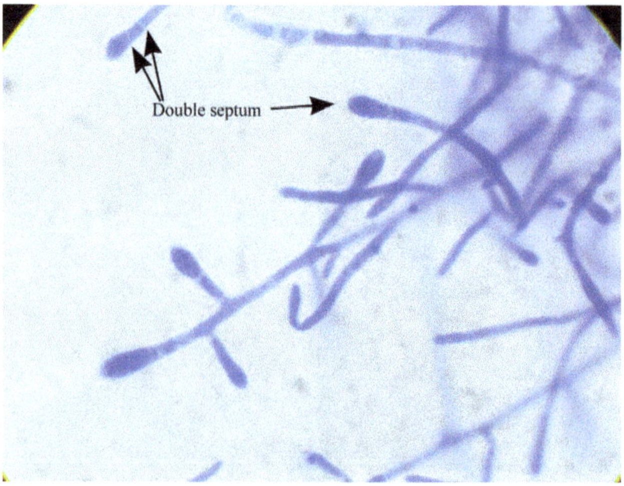

Fig. 8.17.9 *Chrysosporium* showing terminal conidia.

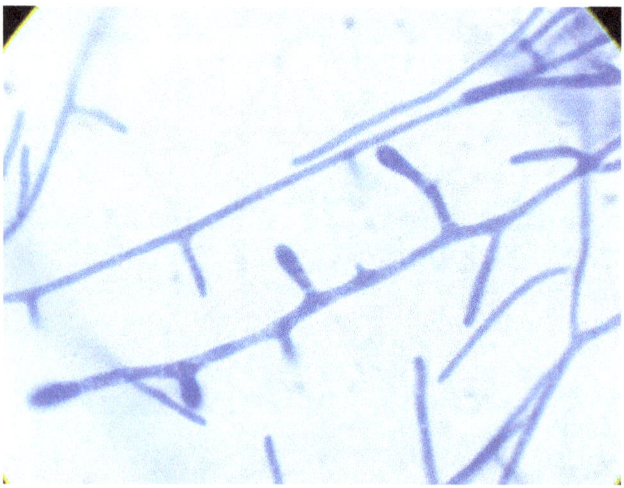

Fig. 8.17.10 *Chrysosporium* showing conidiophore development.

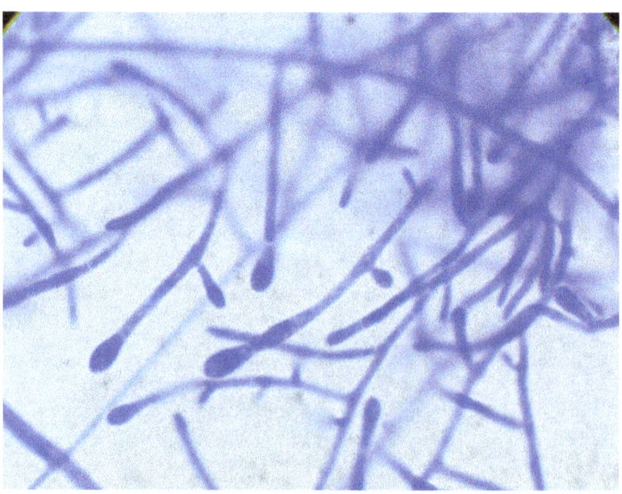

Fig. 8.17.11 *Chrysosporium* terminal conidia.

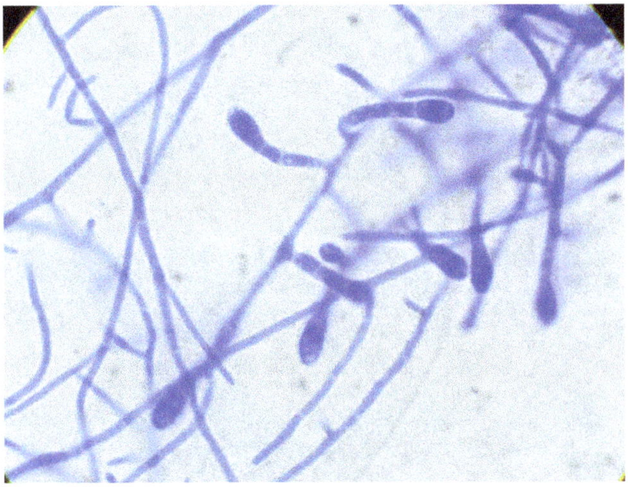

Fig. 8.17.12 *Chrysosporium* showing double septation along conidiophores.

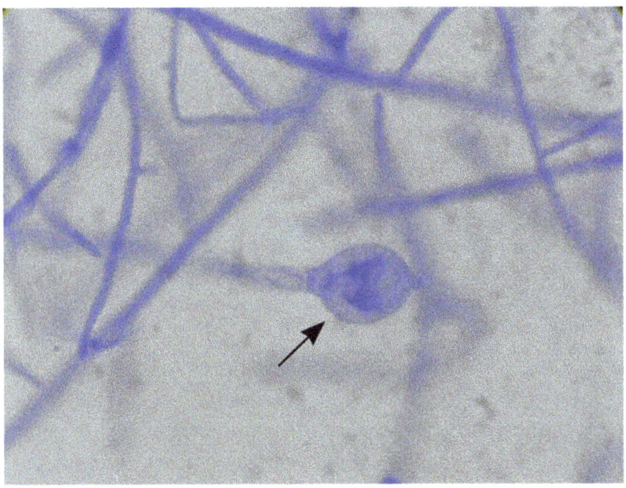

Fig. 8.17.13 With terminal chlamydospore.

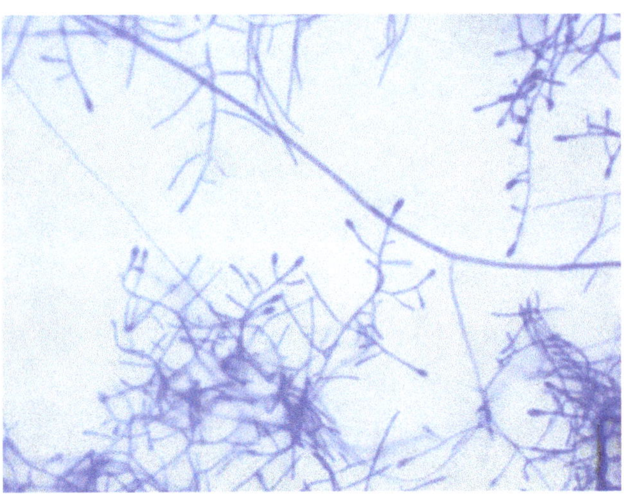

Fig. 8.17.14 *Chrysosporium* may resemble *Sporotrichum* or *Trichophyton* at 100×.

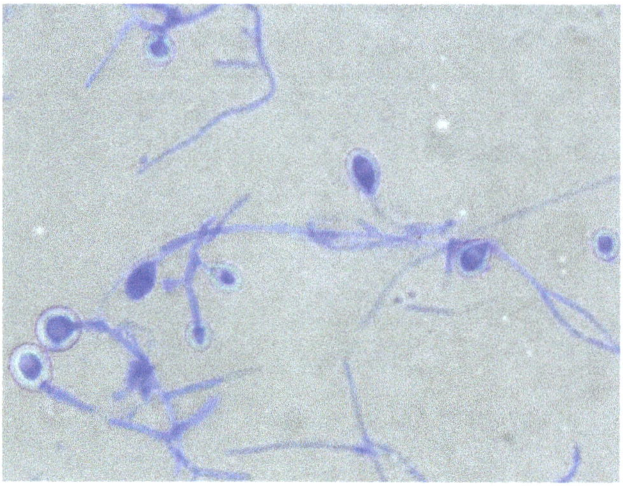

Fig. 8.17.15 Thick-walled conidia.

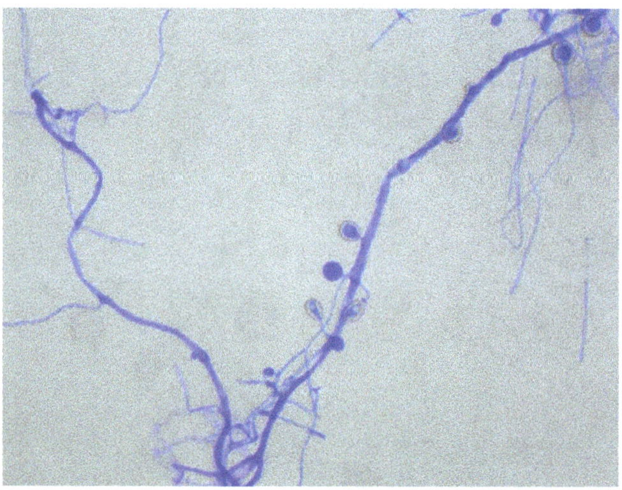

Fig. 8.17.16 Thick-walled conidia.

9

Dematiaceous Fungi (Septate)

9.1 *Acremonium* spp

Colonial morphology: there are approximately 95 species of *Acremonium*. Colonies first appear as small and moist and later become powdery and suede-like. After several days, colonies expand and become wrinkled and convoluted. On blood-containing media or chocolate agar, colonies may stay moist and wrinkled. Colonies first appear white and then may change color to grayish-brown to violet, orange or yellow and in some species, green or dark brown. The reverse color often intensifies in accordance with the top view, depending upon the species.

Microscopic morphology: hyphae are septate, delicate, and thin. Phialides (1.5–3.0 μm × 10–30 μm) arise directly from hyphae and are long and tapering with a pointed end. Phialides have a basal septum. Clusters or chains of conidia form at the apex of phialides. Some phialides may have only one conidium at the tip, especially young cultures. The conidia are small (1.5–2 μm × 3–8 μm), vary in shape and may be curved, columnar or oblong. They are usually one-celled, but can show two-cell septation. Chlamydoconidia are not produced.

Acremonium

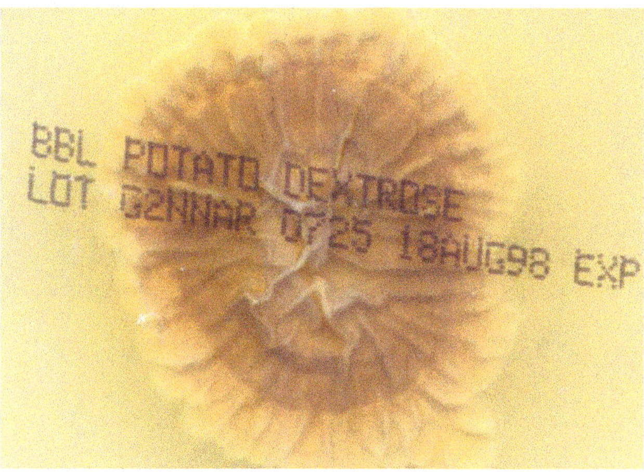

Fig. 9.1.1 *Acremonium* (bottom).

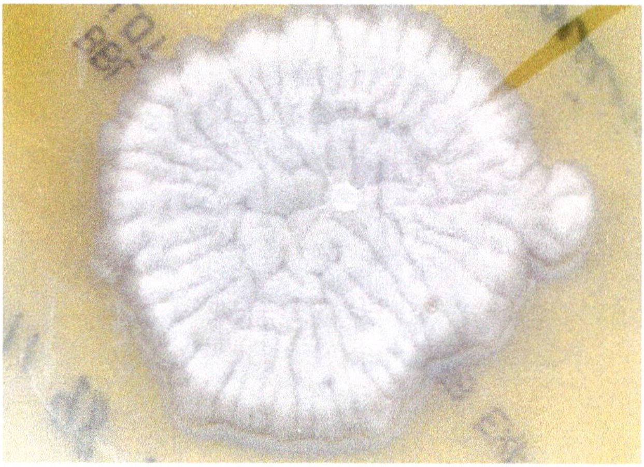

Fig. 9.1.2 *Acremonium* PD.

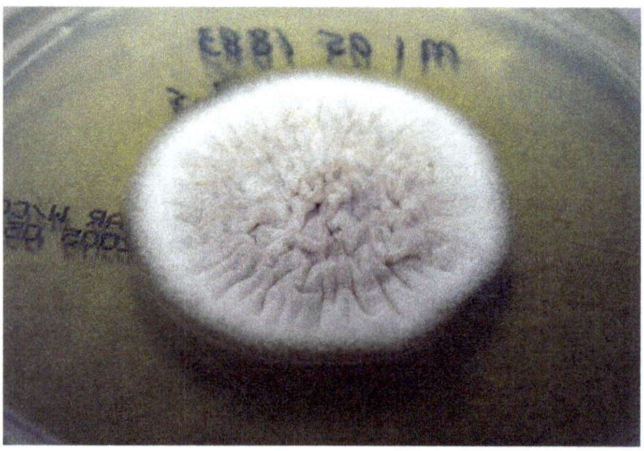

Fig. 9.1.3 *Acremonium* Sab Dx.

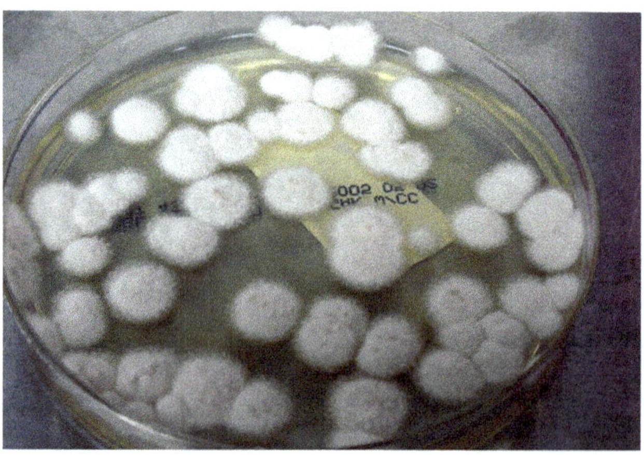

Fig. 9.1.4 *Acremonium* Sab Dx with CC.

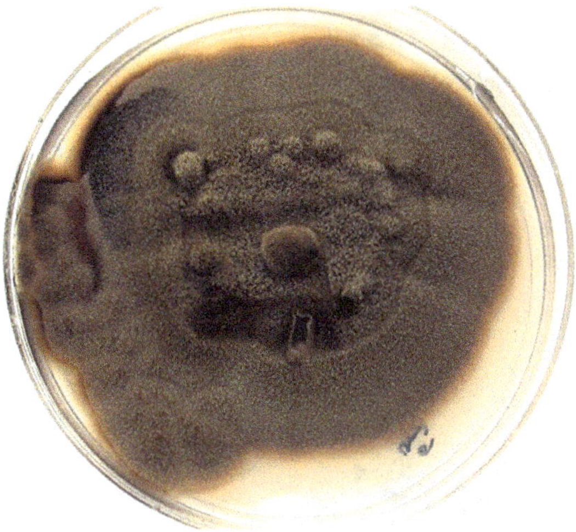

Fig. 9.1.5 *Acremonium* Sab DX (top) dark.

Fig. 9.1.6 *Acremonium* Sab DX (side) dark.

Fig. 9.1.7 *Acremonium* blood agar.

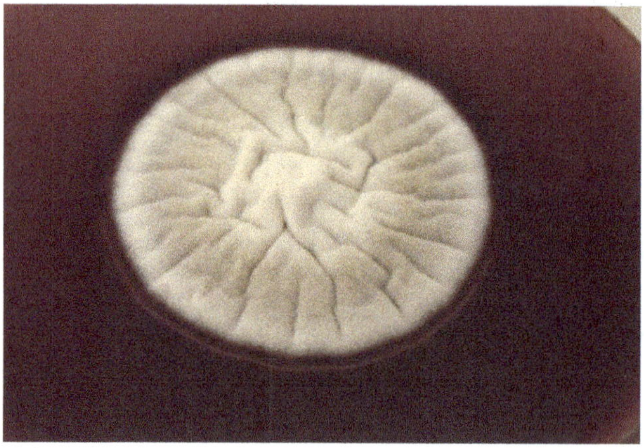

Fig. 9.1.8 *Acremonium* blood agar.

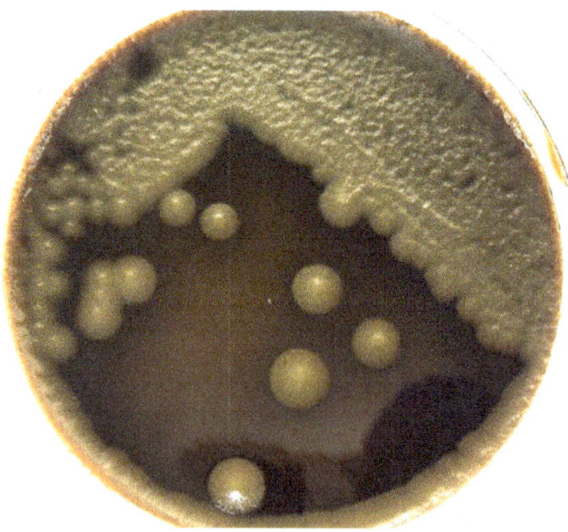

Fig. 9.1.9 *Acremonium* chocolate agar (top).

Fig. 9.1.10 *Acremonium* chocolate agar.

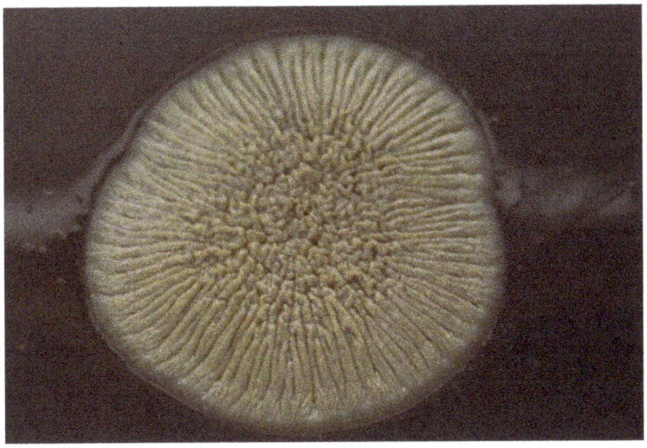

Fig. 9.1.11 *Acremonium* chocolate agar.

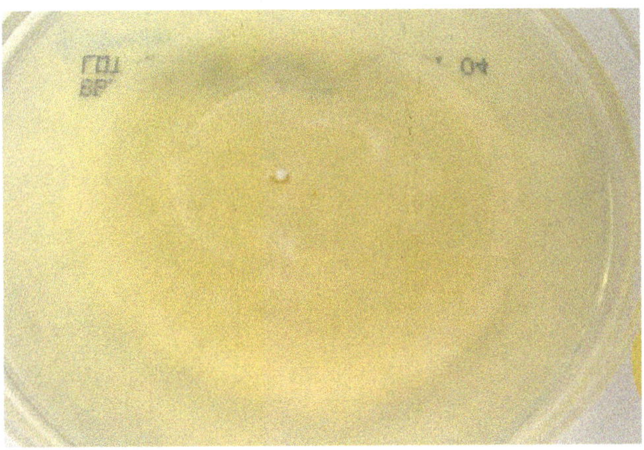

Fig. 9.1.12 *Acremonium* corn meal agar.

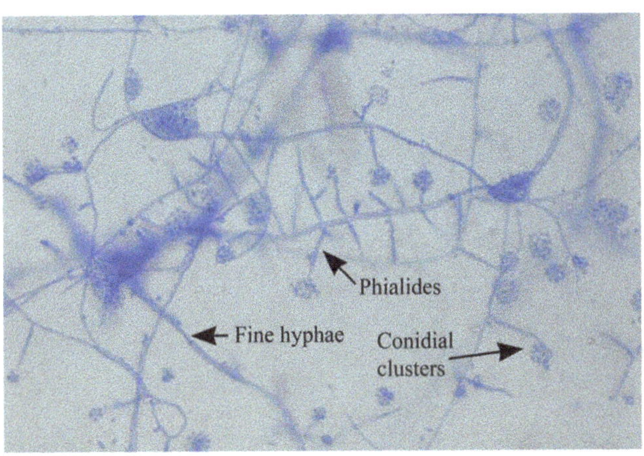

Fig. 9.1.13 Basic morphology, 400×.

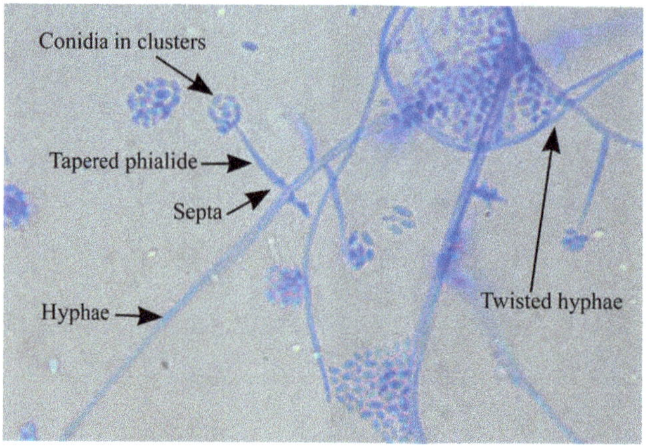

Fig. 9.1.14 *Acremonium*, basic morphology, 1000×.

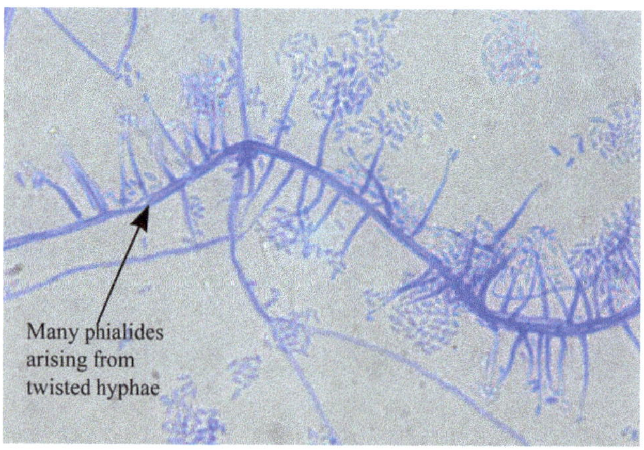

Fig. 9.1.15 Abundant phialides and clusters of conidia.

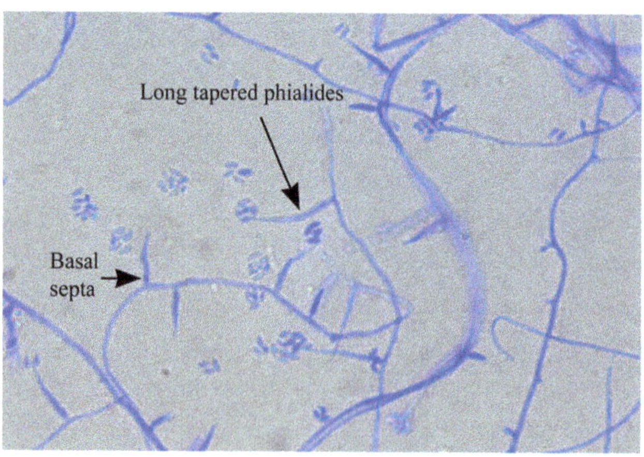

Fig. 9.1.16 Phialide morphology.

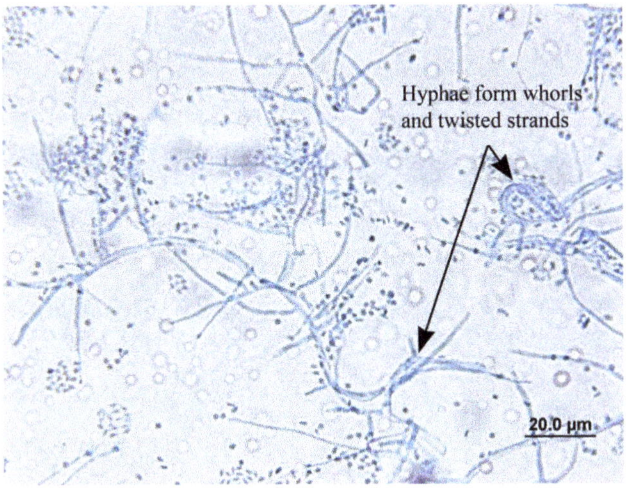

Fig. 9.1.17 *Acremonium*, hyphal morphology.

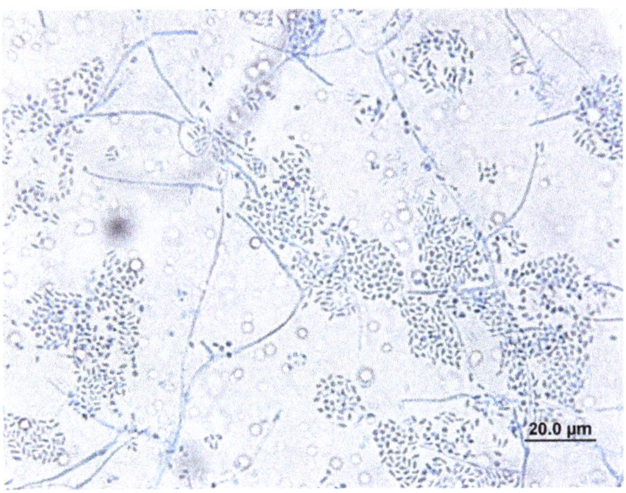

Fig. 9.1.18 Conidia in clusters.

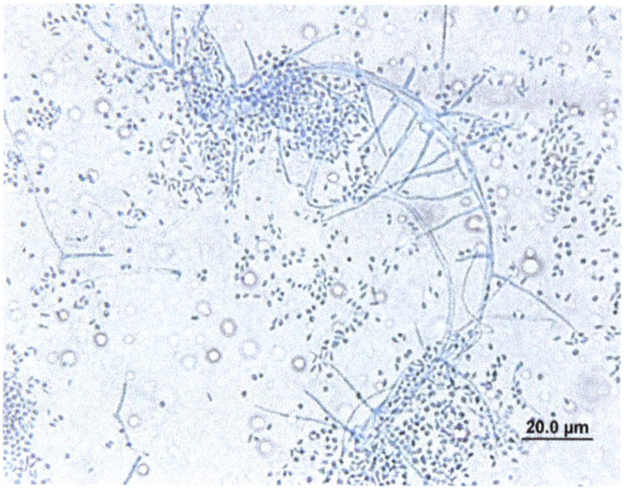

Fig. 9.1.19 *Acremonium*, typical morphology.

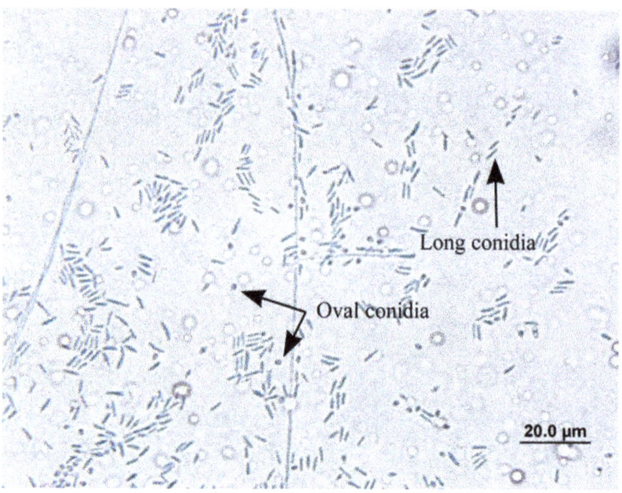

Fig. 9.1.20 *Acremonium*, conidia morphology.

9.2 *Alternaria* spp

Colonial morphology: colonies begin as grayish-white, rapidly turning dark and fluffy. Colonies show concentric circles of white to dark. Dark colors vary from gray, brown, olive, to black. The reverse is usually black. Colonies do not wrinkle but they may form starbursts of colors from the center outward.

Microscopic morphology: hyphae are highly septate, dark walled with dark septa. Conidiophores are produced directly from hyphae with septation. The conidiophores are often foot shaped at the terminal cell, producing conidia in a heel to toe arrangement. Often conidiophores show sympodial (zig-zag) arrangements of conidia production. The conidia are unique, being "snow-shoe" shaped with multiple septations and occurring in chains. Conidia are large and brown (8–16 × 23–50 µm) with transverse and longitudinal septations. Frequently, conidia produce germ tubes that appear as long tails. Young cultures of *Alternaria* may be misidentified as *Stemphylium, Pithomyces* or *Epicoccum*. Often cultures are slow to produce conidia within the first 5–6 days. Cultures may suddenly germinate overnight, showing characteristic conidia at days 7–9. *Alternaria* soon lose their ability to sporulate in culture, so cultures must be watched avidly and patiently.

Alternaria

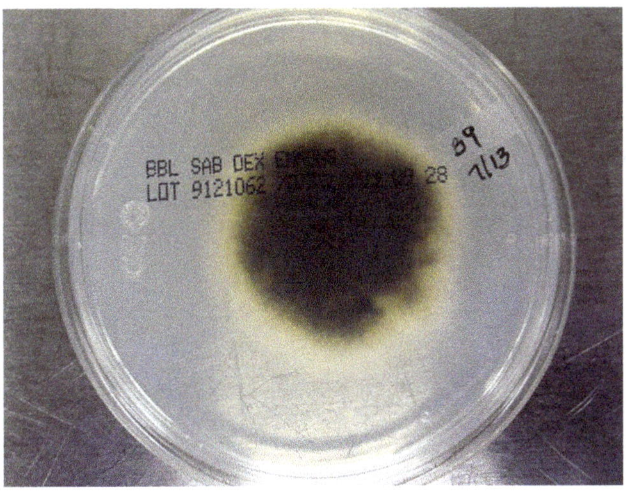

Fig. 9.2.1 *Alternaria* (bottom).

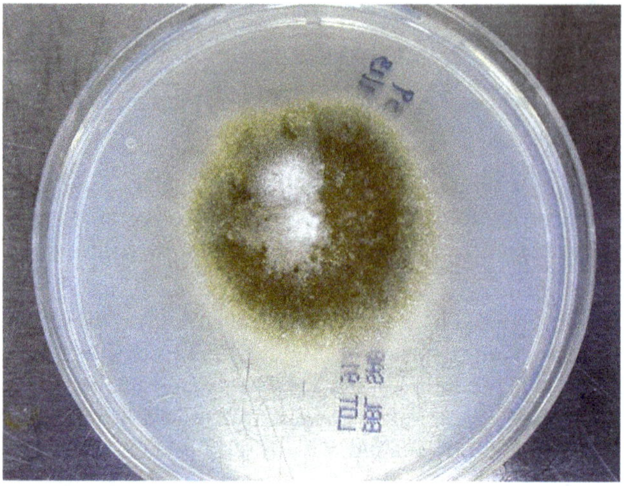

Fig. 9.2.2 *Alternaria* (top).

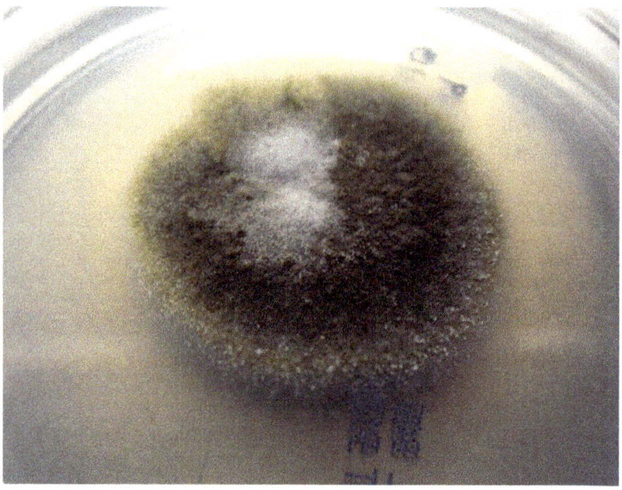

Fig. 9.2.3 *Alternaria* (side).

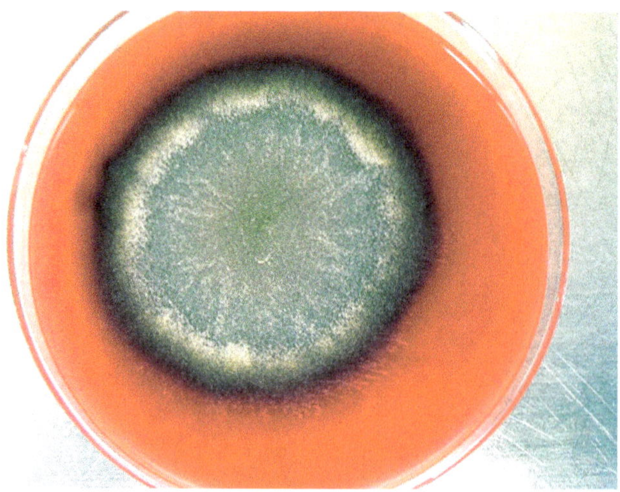

Fig. 9.2.4 *Alternaria* on blood agar.

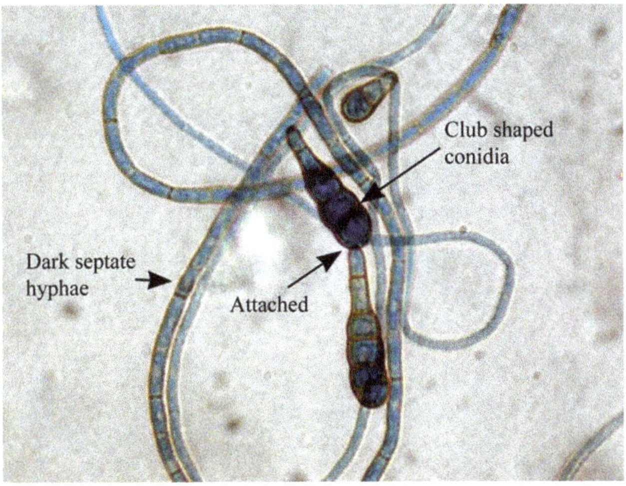

Fig. 9.2.5 *Alternaria*, basic morphology.

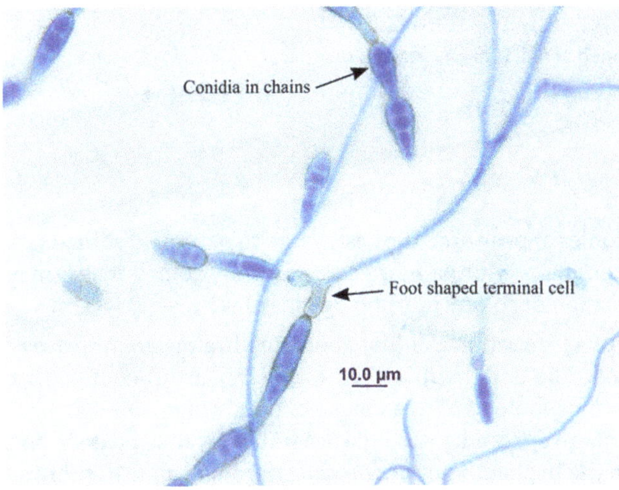

Fig. 9.2.6 *Alternaria*, conidia formation.

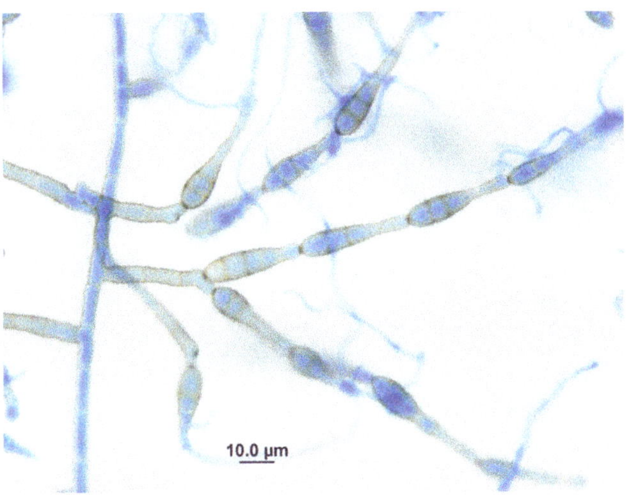

Fig. 9.2.7 Dark septate conidia in chains.

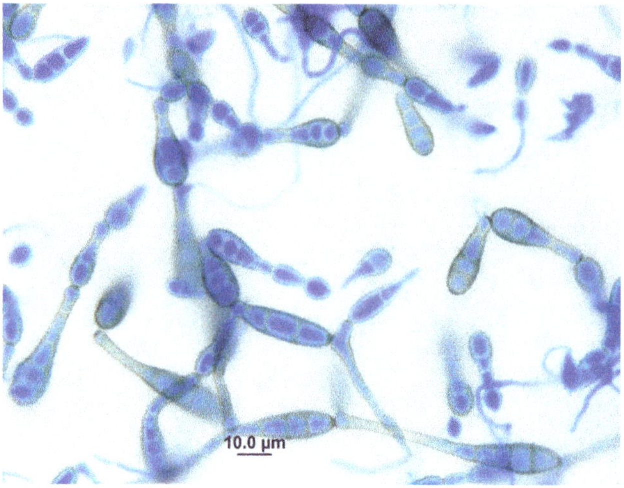

Fig. 9.2.8 *Alternaria*, snow-shoe shaped conidia vary in size and length.

9.3 *Aureobasidium pullulans*

Colonial morphology: the colonies appear first as moist, white to pale pink, sometimes brown to black in color and then mature into a more cottony, fluffy colony which may darken with age. The transformation of colony morphology may take several weeks.

Microscopic morphology: hyphae are septate, at first appearing hyaline, then with age becoming dark and thick walled. The dark hyphae then form large arthroconidia that often have varied non-uniform morphology. These chains of darkly pigmented arthroconidia are commonly called chlamydoconidia. Conidia are round, oval, to pear shaped and are produced directly from the hyphae. They are produced from hyaline to subhyaline hyphae. Several conidia may be produced from a single hyphal cell and may appear as clusters along the hyphae. Conidia may bud much like yeast cells and by some mycologists are considered yeast cells. Hyaline hyphae may also be covered with blebs, giving a rough bumpy appearance to the hyphae.

Aureobasidium

Fig. 9.3.1 *Aureobasidium*.

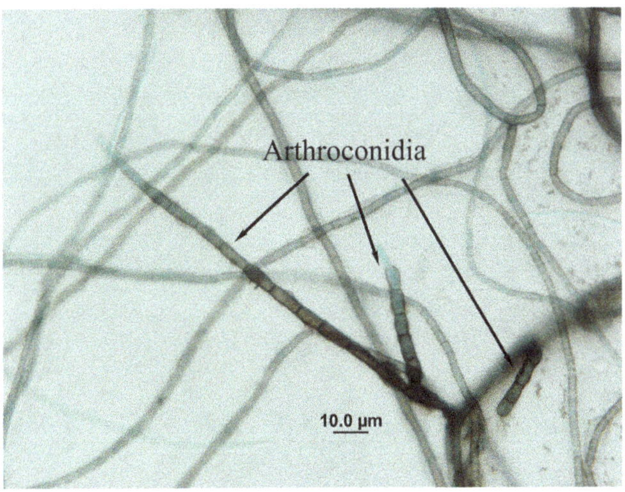

Fig. 9.3.2 Dark arthroconidia.

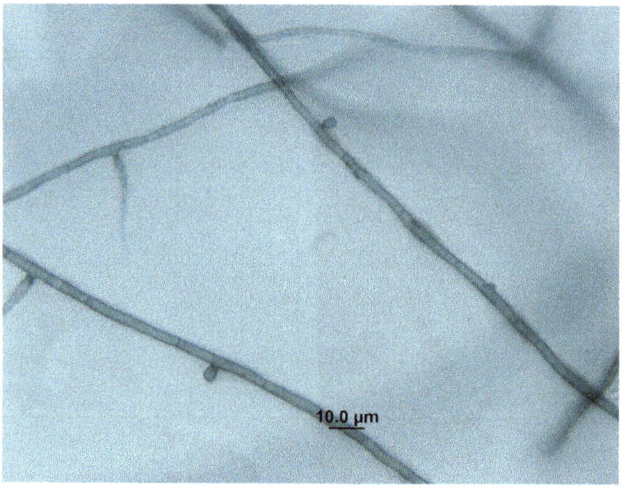

Fig. 9.3.3 *Aureobasidium*, conidia produced directly from hyphae.

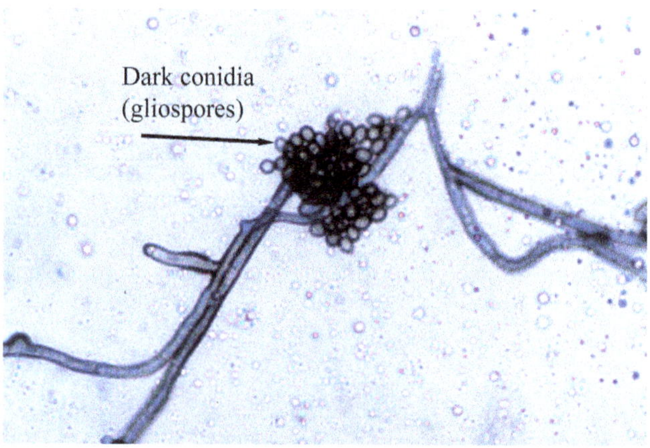

Fig. 9.3.4 Conidia cluster when produced at a single point.

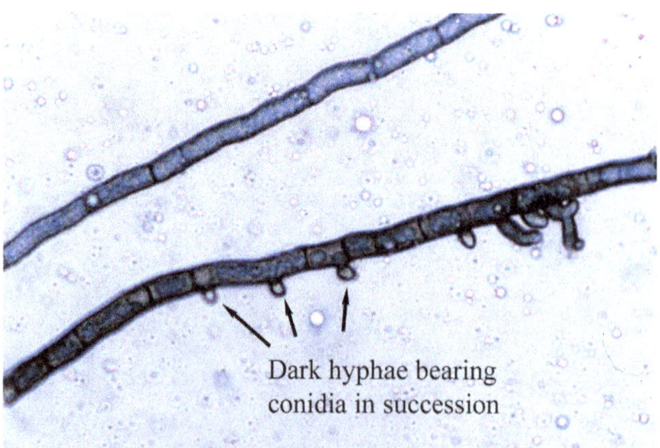

Fig. 9.3.5 Conidia formation.

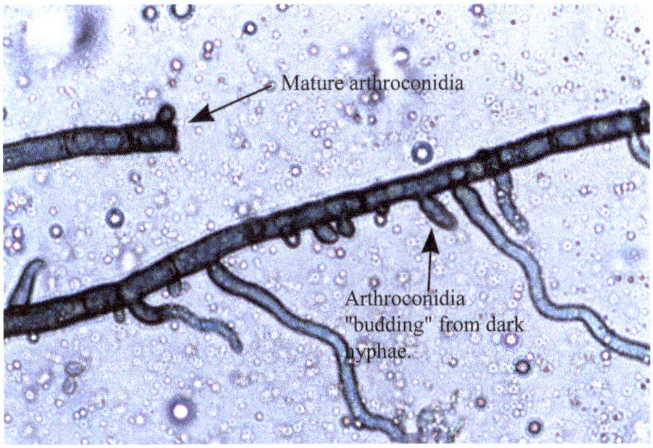

Fig. 9.3.6 *Aureobasidium*, arthroconidia formation.

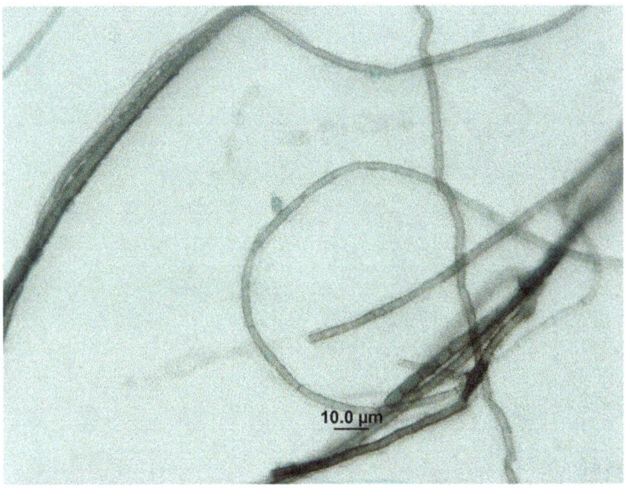

Fig. 9.3.7 *Aureobasidium*, hyphae morphology.

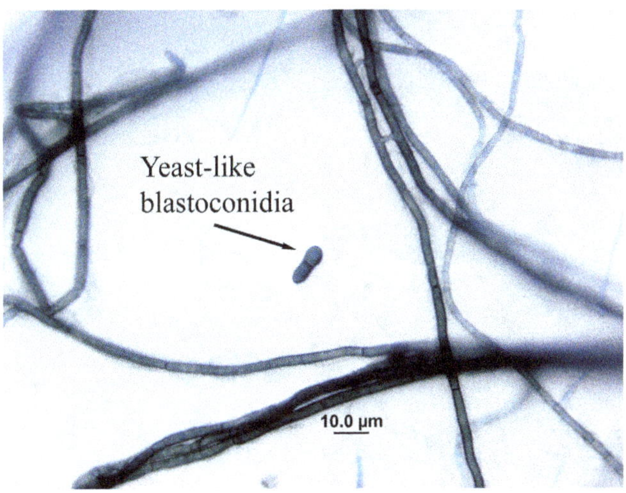

Fig. 9.3.8 *Aureobasidium*, blastoconidia.

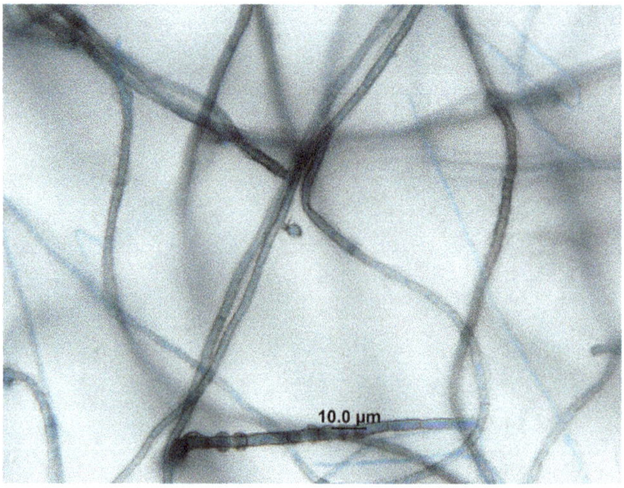

Fig. 9.3.9 *Aureobasidium*, hyphae variation.

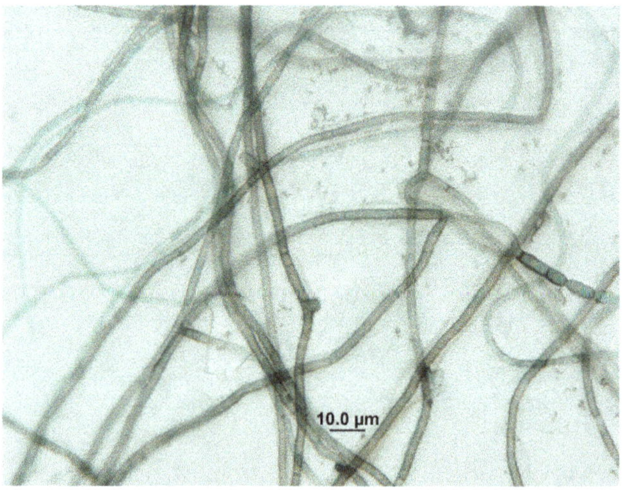

Fig. 9.3.10 Hyphae variation.

Fig. 9.3.11 Rough hyphae.

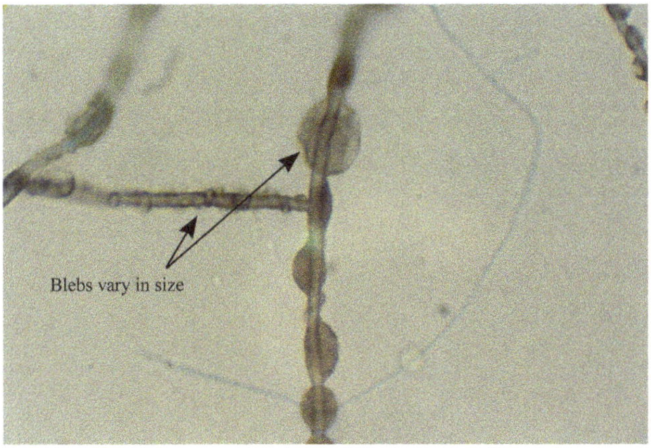

Fig. 9.3.12 *Aureobasidium*, blebs along hyphae.

9.4 *Bipolaris spicifera*, *Curvularia hawaiiensis* (*B. hawaiensis*), and *C. australiensis* (*B. australiensis*)

Colonial morphology: the colony is first grayish-brown with a white center and border. As the colony matures, it becomes darker brown to black with a gray border. It is fuzzy and raised in the center. The reverse is yellow, becoming brown to black with age.

Microscopic morphology: the hyphae are septate, dark and may assume many various shapes. Hyphae may be smooth and bamboo-like or they may swell into structures that resemble chlamydoconidia with or without structures that look like tails. Some mature hyphae darken and give the appearance of arthroconidia which can bud and form other hyphae. Some hyphae may form structures that look like a spiked club. Conidiophores are long and bent, some with sympodial arrangement, and others with a terminal cell that looks like a foot. Large multiseptate (3–5 pseudoseptations) conidia are produced sympodially at each bend of the conidiophores. Conidial septations are not true septations so conidia look like bean pods. Conidia ($8 \times 26\,\mu m$) are rounded at both ends. When conidia germinate, they germinate at both ends and at a 180° angle to the conidium, giving rise to the term "bipolar" for which the genus is named. This feature distinguishes them from *Drechslera* which germinate from intermediate cells within the condidium and usually branch at 90° angles to the conidium. Closely resembling genera are *Drechslera*, *Curvularia*, *Exserohilum*, and *Helminthosporium*. Distinguishing characteristics between these genera are the number of conidia septa, size of the conidia, the protrusion of the hilum, and the germination of the conidia. *B. spicifera* characteristically contains three septa, *B. hawaiiensis* and *B. australiensis* contain 3–6 septa (average of four). *B. australiensis and B. hawaiiensis* were recently reclassified to the genus *Curvularia*.

Bipolaris germ tube formation is accomplished by overlaying a mature culture with sterile distilled water. A sterile applicator stick is then lightly scraped along the colony surface in order to liberate the conidia into suspension. Ten drops of the suspension are then added to any nutrient broth. Almost any broth works, including yeast nitrogen base, tryptic soy broth, RPMI 1640, Sabouraud dextrose broth, thioglycolate broth, brain heart infusion broth, and many others. Incubate the inoculated broth for 2–4 hours at 37 °C. Most species germinate within 2 hours. Germination can be detected in just sterile distilled water if incubated for 24 hours. Prepare a wet prep with a drop of broth and one of lactophenol cotton blue. Bipolar germination, as shown in Figs 9.4.19 and 9.4.20, is unique to *Bipolaris*.

Bipolaris

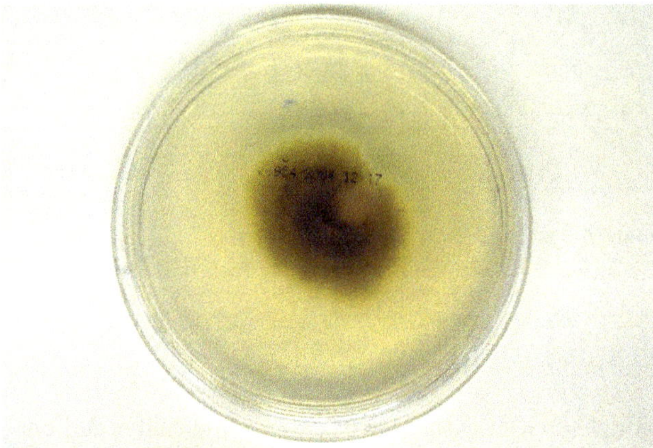

Fig. 9.4.1 *Bipolaris* (bottom).

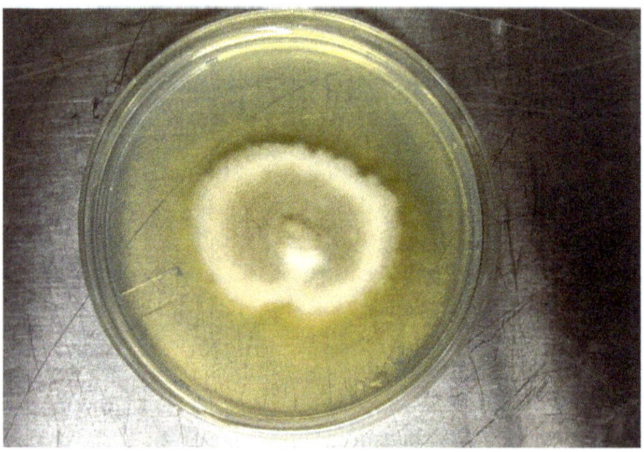

Fig. 9.4.2 *Bipolaris*.

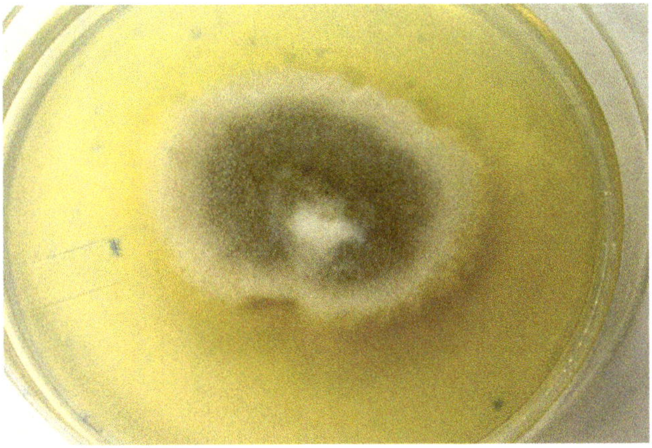

Fig. 9.4.3 *Bipolaris* (side).

Fig. 9.4.4 *Bipolaris* (bottom) at 2 weeks.

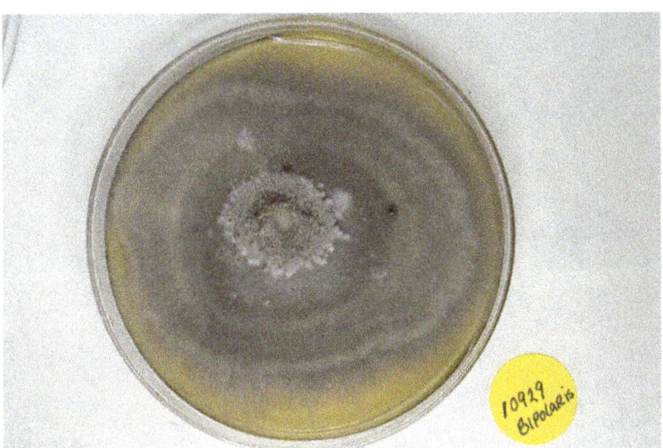

Fig. 9.4.5 *Bipolaris* at 2 weeks.

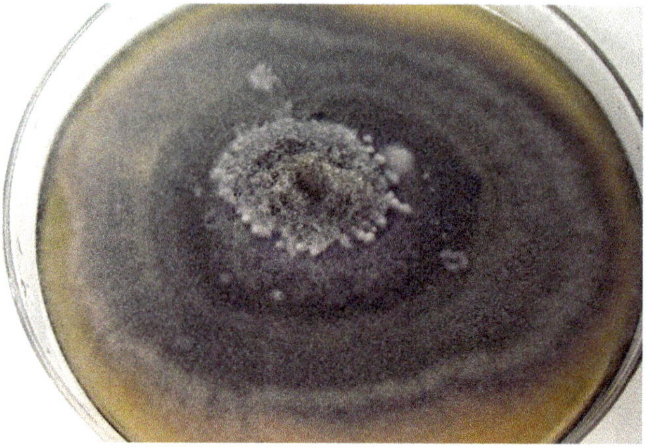

Fig. 9.4.6 *Bipolaris* (side) at 2 weeks.

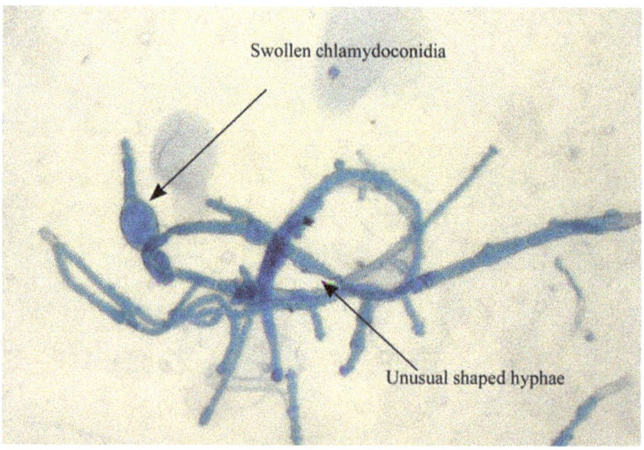

Fig. 9.4.7 *Bipolaris*, varying morphology.

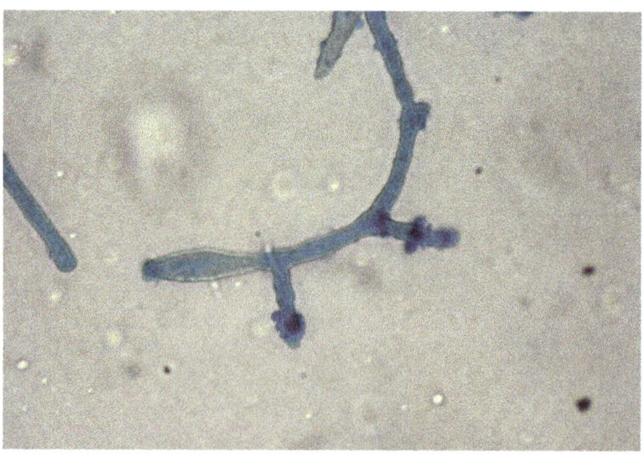

Fig. 9.4.8 *Bipolaris*, varying morphology.

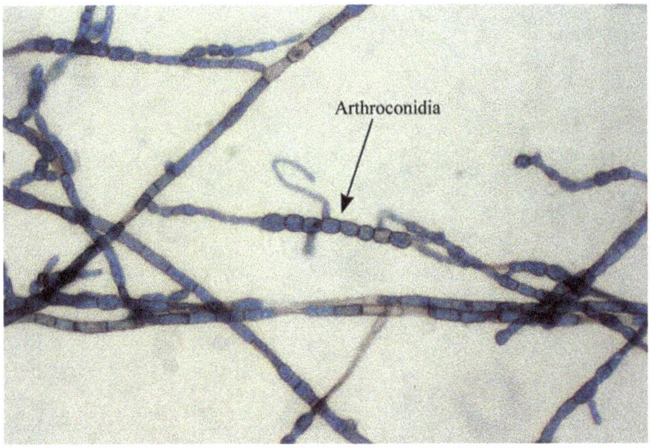

Fig. 9.4.9 *Bipolaris*, arthroconidia.

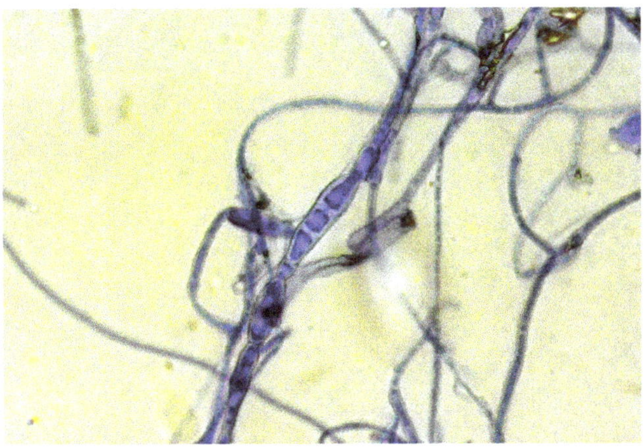

Fig. 9.4.10 *Bipolaris*, varying morphology.

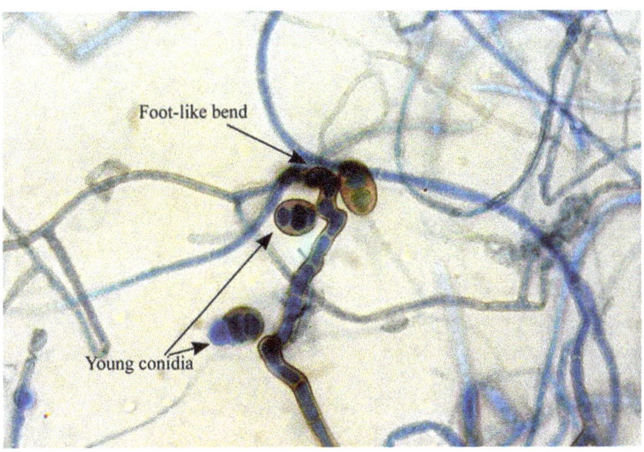

Fig. 9.4.11 *Bipolaris*, early conidia formation, 400×.

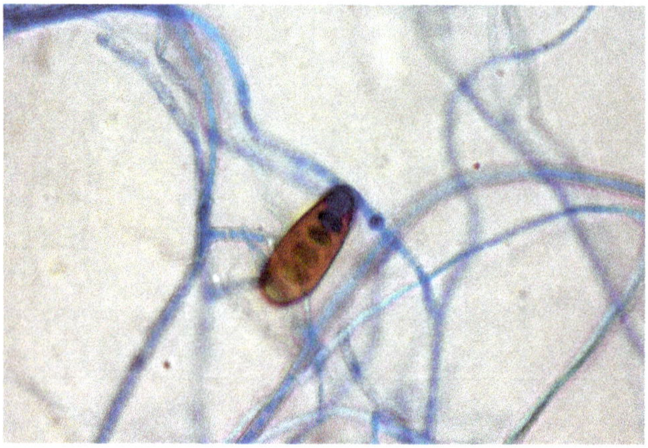

Fig. 9.4.12 *Bipolaris*, early conidia formation, 1000×.

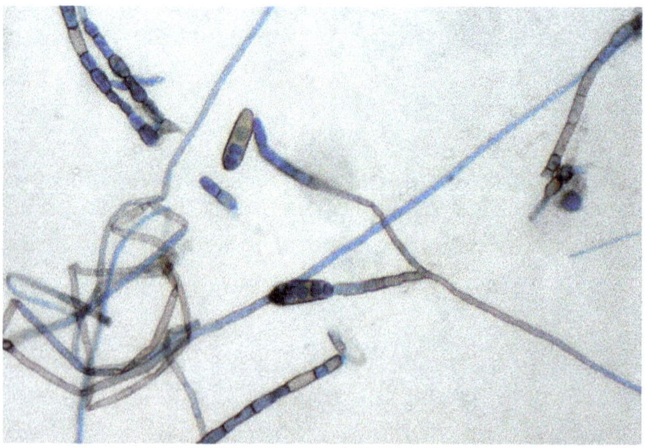

Fig. 9.4.13 *Bipolaris*, conidia and arthroconidia.

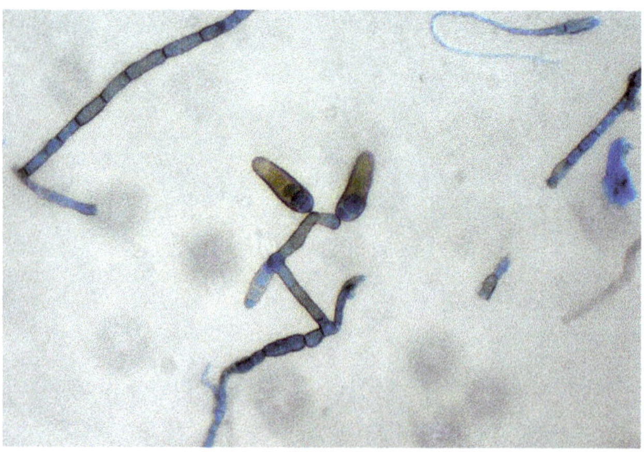

Fig. 9.4.14 *Bipolaris*, early conidia formation.

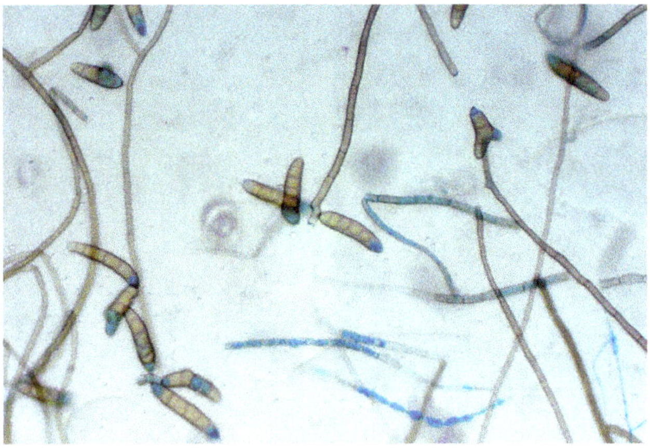

Fig. 9.4.15 *Bipolaris*, mature conidia are dark, 100×.

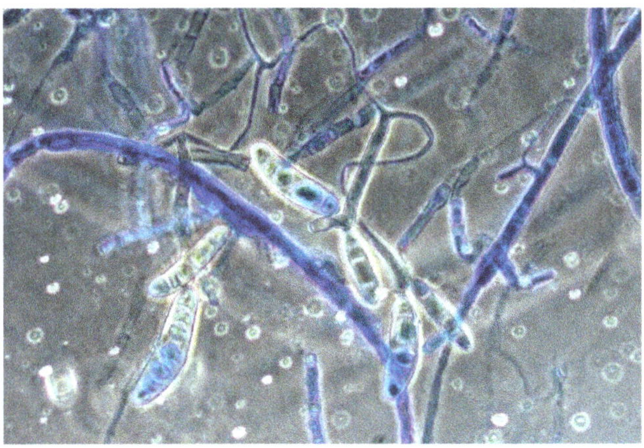

Fig. 9.4.16 *Bipolaris*, phase contrast microscopy, 400×.

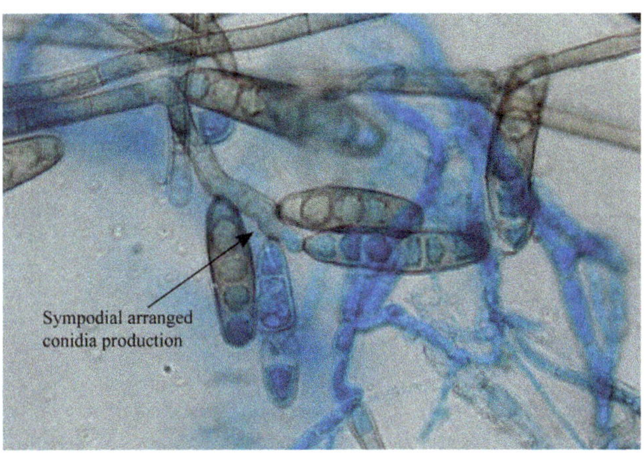

Fig. 9.4.17 *Bipolaris*, mature conidia, 1000×.

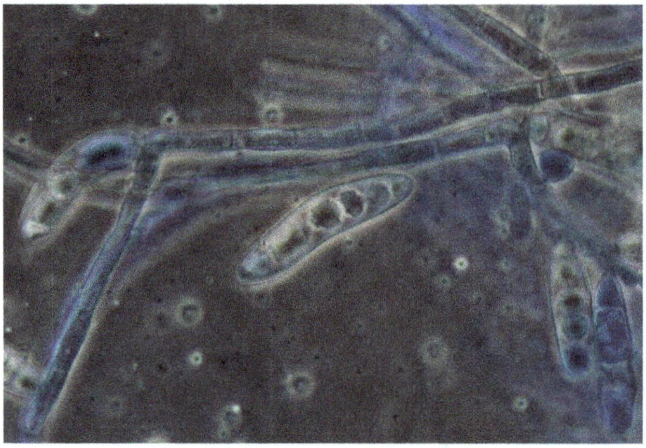

Fig. 9.4.18 *Bipolaris*, mature conidia; phase contrast, 1000×.

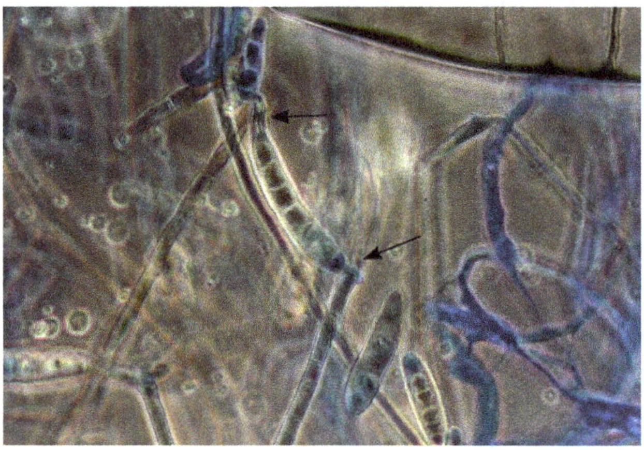

Fig. 9.4.19 Bipolar Germination.

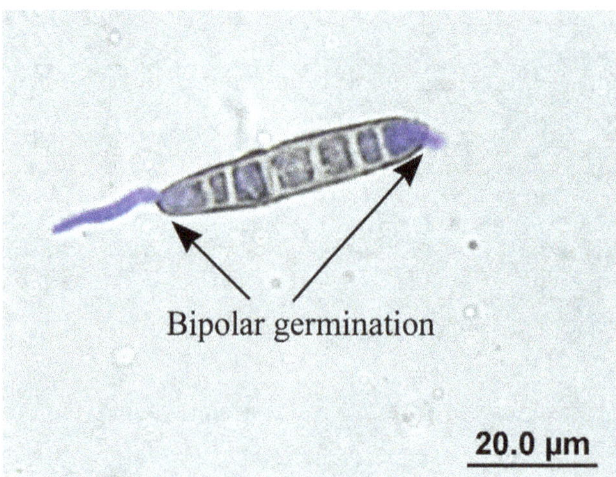

Fig. 9.4.20 *Bipolaris*, bipolar germination.

9.5 *Chaetomium* spp

Colonial morphology: colonies begin as white and cottony, spread, and become dark gray, olive to black with age. The reverse is yellow to orange and sometimes black.

Microscopic morphology: the hyphae are dematiaceous, septate, and wavy. Large (90–170 × 110–250 μm) flask-shaped perithecia are produced. Perithecia are often seen in pairs, with the appearance of a pineapple, with hyphae radiating outward in all directions. The hyphae may be wavy or long and pointed at the ends. Asci are dark, lemon-shaped, single-celled spores that are released from openings in the perithecia. Asci are not always observed.

Chaetomium

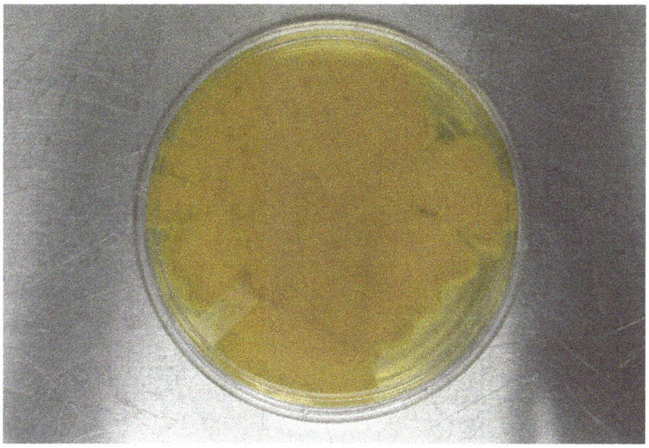

Fig. 9.5.1 *Chaetomium* (bottom).

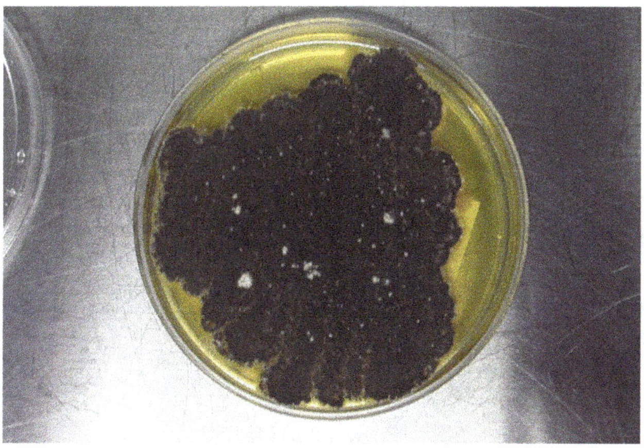

Fig. 9.5.2 *Chaetomium*.

Fig. 9.5.3 *Chaetomium* (side).

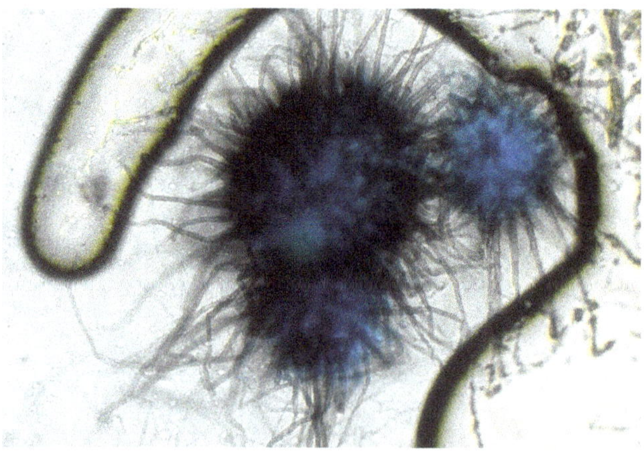

Fig. 9.5.4 Perithecia.

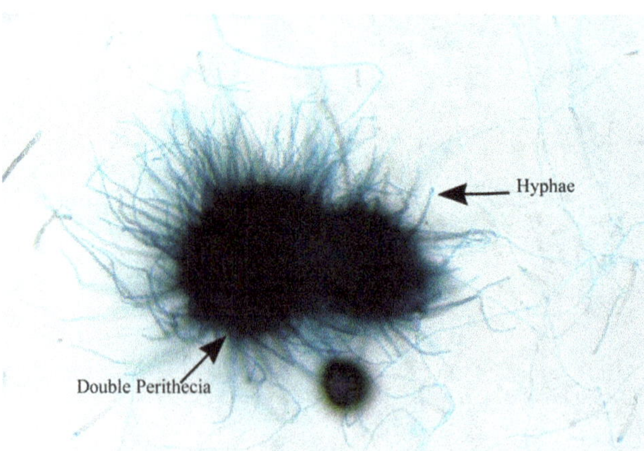

Fig. 9.5.5 Double perithecia.

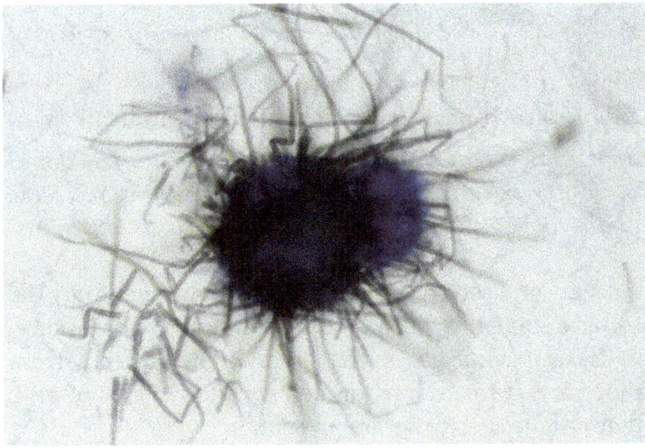

Fig. 9.5.6 *Chaetomium*, "pineapple-like" perithecium with pointed ends.

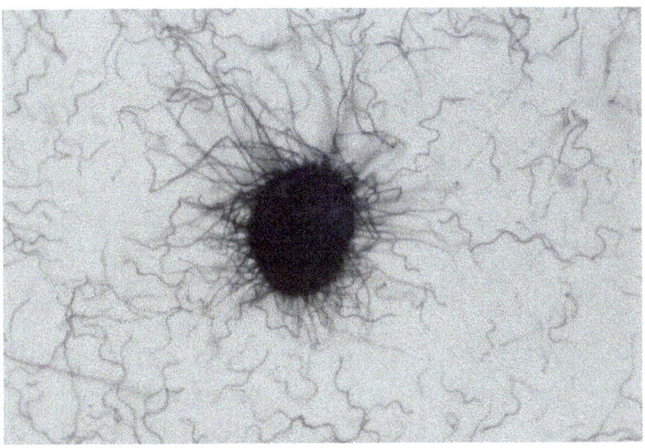

Fig. 9.5.7 Wavy hyphal extensions of perithecium.

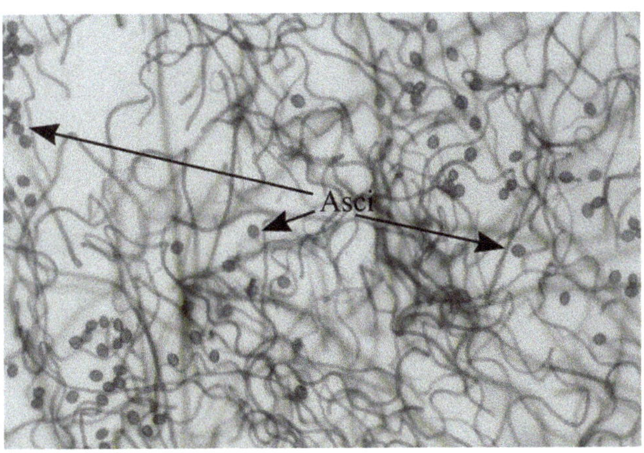

Fig. 9.5.8 Asci released from perithecia.

9.6 *Cladophialophora (Xylohypha) bantiana*

Colonial morphology: the colonies are light green to brown or black with a velvety surface. The reverse is black. Colonies grow slowly and reach maturity within 15 days.

Microscopic morphology: *extreme care is required in handling this fungus. Use BSL-3 precautions and a biosafety hood for making lacto-phenol cotton blue preparations. Slide cultures should not be made. C. bantiana causes cerebral pheohyphomycosis with a respiratory route of infection.*

Hyphae are hyaline and septate, fine, and rare. Most wet preparations reveal only the conidial chains. The hyphae may have some brown color, but are fragile in appearance. They terminate in chains of conidia or in some cases, chains can branch directly from hyphae via short one-cell conidiophores. Mostly, long branching, wavy chains of oval, smooth conidia are seen. No other unique structures are present.

Cladophialophora bantiana

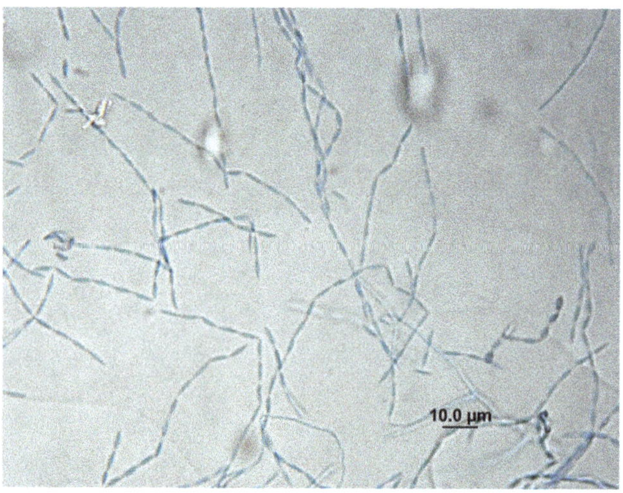

Fig. 9.6.1 *Cladophialophora bantiana.*

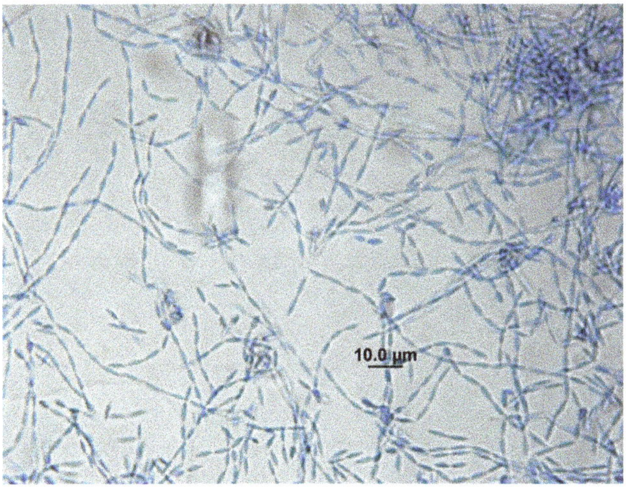

Fig. 9.6.2 *C. bantiana*.

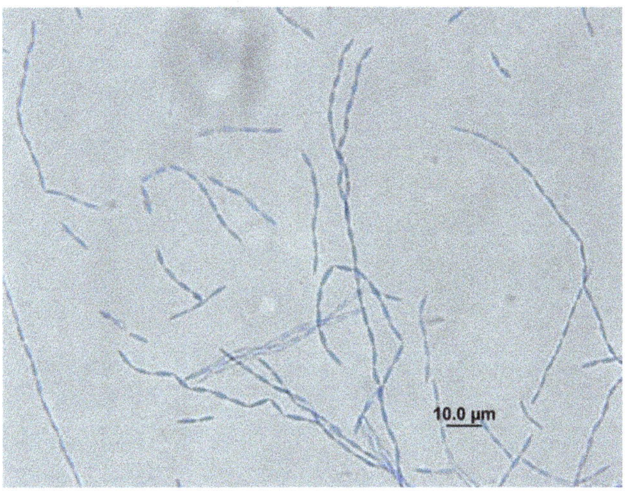

Fig. 9.6.3 *C. bantiana*.

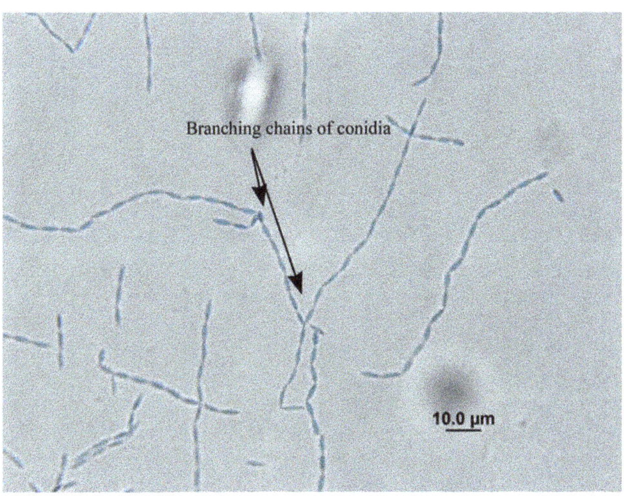

Fig. 9.6.4 *C. bantiana*.

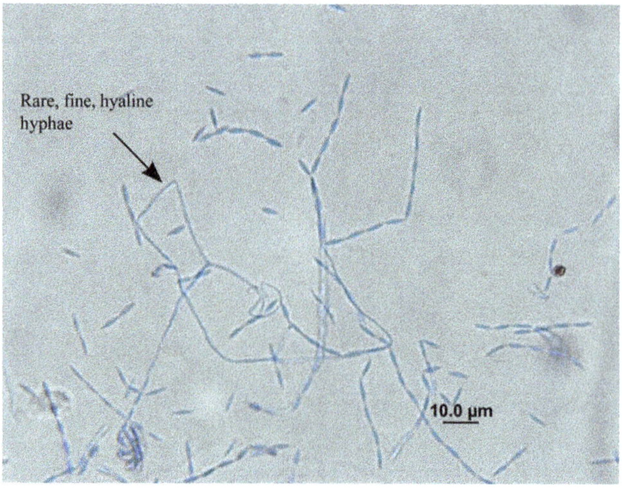

Fig. 9.6.5 *C. bantiana*, hyphae are fine and hyaline.

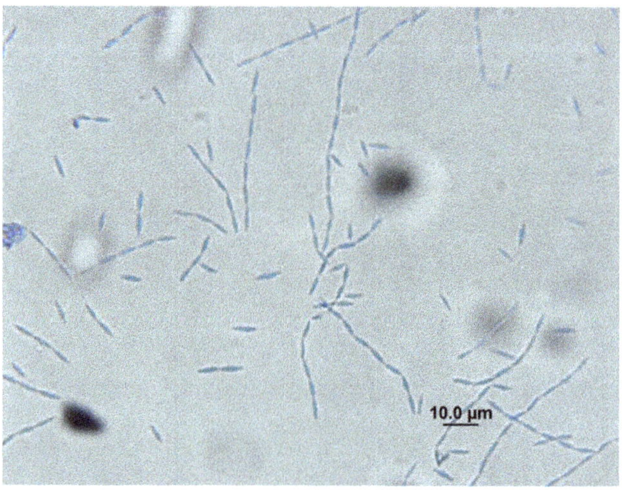

Fig. 9.6.6 Conidial chains.

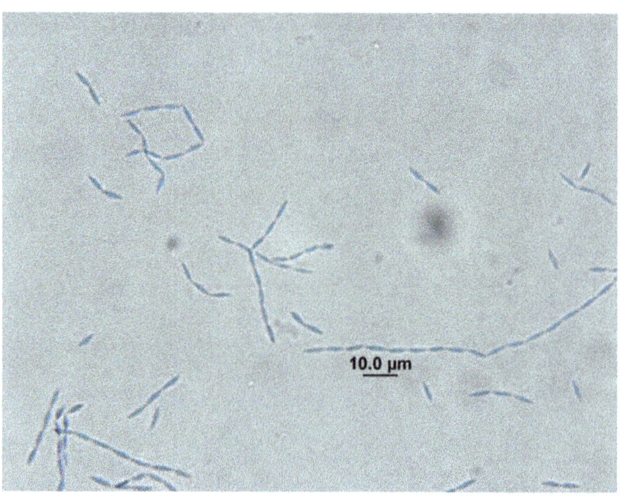

Fig. 9.6.7 Branching conidial chains.

9.7 *Cladophialophora boppi*

Colonial morphology: colonies are velvety black and compact, resistant to cycloheximide. They are mature in 7–10 days at 30 °C.

Microscopic morphology: hyphae are septate, dark, rough, and wavy. Long chains of conidia arise directly from hyphae either by the formation of short conidiophores or differentiation of hyphae into chains of conidia. Conidia are lemon shaped to round with pale attachment scars. Hyphae may form synnema. It grows up to 37 °C but not at 42 °C which differentiates it from *C. bantiana* which grows at 42 °C.

Cladophialophora boppi

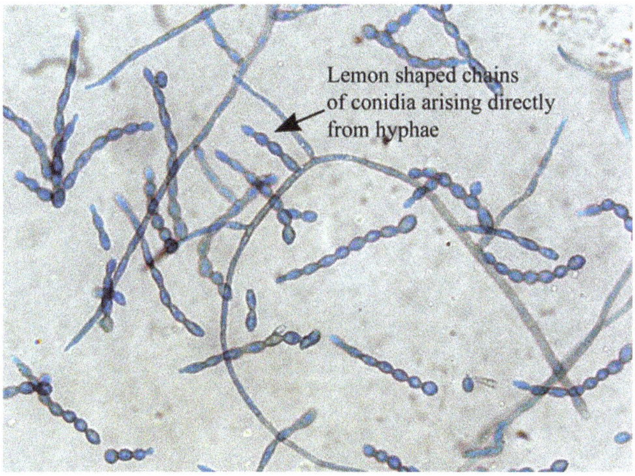

Fig. 9.7.1 *Cladophialophora boppi.*

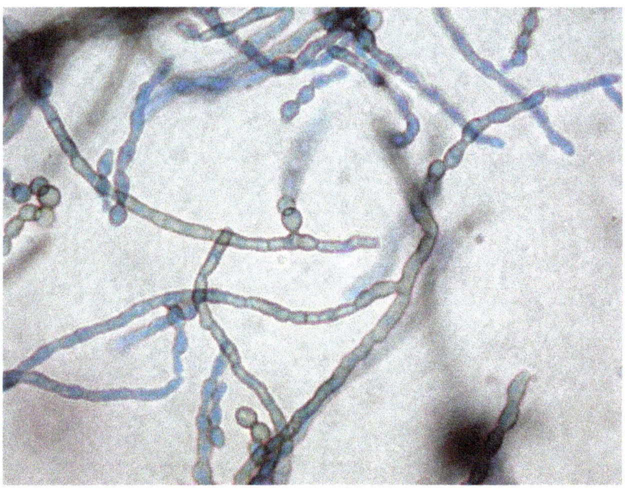

Fig. 9.7.2 *C. boppi.*

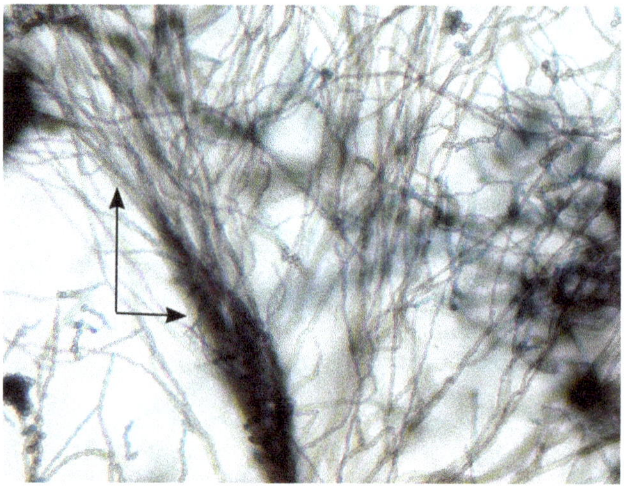

Fig. 9.7.3 *C. boppi*. Synnema (arrows)

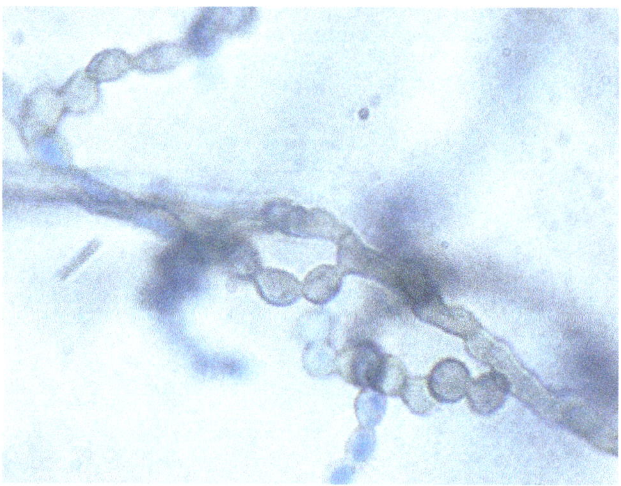

Fig. 9.7.4 *C. boppi*.

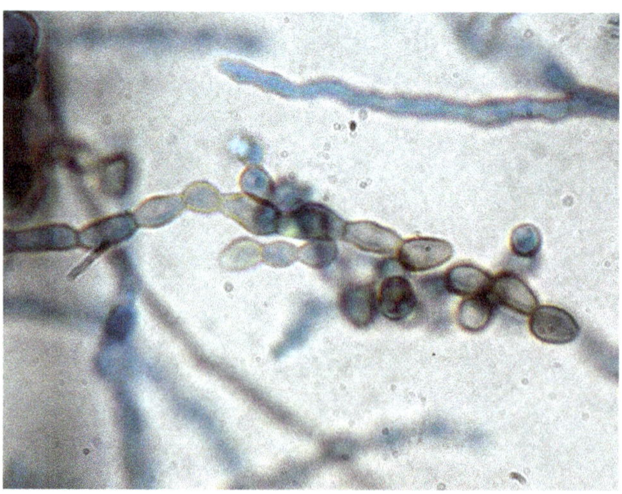

Fig. 9.7.5 *C. boppi*.

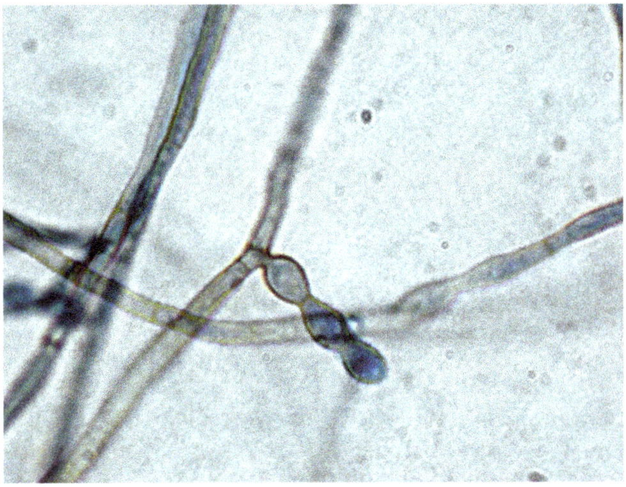

Fig. 9.7.6 *C. boppi*.

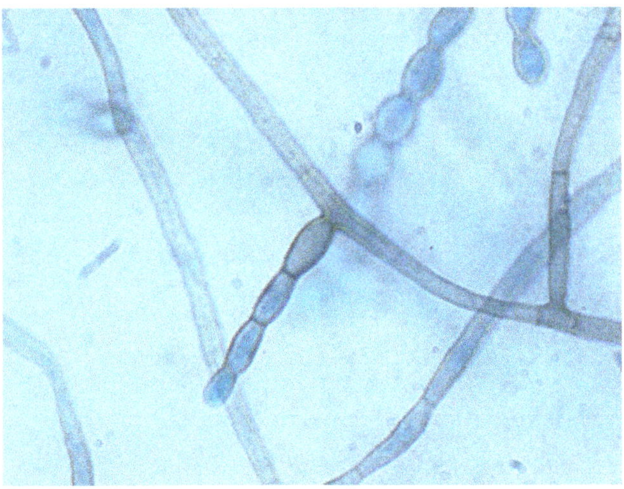

Fig. 9.7.7 *C. boppi*.

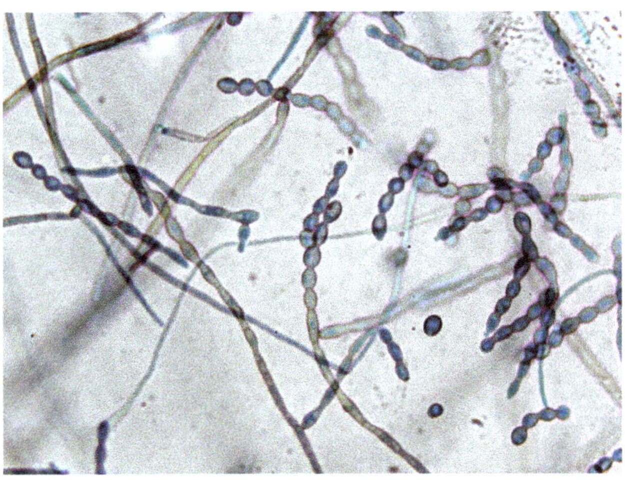

Fig. 9.7.8 *C. boppi*.

9.8 *Cladophialophora carrionii*

Colonial morphology: colonies are first small and white, with a dark center. As colonies mature, they become dark, wrinkled, and covered with a velvety gray or green surface. Reverse is tan with black borders. *C. carrionii* will grow at 37 °C.

Microscopic morphology: hyphae are dark, septate and produce conidiophores, some terminal. Conidiophores produce sterigmata (branching conidia) that also branch into secondary sterigmata prior to producing conidial chains. Secondary sterigmata are referred to as "shield cells" since they have the appearance of a shield. Shield cells that branch three times are rare in *C. carrionii*, thus they often branch only twice, showing two hila (scars). *C. carrionii* demonstrates branching chains of conidia without distinct conidiophores which helps to differentiate it from other fungi in the genera *Cladophialophora* and *Cladosporium*. Conidia chains contain 6–10 conidia whereas other species have 1–6 conidia. Conidia are smooth walled, dark, oval, and pointed at each end. Hila can be seen at each end if images are computer enhanced (as shown), but otherwise they are not as pronounced as they are in other *Cladophialophora* spp.

Cladophialophora carrionii

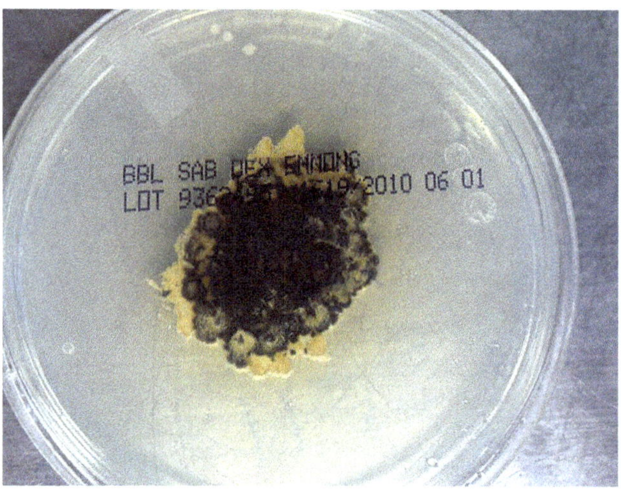

Fig. 9.8.1 *Cladophialophora carrionii*, Sab DX (bottom).

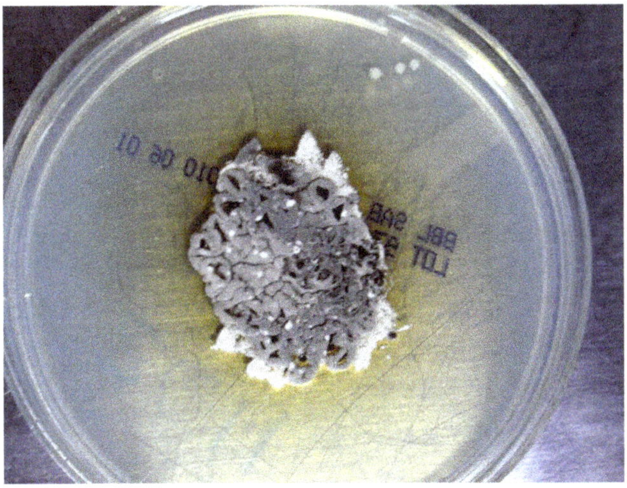

Fig. 9.8.2 *C. carrionii* Sab DX.

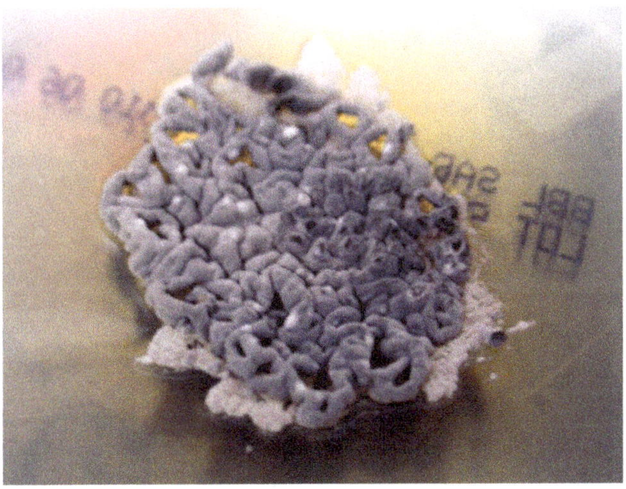

Fig. 9.8.3 *C. carrionii* Sab DX, (side).

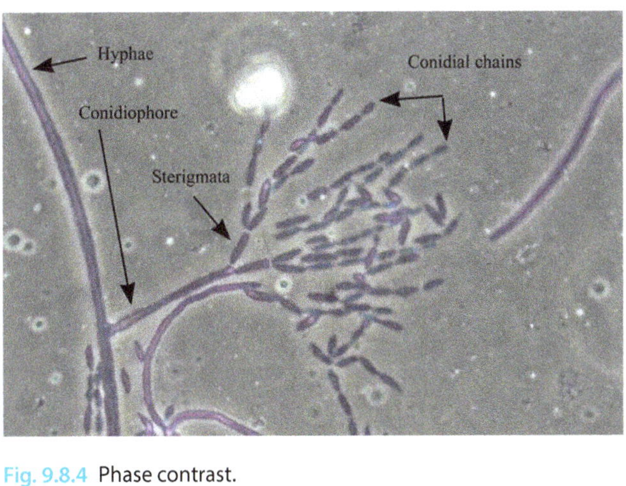

Fig. 9.8.4 Phase contrast.

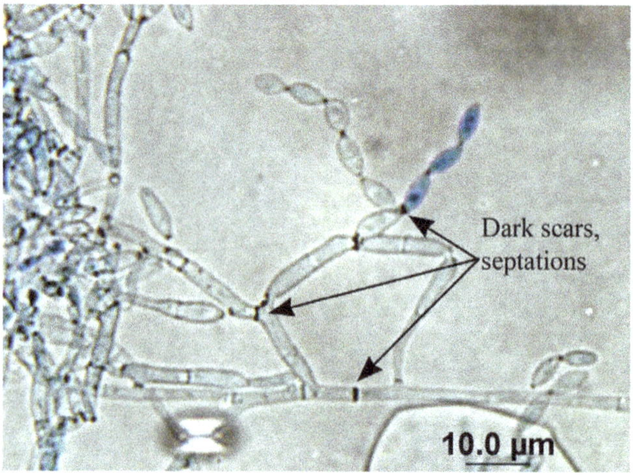

Fig. 9.8.5 *C. carrionii*, dark septations.

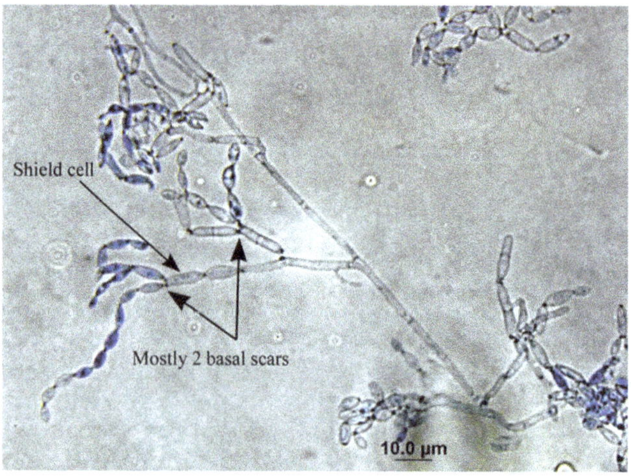

Fig. 9.8.6 Shield cells (sterigmata) with two basal scars.

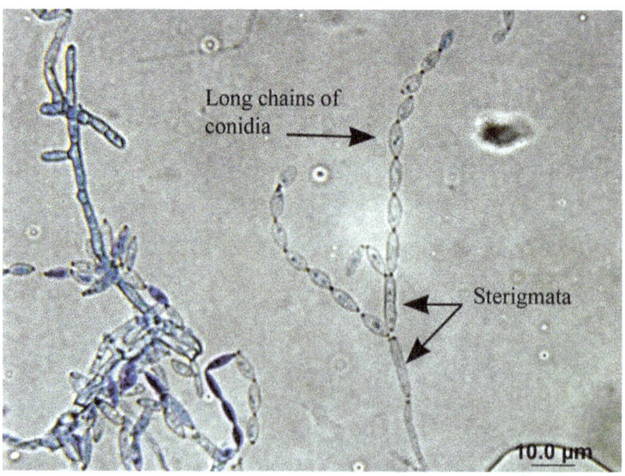

Fig. 9.8.7 Long chains of conidia.

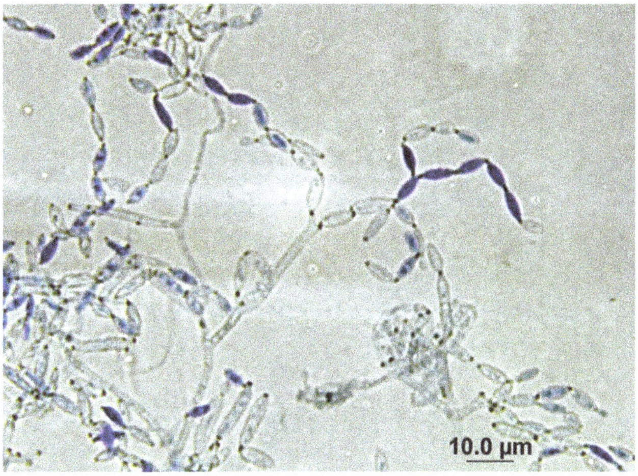

Fig. 9.8.8 *C. carrionii*, common morphology.

9.9 *Cladosporium sphaerospermum*

Colonial morphology: colonies first appear white, then rapidly become dark grass-green. The surface is rough and convoluted, with a smooth to powdery texture. The reverse is dark green to black.

Microscopic morphology: hyphae are hyaline, septate, and rough. Conidiophores are produced from the hyphae, with a large swollen oval-shaped cell from which branching occurs. One but most often two branches occur that develop into flask-shaped cells, which also branch into separate conidiophores bearing flask-shaped phialides. Conidia are produced in chains from these terminal phialides. Conidia are round, remain in distinct chains, and do not bunch together like the conidia of *Gliocladium*. The terminal conidium in the chain is often larger than other conidia in the chain.

Cladosporium sphaerospermum

Fig. 9.9.1 *Cladosporium sphaerospermum* PD.

Fig. 9.9.2 *C. sphaerospermum* CMA

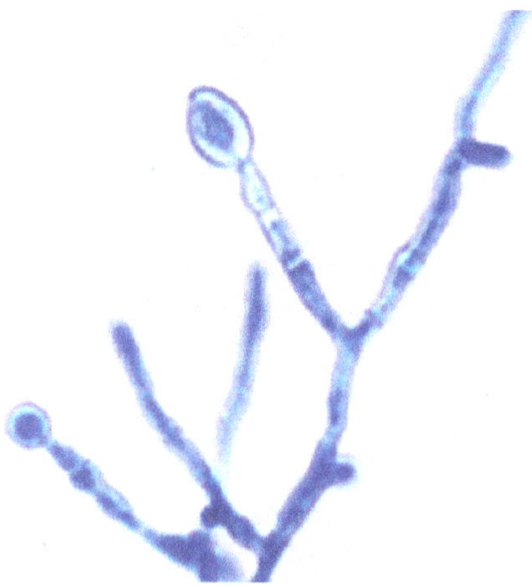

Fig. 9.9.3 *C. sphaerospermum*.

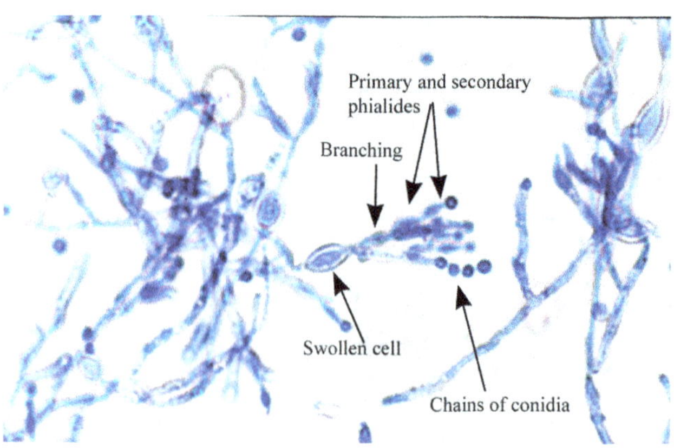

Fig. 9.9.4 *C. sphaerospermum*.

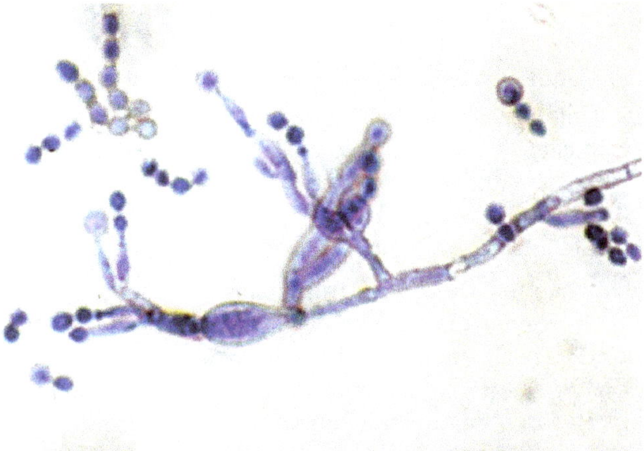

Fig. 9.9.5 *C. sphaerospermum*.

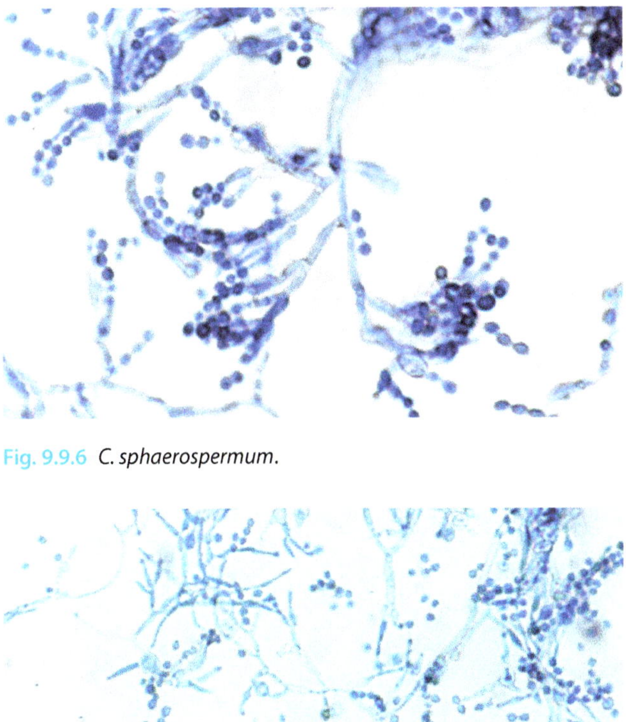

Fig. 9.9.6 *C. sphaerospermum*.

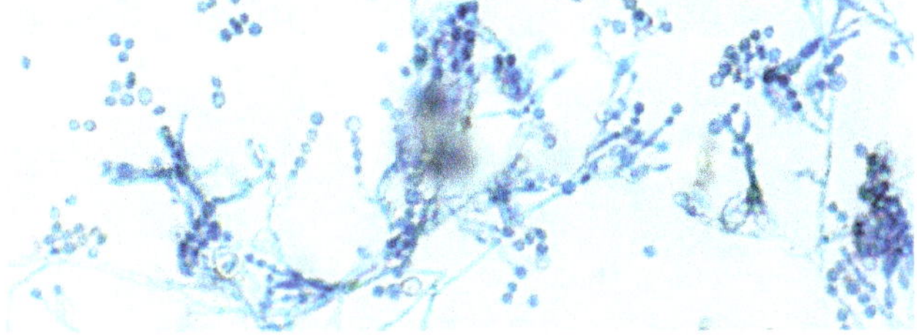

Fig. 9.9.7 *C. sphaerospermum*.

9.10 *Cladosporium* spp

Colonial morphology: colony color varies with species, but they usually begin as green-to-brown fluffy colonies. Colonies retain their color until maturity and may darken and become wrinkled with age. The surface becomes gray to black, velvety or powdery and may show variation in color and texture. Reverse is black. Most strains do not grow at 37 °C.

Microscopic morphology: hyphae are septate and dark. Many branched conidiophores are produced along the hyphae and are distinct from hyphae. Shield cells with 3–4 hila are abundant. Conidia are produced in chains of 1–5, and disarticulate easily when handled. Conidia are round to oval, smooth walled, and when dislodged, hila at each end are easily visible. Sympodial (rachiform) production of conidia is noted for some species. Also, some species may demonstrate unusual sporulation of shield cells giving rise to conidia and hyphae, depending upon growth media and culture conditions.

Cladosporium spp

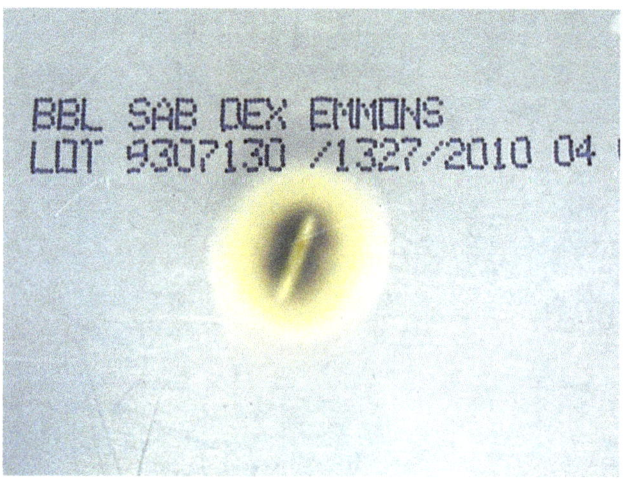

Fig. 9.10.1 *Cladosporium* sp Sab DX (bottom).

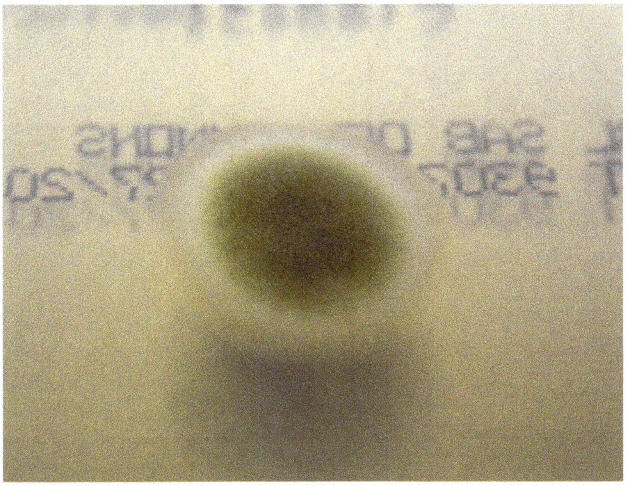

Fig. 9.10.2 *Cladosporium* sp Sab DX.

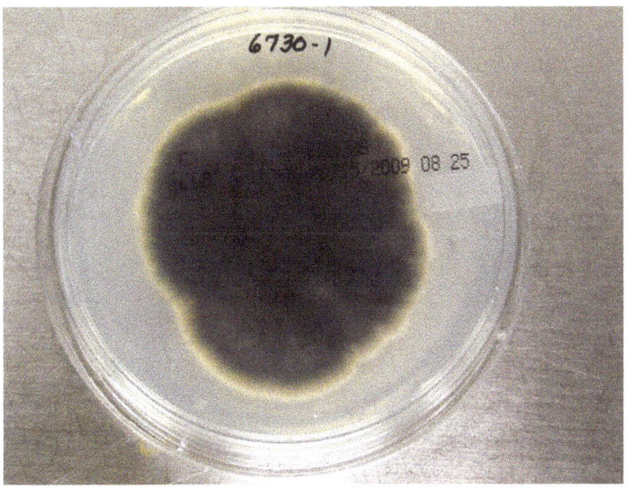

Fig. 9.10.3 *Cladosporium* Sab DX (bottom) at 2 weeks.

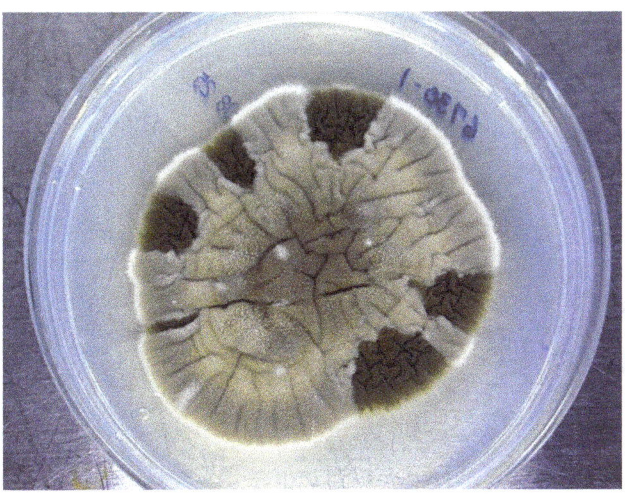

Fig. 9.10.4 *Cladosporium* sp Sab DX at 2 weeks.

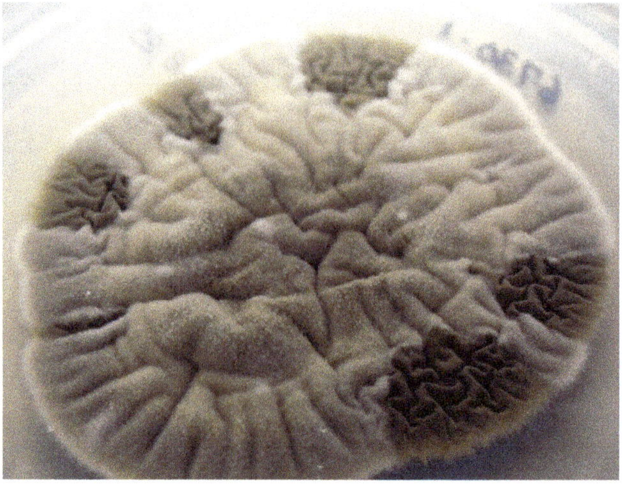

Fig. 9.10.5 *Cladosporium* sp Sab DX (side) at 2 weeks.

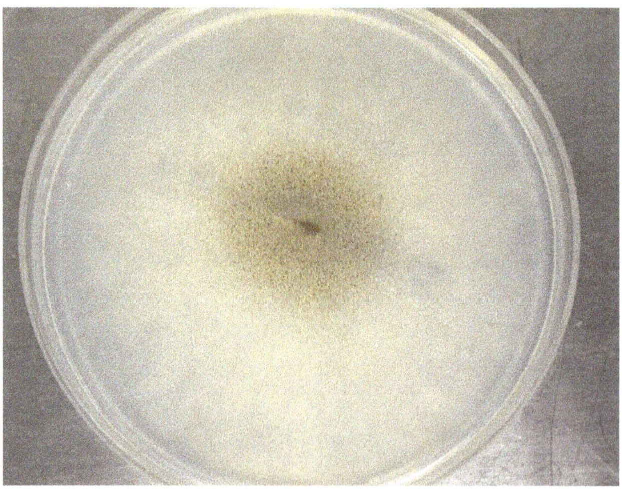

Fig. 9.10.6 *Cladosporium* sp potato DX.

Fig. 9.10.7 *Cladosporium* sp Sab DX (top) dark.

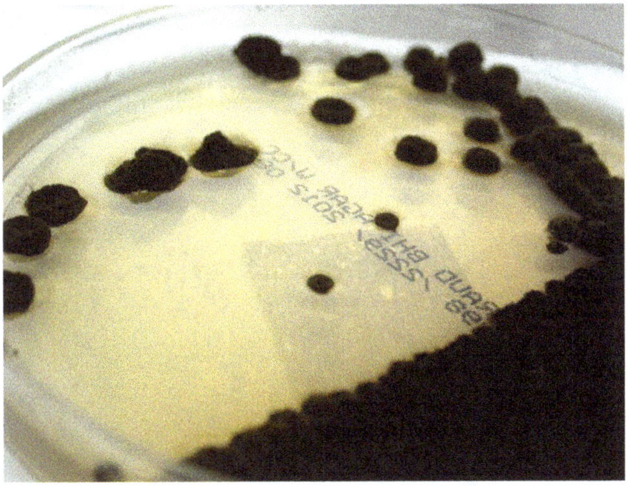

Fig. 9.10.8 *Cladosporium* sp BHI (side).

Fig. 9.10.9 *Cladosporium sp* BHI, (close up), showing agar pitting.

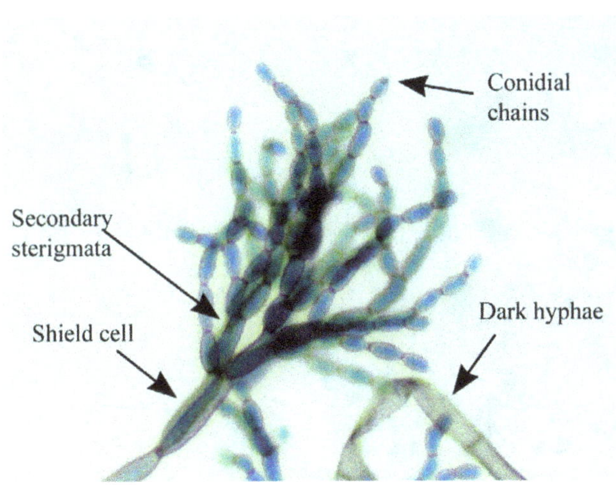

Fig. 9.10.10 *Cladosporium sp*, basic morphology.

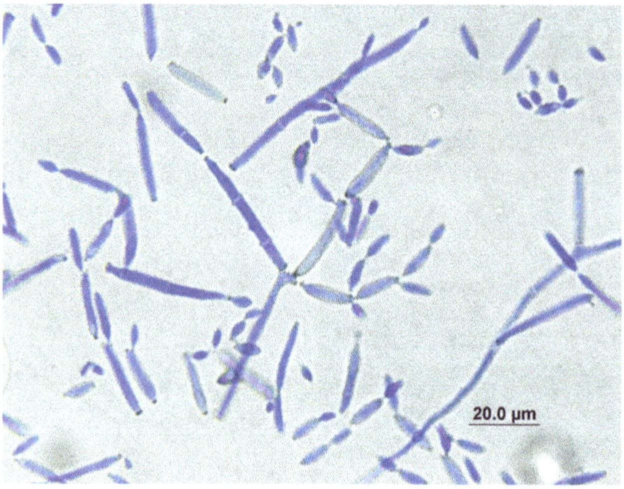

Fig. 9.10.11 Broken cells often seen.

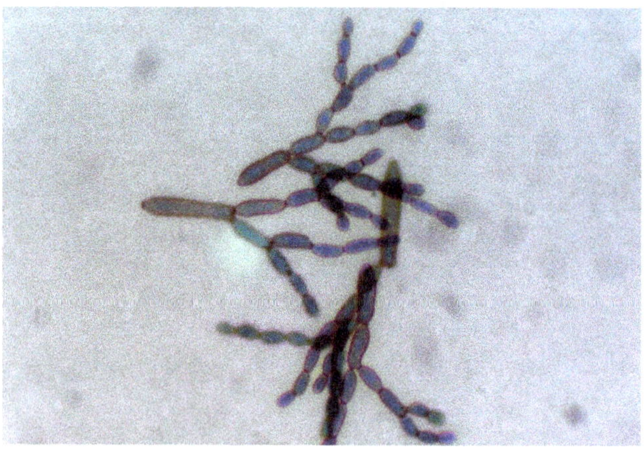

Fig. 9.10.12 *Cladosporium sp*, multiple branching.

Fig. 9.10.13 Profuse branching.

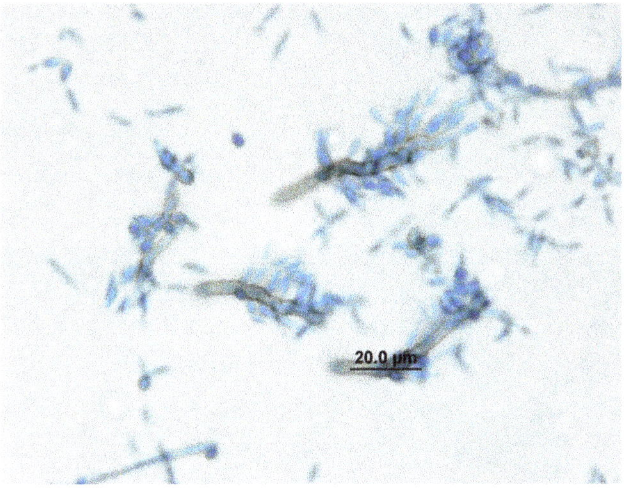

Fig. 9.10.14 Sympodial production of conidia.

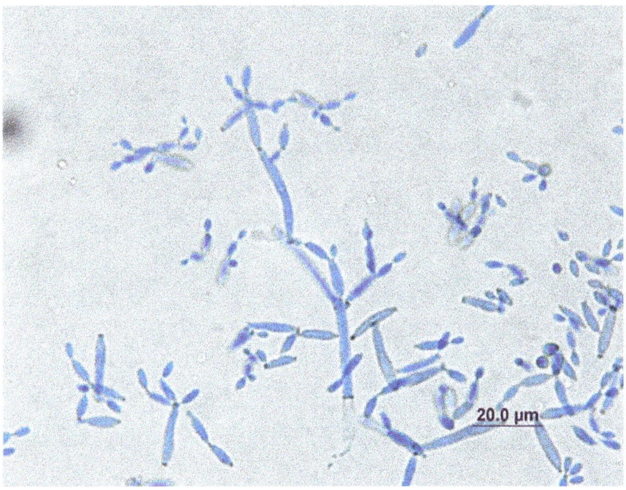

Fig. 9.10.15 *Cladosporium sp*, broken up cell morphology.

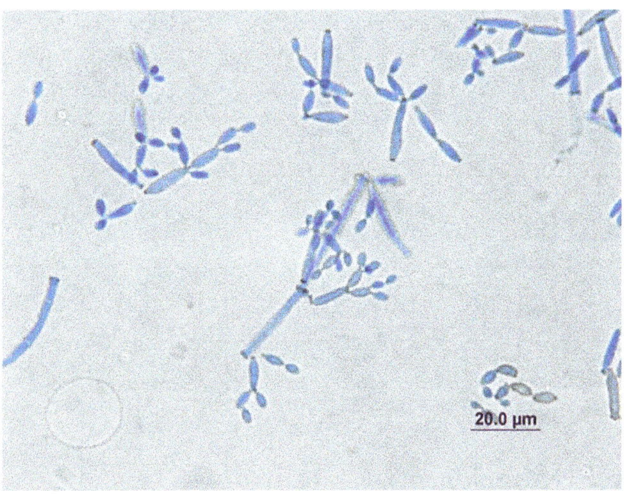

Fig. 9.10.16 *Cladosporium sp*, broken up cell morphology.

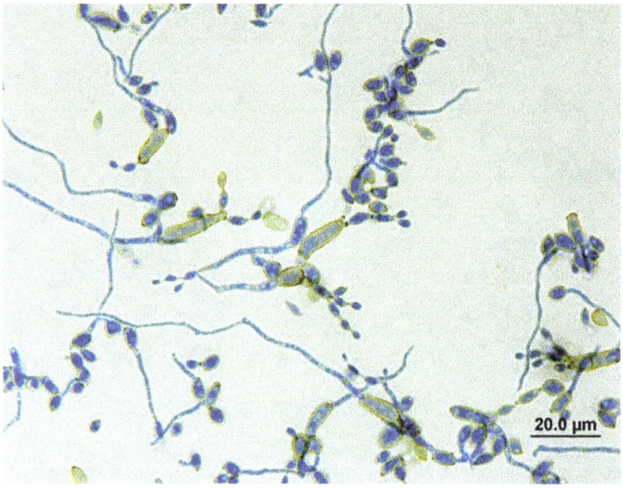

Fig. 9.10.17 Unusual forms.

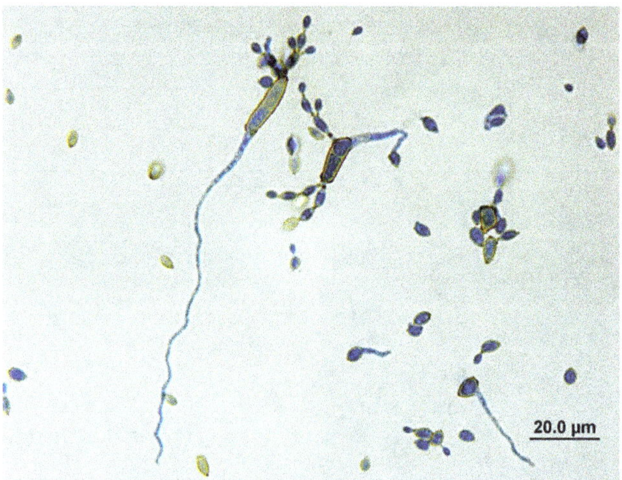

Fig. 9.10.18 Unusual forms.

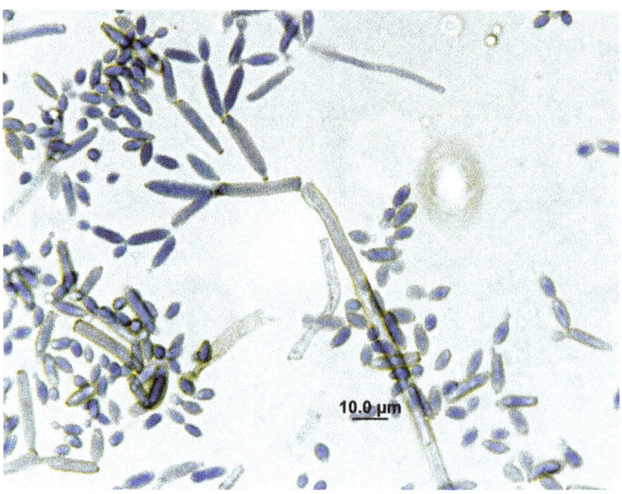

Fig. 9.10.19 *Cladosporium sp*, broken up cell morphology.

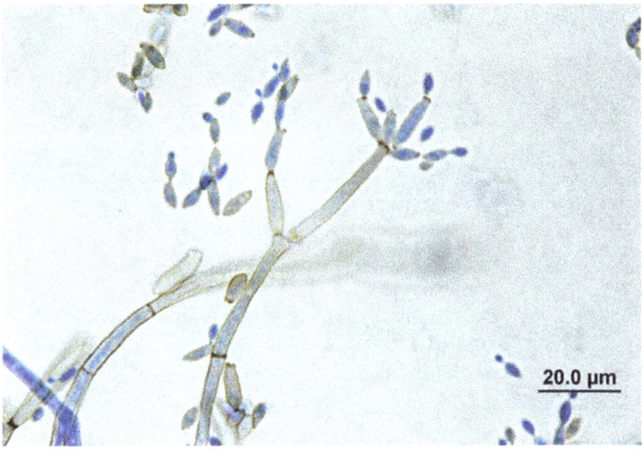

Fig. 9.10.20 *Cladosporium sp*, typical morphology.

9.11 *Curvularia* spp

Colonial morphology: on Sabouraud dextrose agar, colonies are brown to black, with a black reverse. On potato dextrose, colonies first appear as dark green to brown and later become dark brown with a green reverse. The surface of colonies is wrinkled, suede-like to downy.

Microscopic morphology: hyphae are septate and dark. Conidiophores are produced perpendicular to hyphae. They develop dark walls and are highly septate. Conidiophores curve at the end, in a shape resembling a foot. Conidia are produced from the heel of the foot. Some conidiophores are sympodially branched with conidia produced at each bend of the conidiophore. Multiple conidia (1–8) can be produced from one conidiophore. Conidia are characteristically bent or curved, dark, and usually contain four cells. One of the four cells is enlarged, causing the curvature of the conidia. Conidia become dark brown with age. They germinate bipolar, and some may have a prominent hilum at the terminus. Hyphae have swollen cells along the axis of the hyphae, looking much like clubs linked end to end. This can easily be confused with *Alternaria*, but the club-like hyphae do not have septations into multiple cells like *Alternaria*.

Curvularia

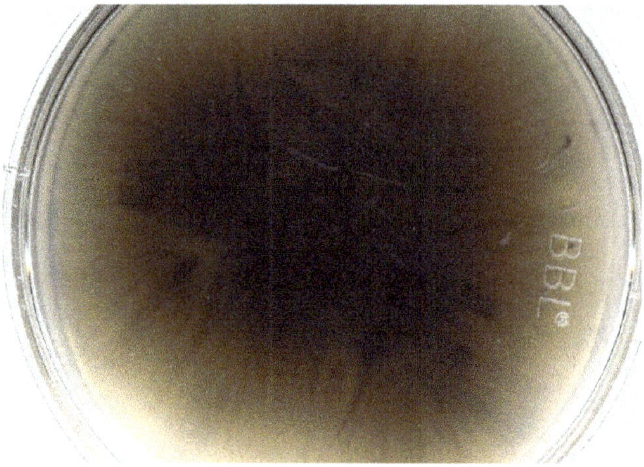

Fig. 9.11.1 *Curvularia* corn meal agar (top).

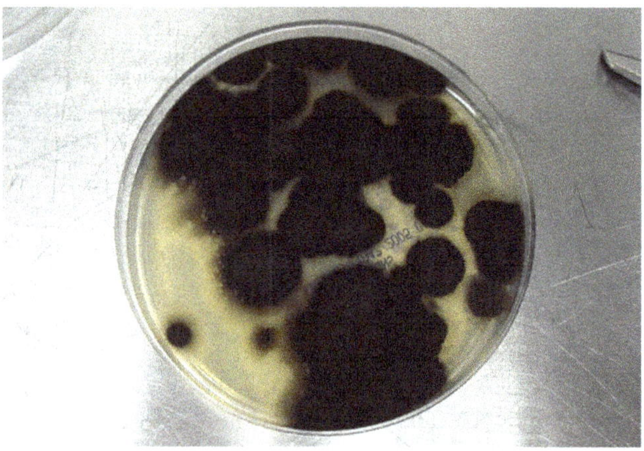

Fig. 9.11.2 Sab DX.

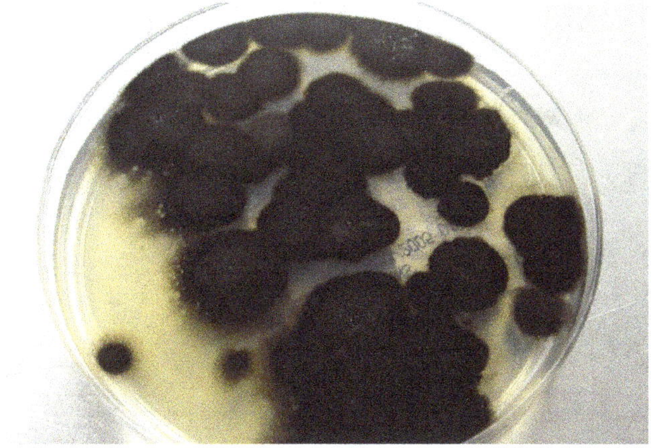

Fig. 9.11.3 Sab DX (side).

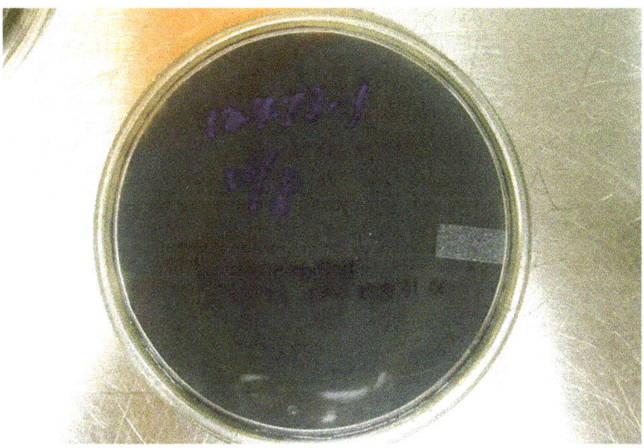

Fig. 9.11.4 Potato DX (bottom).

Fig. 9.11.5 Potato DX.

Fig. 9.11.6 *Curvularia*, curved conidiophore.

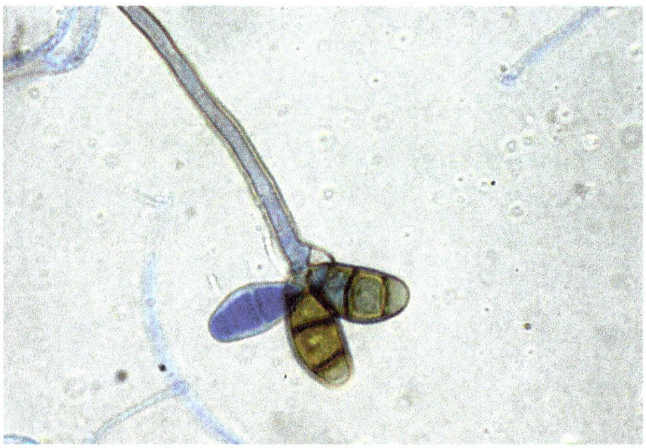

Fig. 9.11.7 *Curvularia*, conidia.

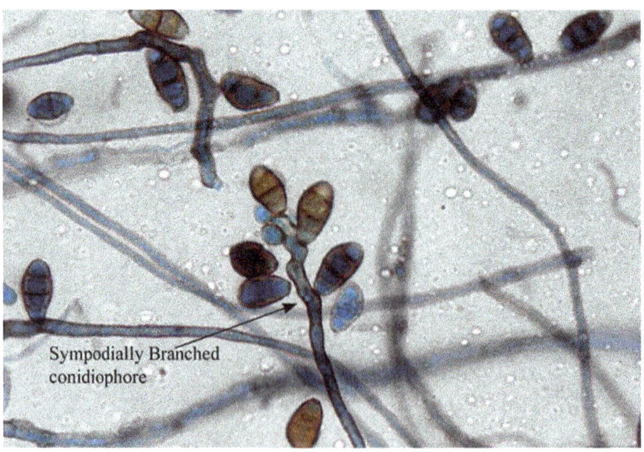

Fig. 9.11.8 *Curvularia*, typical morphology.

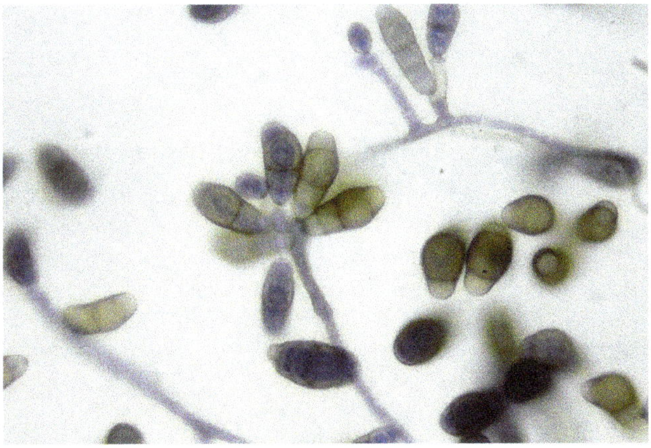

Fig. 9.11.9 *Curvularia*, varying stages of conidia development.

Fig. 9.11.10 *Curvularia*, varying stages of conidia development.

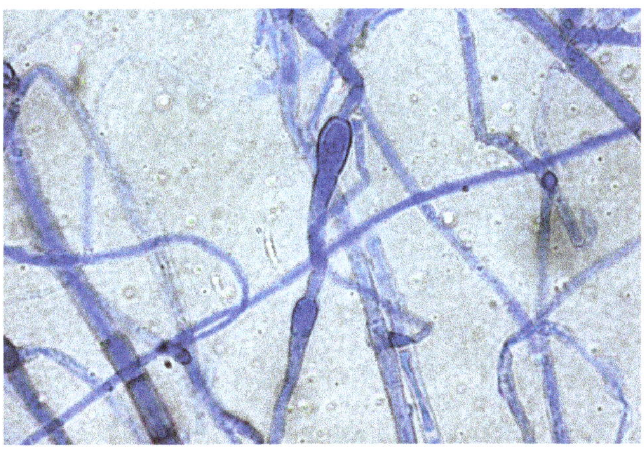

Fig. 9.11.11 *Curvularia*, club-like hyphae resemble *Alternaria*.

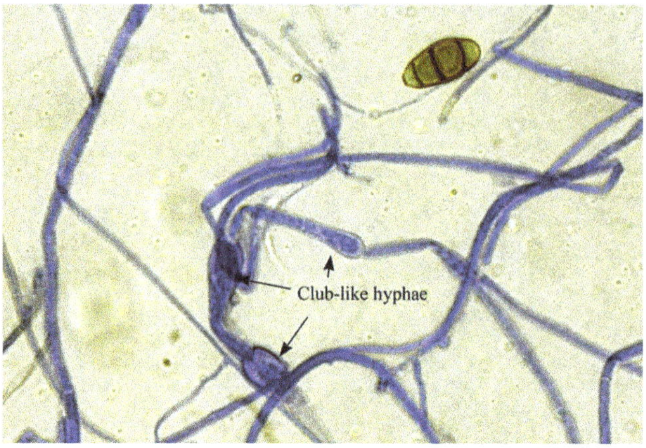

Fig. 9.11.12 *Curvularia*, club-like hyphae resemble *Alternaria*.

9.12 *Epicoccum* spp

Colonial morphology: colonies grow rapidly, mature within 7 days, are suede-like to downy, with a strong yellow to orange-brown diffusible pigmentation. Colony reverse is brown to black. Colonies grow more slowly on BHI blood agar and are white to cream, wrinkled and indent the agar.

Microscopic morphology: hyphae are septate, hyaline to brown. Dark sporodochia (groups of conidiophores) develop prior to sporulation. Once they germinate, sporodochia on the surface of the colony become visible to the naked eye as dark spots on the surface. Dark brown conidia are produced on short conidiophores arising singly or in clusters from the hyphae. Conidia are round to pear shaped, mostly 15–25 μm diameter with a conical base and broad attachment scar. Conidia are single, dark brown, and multicellular, with a rough, bumpy, verrucose surface, cross-striated with transverse and longitudinal septations. Conidia can be seen at all stages of development along the hyphae. They can be produced either singly or in clusters. Chlamydospores (1–3-celled) may be present. Sometimes brush-like or finger-like terminal hyphal elements can be seen. Cultures may be sterile for several days, then overnight seem to germinate their conidia in abundance. Sometimes *Epicoccum* are difficult to differentiate from *Pithomyces*. The mature conidia of *Epicoccum* are more rounded as opposed to elongate with *Pithomyces*. A key feature is that young conidia of *Pithomyces* are elongate and echinulate prior to developing septation.

Epicoccum

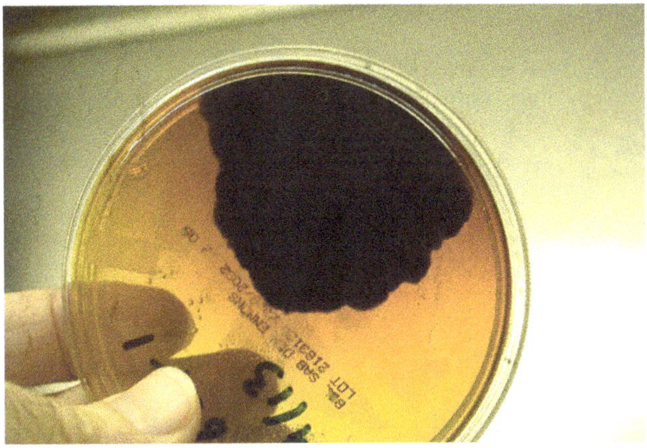

Fig. 9.12.1 *Epicoccum* (bottom).

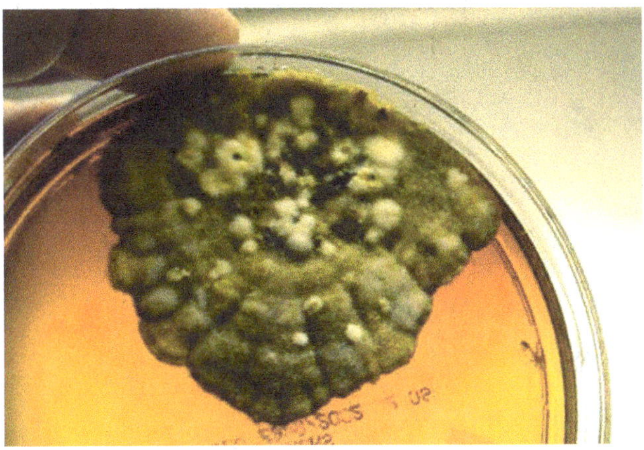

Fig. 9.12.2 *Epicoccum* (top).

Fig. 9.12.3 *Epicocccum* BHI blood at 5 weeks.

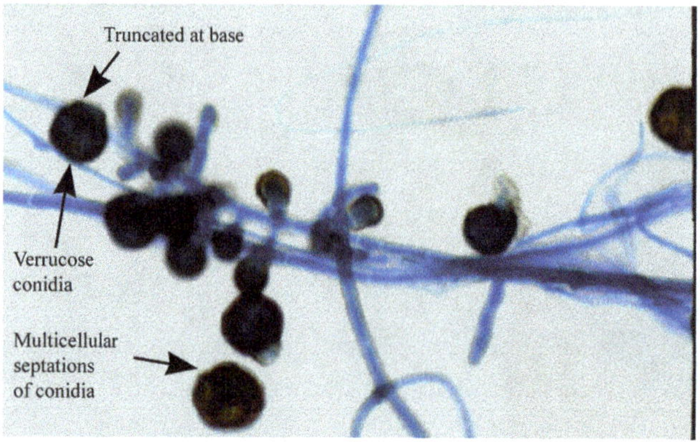

Fig. 9.12.4 *Epicoccum*, general morphology.

Fig. 9.12.5 Conidia with longitudinal and transverse septations.

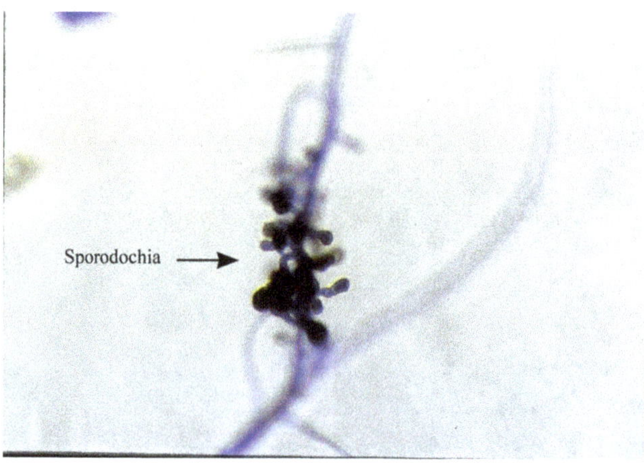

Fig. 9.12.6 Sporodochia with developing conidia.

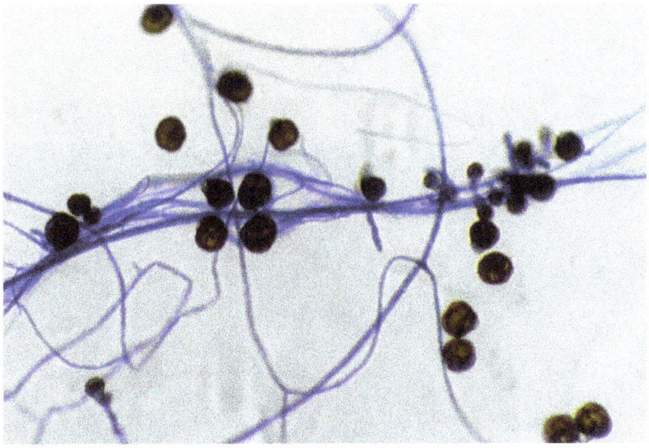

Fig. 9.12.7 *Epicoccum*, developing conidia along hyphae.

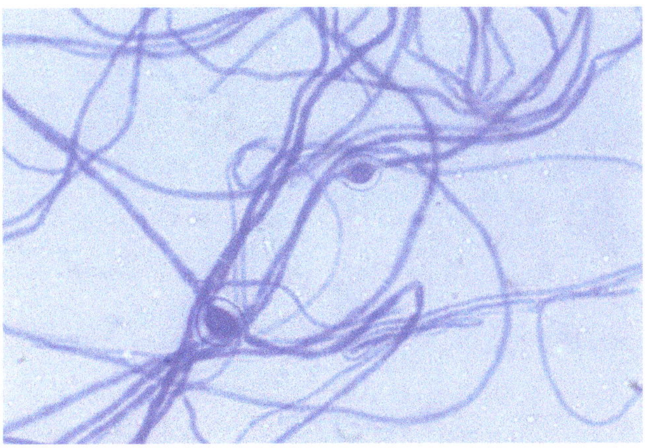

Fig. 9.12.8 Single-cell chlamydospores.

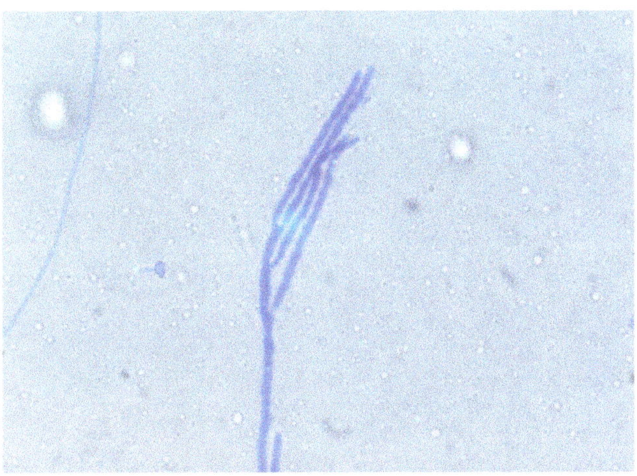

Fig. 9.12.9 Brush-like hyphae.

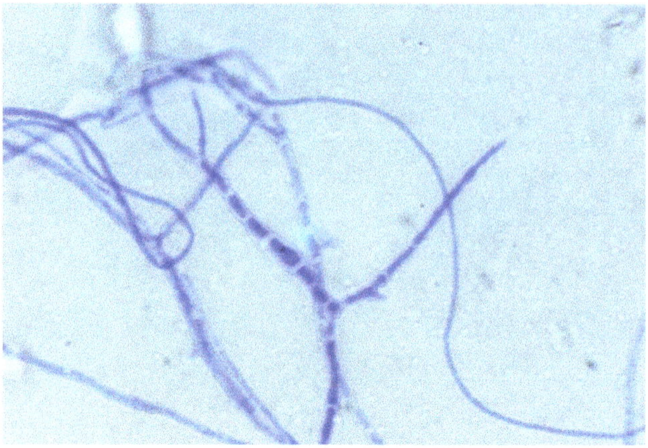

Fig. 9.12.10 Coarse irregular hyphae.

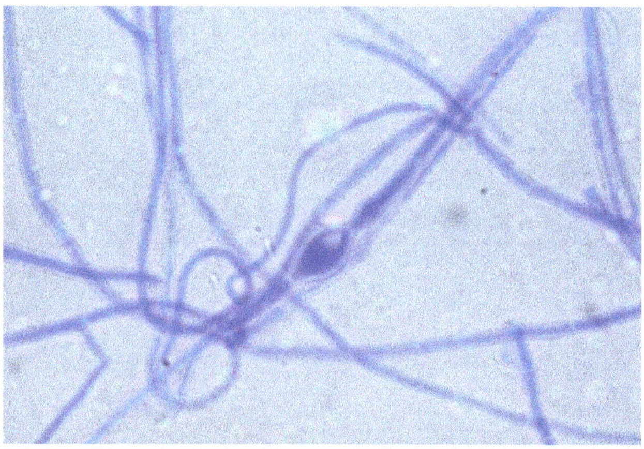

Fig. 9.12.11 Chlamydospore formation.

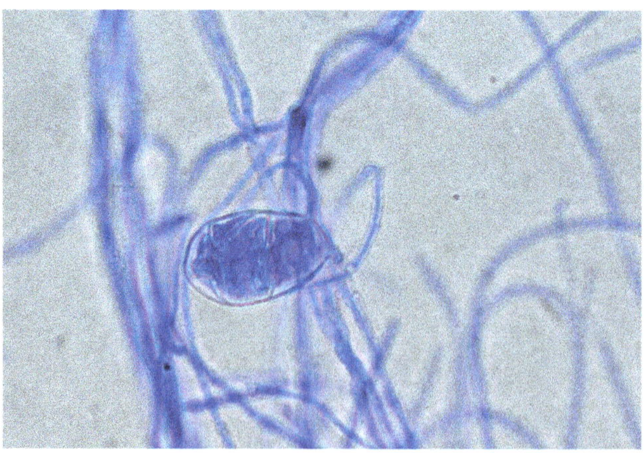

Fig. 9.12.12 Three-cell chlamydospore.

9.13 *Exserohilum* spp

Colonial morphology: colonies are velvety, white to gray (cream reverse) on Sabouraud dextrose agar and brown to black (black reverse) on potato dextrose agar.

Microscopic morphology: hyphae are dark and septate. They may be ribbon-like or firm with a diameter of 3–6 µm. Conidiophores show sympodial, geniculate (bent) conidium production. The conidia are brown and rounded at the apex, averaging 15–80 µm in size. They are multicellular (4–9 divisions) and have a dark, protruding hilum at the base. Cells within the conidium are "plumb-bob" shaped. Terminal cells within the mature conidia have a dark, distinct septum at each end (Fig. 9.13.16). The number of divisions within conidia of *Exserohilum*, a poroconidium with protuberant hilum, and the darker septum at each end of the conidium distinguish *Exserohilum* from *Bipolaris* (3–6 divisions).

Exserohilum rostratum

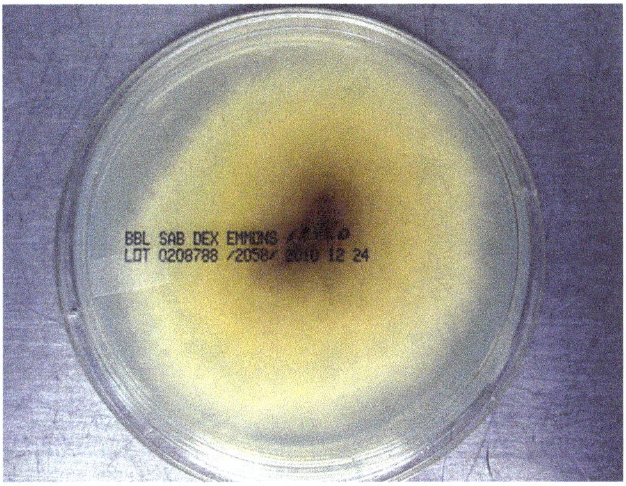

Fig. 9.13.1 *Exserohilum* on SABDX (bottom).

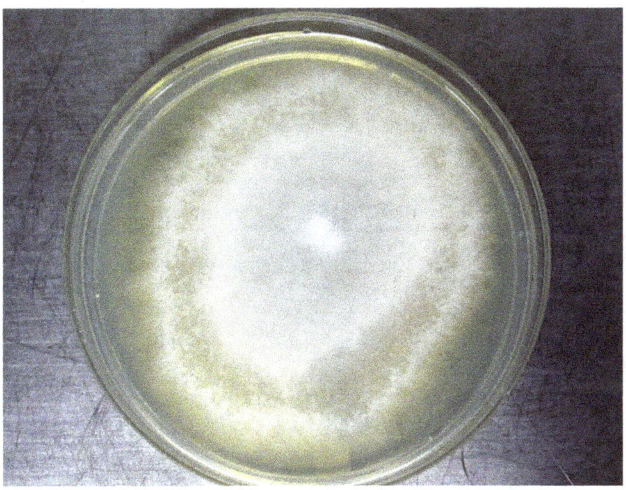

Fig. 9.13.2 *Exserohilum* on SABDX (top).

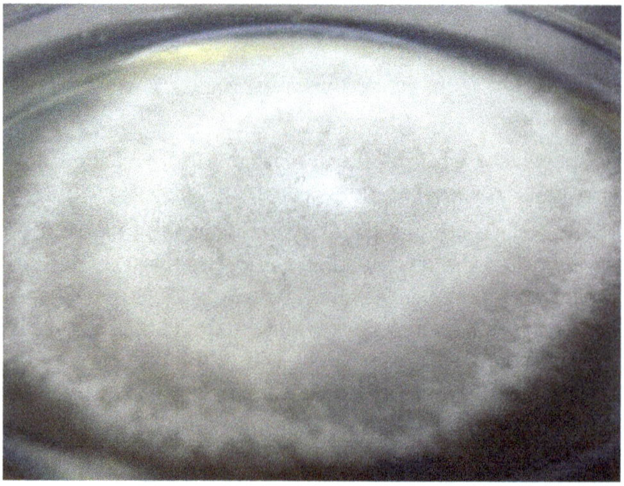

Fig. 9.13.3 *Exserohilum* on SABDX (side).

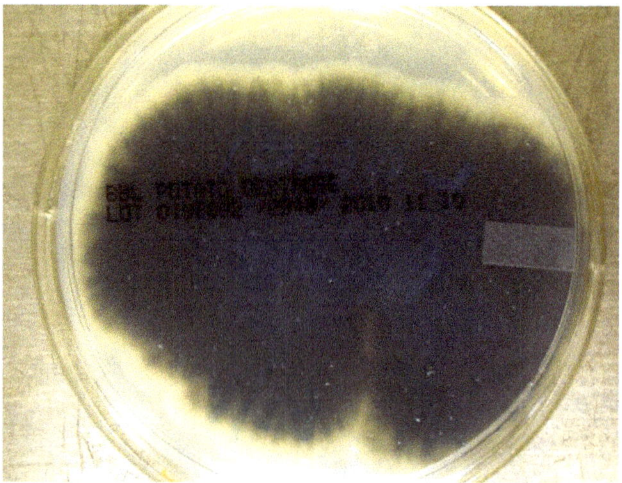

Fig. 9.13.4 *Exserohilum* PDA (bottom).

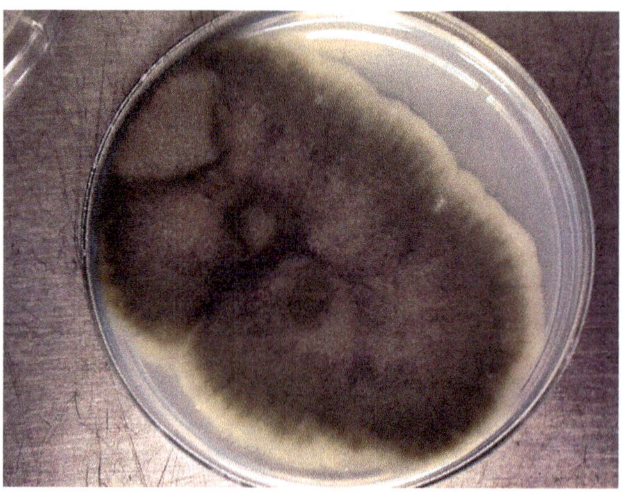

Fig. 9.13.5 *Exserohilum* PDA.

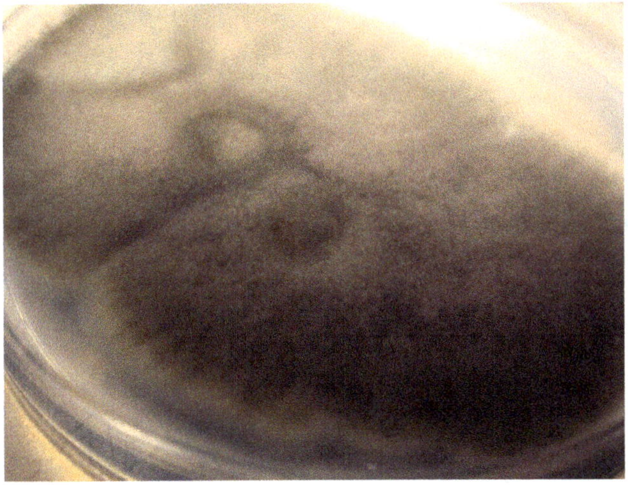

Fig. 9.13.6 *Exserohilum* PDA (side).

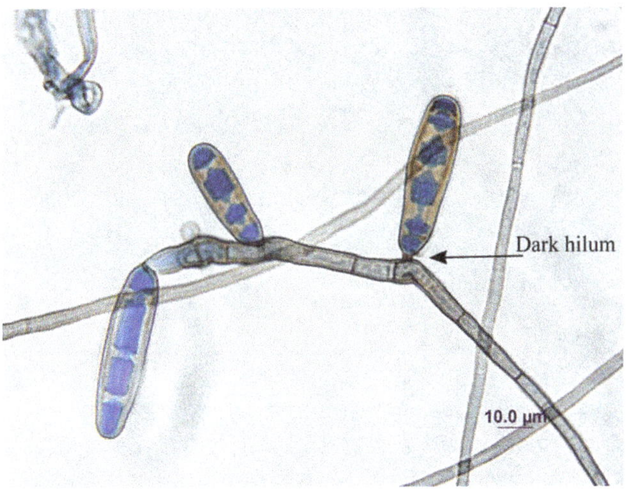

Fig. 9.13.7 Poroconidia produced along conidiophore.

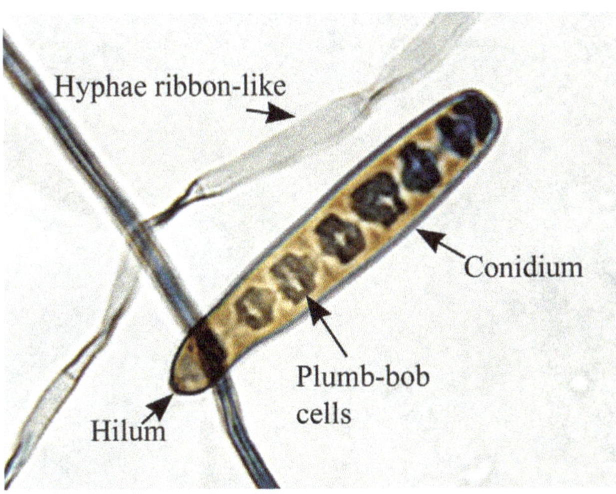

Fig. 9.13.8 Conidium morphology.

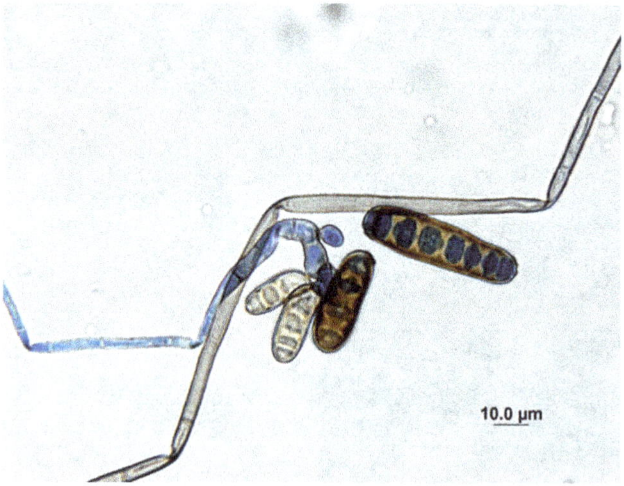

Fig. 9.13.9 *Exserohilum*, different stages of conidia development.

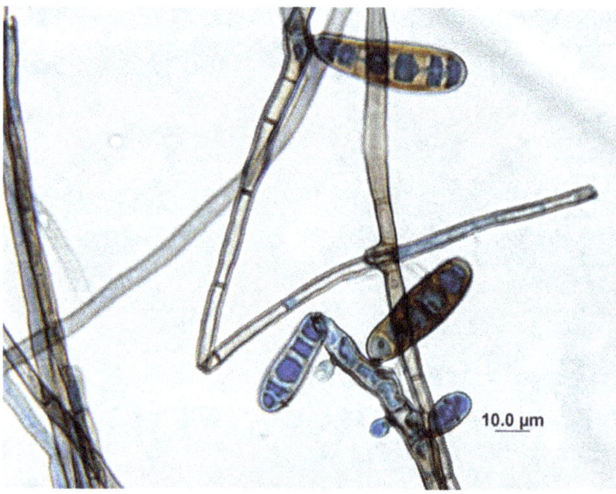

Fig. 9.13.10 *Exserohilum*, poroconidia.

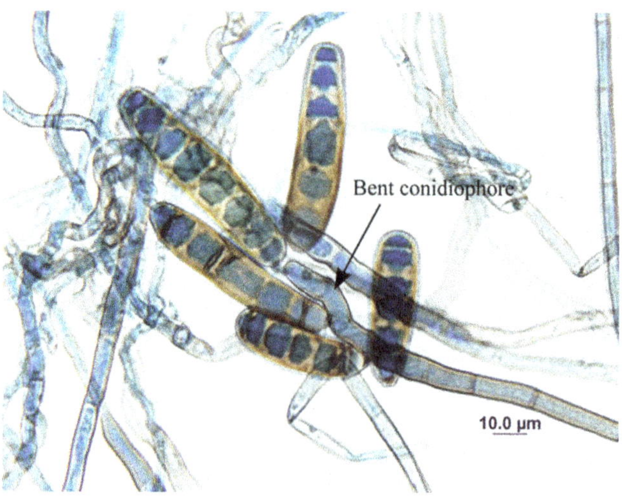

Fig. 9.13.11 Mature conidia with bent conidiophore.

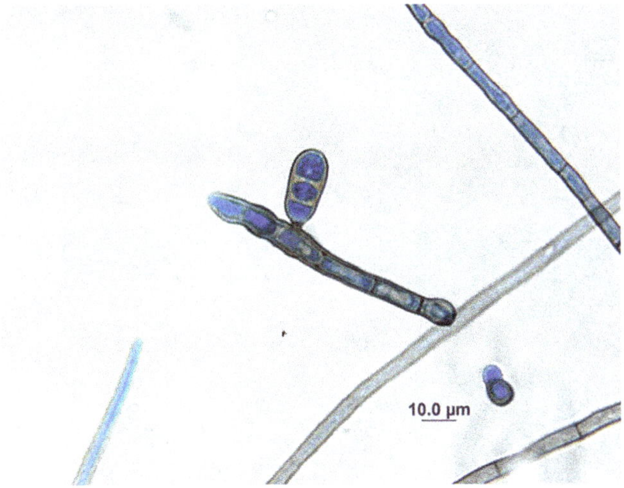

Fig. 9.13.12 *Exserohilum*, conidium produced from fertile hyphae.

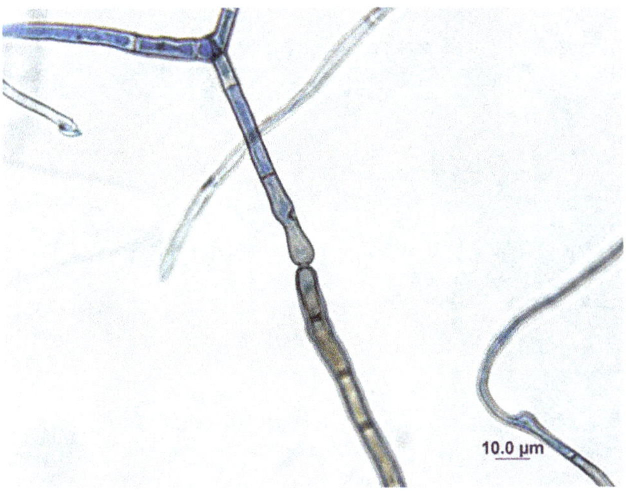

Fig. 9.13.13 Irregular hyphae.

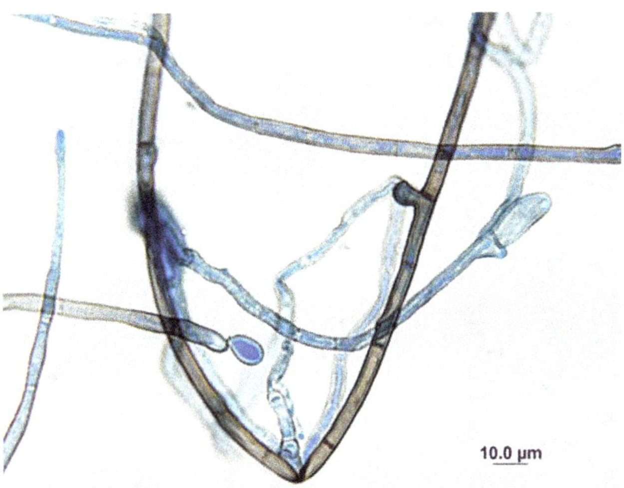

Fig. 9.13.14 Varying morphology of hyphae.

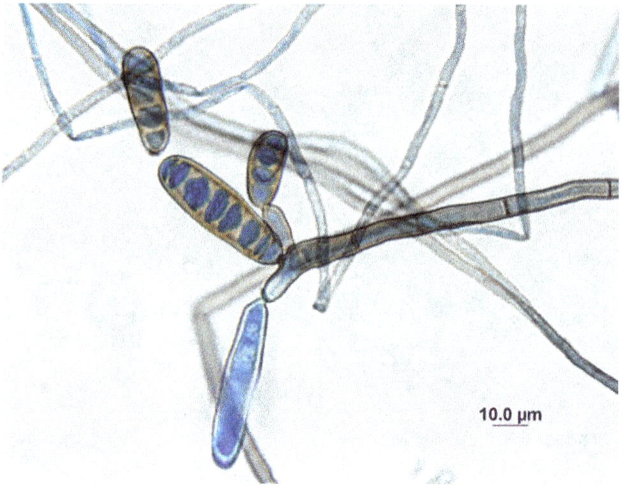

Fig. 9.13.15 *Exserohilum*, varying degrees of conidial septation.

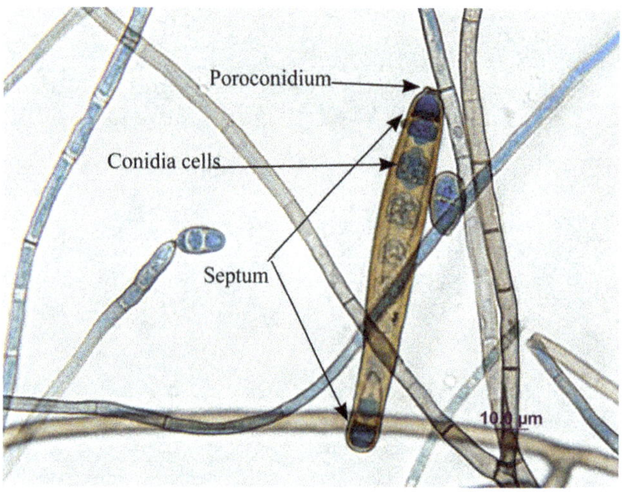

Fig. 9.13.16 Morphology of poroconidia.

9.14 *Fonsecaea*

Fonsecaea consists of two pathogenic species: *F. pedrosoi* and *F. compactum*. *Fonsecaea* species have various morphological appearances.

Colonial morphology: colonies of *F. pedrosoi* are slow growing, and grow best on potato dextrose agar > corn meal agar > Sabouraud dextrose agar. *F. compactum* grows well on Sabouraud dextrose. On potato dextrose, colonies develop entire, flat, dark green or dark brown to black and then become wrinkled and velvety. Colonies become embedded in the culture medium as they wrinkle. Reverse is black. On corn meal agar, colonies are olive to dark green, later developing a brown-black center. Colonies are radiate and fluffy on corn meal agar with aerial hyphae forming a silver-gray frost in the center of the colony.

Microscopic morphology: hyphae are septate, branched, hyaline to dark brown, showing dark brown septations. Conidiophore formation is discussed below.

There are four basic types of conidia formation in *Fonsecaea*.

- *Rhinocladiella* type (raduliform) (Fig. 9.14.11): primary conidia are produced by sympodial blastoconidial formation. The conidiophores develop from the hyphae either by branching or at the end of the hyphae. Conidiophores usually are composed of 2–3 cells arising from the hyphae. The conidiogenous zone is confined to the upper portion of the conidiophore. Secondary, 1–2-celled conidiophores develop sympodially along the primary conidiophore showing dark brown zig-zag septations. Single, oval conidia develop at the apex of the secondary conidiophores, and aggregate along the conidiophore.
- *Fonsecaea* type (Fig. 9.14.6): conidiophores are long and have multiple cells, which resemble long, septate mycelia. The conidiophores form multiple, swollen denticles at the termini. Denticles develop single-cell conidia at the termini. At each point of septation along the conidiophores, denticles also develop, which then produce conidia. In addition, prolific sites may develop along the conidiophores and make several elongate conidia, resembling asterisk-like projections (Fig. 9.14.14).
- *Cladosporium* type (Fig. 9.14.10): conidiophores are dark, long, erect and terminate in primary, shield cell conidia. Secondary conidia develop in short branching chains of lemon-shaped conidia. In some cases and often with *F. compactum*, tertiary conidia develop from secondary shield-type conidia. These fruiting heads are very delicate and regularly fragment in lacto-phenol cotton blue touch preparations. It is rare to capture an intact fruiting head by direct Scotch tape preparation. Alternatively, slide cultures are best to reveal this type of conidiation. *Fonsecaea* are often mistaken for *Cladosporium* because of this tendency to fragment, but *Cladosporium* usually does not fragment as easily and some intact structures can be identified. Sometimes, single phialides may be seen which can be confused with *Phialophora* (Fig. 9.14.13). It is important to review multiple preparations at different daily intervals to avoid misidentification.
- *Phialophora* type (Figs 9.14.12, 9.14.13): phialides are produced directly from mycelia. Phialides form single conidia at the apices; often, a cup-like collarette can be seen. These are rare and conidiation is sparse.

9.14.1 Differences between species

Cultural and microscopic appearances of both species are similar. The distinguishing characteristic of *F. compactum* is that when *Fonsecaea* conidiation is present (predominant form), short chains of round conidia form (versus oval, lemon-shaped conidia of *F. pedrosoi*), with wide septa occurring between conidia giving them the appearance of "strands of pearls." The fruiting heads appear more "compact" than with *F. pedrosoi*. In tissue, they have a characteristic appearance of dark, brown, oval-round septate mycelia (Fig. 9.14.11).

Fonsecaea

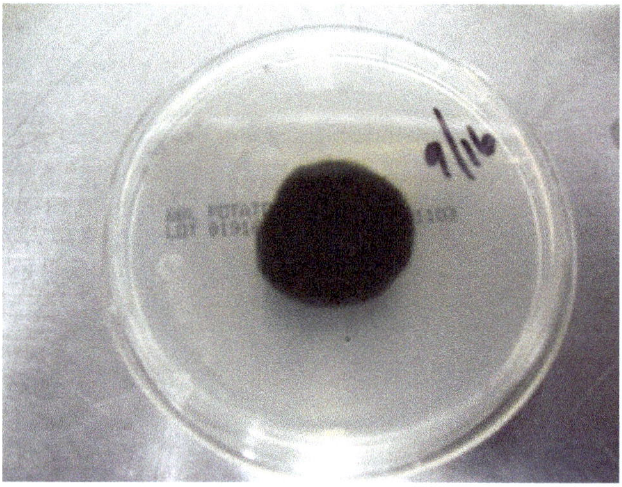

Fig. 9.14.1 *Fonsecaea pedrosoi*, PDA (bottom).

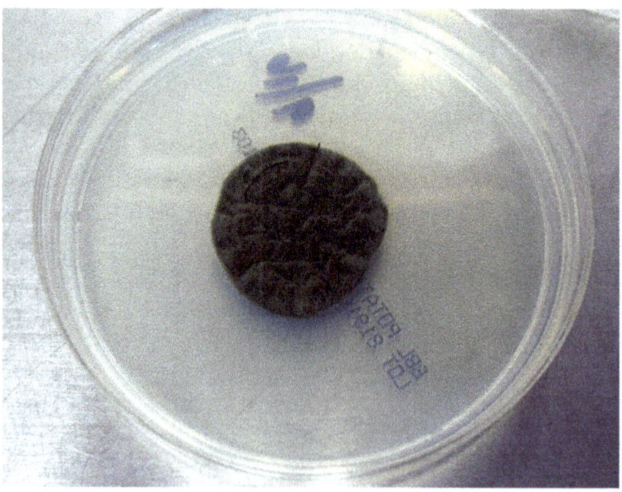

Fig. 9.14.2 *F. pedrosoi*, PDA (top).

Fig. 9.14.3 *F. pedrosoi*, PDA (side).

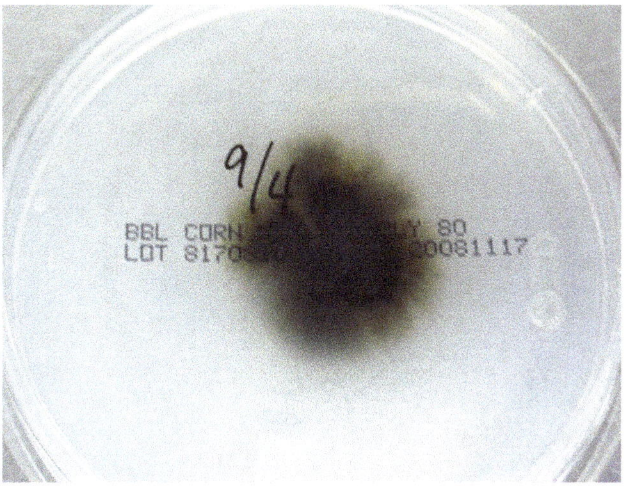

Fig. 9.14.4 *F. pedrosoi*, CMA (bottom).

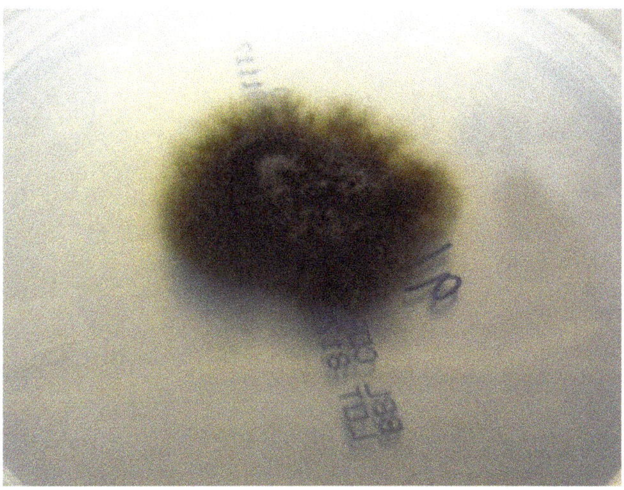

Fig. 9.14.5 *F. pedrosoi*, CMA (side).

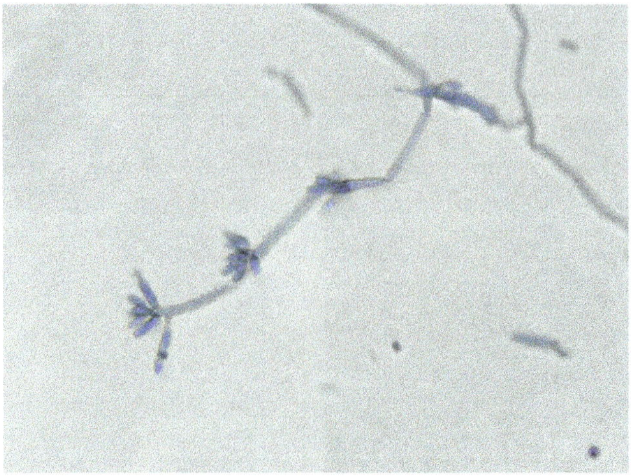

Fig. 9.14.6 *F. pedrosoi*, *Fonsecaea* type conidiation.

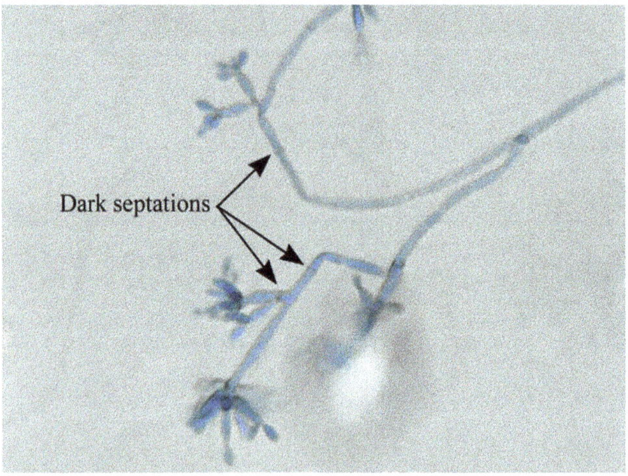

Fig. 9.14.7 *F. pedrosoi*.

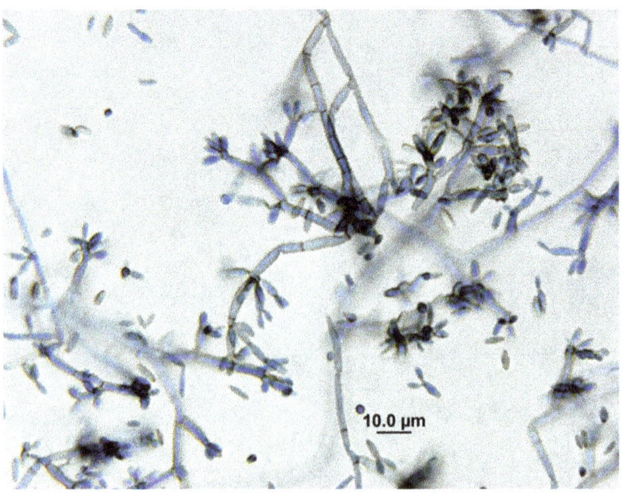

Fig. 9.14.8 *F. pedrosoi*.

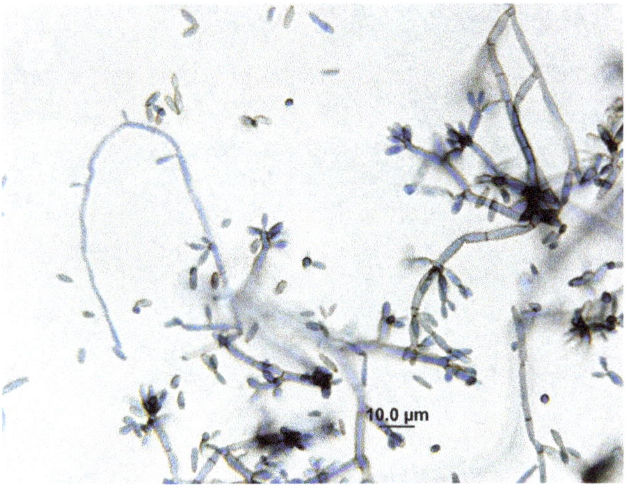

Fig. 9.14.9 *F. pedrosoi*.

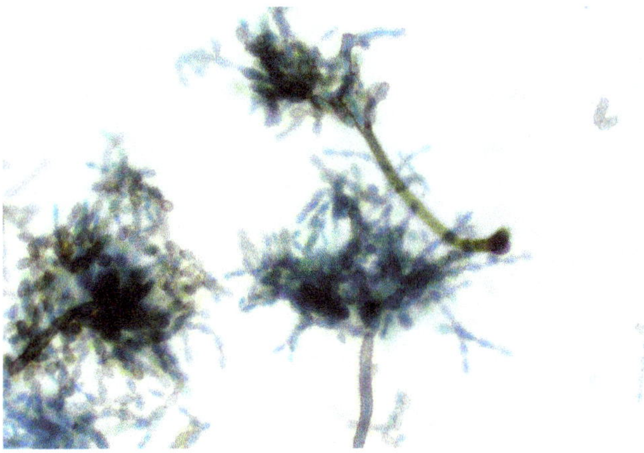

Fig. 9.14.10 *F. pedrosoi*, *Cladosporium* type conidiation.

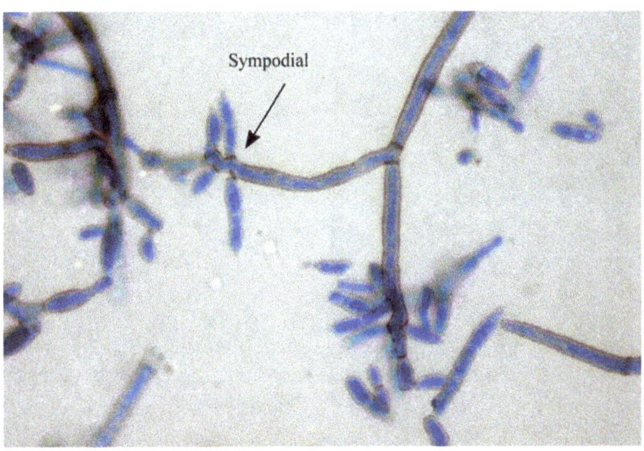

Fig. 9.14.11 *F. pedrosoi*, *Rhinocladiella* type conidiation.

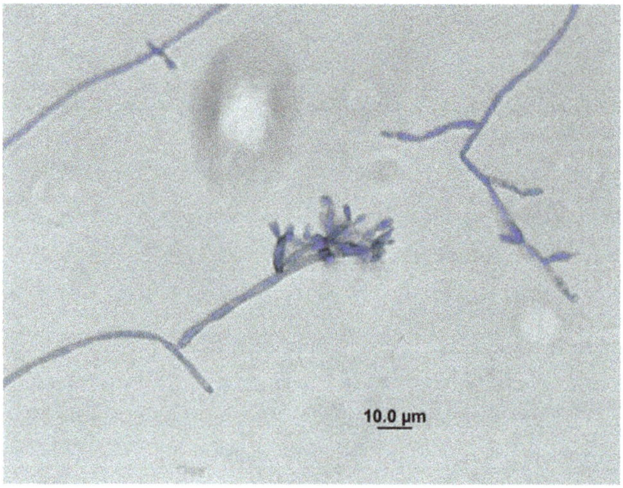

Fig. 9.14.12 *F. pedrosoi.*

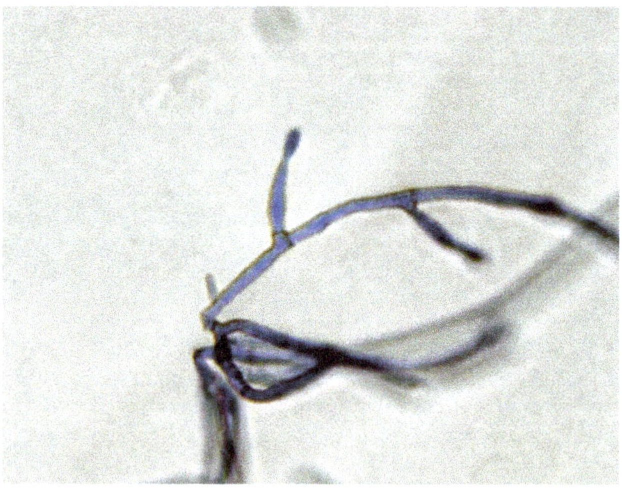

Fig. 9.14.13 *F. pedrosoi*, single phialides.

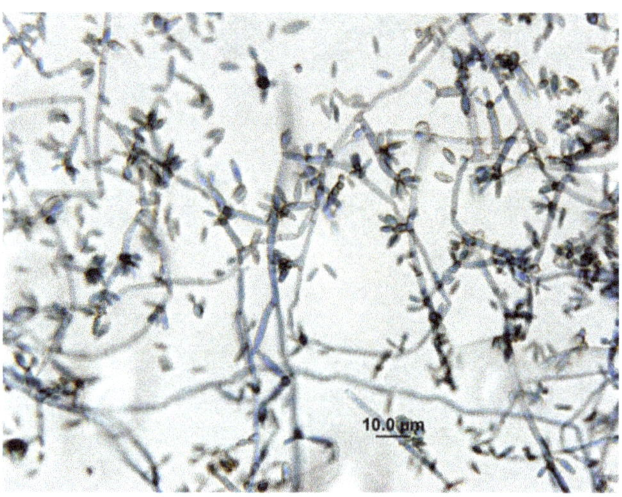

Fig. 9.14.14 *F. pedrosoi*, asterisk-like projections.

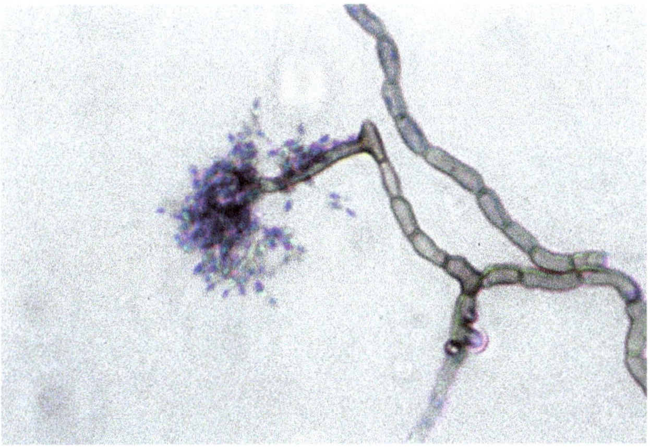

Fig. 9.14.15 *F. compactum*.

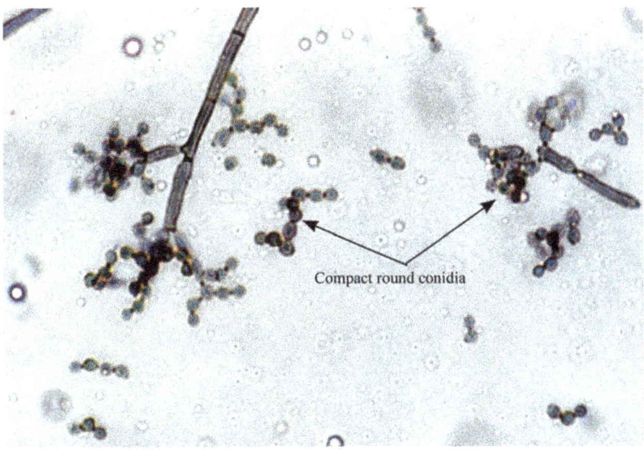

Fig. 9.14.16 *F. compactum*.

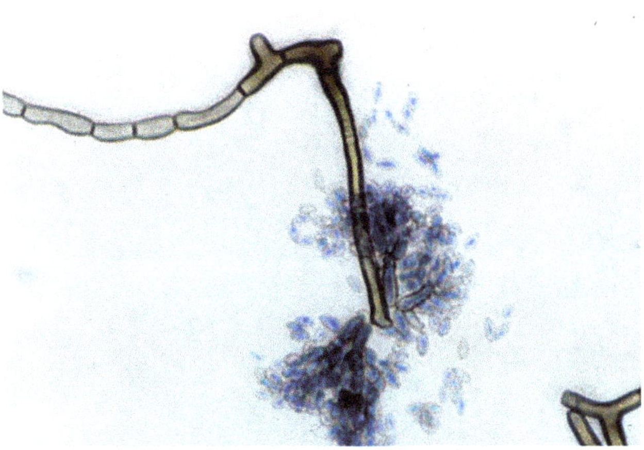

Fig. 9.14.17 *F. compactum*.

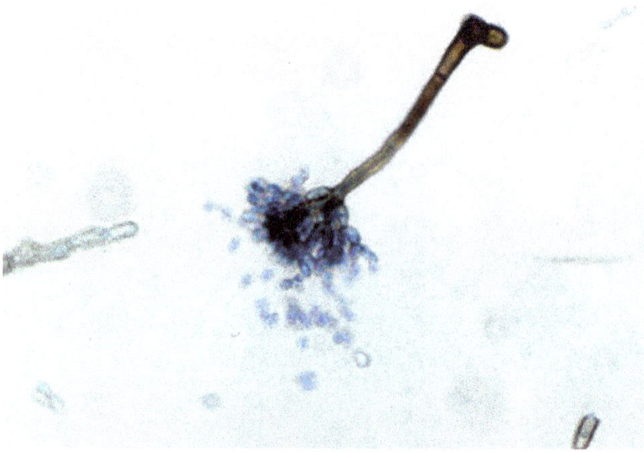

Fig. 9.14.18 *F. compactum*.

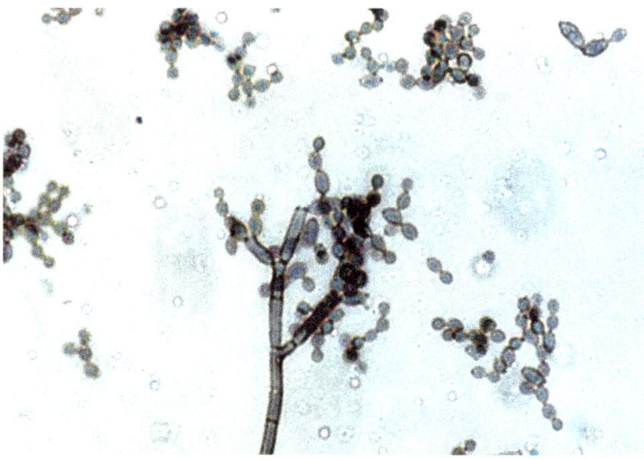

Fig. 9.14.19 *F. compactum*.

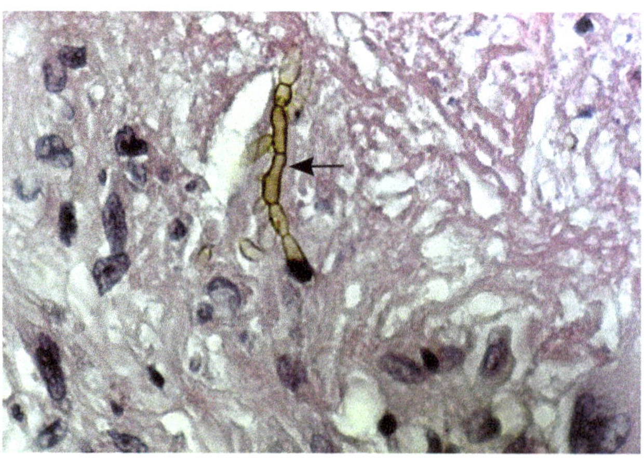

Fig. 9.14.20 *F. compactum* in tissue.

9.15 *Hormonema dematioides*

Colonial morphology: white to cream colored colonies appear rapidly within 3–5 days. With age, colonies darken and become black-gray in the center surrounded by a white border. Colony texture is wooly or chalk-like. Although colonies mature rapidly, several weeks may pass before fruiting structures appear.

Microscopic morphology: hyphae are hyaline and dematiaceous, similar to *Scytalidium dimidiatum* and *Aureobasidium pullulans.* Blastoconidia are produced directly from hyaline hyphae and darker arthroconidia. Blastoconidia are produced asynchronously along the hyphae, meaning that conidia are produced in succession, individually, from a single prolific point, at many points along the hyphae. Conidia will continue to bud once they are released. In *Aureobasidium*, blastoconidia are produced synchronously, each from a single prolific point (sometimes clustered) so that clusters of blastoconidia can be seen along the hyphae. In *Scytalidium* no conidiophores or blastoconidia are produced, only arthroconidia. Young cultures of *Hormonema* produce yeast-like cells that are single celled, round and stain poorly Gram negative.

Hormonema

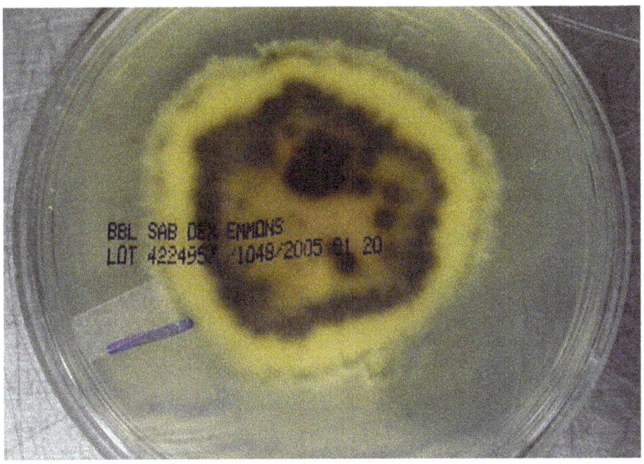

Fig. 9.15.1 *Hormonema* (bottom).

Fig. 9.15.2 *Hormonema* (top).

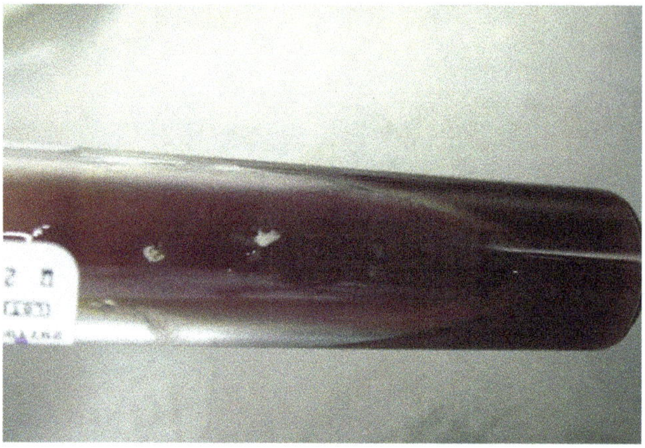

Fig. 9.15.3 *Hormonema* yeast.

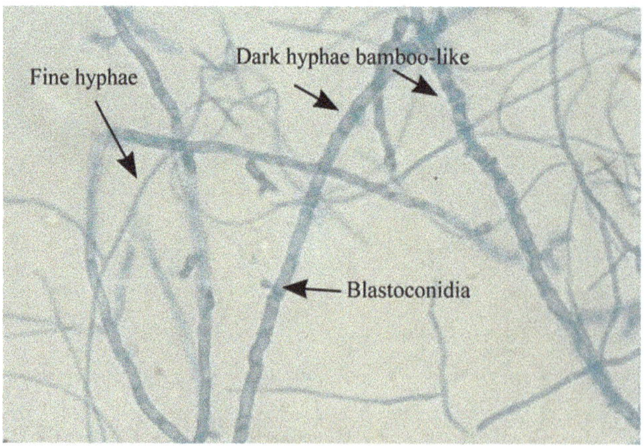

Fig. 9.15.4 Varying morphology of hyphae and blastoconidia formation.

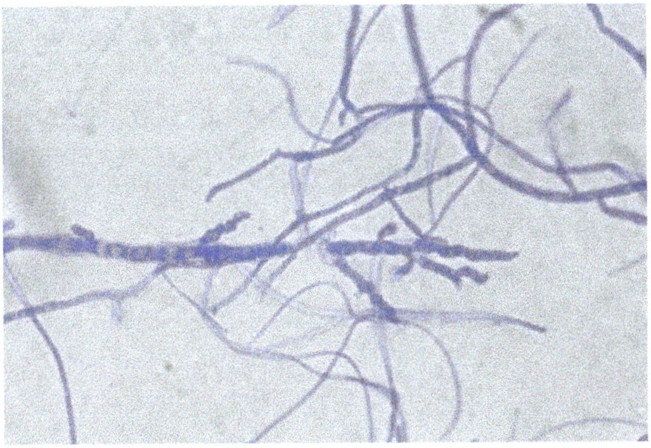

Fig. 9.15.5 *Hormonema*, broad irregular hyphae.

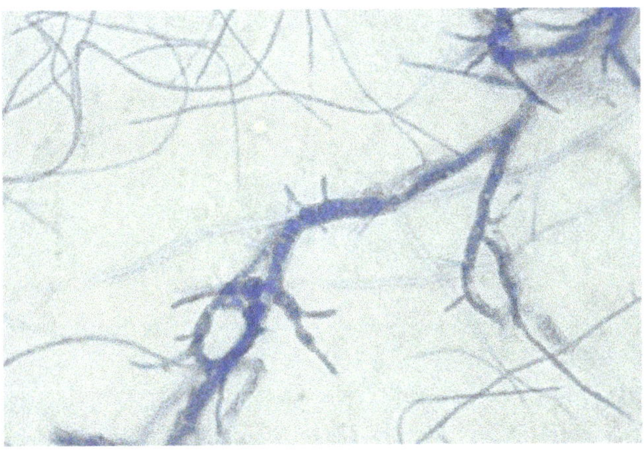

Fig. 9.15.6 Broad irregular hyphae.

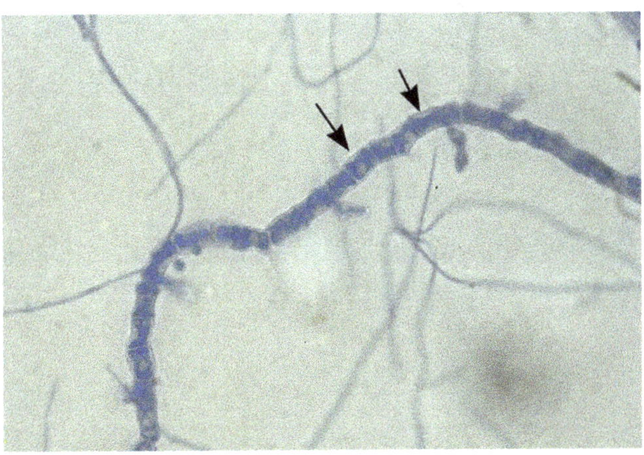

Fig. 9.15.7 Arthroconidia.

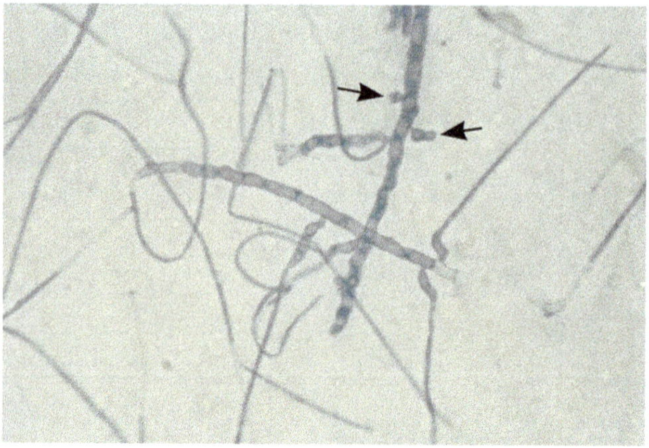

Fig. 9.15.8 Showing blastoconidia production.

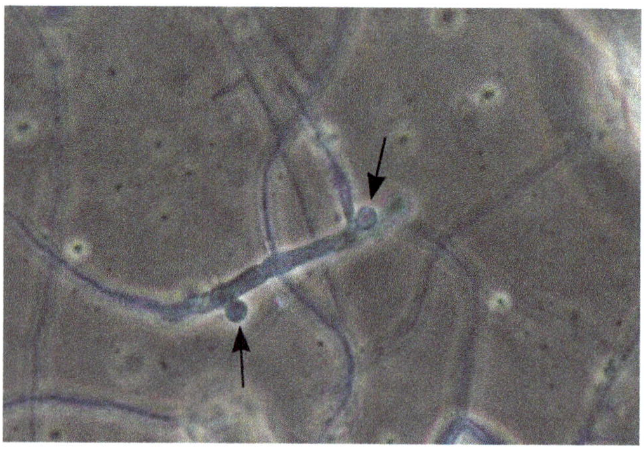

Fig. 9.15.9 Phase contrast showing blastoconidia.

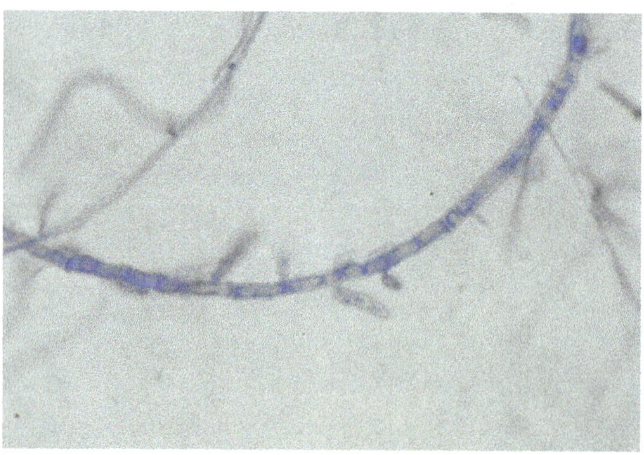

Fig. 9.15.10 Arthroconidia formation.

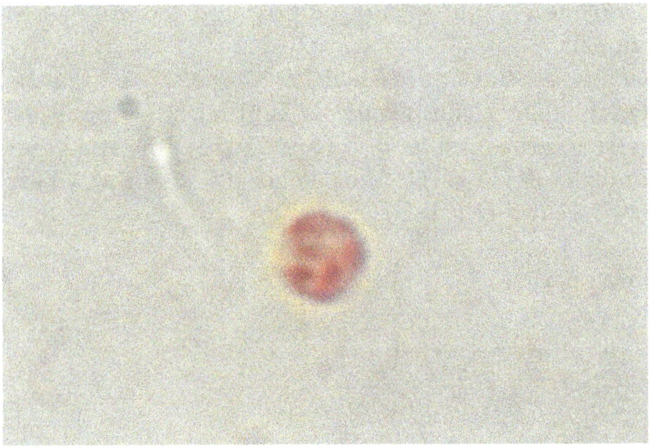

Fig. 9.15.11 Gram stain of yeast cell.

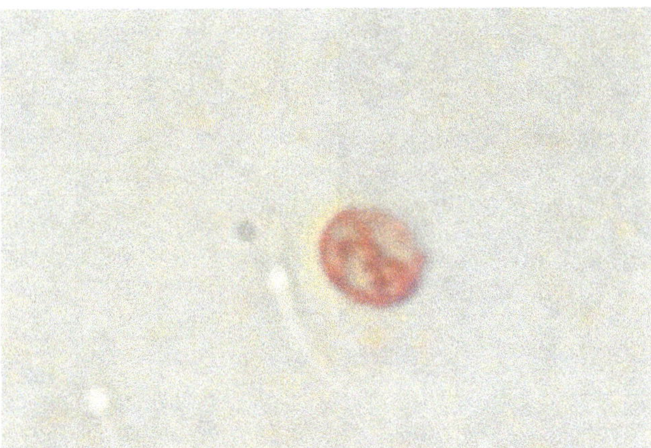

Fig. 9.15.12 Gram stain of yeast cell.

9.16 *Hortaea (Phaeoannellomyces, Exophiala) werneckii*

Colonial morphology: colonies appear yeast-like, smooth, and slow growing. They may be initially white or black but then become dark olive or black in older cultures. Aerial mycelia develop in mature cultures.

Microscopic morphology: this is a polymorphic fungus beginning yeast-like and becoming filamentous. At first, dark hyphae are mixed with biclavate yeast-like cells (conidia) that are dark and extensively occur in groups. The yeast-like biclavate (two-celled) conidia are often tapered at one end and possibly annellide. They contain a dark septum. Mature cultures develop into filamentous hyphae to complete the anamorph life cycle, developing broad hyphae on which swollen phialides produce conidia at the annellide tip. They are abundant and often confused with *Cladosporium*. Some conidia may be produced directly from phialides on dark, extensively septate hyphae. Other structures such as favic chandeliers and chlamydospores may be seen. Chlamydospores may be round and produced on the ends of hyphae or produced intercalary. No teleomorph stage has been identified.

Since morphological characteristics are similar to *Exophiala (Wangiella) dermatitidis*, this fungus was at one time placed into the genus *Exophiala* but later, also incorrectly, transferred to *Phaeoannellomyces*. *Hortaea* can be distinguished from the two other genera by the ability to grow in high salt (15–30%). The conidia of *E. dermatitidis* are single-celled whereas *H. werneckii* are two-celled.

Hortaea (Phaeoannellomyces) werneckii

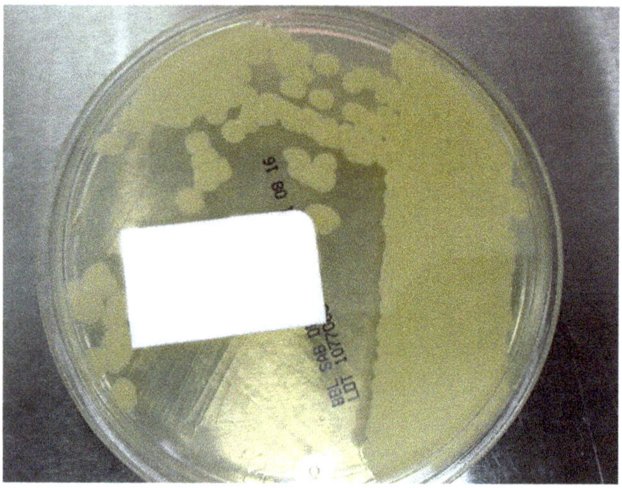

Fig. 9.16.1 *Hortaea werneckii* (bottom).

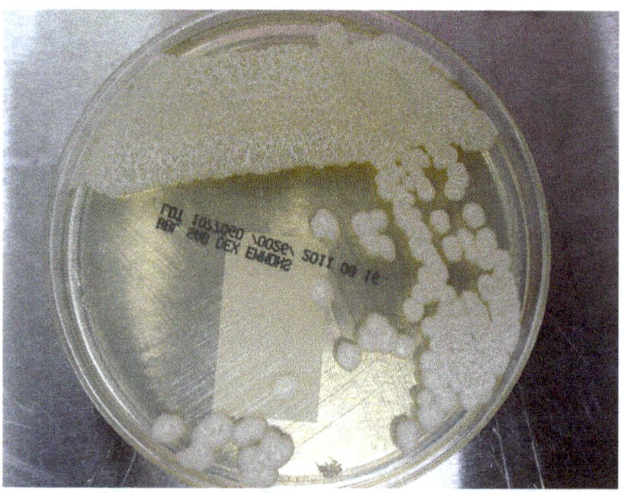

Fig. 9.16.2 *H. werneckii* (top).

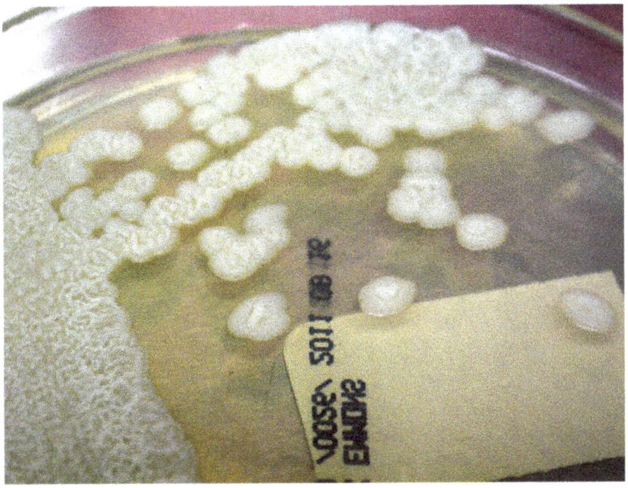

Fig. 9.16.3 *H. werneckii* (side).

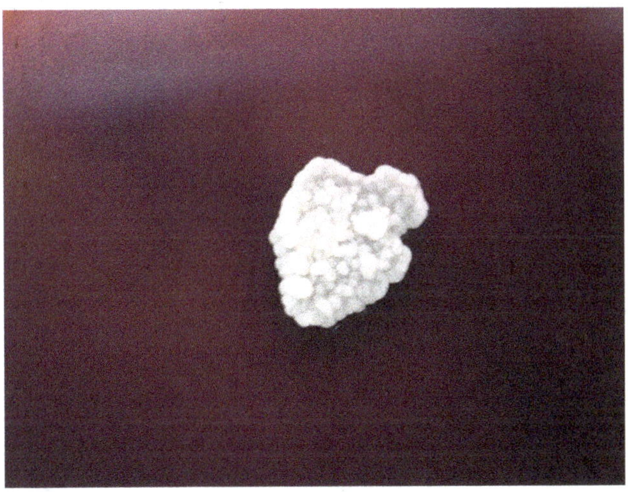

Fig. 9.16.4 *H. werneckii* blood agar (top).

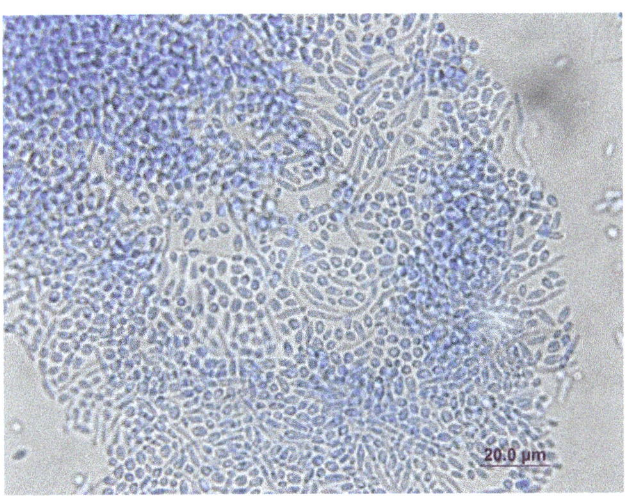

Fig. 9.16.5 *H. werneckii* early yeast-like phase.

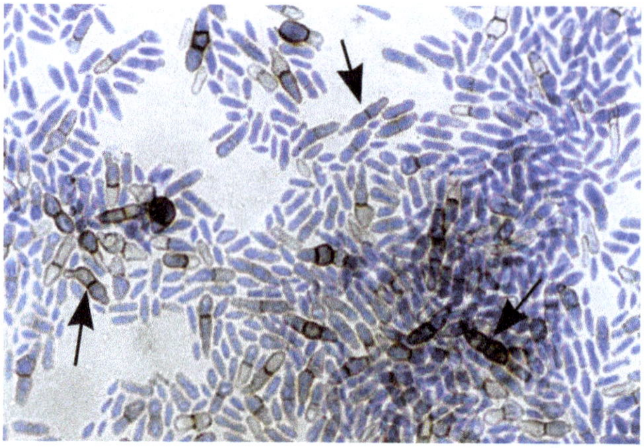

Fig. 9.16.6 *H. werneckii* mature, dark, biclavate conidia.

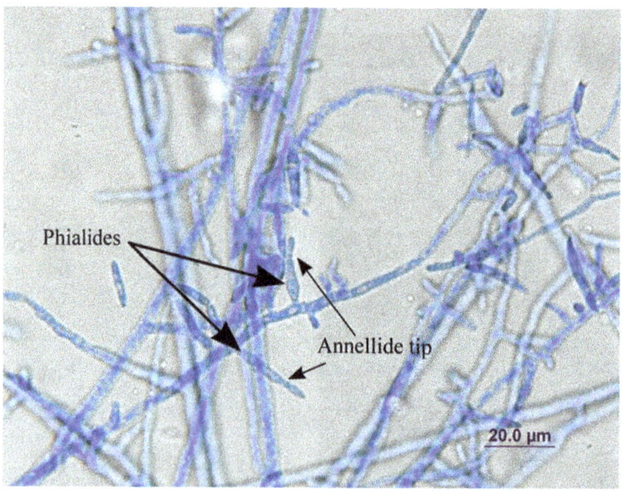

Fig. 9.16.7 Single annellide phialides.

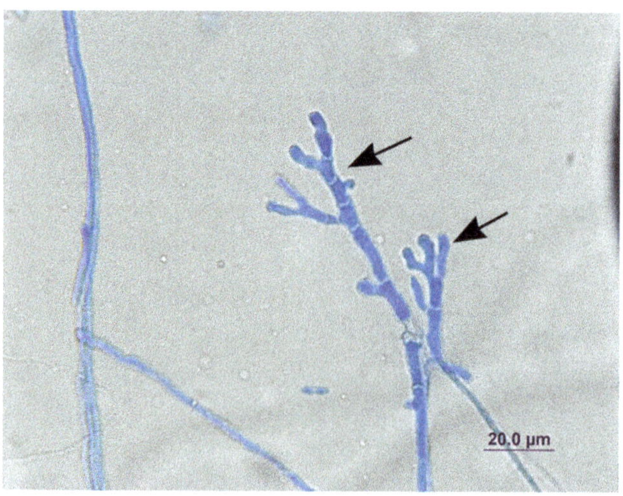

Fig. 9.16.8 Favic chandeliers.

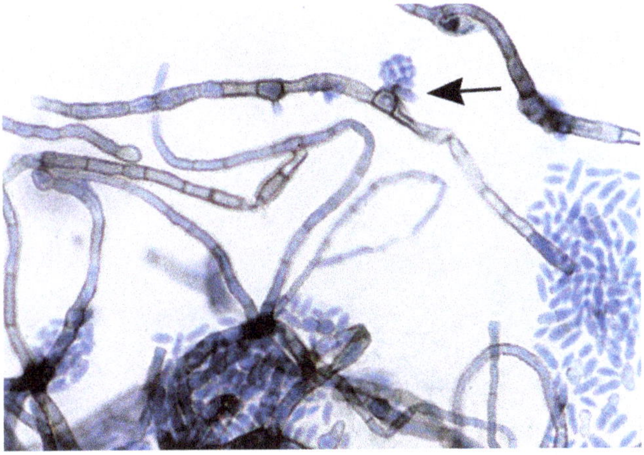

Fig. 9.16.9 Accumulation of conidia at annellidic points along hyphae.

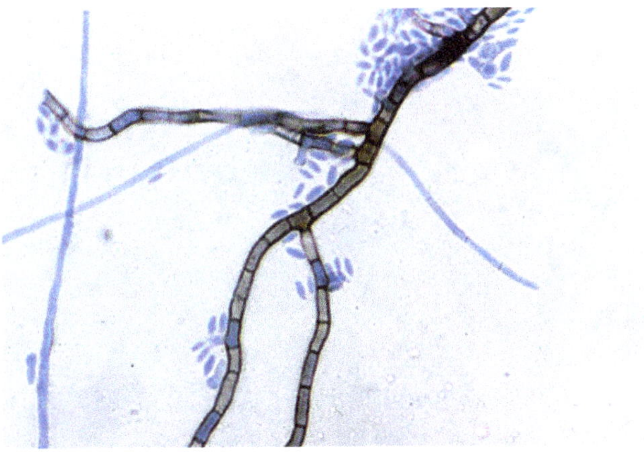

Fig. 9.16.10 Dark dematiaceous hyphae.

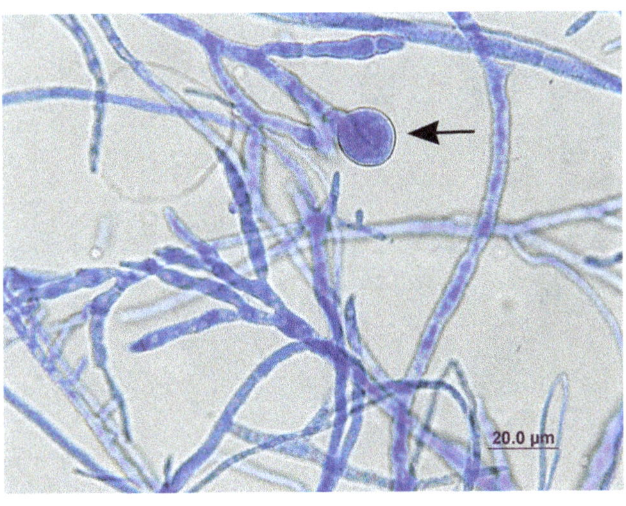

Fig. 9.16.11 Chlamydospore formation at hyphal tip.

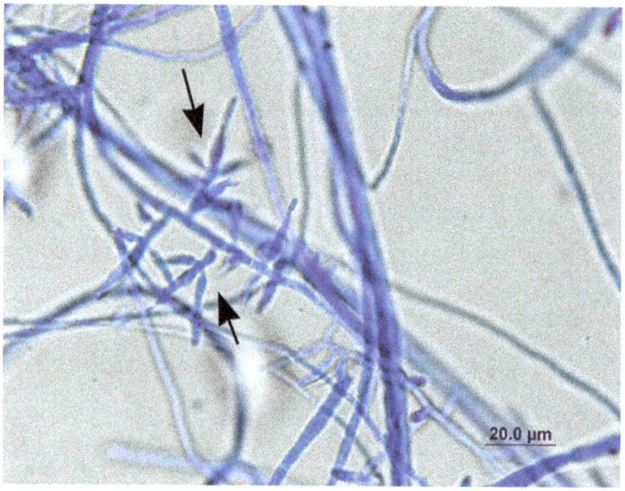

Fig. 9.16.12 *Cladosporium*-like phialide production.

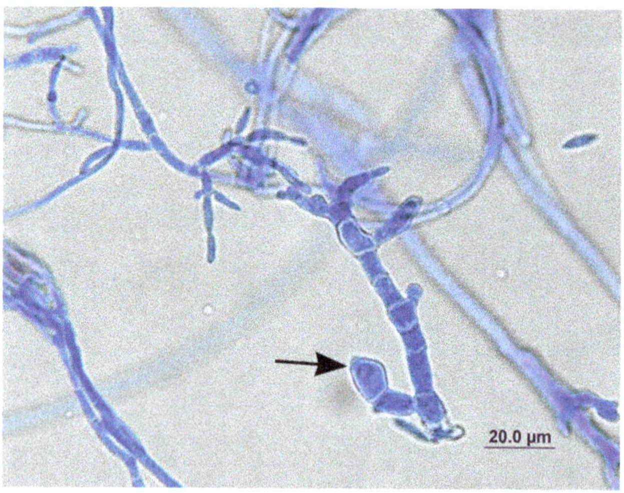

Fig. 9.16.13 Rough large-celled chlamydospores.

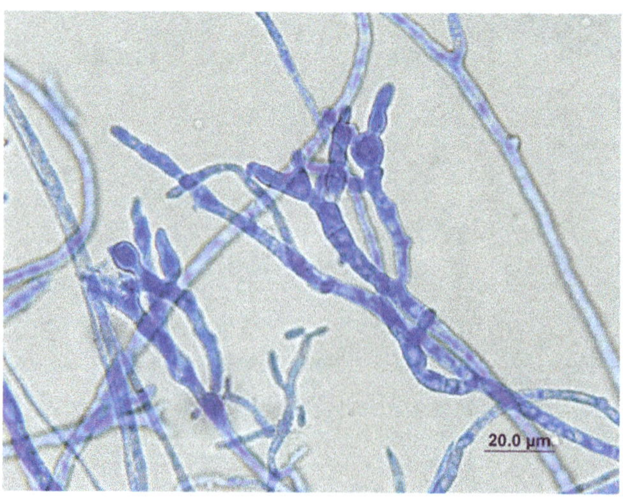

Fig. 9.16.14 Irregular hyphae.

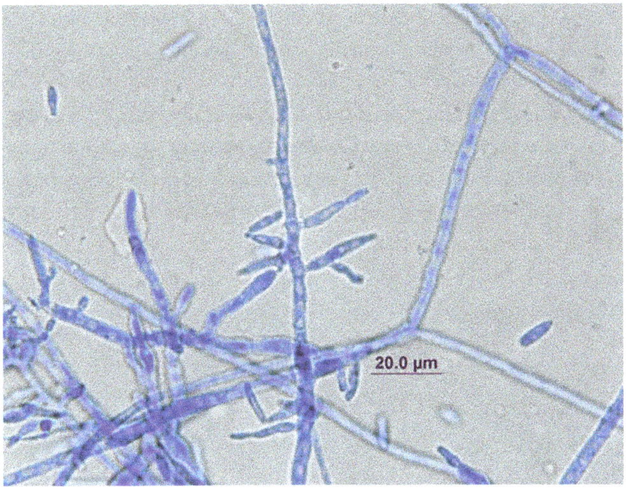

Fig. 9.16.15 Phialides along hyphae.

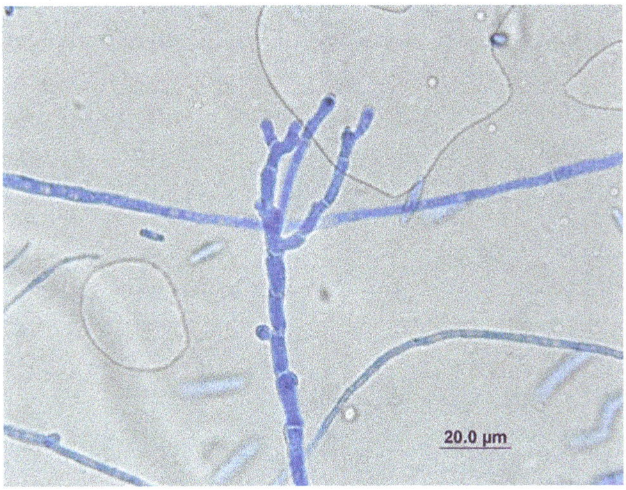

Fig. 9.16.16 Branching irregular hyphae.

9.17 *Madurella* spp

Colonial morphology: colony morphology varies greatly, and may be dark gray, "mouse fur-like" (Figs 9.17.2, 9.17.3) or white or yellow-brown. The surface may also be smooth, shiny, and folded. Reverse of gray colonies is yellow (Fig. 9.17.1), while the reverse of yellow-brown colonies is dark brown. Some species secrete a dark brown pigment into the agar.

Microscopic morphology: mycelia are dark, highly septate, and toruloid. *M. mycetomatis* grown on Sabouraud dextrose agar forms numerous chlamydoconidia. Along the mycelia and perpendicularly produced are multicellular hyphae. These multicellular hyphae are highly septate, condensed, rounded, bumpy, and end either in chlamydoconidia or structures that appear like fingers that resemble "favic chandeliers" or both structures combined (Fig. 9.17.9). Older cultures form dark mycelia with normal septations, but more intercalary chlamydoconidia are present. Sclerotia also form in older cultures (Fig. 9.17.13). It has been noted that *M. mycetomatis* grown on corn meal agar will produce, directly from mycelia, phialides that produce round conidia. *M. grisea* forms only sterile, toruloid (bead-like) mycelia and does not grow at 37 °C, whereas *M. mycetomatis* does. *Madurella grisea* was recently reclassified to the genus *Trematosphaeria*.

Madurella grisea

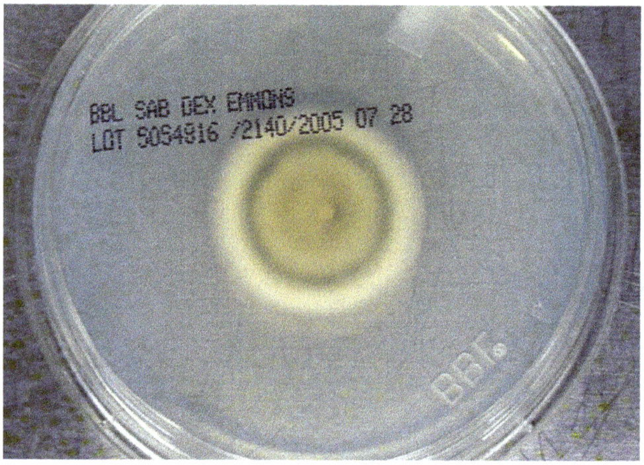

Fig. 9.17.1 *Madurella grisea* (bottom).

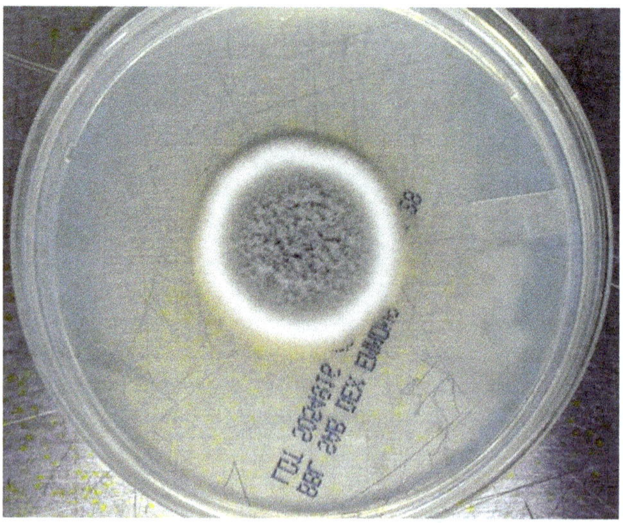

Fig. 9.17.2 *Madurella* (top).

Fig. 9.17.3 *Madurella* (side).

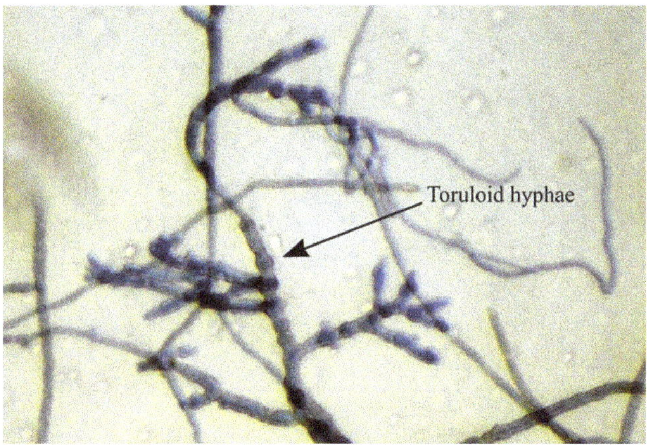

Fig. 9.17.4 Bead-like toruloid hyphae.

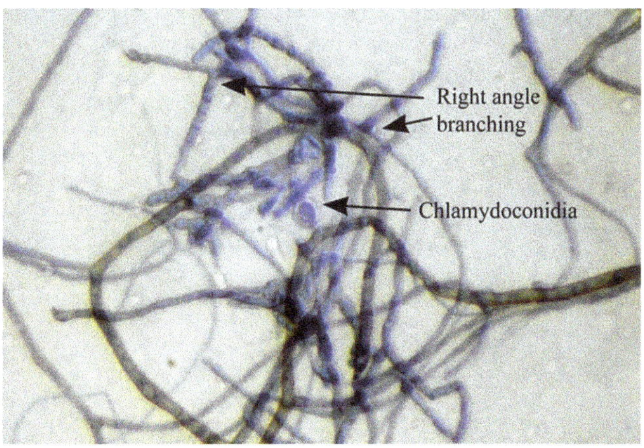

Fig. 9.17.5 Chlamydoconidia.

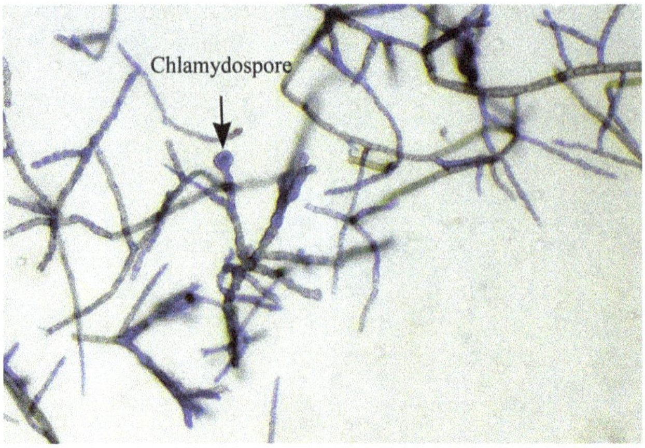

Fig. 9.17.6 Toruloid hyphae with chlamydospore.

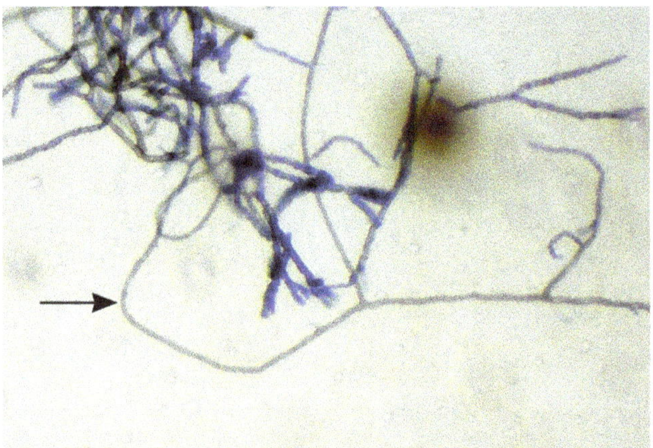

Fig. 9.17.7 Fine hyphae present.

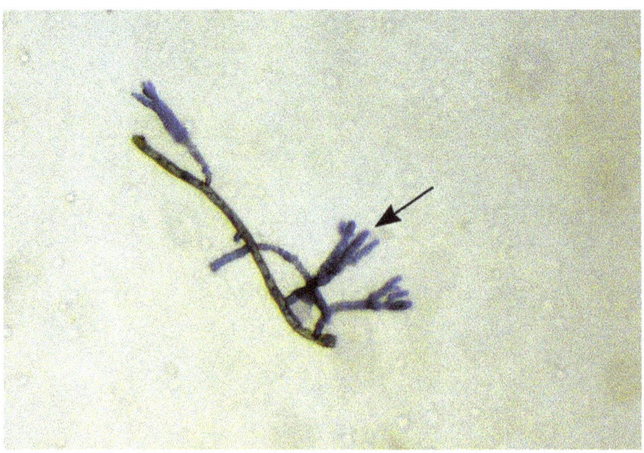

Fig. 9.17.8 Finger-like projections.

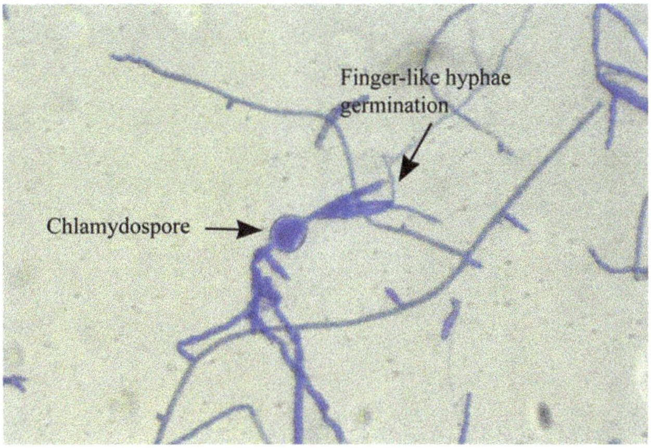

Fig. 9.17.9 *Madurella*, chlamydospore with finger like projections.

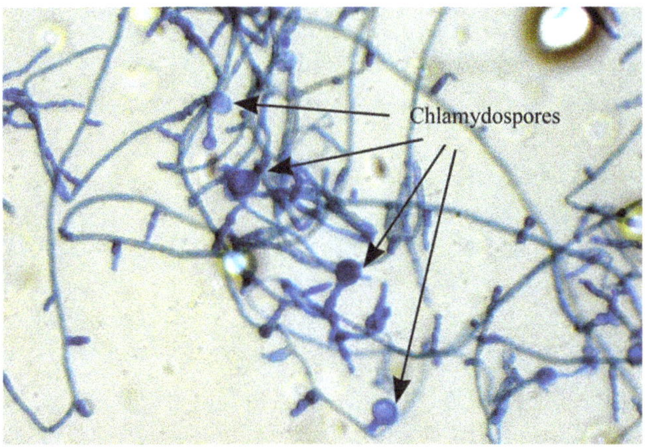

Fig. 9.17.10 Chlamydospores.

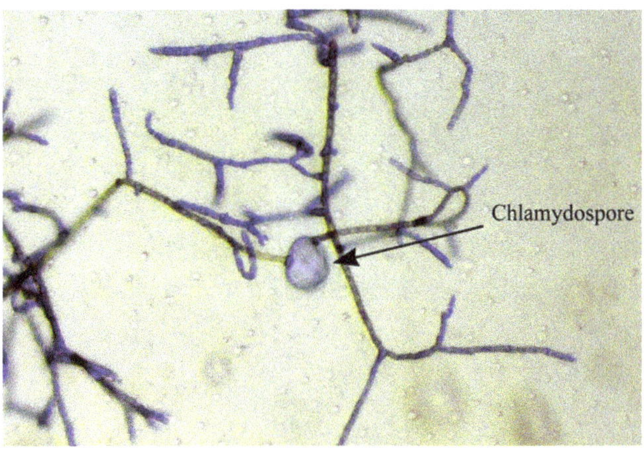

Fig. 9.17.11 Chlamydospore.

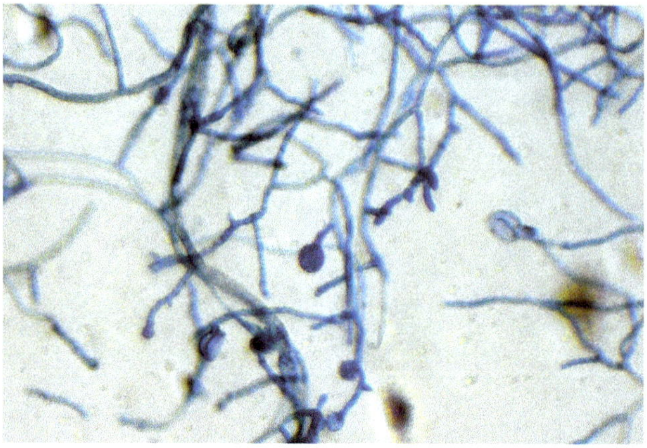

Fig. 9.17.12 *Madurella*, frequent chlamydospores.

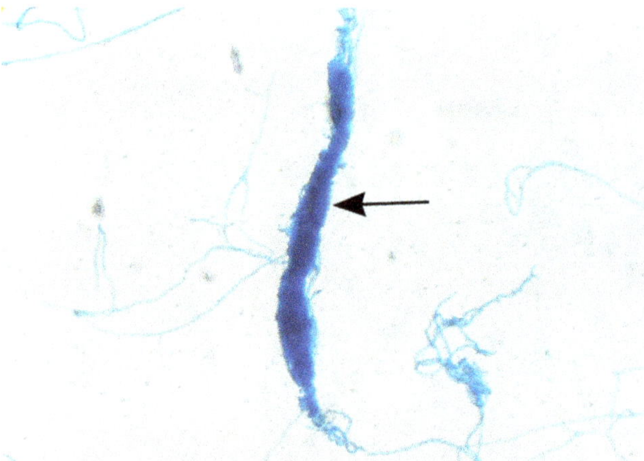

Fig. 9.17.13 Sclerotia 100× magnification.

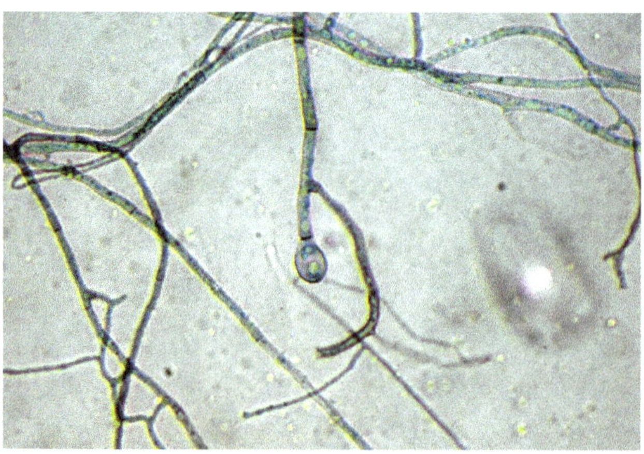

Fig. 9.17.14 Dark hyphae with chlamydospore.

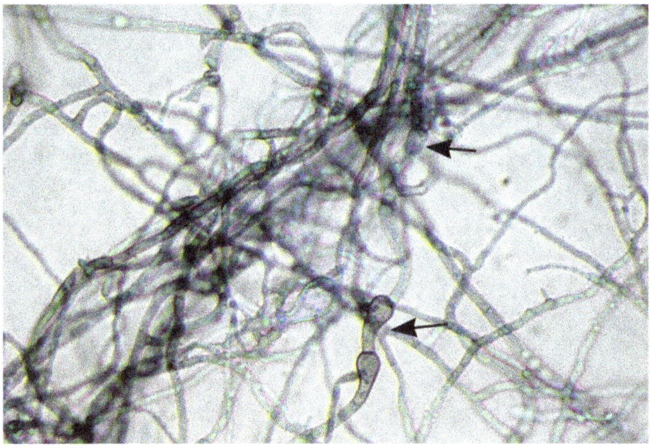

Fig. 9.17.15 Racquet hyphae.

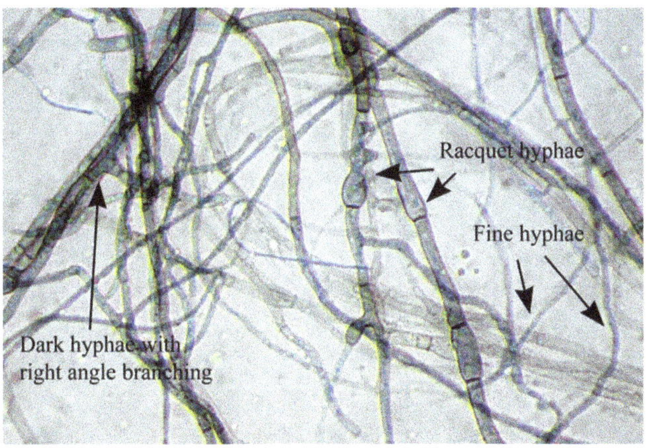

Fig. 9.17.16 Varying types of hyphae.

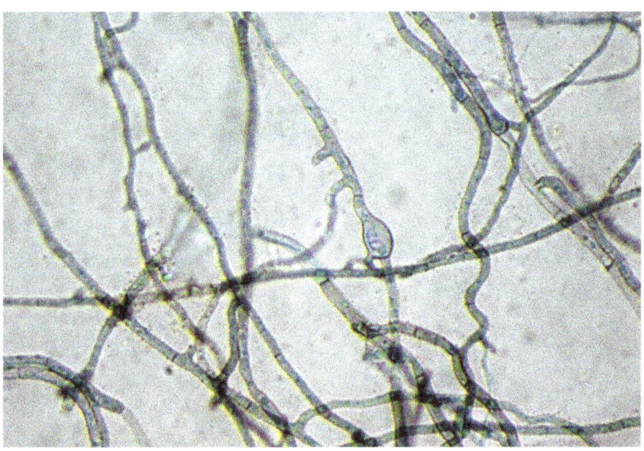

Fig. 9.17.17 Intercalary chlamydospore.

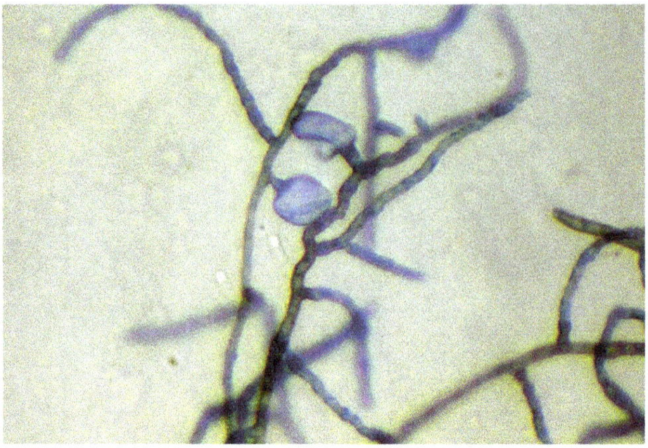

Fig. 9.17.18 *Madurella*, collapsed chlamydospores.

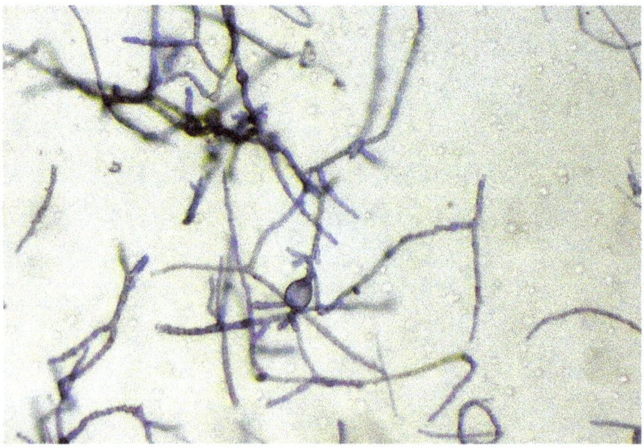

Fig. 9.17.19 Toruloid hyphae with chlamydospore.

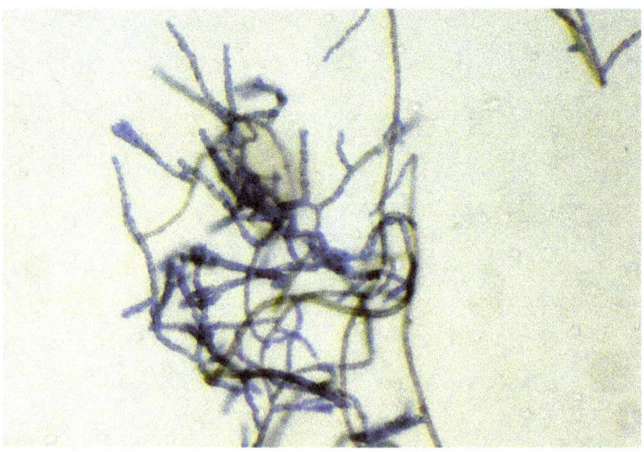

Fig. 9.17.20 Toruloid hyphae with finger-like projections.

9.18 *Neoscytalidium* spp

Colonial morphology: growth is rapid, filling the plate within 3–4 days. Colonies are wooly, gray to brown with dark reverse (*N. dimidiatum*) or white to cream, maybe gray with a cream to yellow reverse (*N. hyalinum*). The center of the colony is differentiated from the border by a change in color or indentation. Black (*N. dimidiatum*) or yellow (*N. hyalinum*) liquid droplets may form in the center of the colony. Growth is inhibited by cycloheximide.

Microscopic morphology: branched, septate hyphae are both hyaline and dark. Dark brown hyphae develop into arthroconidia that have flattened ends and are consecutively produced with no blank divisions between (distinguishing them from *Coccidioides*). Arthroconidia (4–6 × 4–16 µm) are beaded inside, appearing falsely septate (bicellular), but when broken apart, lack a flange. Arthroconidia of *N. hyalinum* are hyaline. Pycnidia may also be produced with age. The pycnidial state is referred to as *Nattrassia mangiferae,* formerly *Hendersonula toruloidea.* A key characteristic is that both fine hyaline hyphae and broad large hyphae (that fragment into arthroconidia) are present.

Neoscytalidium species

Fig. 9.18.1 *Neoscytalidium dimidiatum* (bottom).

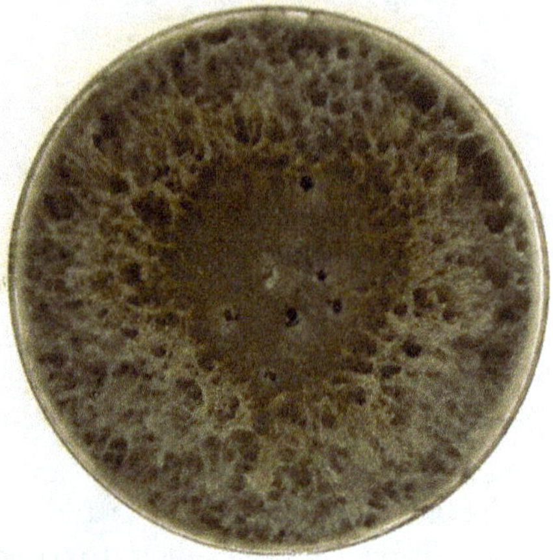

Fig. 9.18.2 *N. dimidiatum* (top).

Fig. 9.18.3 *N. dimidiatum* (side).

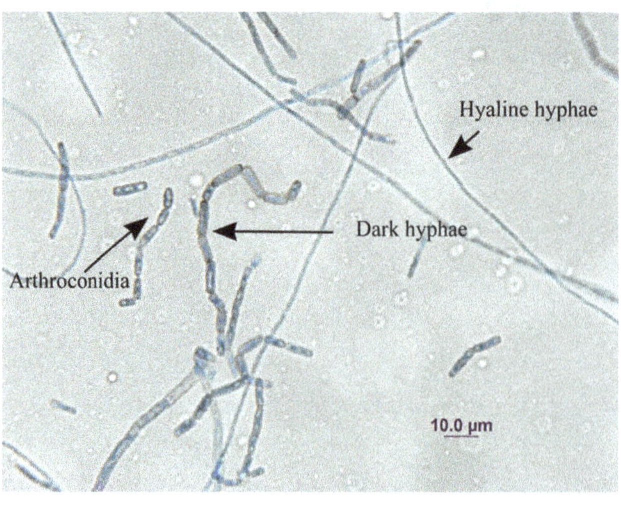

Fig. 9.18.4 *N. dimidiatum*, 1–2-cell arthroconidia.

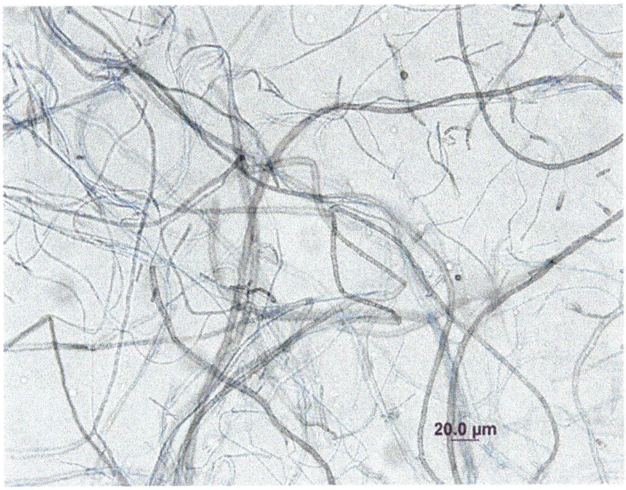

Fig. 9.18.5 *N. dimidiatum*, dark and hyaline hyphae.

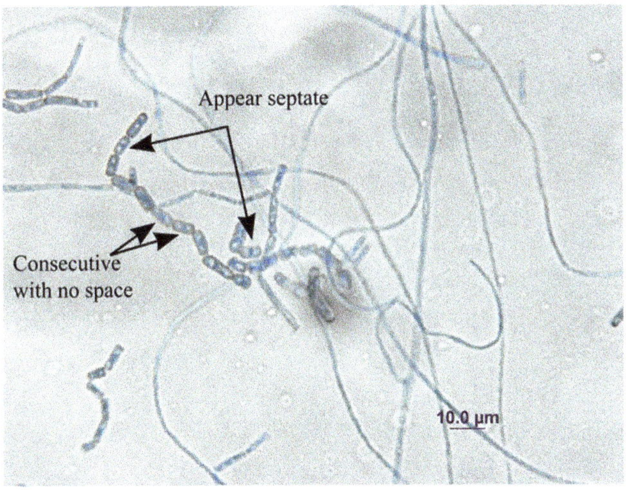

Fig. 9.18.6 *N. dimidiatum*.

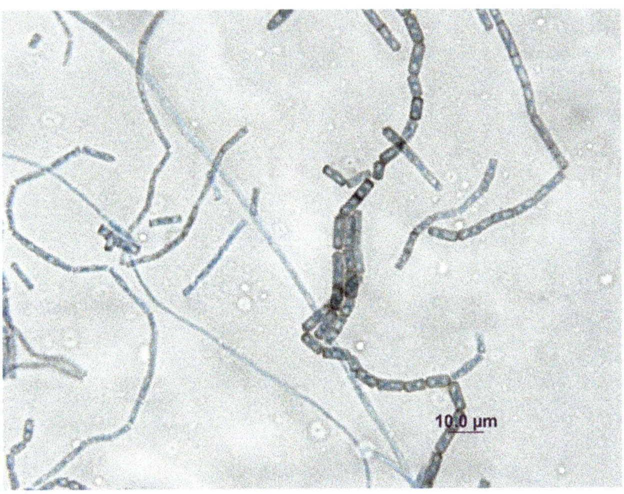

Fig. 9.18.7 *N. dimidiatum*.

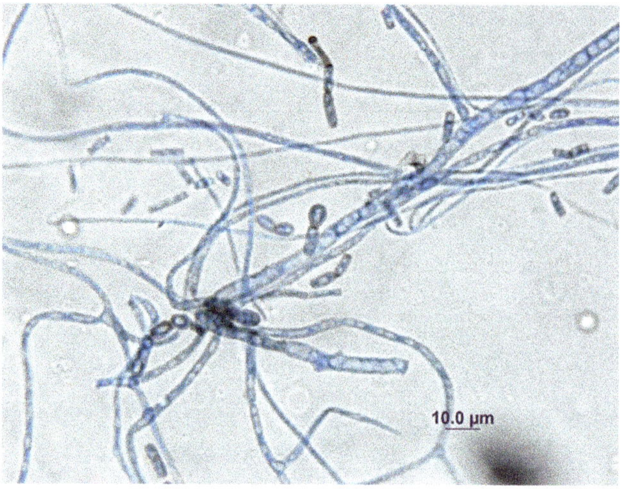

Fig. 9.18.8 *N. dimidiatum*.

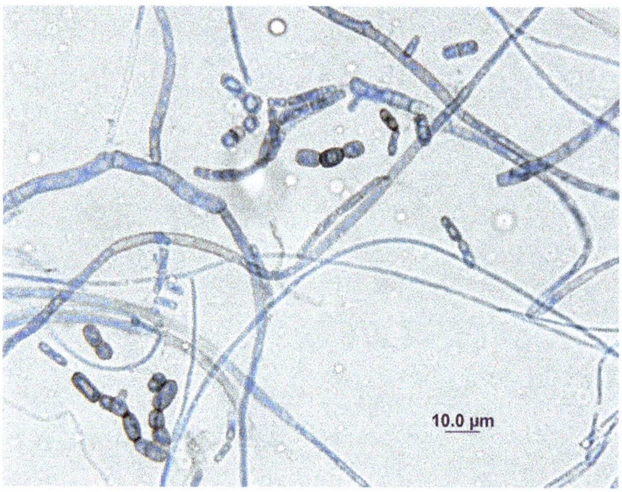

Fig. 9.18.9 *N. dimidiatum*.

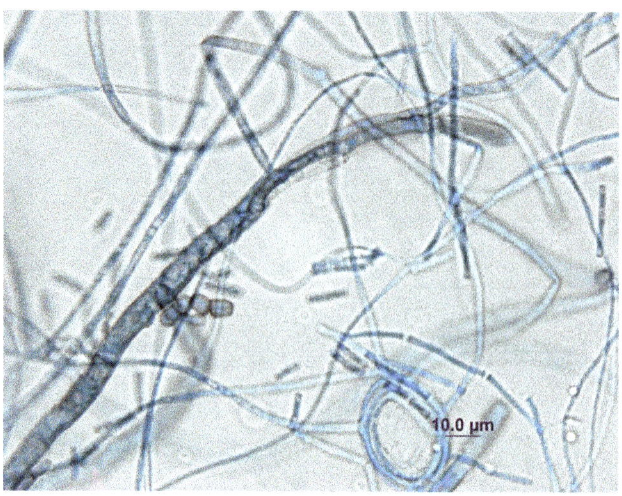

Fig. 9.18.10 *N. dimidiatum*.

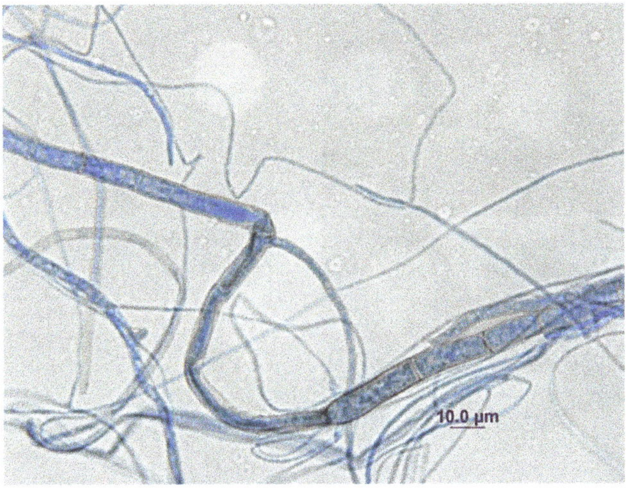

Fig. 9.18.11 *N. dimidiatum*, broad, dark hyphae.

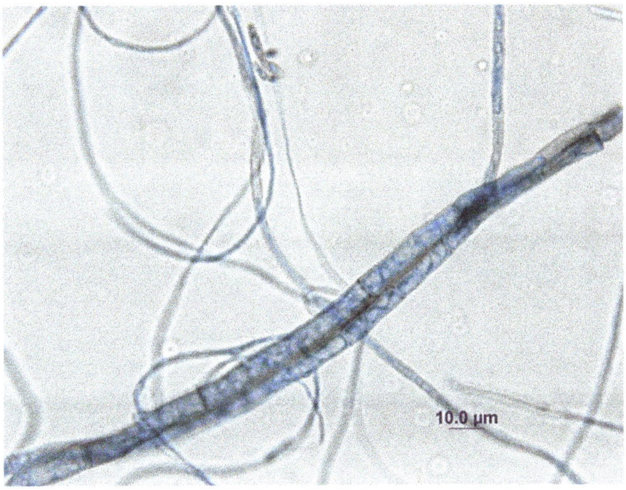

Fig. 9.18.12 *N. dimidiatum*.

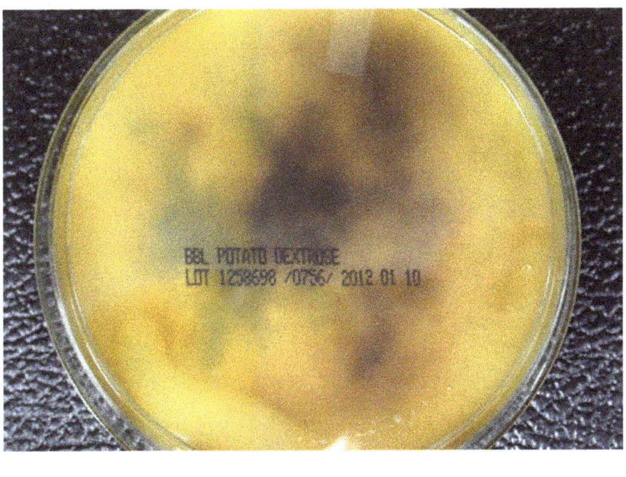

Fig. 9.18.13 *N. hyalinum* (bottom).

Fig. 9.18.14 *N. hyalinum* (top).

Fig. 9.18.15 *N. hyalinum* (side).

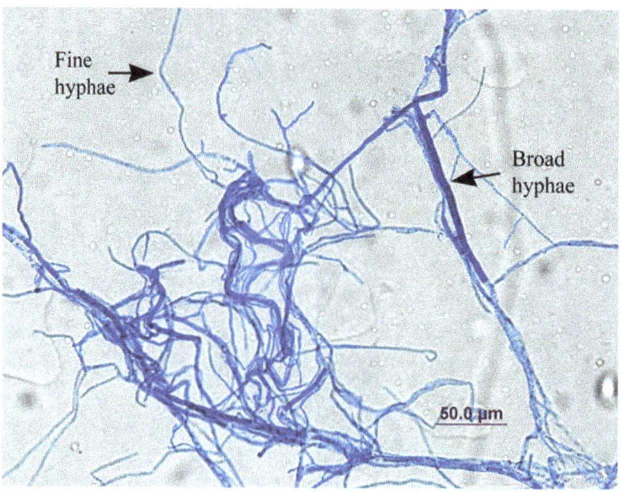

Fig. 9.18.16 *N. hyalinum*.

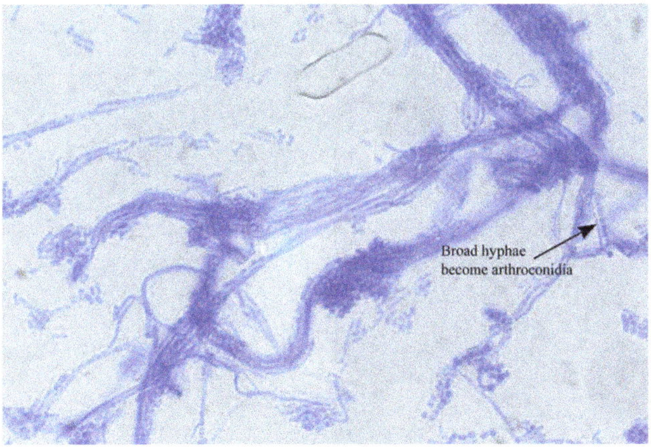

Fig. 9.18.17 *N. hyalinum*.

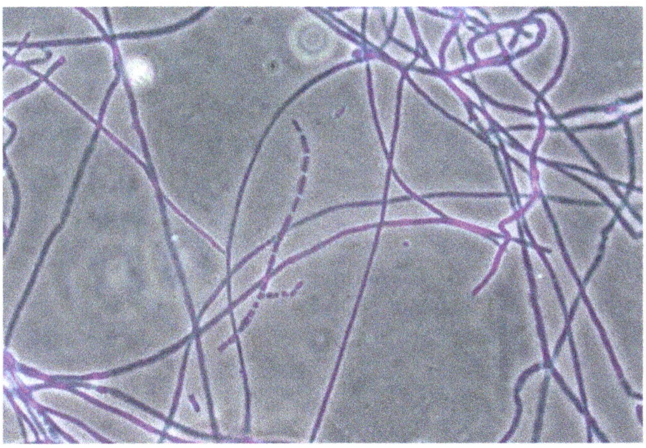

Fig. 9.18.18 *N. hyalinum*, phase contrast.

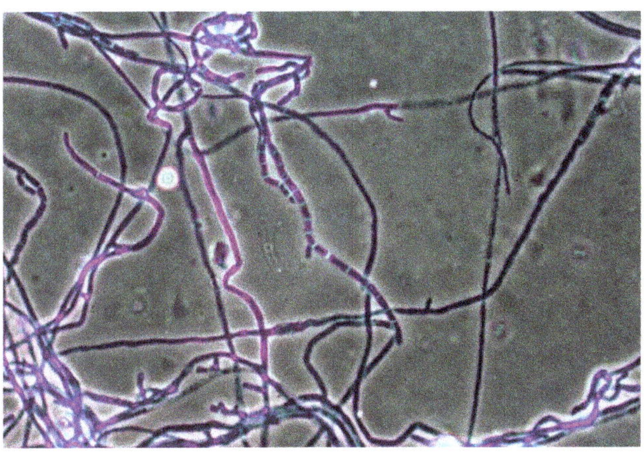

Fig. 9.18.19 *N. hyalinum*, phase contrast.

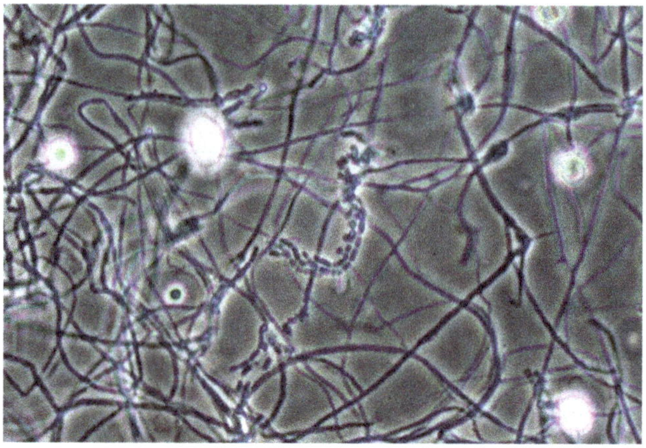

Fig. 9.18.20 *N. hyalinum*, phase contrast.

9.19 *Nigrospora* spp

Colonial morphology: colonies begin as white and wooly, then become gray with age. Reverse is tan to black. Cultures mature within 3–4 days. Older cultures develop black spots of conidiation on the surface.

Microscopic morphology: hyphae are hyaline, septate, and bamboo-like in appearance. Different widths of hyphae occur, from fine to broad in diameter. Some are coarse and give the appearance of arthroconidia but *Nigrospora* does not make arthroconidia. Thus, differentiation from *Neoscytalidium* is important because some strains of *Nigrospora* may take 3 weeks to sporulate. Conidiophores are hyaline, short (1–4 cells), and swollen with a descending end that supports the large conidia, giving the overall appearance of small mushrooms attached to the hyphae or extending from the ends of conidiophores. Conidia are large (14–20 µm in diameter), black, oval to round and may be slightly horizontally flattened on one side. Mature conidia have no internal structure, just solid black.

Nigrospora

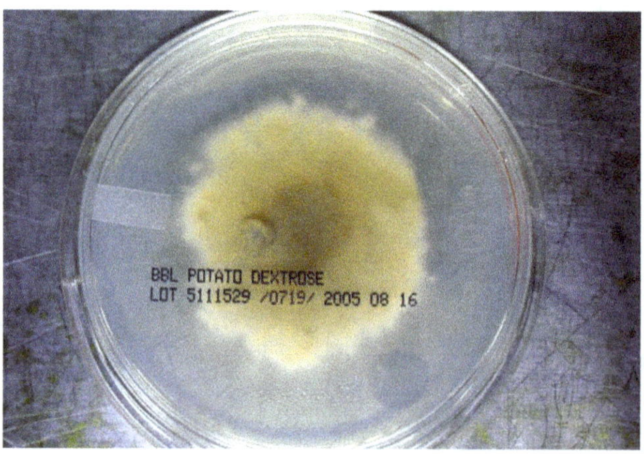

Fig. 9.19.1 *Nigrospora* (bottom).

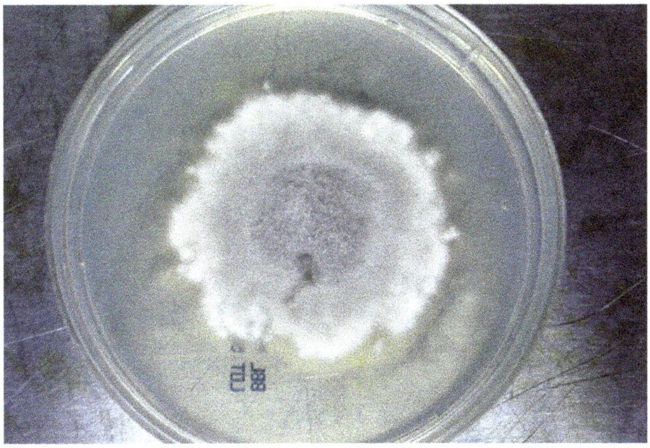

Fig. 9.19.2 *Nigrospora* (top).

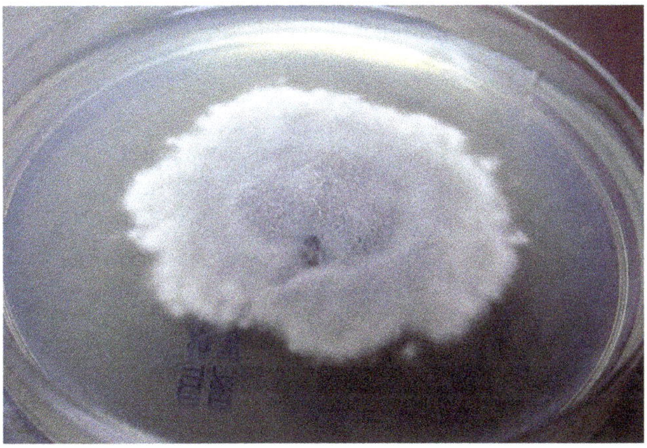

Fig. 9.19.3 *Nigrospora* (side).

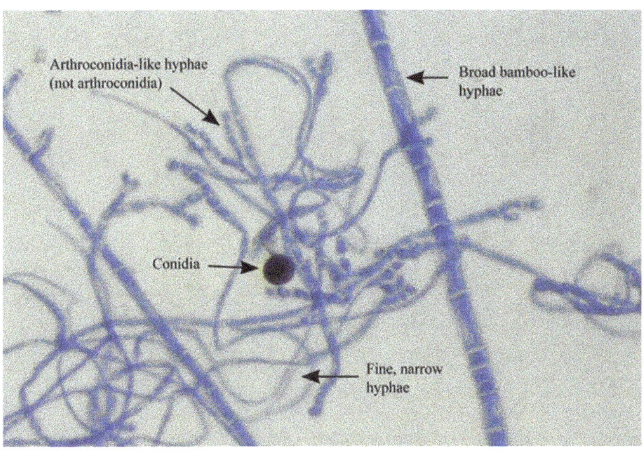

Fig. 9.19.4 Broad and narrow hyphae.

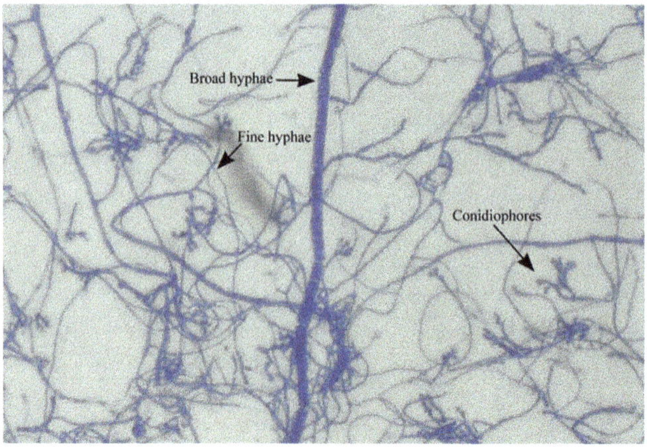

Fig. 9.19.5 *Nigrospora*, 100×.

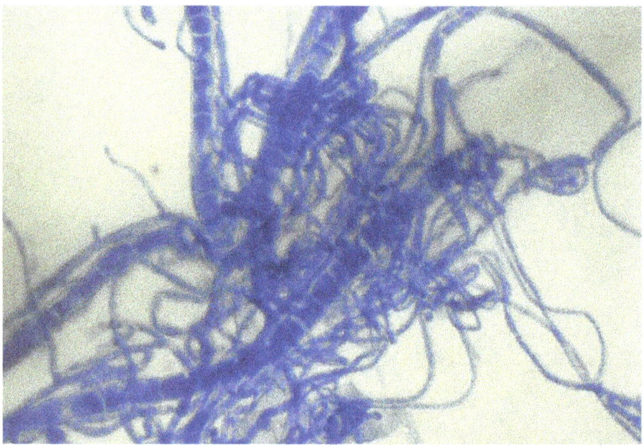

Fig. 9.19.6 Dark bamboo-like hyphae.

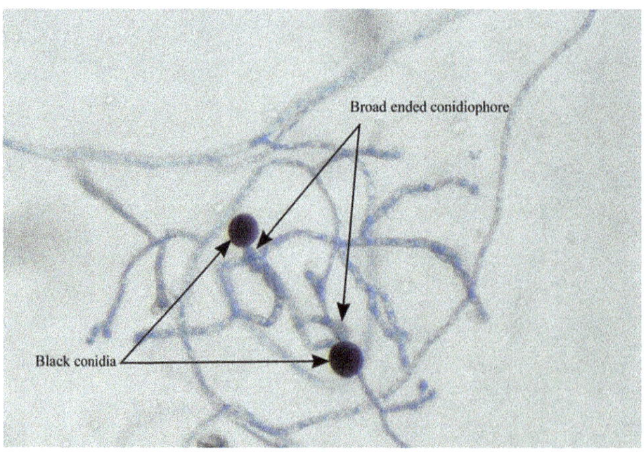

Fig. 9.19.7 Conidiophore is broad ended with black round conidia.

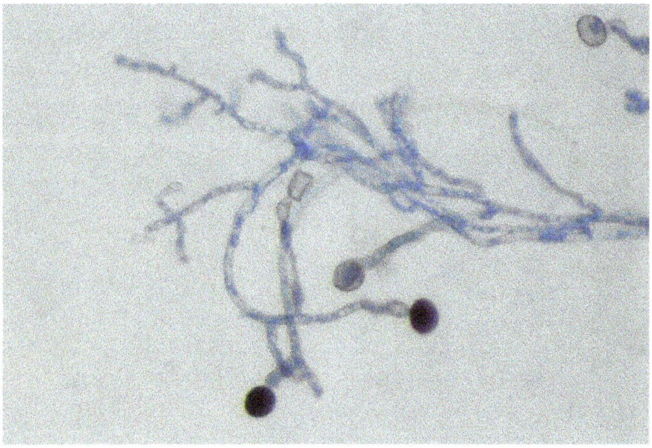

Fig. 9.19.8 Black, round to flattened conidia.

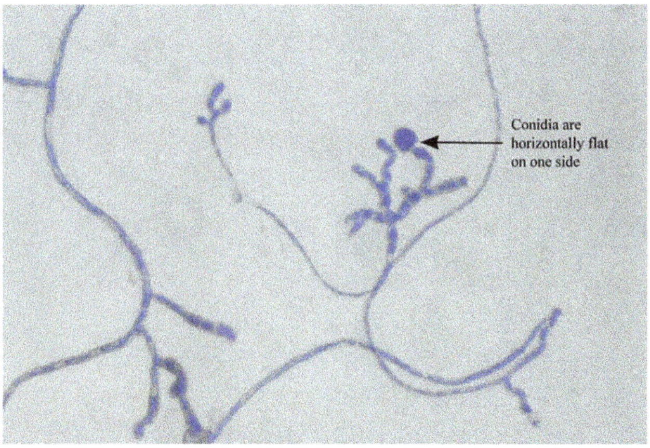

Fig. 9.19.9 *Nigrospora*.

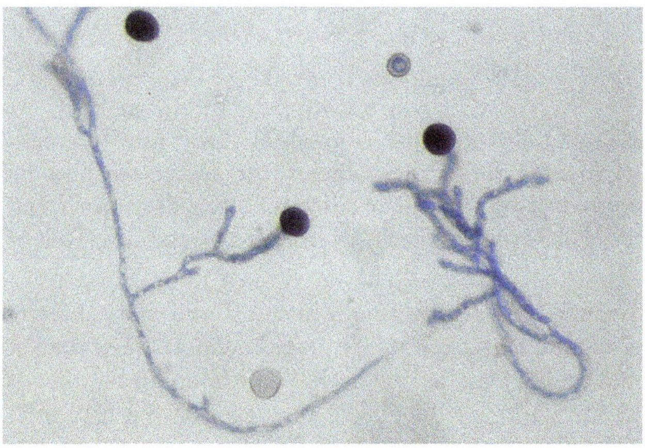

Fig. 9.19.10 *Nigrospora*, 400×.

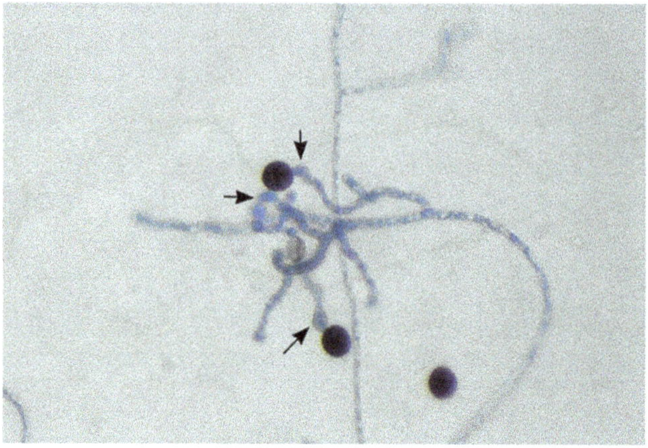

Fig. 9.19.11 Swollen conidiophores.

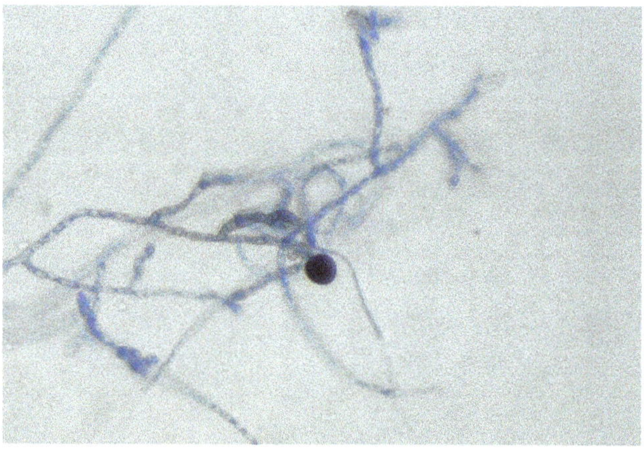

Fig. 9.19.12 *Nigrospora*, hyphae are highly septate.

9.20 *Ochroconis (Dactylaria)* spp

Colonial morphology: on Sabouraud dextrose agar, colonies are furry gray to greenish-brown, developing a dark pigmented border that radiates from the colony into the agar as a red halo. Mature colonies develop concentric circles of black, dark green, and gray with a bumpy, wavy surface. The reverse is black, surrounded by a diffused red or brown halo.

Microscopic morphology: hyphae are septate, both hyaline and dark brown. Conidiophores are produced along the dark hyphae. The conidiophores are septate at the base, swollen (5 × 10 μm), and have a long narrow neck (denticle) that easily bends (collapses). Conidia (3 × 10 μm) are dark, "corndog"-like in shape, having a distinct central septum (biclavate). As conidia separate from the conidiophores, the denticle sometimes remains attached, further giving the appearance of a "corndog" on a stick.

The central septum shows constriction of the sides of the conidia. Intercalary conidia are present but unusual. *Ochroconis gallopava* was recently reclassified to the genus *Verruconis*.

Ochroconis

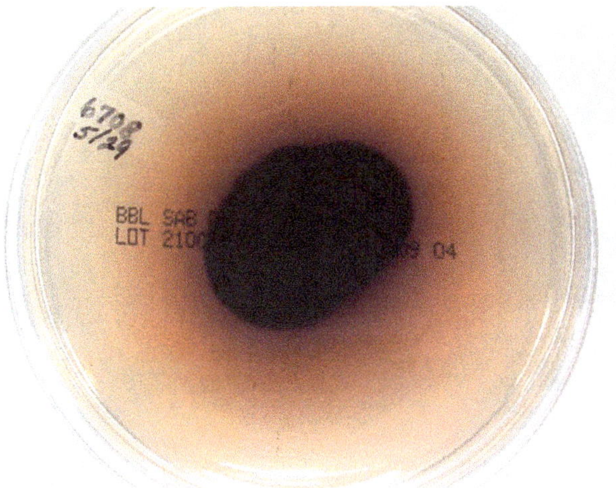

Fig. 9.20.1 *Ochroconis* (bottom).

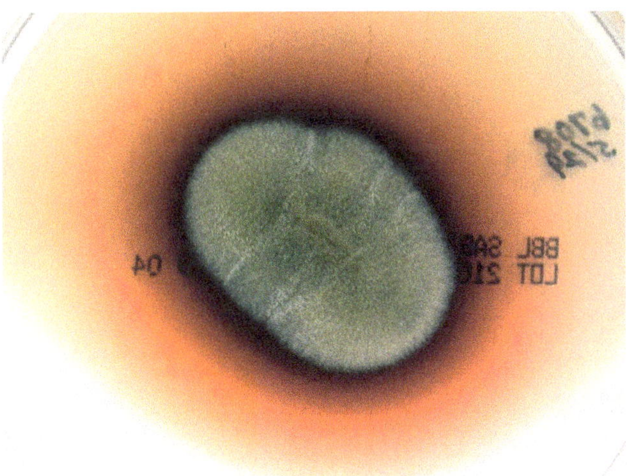

Fig. 9.20.2 *Ochroconis* (top).

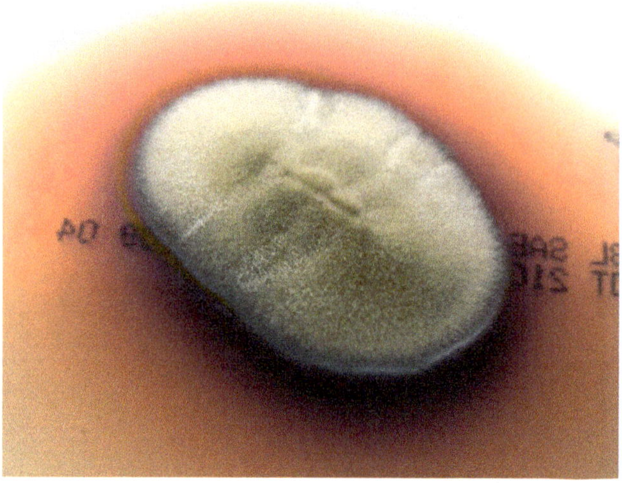

Fig. 9.20.3 *Ochroconis* (side).

Fig. 9.20.4 *Ochroconis* (side) mature.

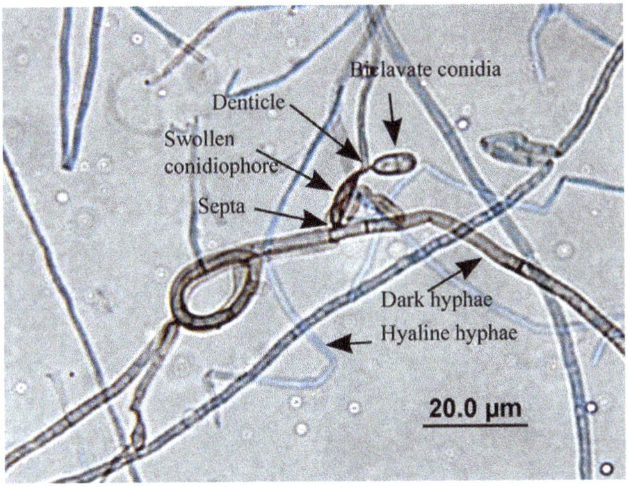

Fig. 9.20.5 Key features.

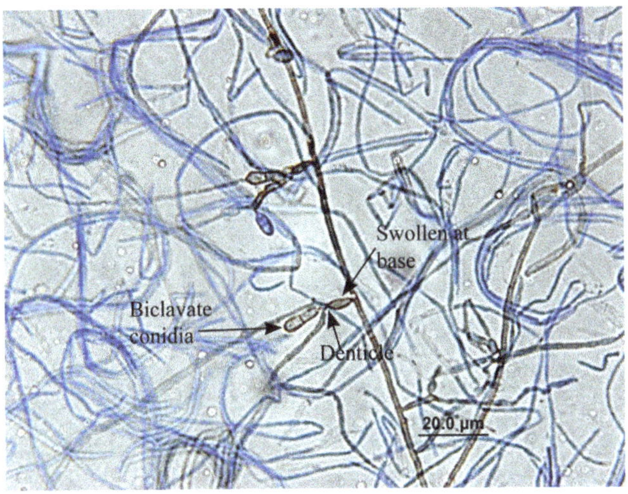

Fig. 9.20.6 General morphology.

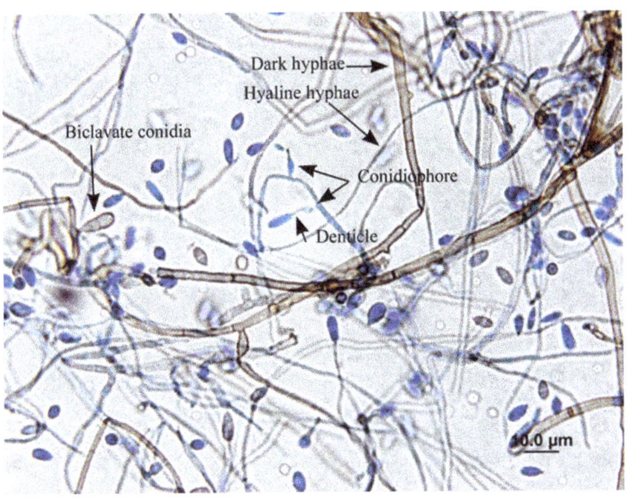

Fig. 9.20.7 General morphology.

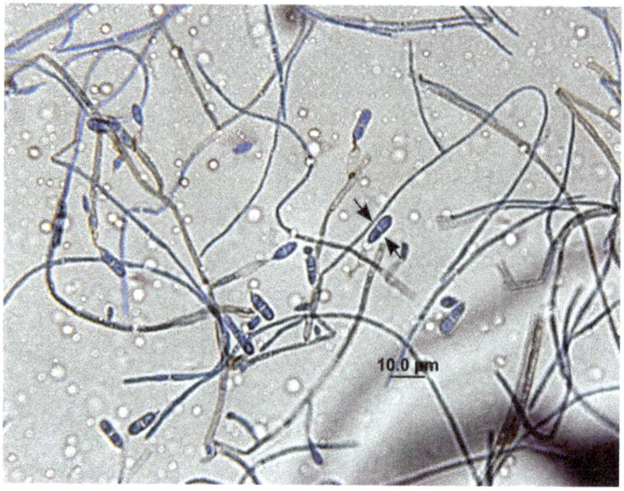

Fig. 9.20.8 Conidia constricted at center.

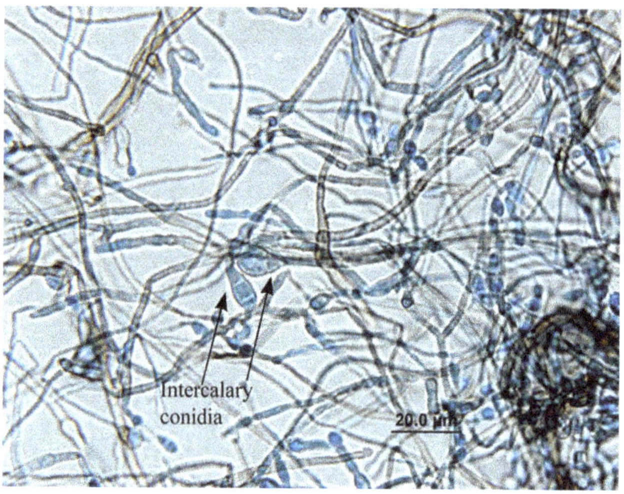

Fig. 9.20.9 Intercalary conidia.

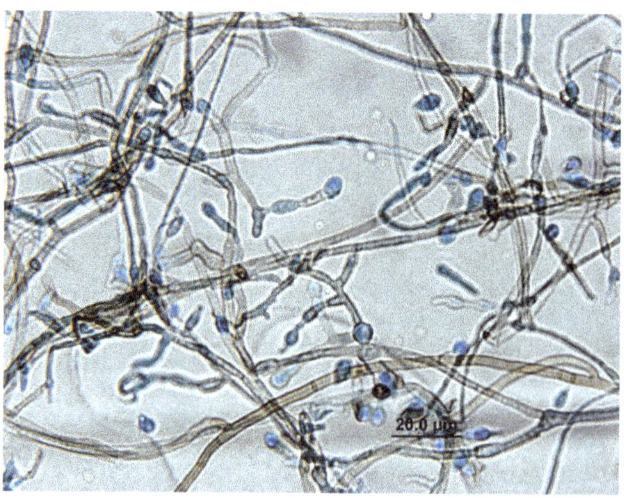

Fig. 9.20.10 *Ochroconis*.

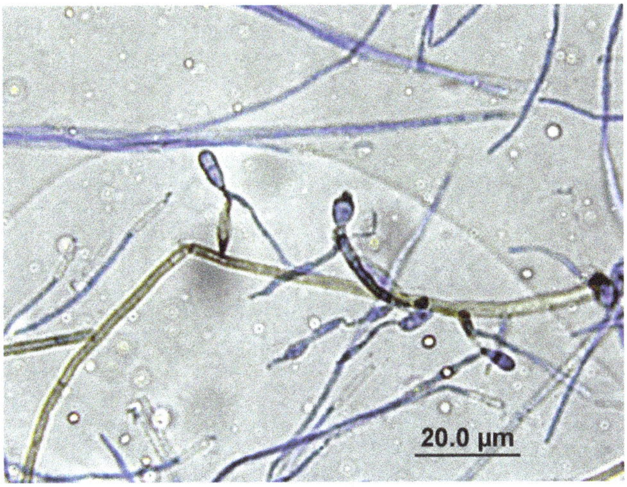

Fig. 9.20.11 *Ochroconis*, "corn dog-like" structures.

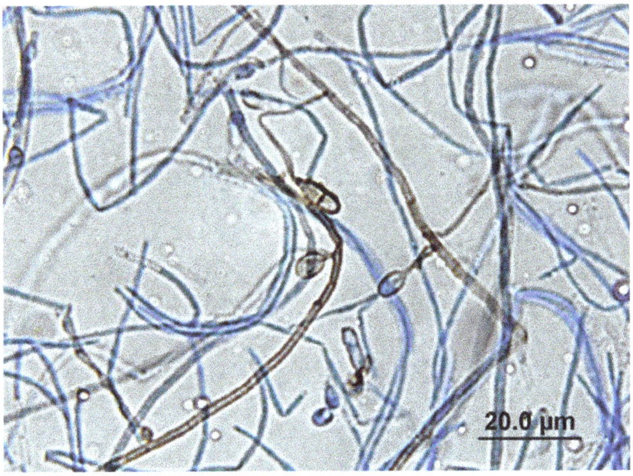

Fig. 9.20.12 *Ochroconis*.

9.21 *Phaeoacremonium parasiticum*

Colonial morphology: colonies are first cream, then convert to olive-gray and with age turn dark brown or black. Reverse is black. Surface texture is velvety or suede.

Microscopic morphology: hyphae are septate, first hyaline then become brown. Hyphae often occur in bundles (fascicles) or in circular whorls. Phialides (2–3 × 12–30 µm) are produced along hyphae singularly or on branched or unbranched conidiophores that become rough at the base. Phialides are brown, have distinct brown basal septation and descend toward the tip, which may terminate in a funnel-shaped collarette. Conidia are hyaline or brown, avocado shaped, and are formed in chains, which may cluster at the tip of phialides. The occurrence of circular whorls of hyphae with phialides that radiate outward in a pinwheel formation (Figs 9.21.5, 9.21.12) is distinguishing from *Phialemonium* and *Acremonium*.

Phaeoacremonium parasiticum

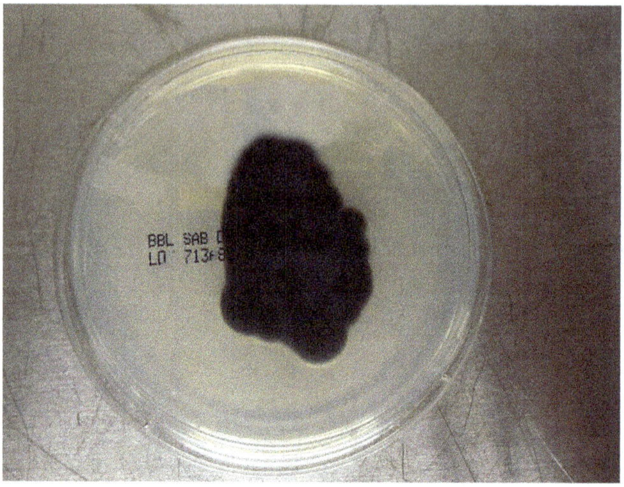

Fig. 9.21.1 *Phaeoacremonium parasiticum* (bottom).

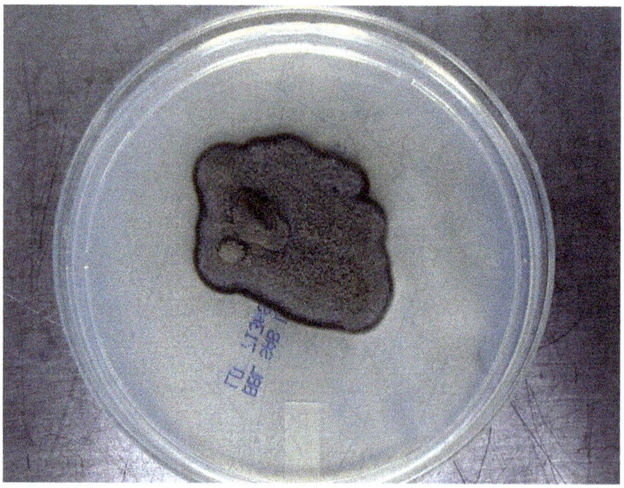

Fig. 9.21.2 *P. parasiticum* (top).

Fig. 9.21.3 *P. parasiticum* (side).

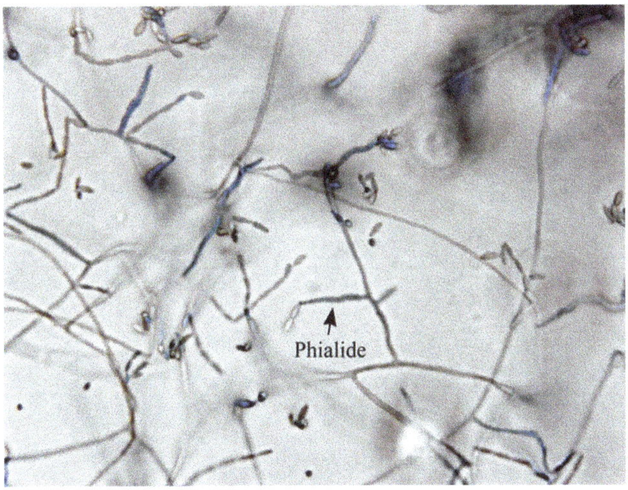

Fig. 9.21.4 Conidia formed in chains versus clusters.

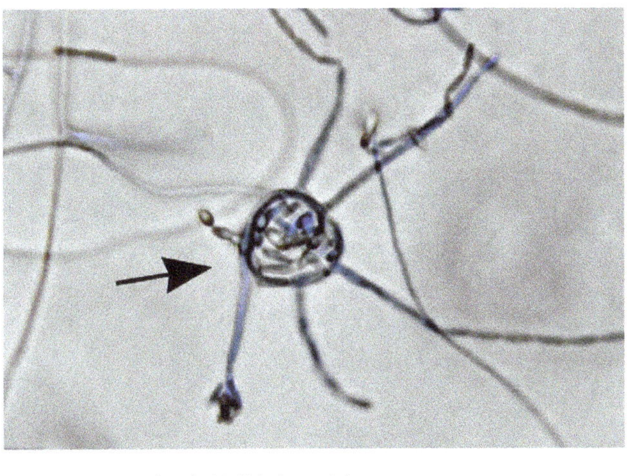

Fig. 9.21.5 Pinwheel phialide branching.

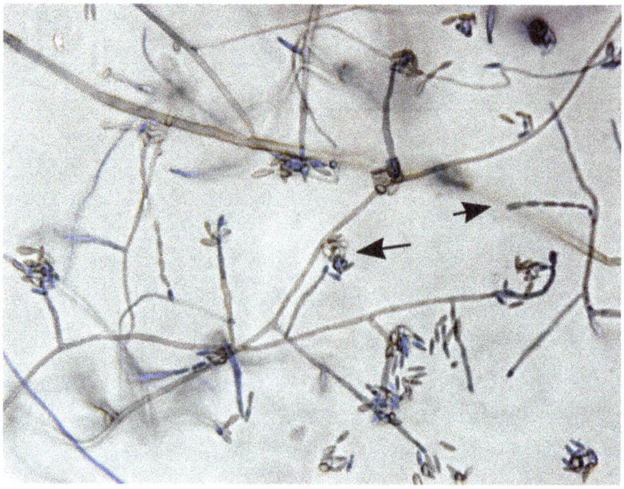

Fig. 9.21.6 Conidia in chains and clusters.

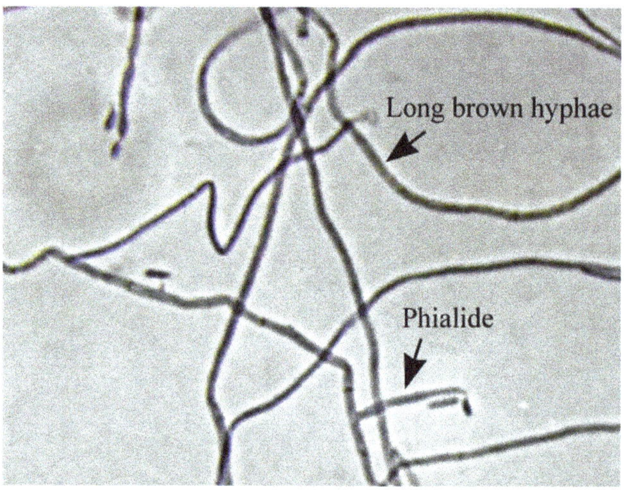

Fig. 9.21.7 Dark brown hyphae and slender phialide.

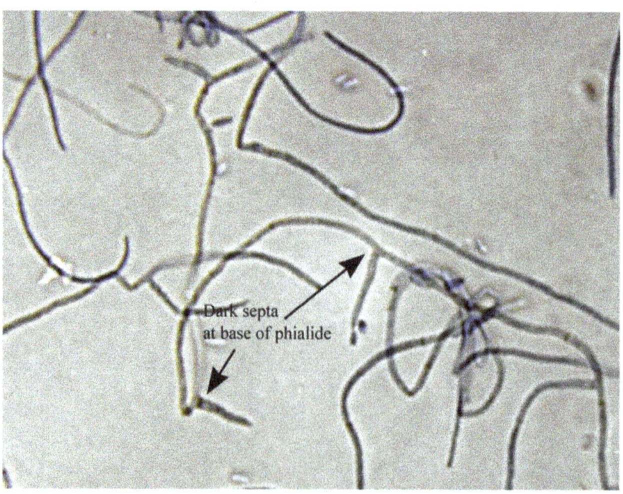

Fig. 9.21.8 Hyaline and dark hyphae.

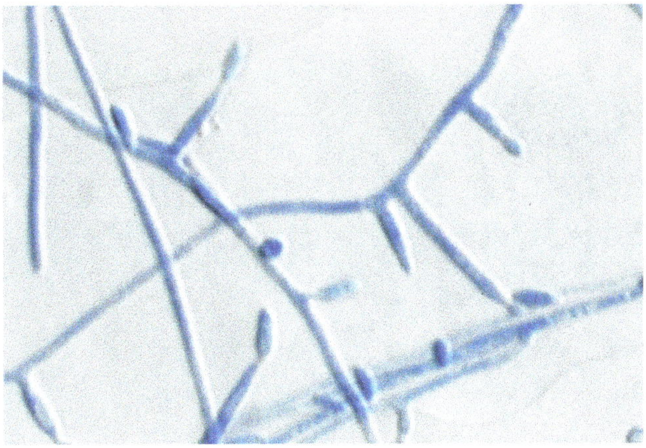

Fig. 9.21.9 Phialides with septum at base.

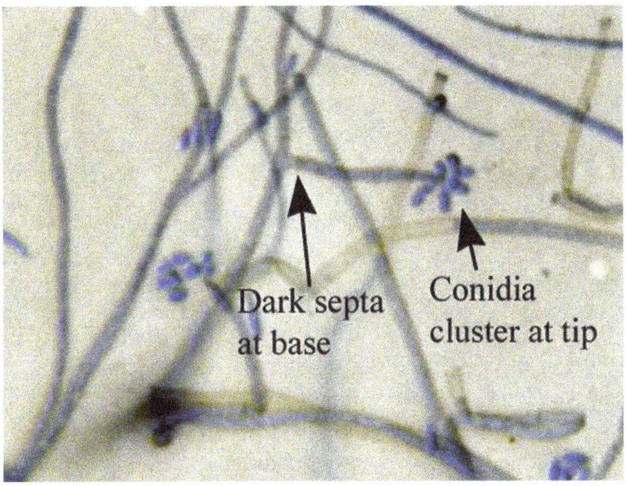

Fig. 9.21.10 Phialides with basal septum; clustered conidia at tip. 1000×.

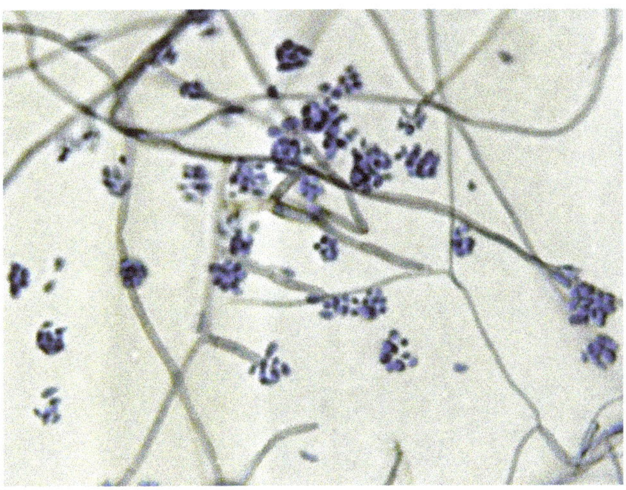

Fig. 9.21.11 Conidia in clusters at phialide tips.

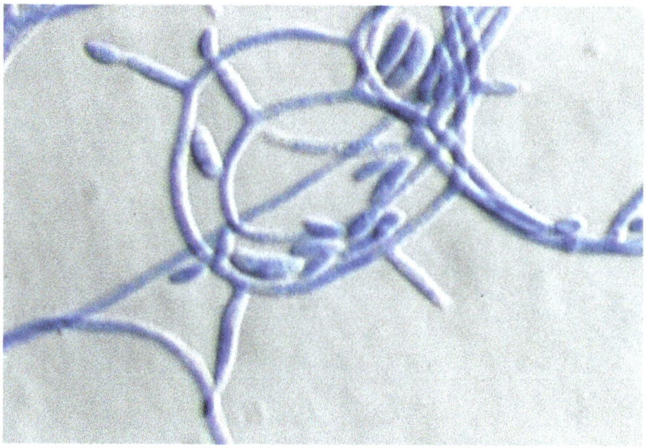

Fig. 9.21.12 Pinwheel phialide branching.

9.22 *Phialemonium* spp

Colonial morphology: the colony is at first white and suede-like, becoming tan to yellow-brown and developing a cottony surface overlay. Some species may develop a green to slight green-yellow pigment with age. Sometimes the pigment diffuses into the agar. Reverse is yellow to brown.

Microscopic morphology: hyphae are hyaline and septate. They become compactly aggregated into fascicles or synnema (Figs 9.22.7, 9.22.10). Phialides are produced perpendicular to the hyphae. The phialides are single, and may be long or short, but do not usually have a visible septum at the base. They appear long and descending or cylindrical, but not swollen at the base. Conidia (single celled) are produced at the terminal end of phialides and cluster in large groups. *P. obovatum* produces chlamydospores and conidia that are mostly oval or tear shaped whereas *P. curvatum* produces conidia that are mostly curved or long.

Phialemonium obovatum

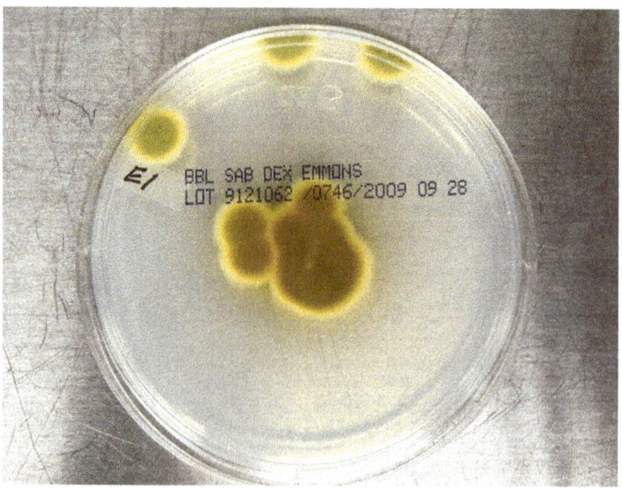

Fig. 9.22.1 *Phialemonium obovatum* (bottom).

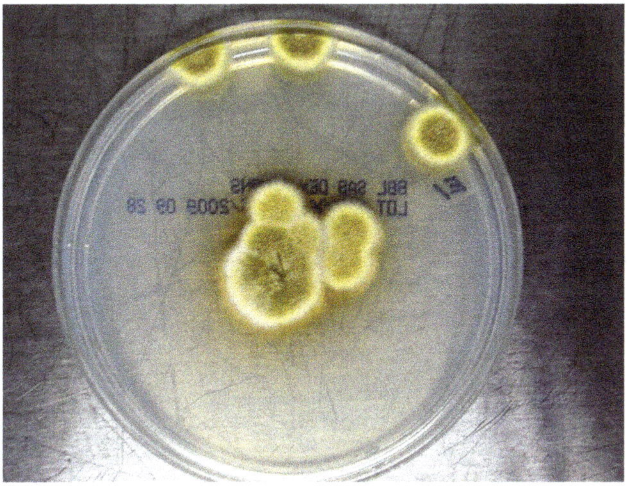

Fig. 9.22.2 *P. obovatum* (top).

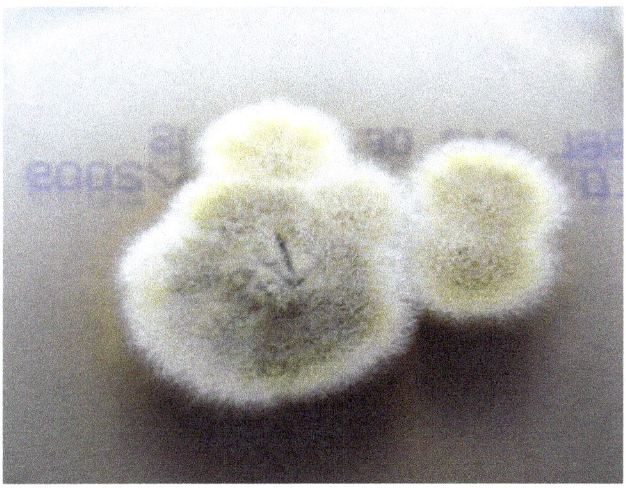

Fig. 9.22.3 *P. obovatum* (side).

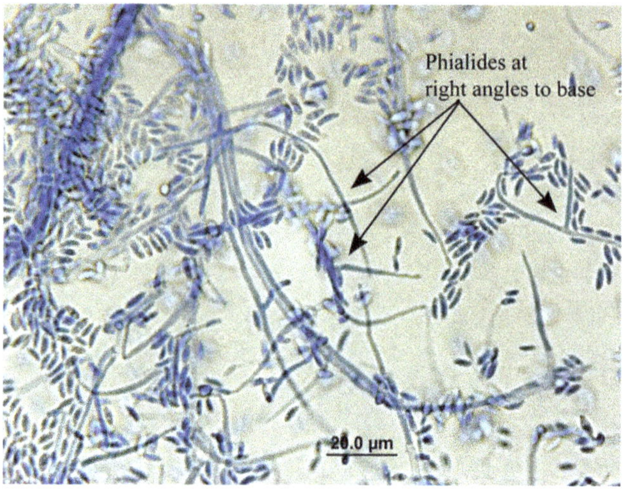

Fig. 9.22.4 *P. obovatum*, phialide production.

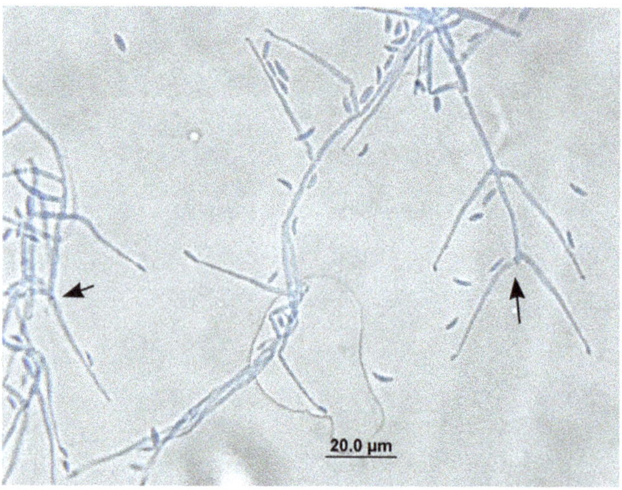

Fig. 9.22.5 Phialides without dark basal septum.

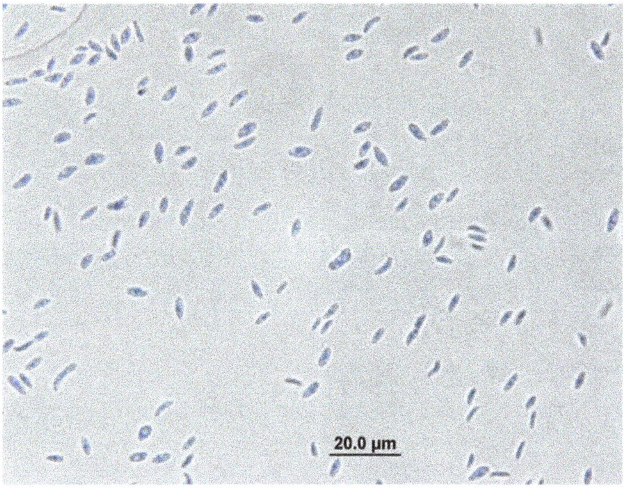

Fig. 9.22.6 Hyaline single cell conidia.

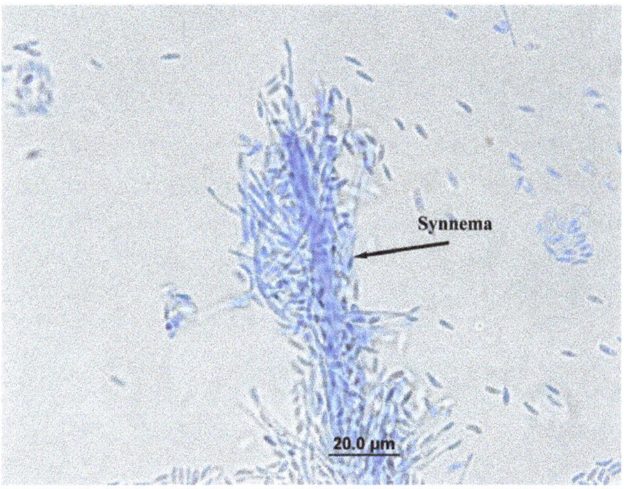

Fig. 9.22.7 *P. obovatum*, synnema with multiple phialides.

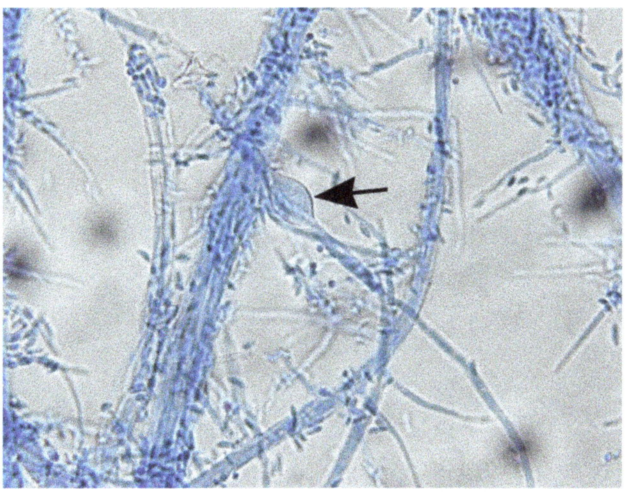

Fig. 9.22.8 Chlamydoconidia.

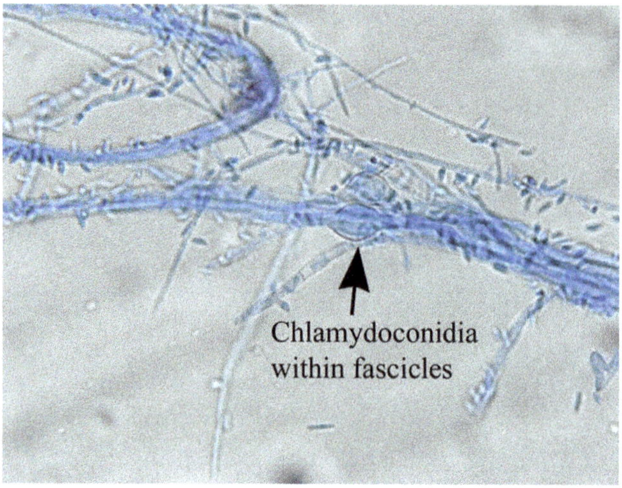

Fig. 9.22.9 *P. obovatum*, chlamydoconidia within fascicles.

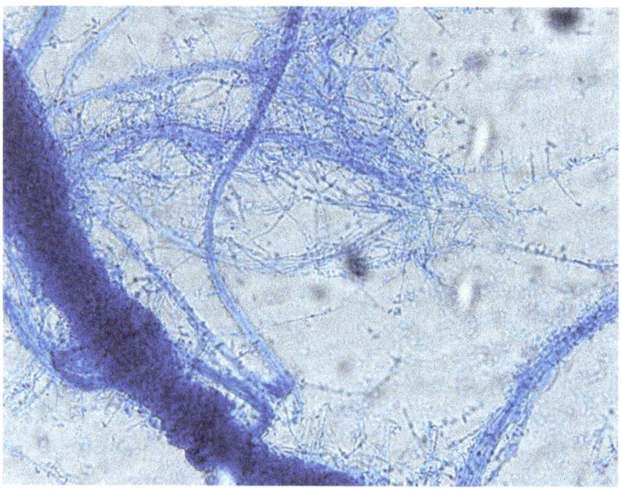

Fig. 9.22.10 Fascicles.

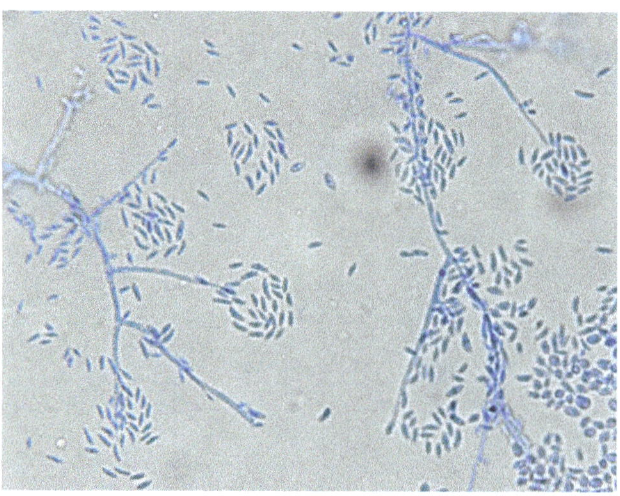

Fig. 9.22.11 Conidia clusters at ends of phialides.

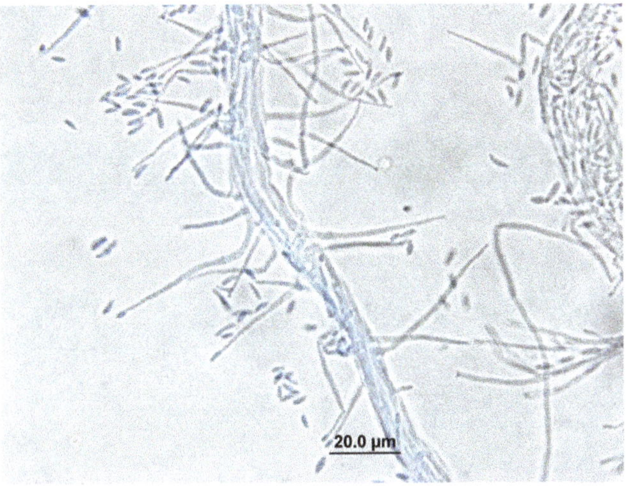

Fig. 9.22.12 *P. obovatum*, phialide morphology.

9.23 *Phialophora* spp

Colonial morphology: colonies are dome shaped, dark green, gray to black, with a dense suede surface. Growth is elevated above the agar and may develop variations in color (white spots) with age. Reverse is black.

Microscopic morphology: hyphae are brown and septate. Phialides are true vase shaped with a wide collarette resembling a cup (*P. verrucosa*) or saucer (*P. richardsiae*). Many phialides are produced singly, in pairs or in triplets directly from hyphae or on conidiophores. Single-celled phialoconidia are produced with the youngest at the base, consecutively, and stick together forming clusters at the end of the phialide. *P. verrucosa* produces hyaline conidia (3.0–5.0 × 1.5–3.0 µm) that are oval, sometimes with a flattened base, whereas those of *P. richardsiae* are round. *P. richardsiae* has recently been transferred to the genus *Pleurostomophora richardsiae*.

Phialophora verrucosa

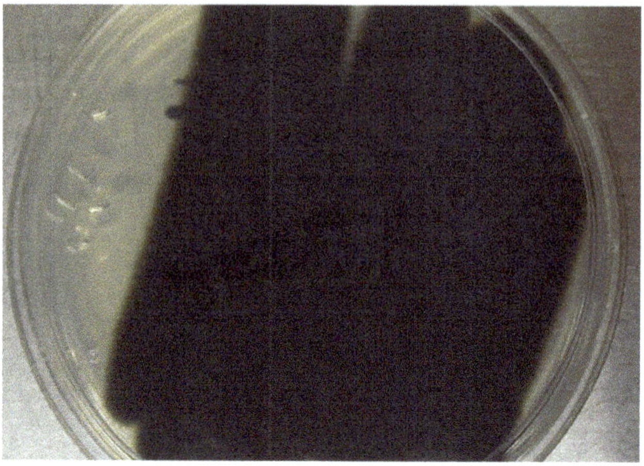

Fig. 9.23.1 *Phialophora verrucosa* (bottom).

Fig. 9.23.2 *P. verrucosa* (top).

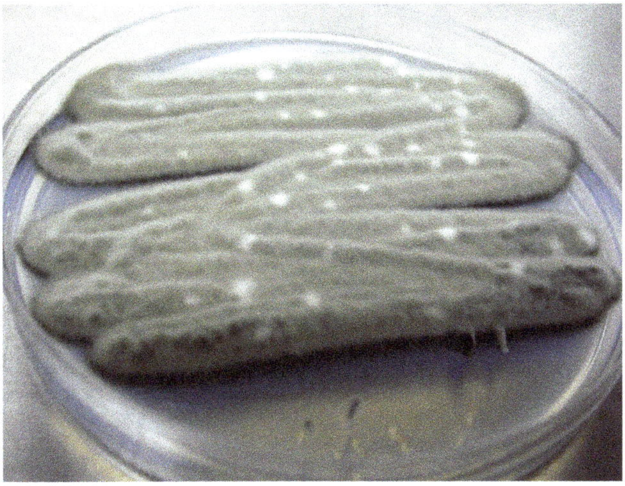

Fig. 9.23.3 *P. verrucosa* (side).

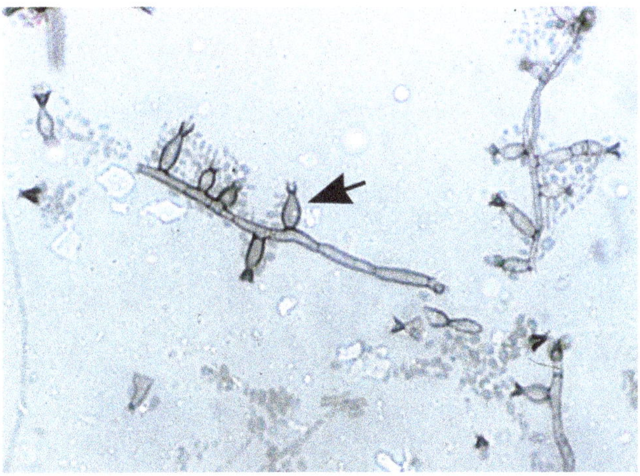

Fig. 9.23.4 Vase-shaped phialides with cup collarette.

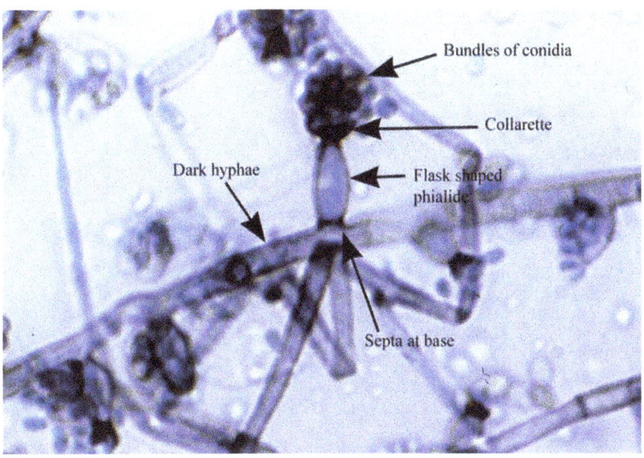

Fig. 9.23.5 Morphological features.

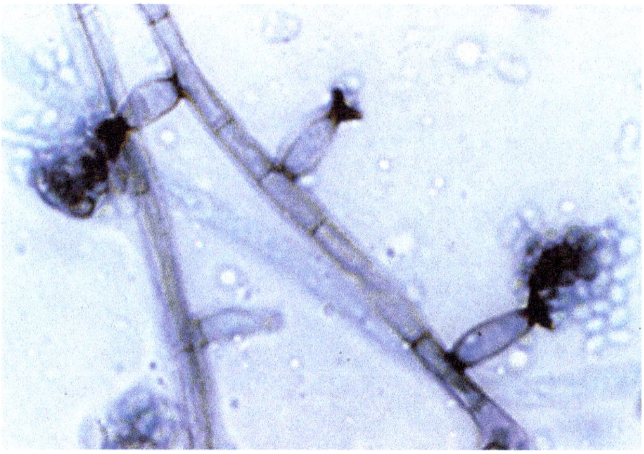

Fig. 9.23.6 Flask shaped phialides.

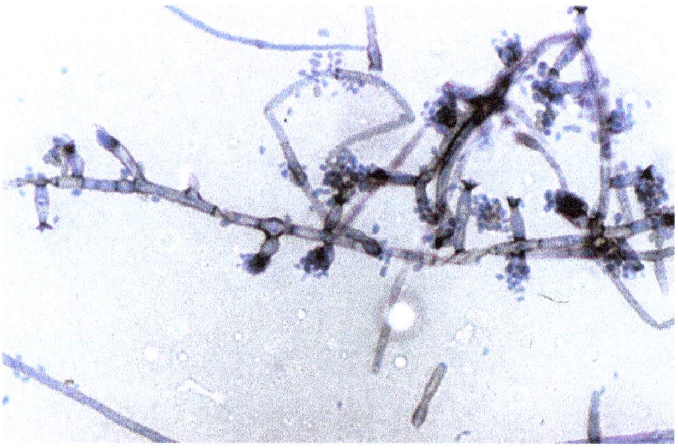

Fig. 9.23.7 Septum at base of phialides.

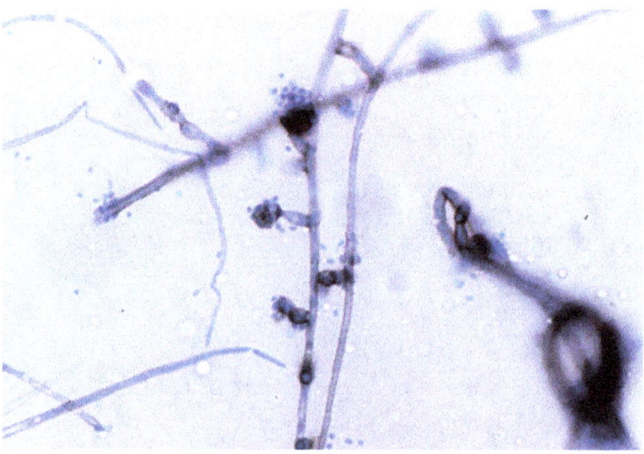

Fig. 9.23.8 Conidia cluster at tip of phialides.

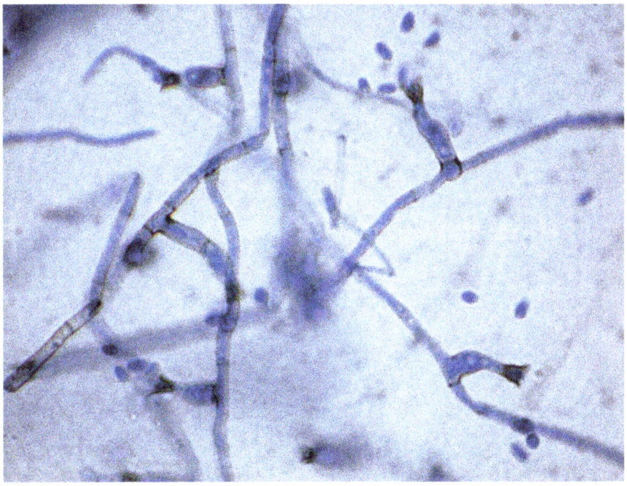

Fig. 9.23.9 Phialide morphology.

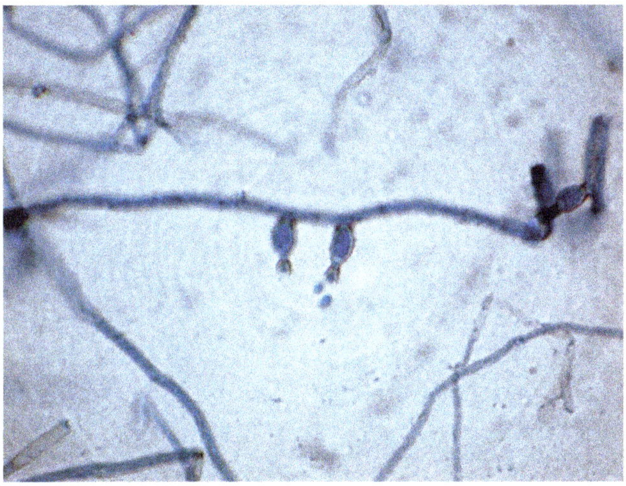

Fig. 9.23.10 Phialide morphology.

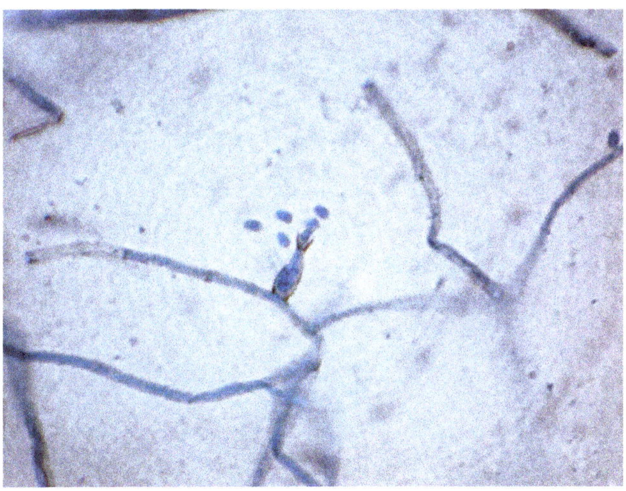

Fig. 9.23.11 Phialide morphology.

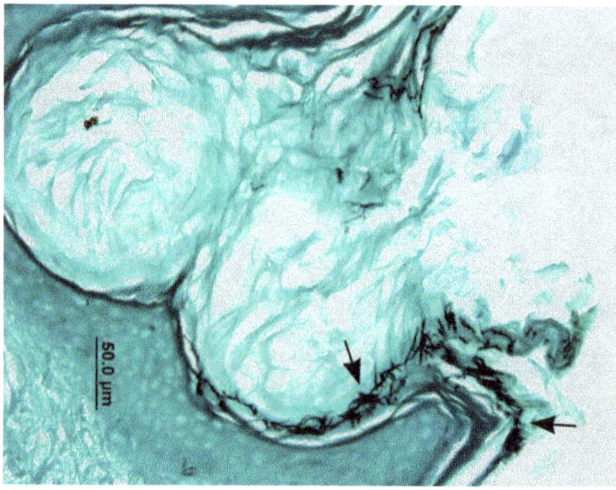

Fig. 9.23.12 *P. verrucosa* in human tissue, GMS stain.

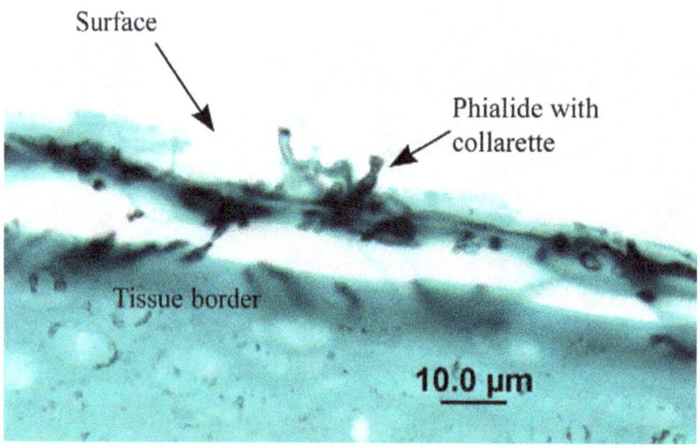

Fig. 9.23.13 *P. verrucosa* in tissue, GMS stain.

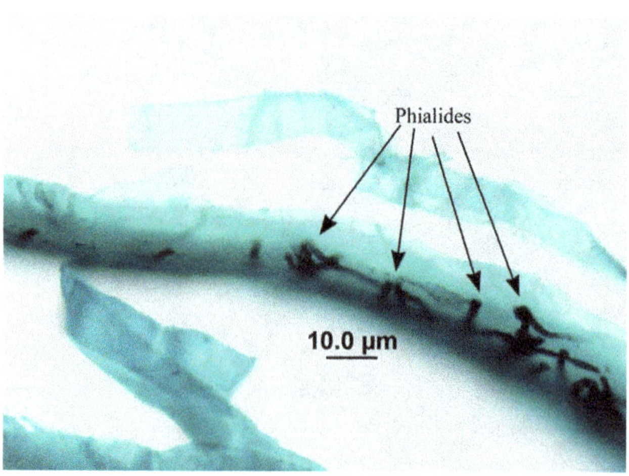

Fig. 9.23.14 *P. verrucosa* in tissue, GMS stain.

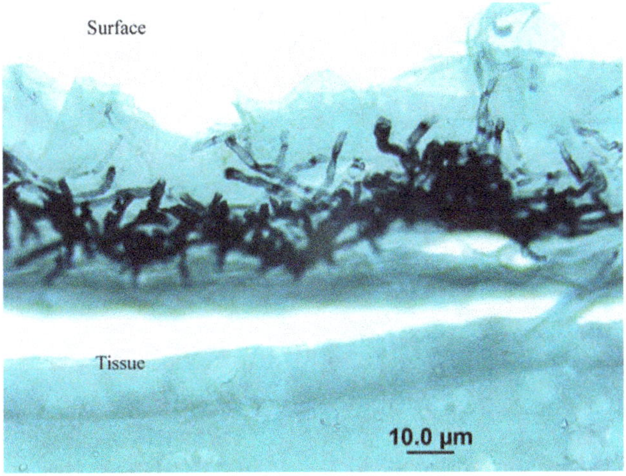

Fig. 9.23.15 *P. verrucosa* in tissue.

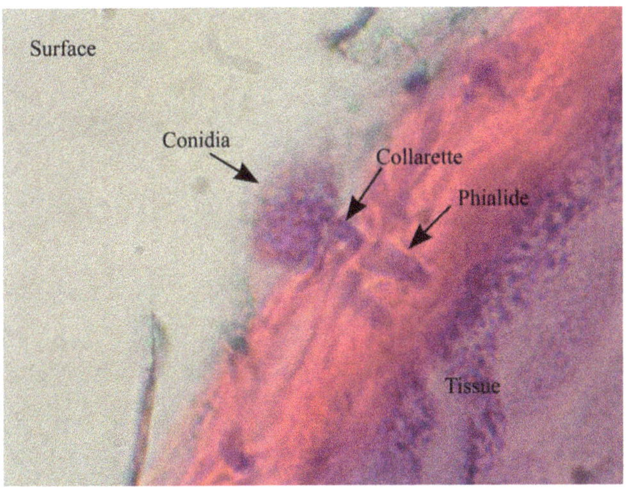

Fig. 9.23.16 *P. verrucosa* in human tissue, H&E stain.

9.24 *Pithomyces* sp

Colonial morphology: colonies on Sabouraud dextrose agar are white to cream, suede-like, and with age may develop a flocculent center with a brown hue. Reverse is golden with a dark brown center. Colonies on potato dextrose agar may be gray.

Microscopic morphology: hyphae are dark brown and highly septate. Conidiophores are produced directly from hyphae with no specific conidiophore structure. *Pithomyces* is differentiated from *Alternaria* and *Ulocladium* by its poor (or lack of) differentiation of conidiophores. Conidia are dark brown, multiseptate, egg shaped or pear shaped or oblong, and frequently spiny or warty. They contain 2–3 cross-septations and a transverse septum giving the appearance of the letters "H" or "IH." As conidia separate from conidiophores, an annular frill is attached at the base. Conidia are never produced in chains. Hyphae sometimes darken, swell and form dumbbell-shaped chlamydoconidia (Figs 9.24.13, 9.24.14).

Pithomyces

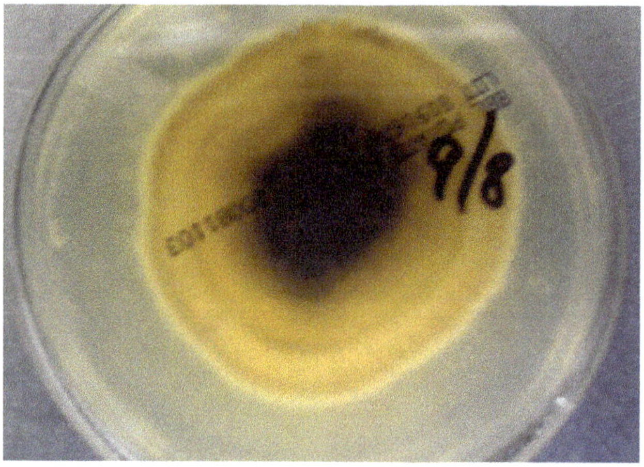

Fig. 9.24.1 *Pithomyces*, PDA (bottom).

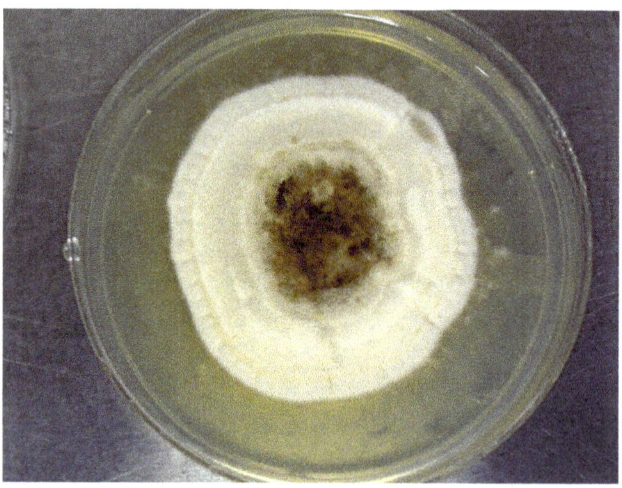

Fig. 9.24.2 *Pithomyces*, PDA (top).

Fig. 9.24.3 *Pithomyces*, PDA (side).

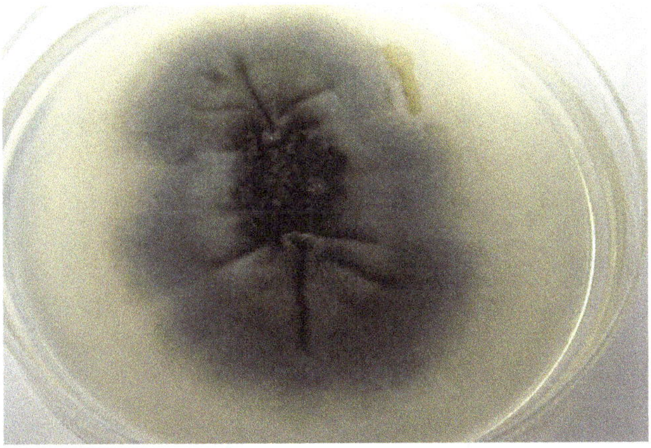

Fig. 9.24.4 *Pithomyces*, CMA (side).

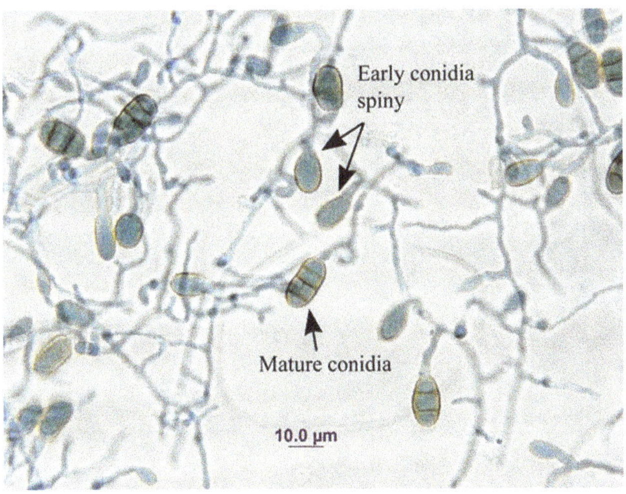

Fig. 9.24.5 General morphology.

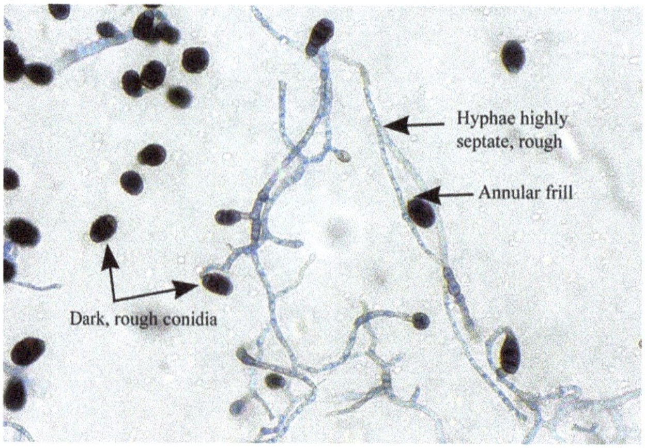

Fig. 9.24.6 *Pithomyces*, general morphology.

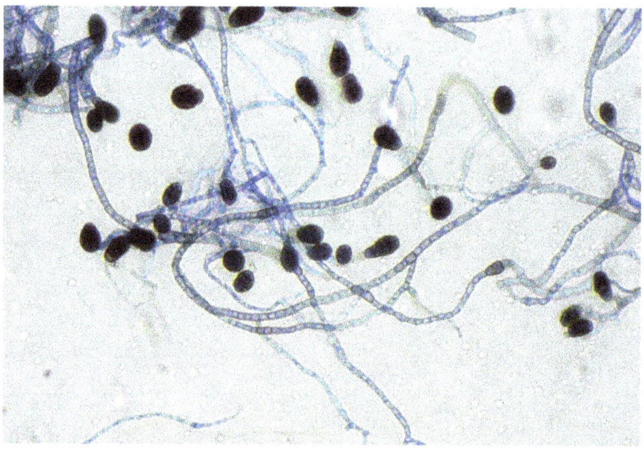

Fig. 9.24.7 General morphology.

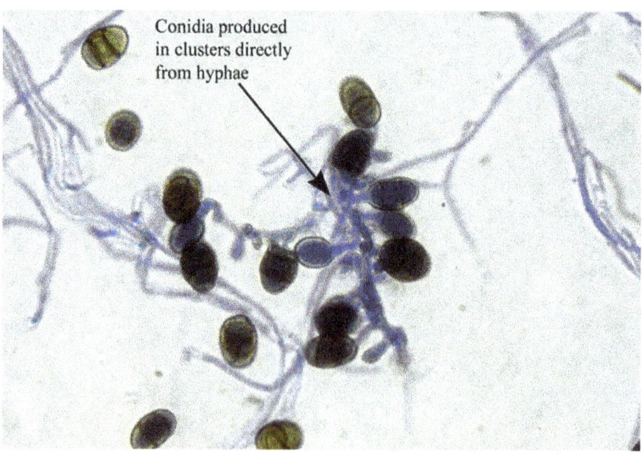

Fig. 9.24.8 Conidia production.

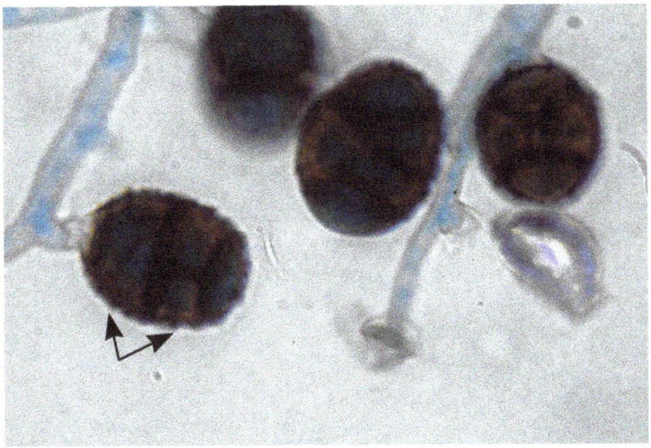

Fig. 9.24.9 Rough-walled conidia.

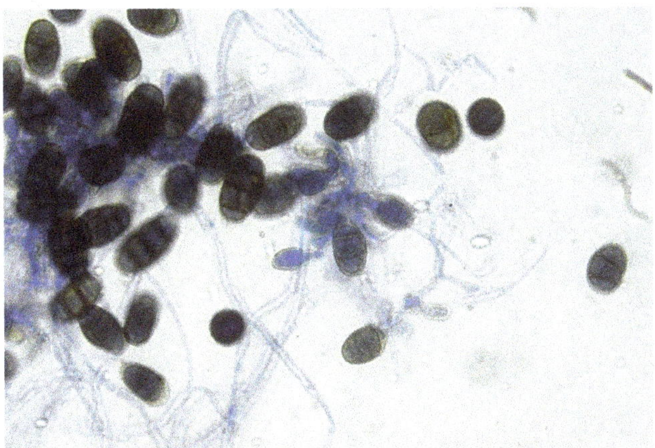

Fig. 9.24.10 *Pithomyces*, conidia showing cross-septal walls.

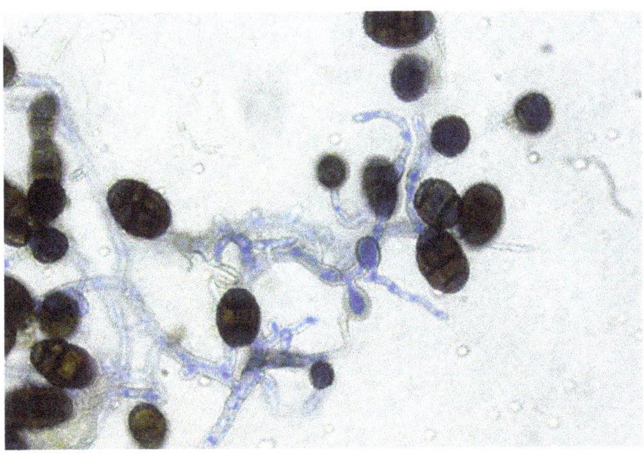

Fig. 9.24.11 Cross-septations appearing as "IH" or "H" divisions.

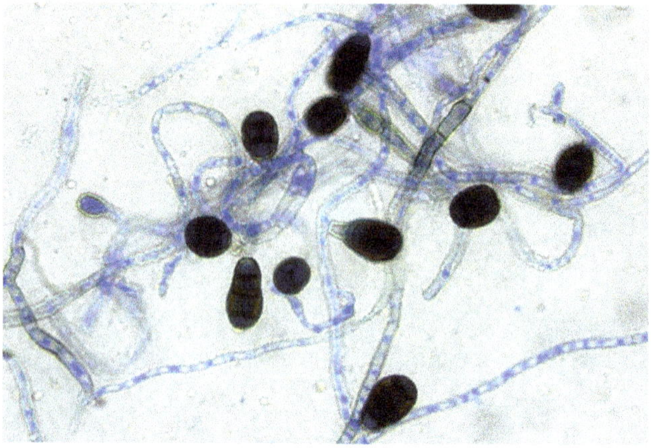

Fig. 9.24.12 *Pithomyces*, hyphae are multi-septate.

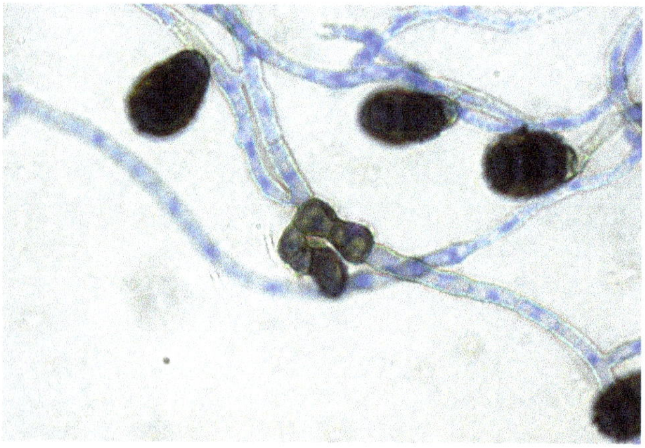

Fig. 9.24.13 Odd shaped chlamydoconidia are present.

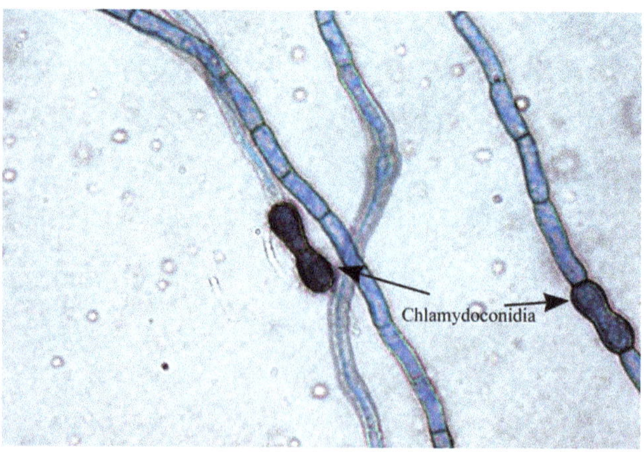

Fig. 9.24.14 Intercalary chlamydoconidia.

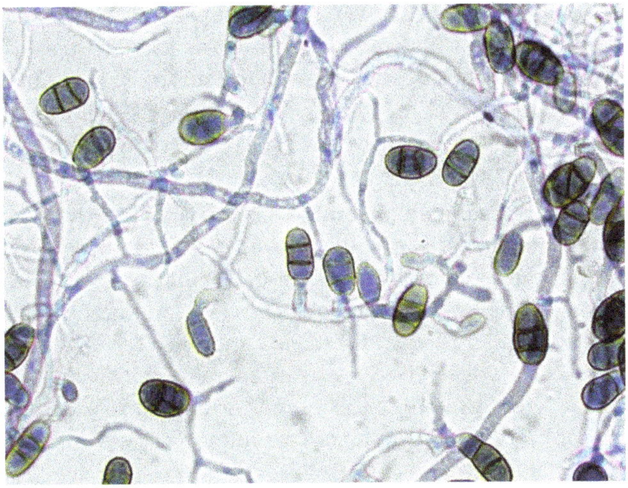

Fig. 9.24.15 *Pithomyces*, conidia produced on short or medium conidiophores.

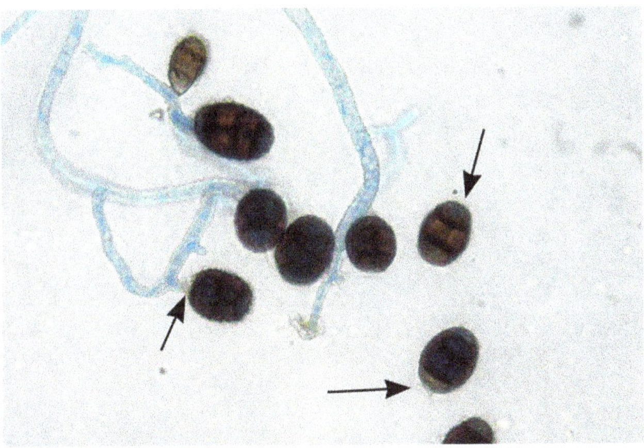

Fig. 9.24.16 Annular frill at base of conidia.

9.25 *Pseudallescheria boydii*

The taxonomy of both the teleomorph (sexual or perfect state) and anamorph (asexual or conidial state) phases of this fungus has undergone recent name changes. The name for the teleomorph phase is *Pseudallescheria boydii* (the names *Petriellidium boydii* and *Allescheria boydii* are no longer used). The name for the anamorph phase is *Scedosporium apiospermum* (the name *Monosporium apiospermum* is no longer used). All species of *Pseudallescheria* have *Scedosporium* anamorphs, therefore, the teleomorph name is restricted to description of this organism only when cleistothecia are present.

Colonial morphology: growth is described from corn meal agar because cleistothecia formation is best stimulated on corn meal agar or other nutrient-deficient media. Many isolates fail to produce cleistothecia regardless of growth conditions. The surface is first white and fluffy which then turns gray or brown. The reverse is first white but becomes gray or black. Colonies spread over the entire plate and dark brown/black dots may be seen; cultures often look as if someone has sprinkled pepper on the surface. These dots are cleistothecia (non-ostiolate ascocarps) that are yellow-brown to black, spherical, mostly submerged in the agar and are composed of irregularly interwoven brown hyphae. Inside cleistothecia are many ascospores.

Microscopic morphology: hyphae are hyaline and septate. Conidia are mostly pear shaped, non-clavate, pale brown or, rarely, broadly clavate elongated conidia ($4–9 \times 6–10\,\mu m$), that are rounded at the top and flat at the base. Conidia are produced singly or in clusters on long, simple or branched conidiophores or perpendicular along the hyphae. They are very similar to conidia of *Blastomyces dermatitidis*. *Graphium*-type conidiation may also be observed in slide preparations. Due to the variability of conidia size and shape and production, this fungus is difficult to identify. See *Scedosporium apiospermum* for more photos of the most common presentation of this fungus.

Pseudallescheria boydii culture photos

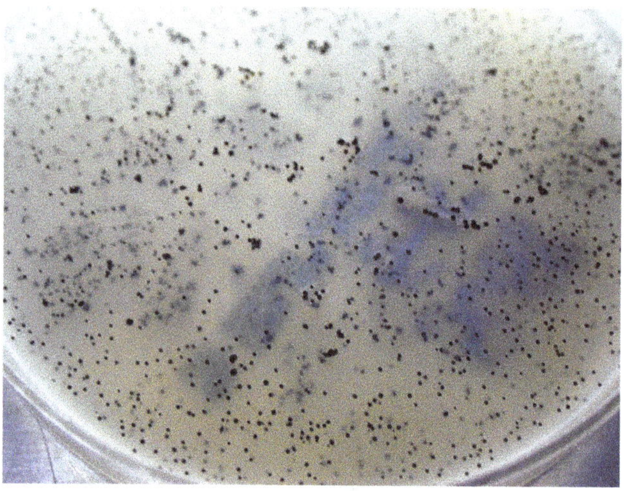

Fig. 9.25.1 *Pseudallescheria boydii*, CMA (top).

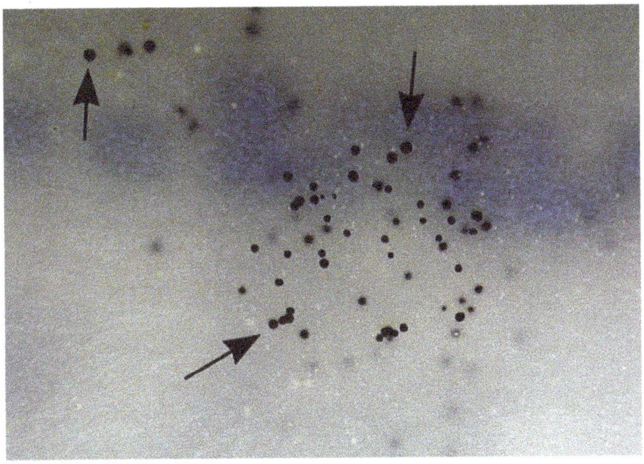

Fig. 9.25.2 *P. boydii*, cleistothecia on CMA (close).

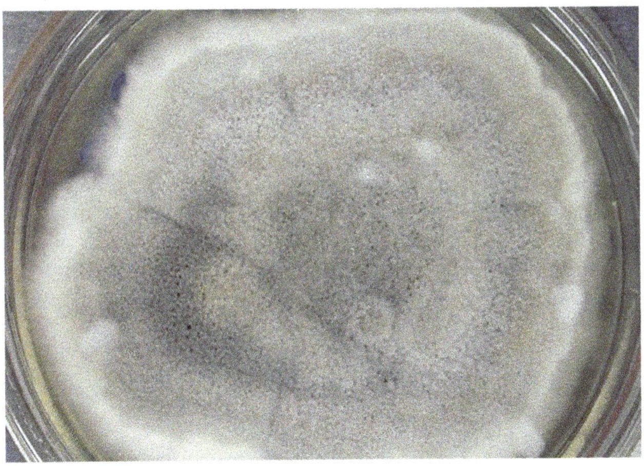

Fig. 9.25.3 *P. boydii*, SABDX (top).

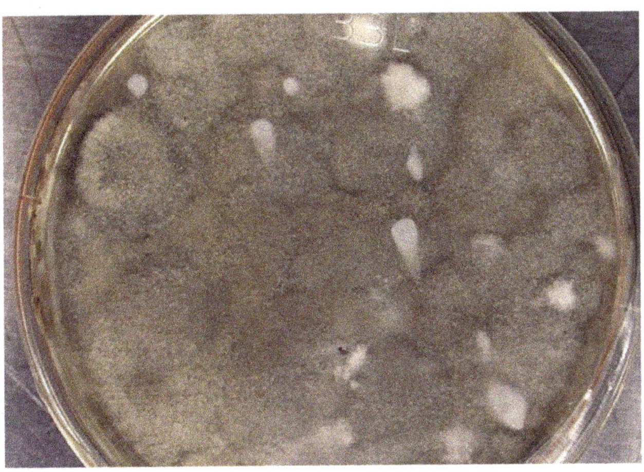

Fig. 9.25.4 *P. boydii*, SAB BHI with CC (top).

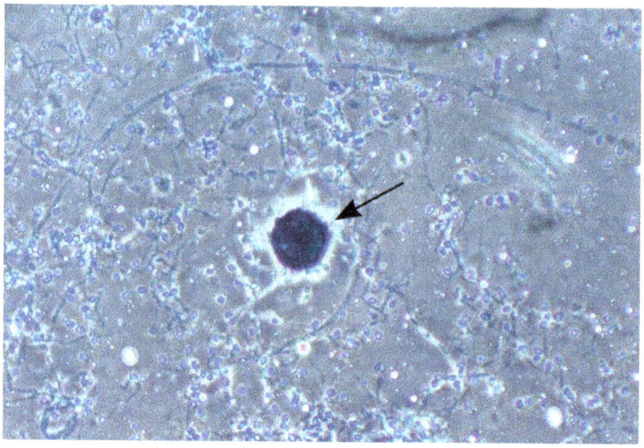

Fig. 9.25.5 Showing cleistothecia.

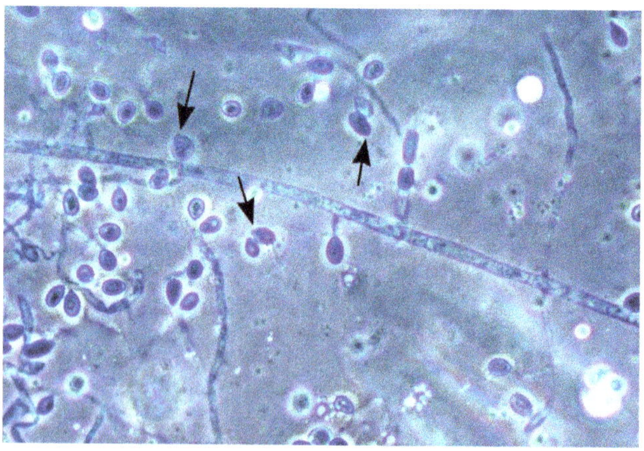

Fig. 9.25.6 Truncate conidia.

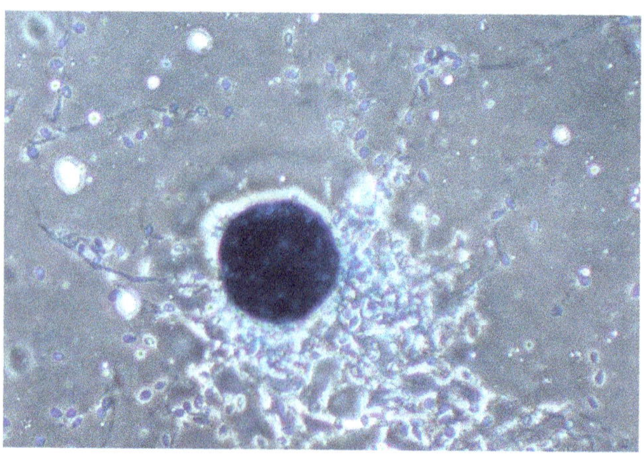

Fig. 9.25.7 Cleistothecium, 400×.

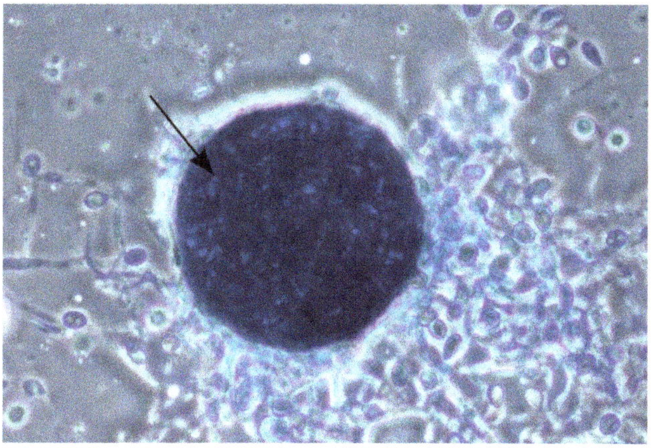

Fig. 9.25.8 Ascospores within cleistothecia.

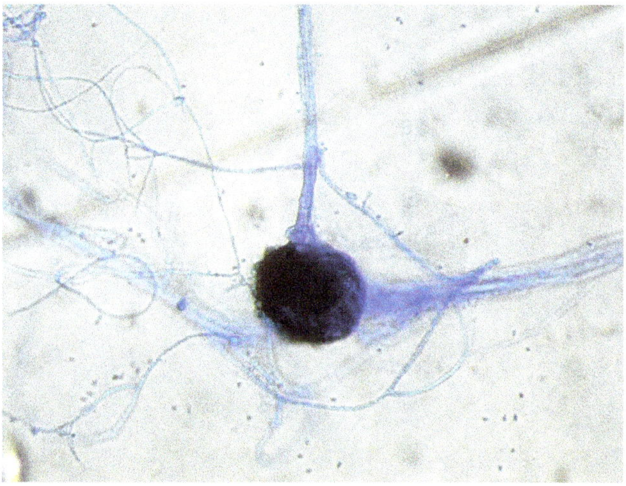

Fig. 9.25.9 Cleistothecium attached to clustered hyphae.

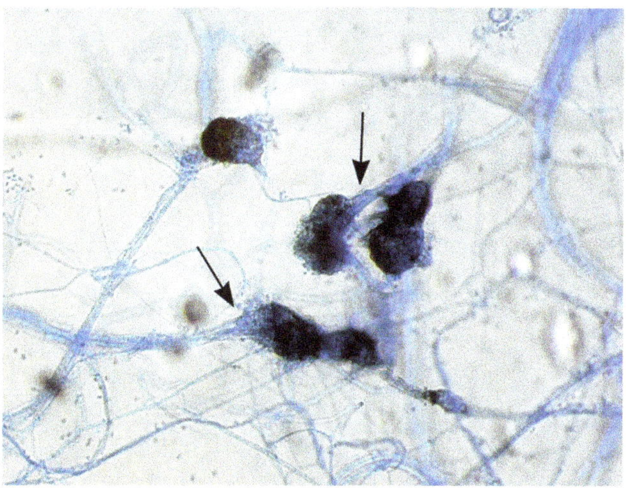

Fig. 9.25.10 *Graphium*-like clustering of hyphae with atypical cleistothecia formation.

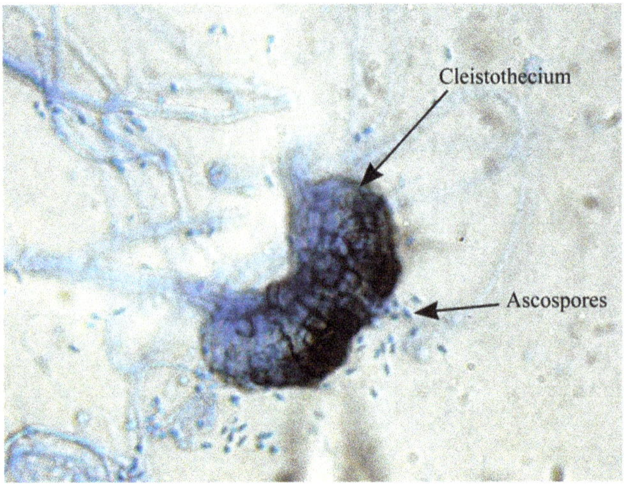

Fig. 9.25.11 Ascopores released from cleistothecium.

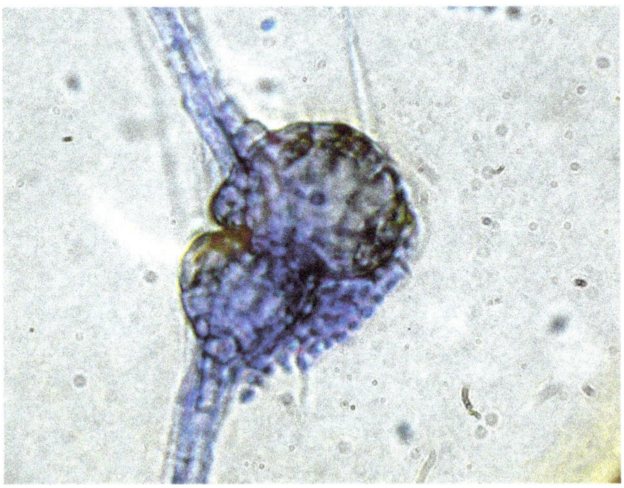

Fig. 9.25.12 Cleistothecium, 1000×; with released ascospores clustered down the side.

9.26 *Scedosporium*

9.26.1 *Scedosporium apiospermum*

Teleomorph = *Pseudallescheria boydii*

Colonial morphology: colonies are wooly to cottony. Colonies are at first white, then produce aerial hyphae that later become gray-brown. The reverse is white-yellow and may turn black with age. It is resistant to cycloheximide whereas the teleomorph may be inhibited.

Microscopic morphology: hyphae are hyaline and septate (2–4 µm in diameter). Conidia are produced singly or in small groups on long, simple or branched conidiophores or produced laterally on hyphae. Conidia are single-celled, pale brown, broad to egg shaped, 4–9 × 6–10 µm in size and rounded on the far end, much like balloons. The base of the conidium is truncate (flat) and just below the point of attachment, annellations can be observed on the conidiophores. This appears as concentric ring-like structures along the apical ends of the conidiophores. Rarely, conidiophores may assimilate together forming synnema (*Graphium* state). For further information on the teleomorph state, see *Pseudallescheria boydii*.

Note: Microscopic morphology of this genus is variable depending upon the stage of growth and culture media used. Growth on BHI agar often shows colonies resembling *Blastomyces dermatitidis*. Paired with microscopic morphology, it can easily be misidentified as *Blastomyces*. Yeast conversion at 37 °C is positive for *Blastomyces* and negative for *Scedosporium* spp. When in doubt, gene probes for *B. dermatitidis* should be performed. Microscopic morphology may also resemble that of *Chrysosporium*, *Emmonsia parva*, and *Trichophyton* spp.

Scedosporium apiospermum

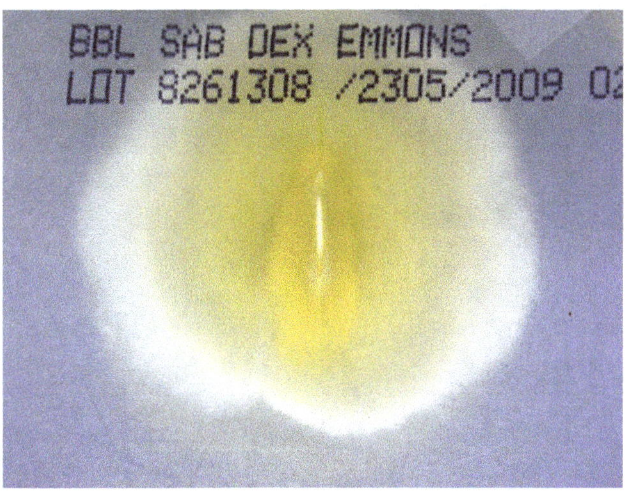

Fig. 9.26.1 *Scedosporium apiospermum* (bottom).

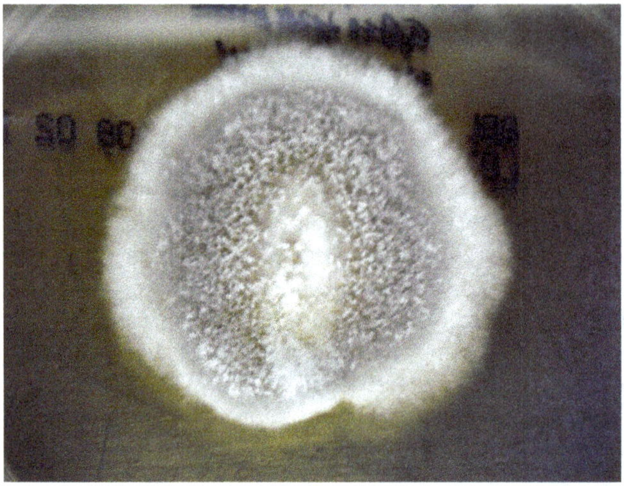

Fig. 9.26.2 *S. apiospermum* (top).

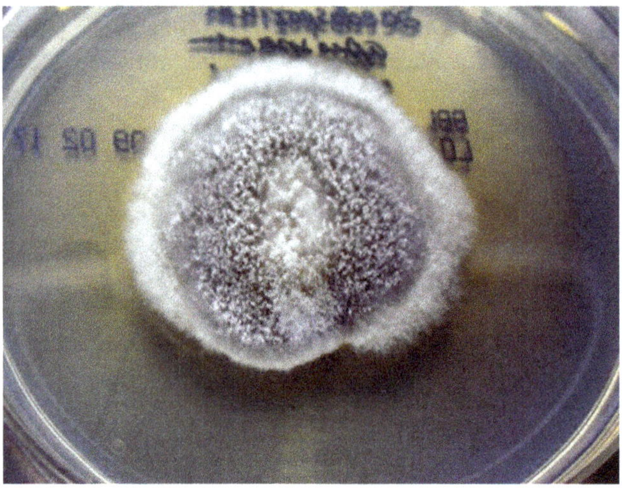

Fig. 9.26.3 *S. apiospermum* (side).

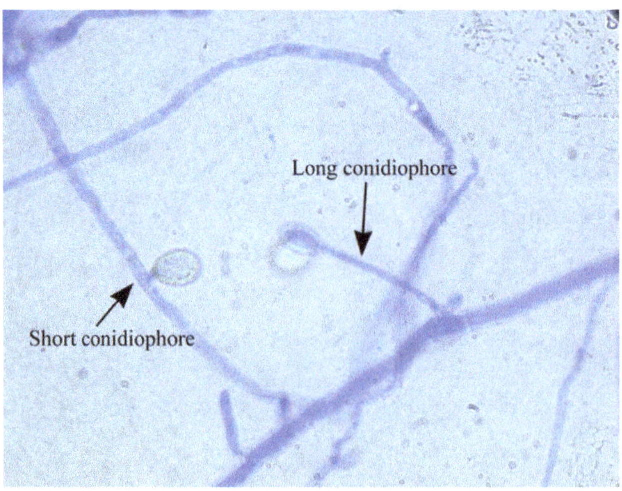

Fig. 9.26.4 *S. apiospermum*, conidiophores.

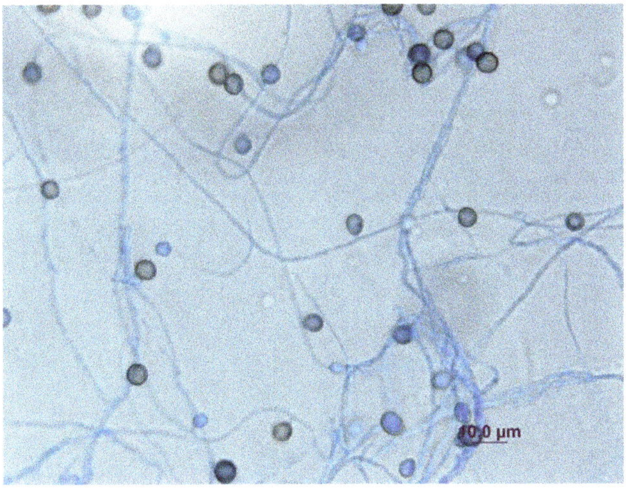

Fig. 9.26.5 *S. apiospermum*, conidia.

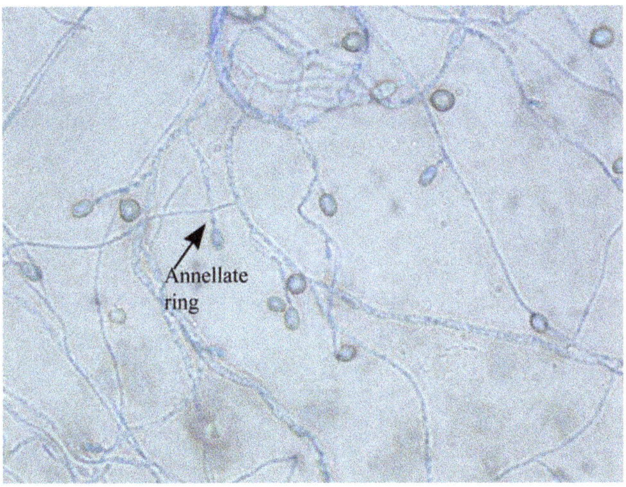

Fig. 9.26.6 *S. apiospermum*, conidiophores with annellate ring.

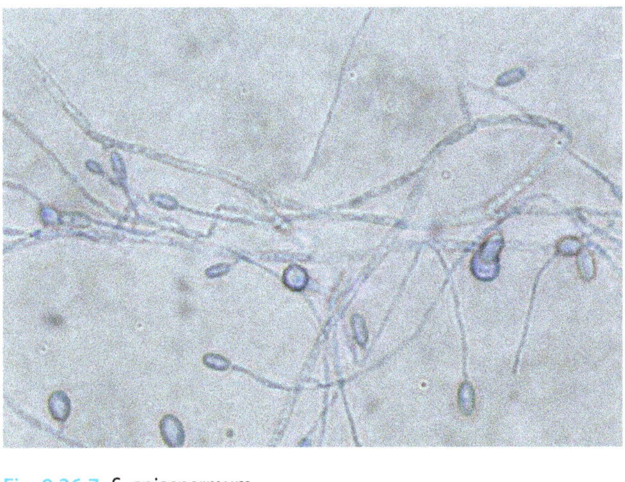

Fig. 9.26.7 *S. apiospermum*.

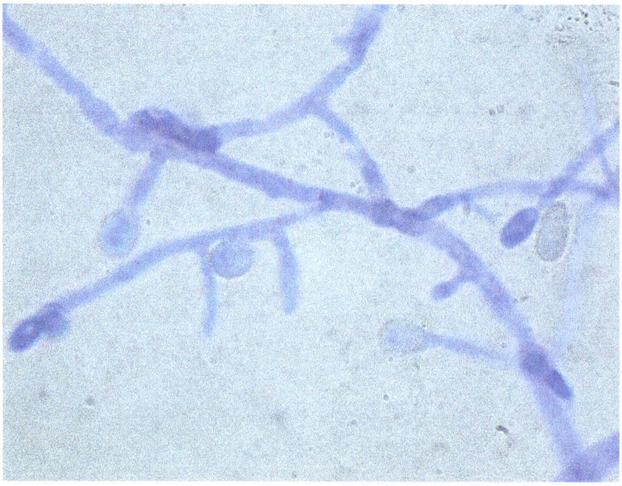

Fig. 9.26.8 *S. apiospermum*, 1000×.

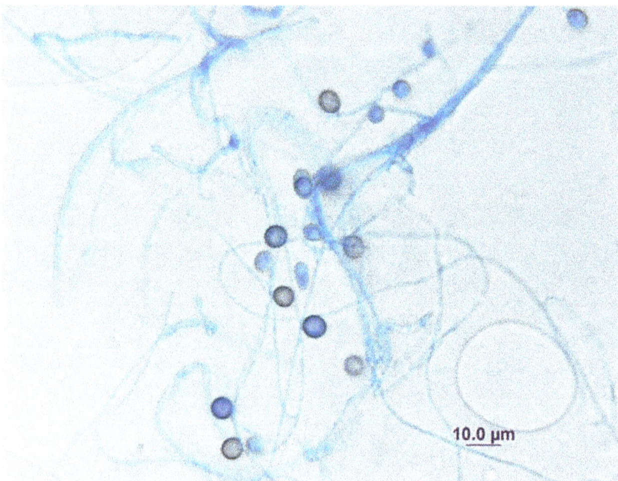

Fig. 9.26.9 *S. apiospermum*.

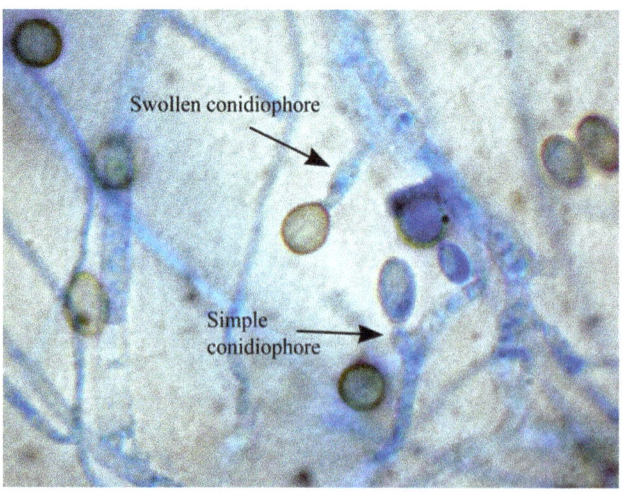

Fig. 9.26.10 *S. apiospermum*, conidiophores, 1000×.

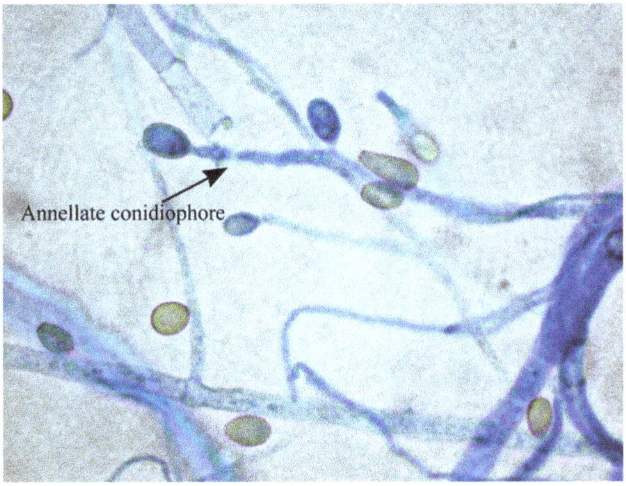

Fig. 9.26.11 *S. apiospermum*, conidiophores.

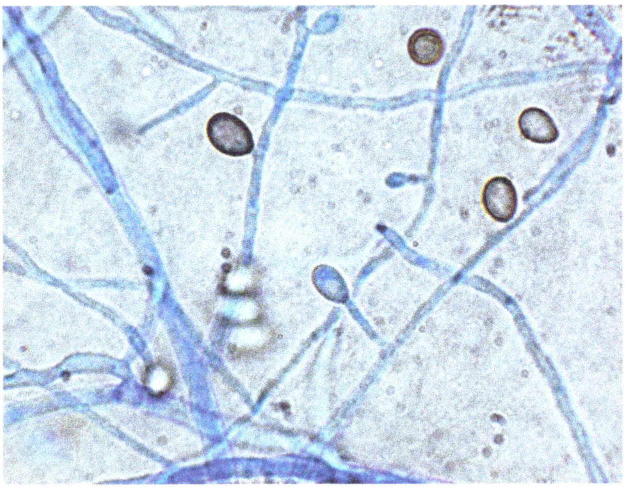

Fig. 9.26.12 Dark conidia.

9.26.2 *Scedosporium prolificans (inflatum)*

Scedosporium prolificans has no known sexual state. Culture growth and microscopic morphology are very similar to *S. apiospermum*. It is differentiated from *S. apiospermum* by growth at 45 °C, inhibition by cycloheximide and microscopic morphology. The conidiophores of *S. prolificans* are inflated and broad as they arise from the mycelium. They then narrow at the annellide end where conidia attach. Thus, they resemble phialides and can be mistaken for *Phialophora*.

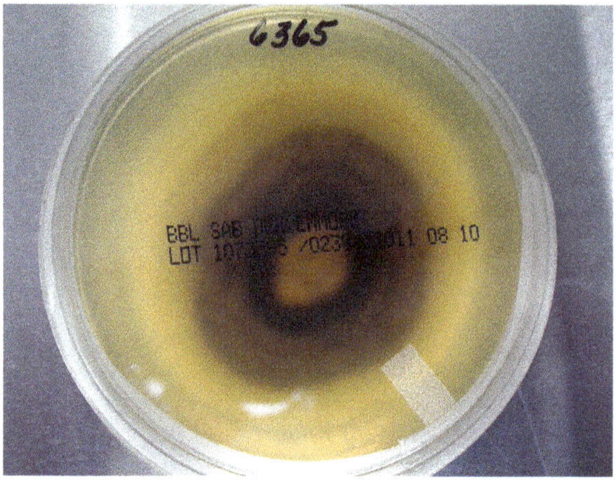

Fig. 9.26.13 *Scedosporium prolificans* (bottom).

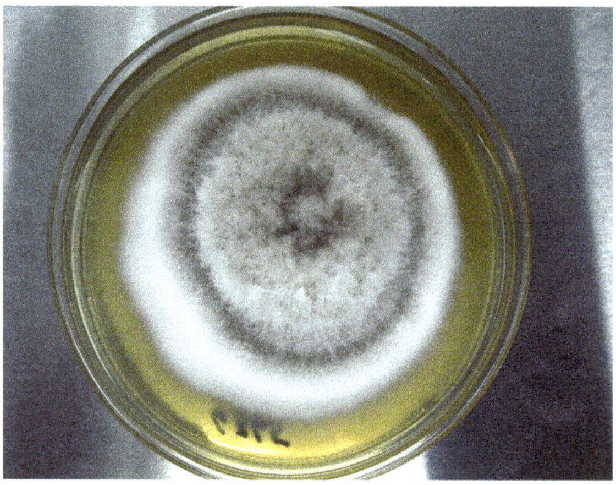

Fig. 9.26.14 *S. prolificans* (top).

Fig. 9.26.15 *S. prolificans* (side).

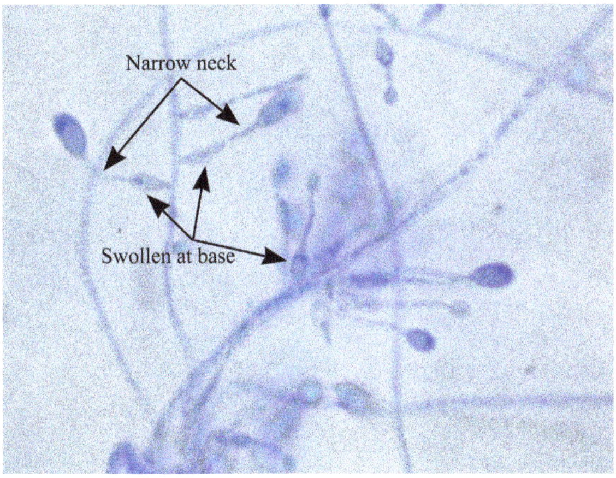

Fig. 9.26.16 *S. prolificans*, conidiophore morphology.

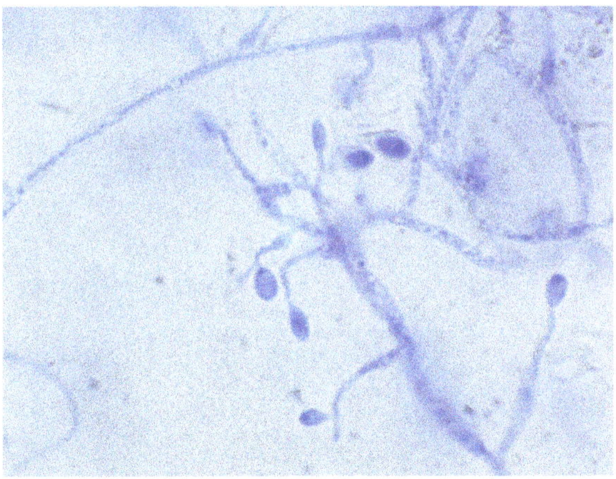

Fig. 9.26.17 Conidiophore size variation.

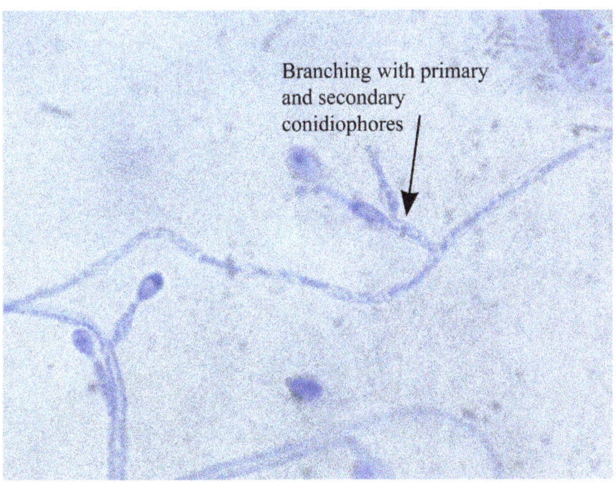

Fig. 9.26.18 *S. prolificans*, conidiophore morphology.

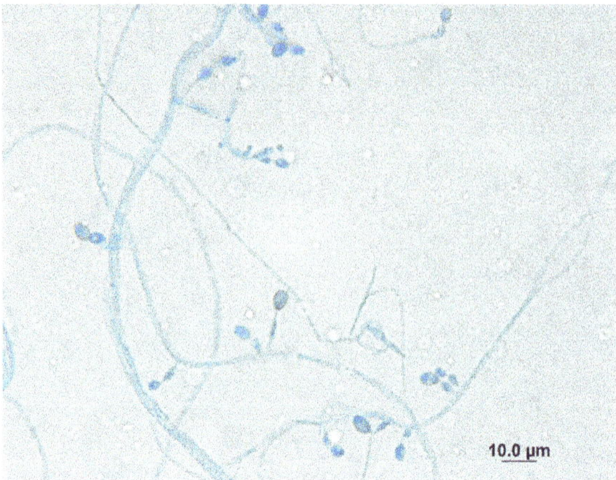

Fig. 9.26.19 Mature conidiophores.

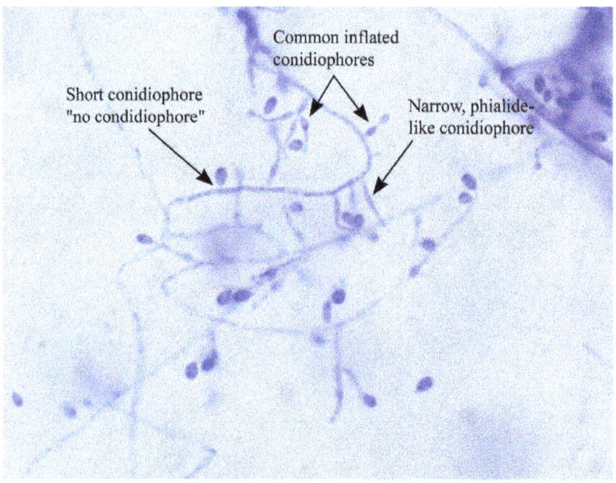

Fig. 9.26.20 Types of conidiophores.

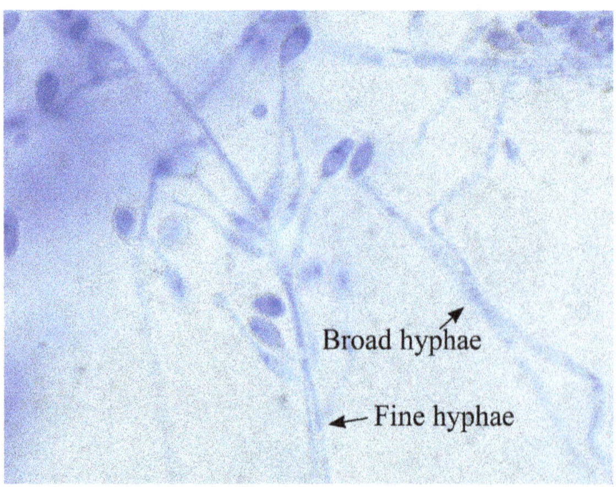

Fig. 9.26.21 Hyphae.

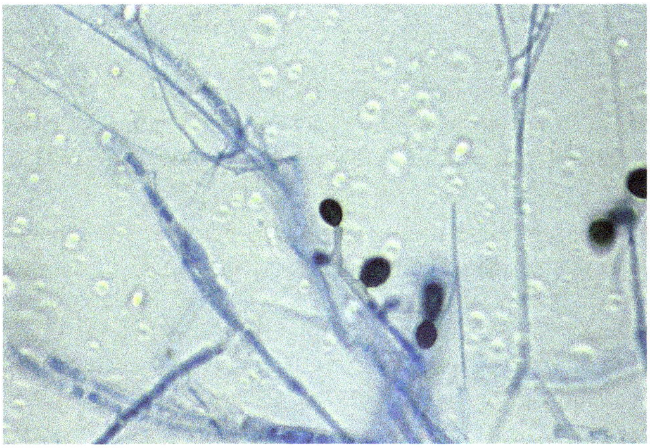

Fig. 9.26.22 Stages of conidial development.

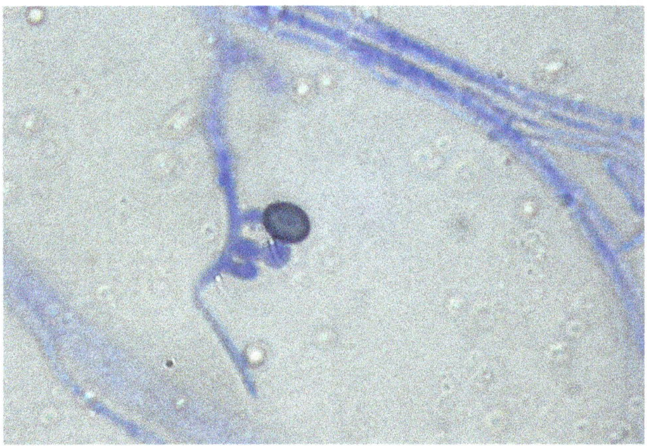

Fig. 9.26.23 Conidia development.

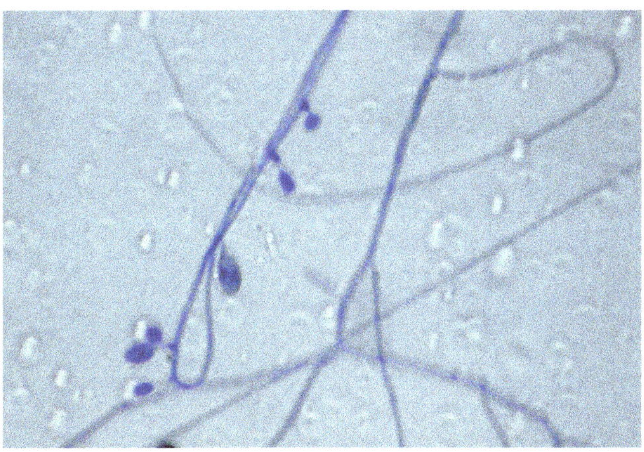

Fig. 9.26.24 Conidia development.

9.27 *Stachybotrys chartarum*

Colonial morphology: colonies grow rapidly in 3–5 days and fill the plate. They are at first white, and then become dark brown to black. Colonies are fuzzy and may have concentric rings of dark color. The reverse is light and then becomes dark brown to black with age.

Microscopic morphology: hyphae are hyaline and septate but conidiophores are dark. Conidiophores are narrow when young, but become thick, septate and sometimes branched. Hyphae and conidiophores may sometimes fuse together when they touch (Fig. 9.27.3). Conidiophores produce clusters (3–8) of phialides from which dark brown longitudinal, oval conidia (4 × 9 μm) are produced in clusters. Young conidia are smooth and hyaline while older conidia become rough, warty or spiny. Conidia are single celled, but some may give the appearance of being septate. Rarely, round chlamydospores may be produced directly from hyphae or at the tips of hyphae on septate chlamydophores (Fig. 9.27.12).

Stachybotrys is an environmental fungus associated with "sick building syndrome." It produces mycotoxins, specifically tricothecenes. Although it is not considered a pathogen, the potent toxins are harmful to humans and can cause a variety of serious and sometimes deadly symptoms.

Stachybotrys chartarum

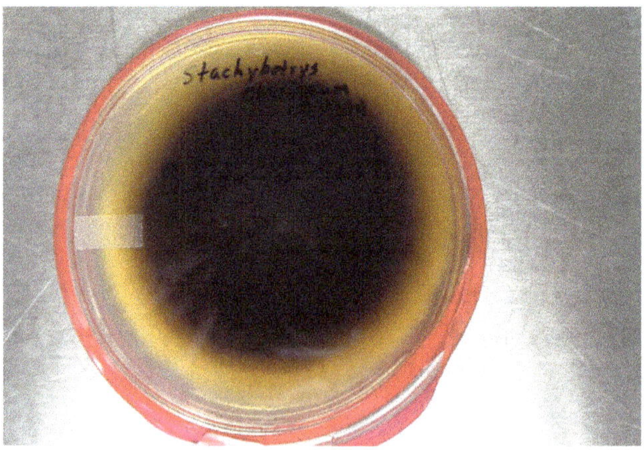

Fig. 9.27.1 *Stachybotrys chartarum* (bottom).

Fig. 9.27.2 *S. chartarum* (top).

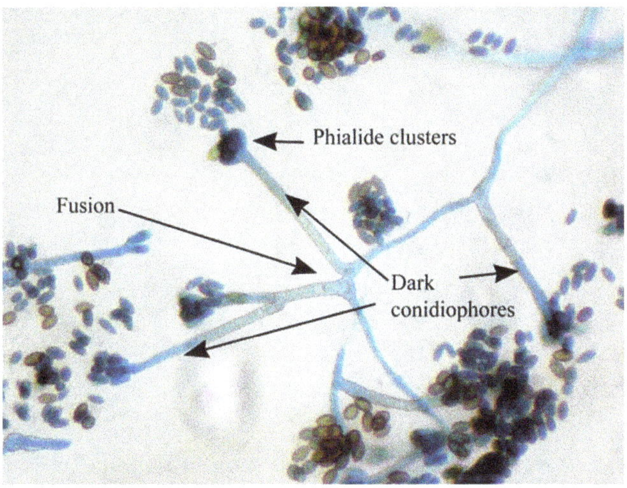

Fig. 9.27.3 *S. chartarum*, basic morphology.

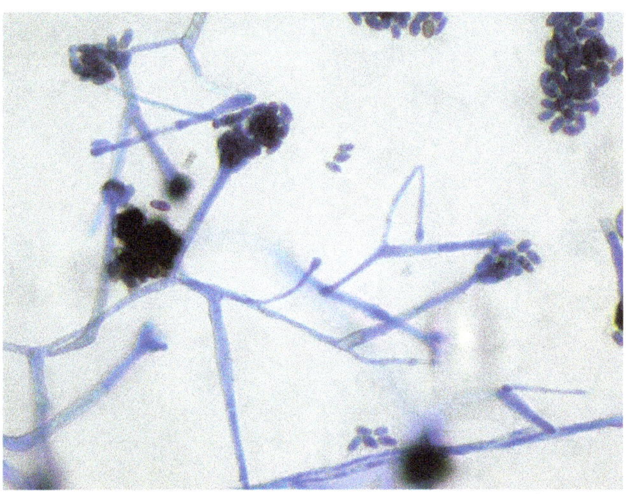

Fig. 9.27.4 Stages of conidiophore development.

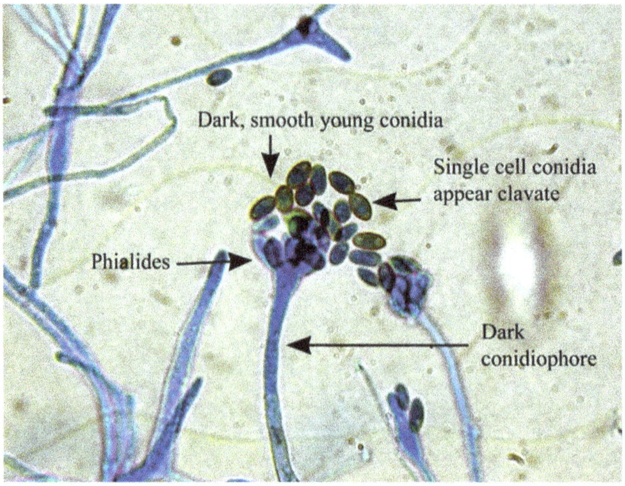

Fig. 9.27.5 *S. chartarum*, basic morphology.

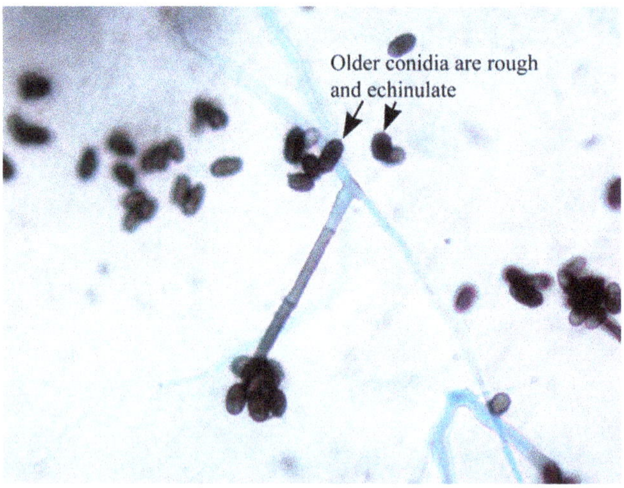

Fig. 9.27.6 *S. chartarum*, with rough echinulate conidia.

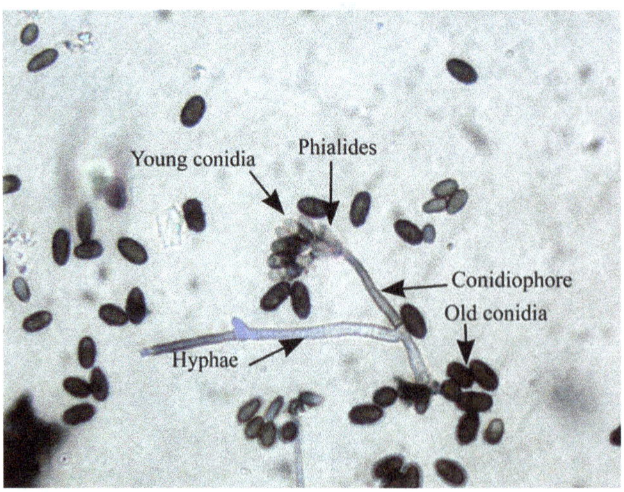

Fig. 9.27.7 *S. chartarum*, basic morphology.

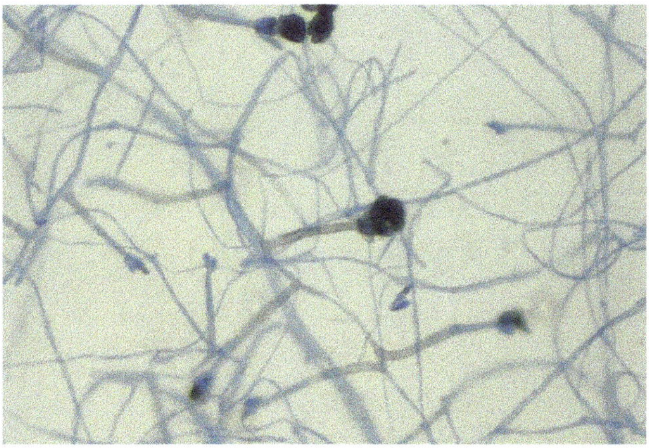

Fig. 9.27.8 Light stain showing dark conidiophore.

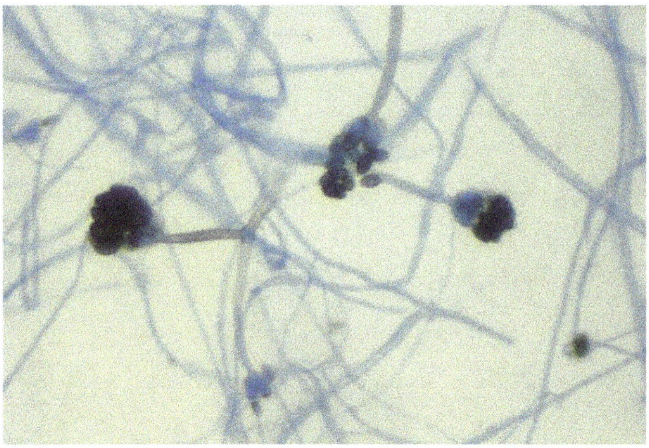

Fig. 9.27.9 Light stain showing dark conidia.

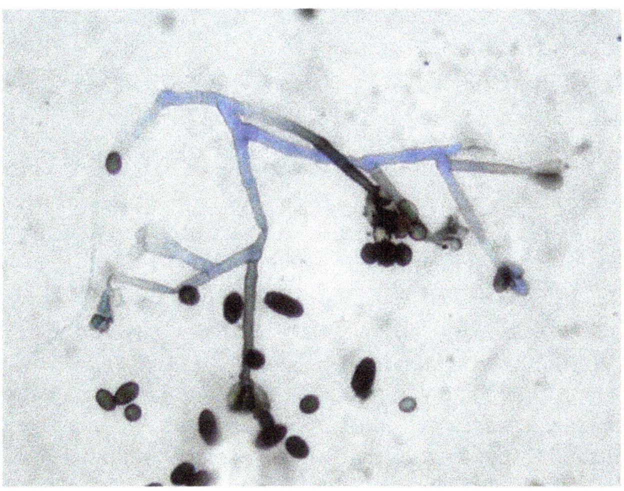

Fig. 9.27.10 Smooth- and rough-walled conidia.

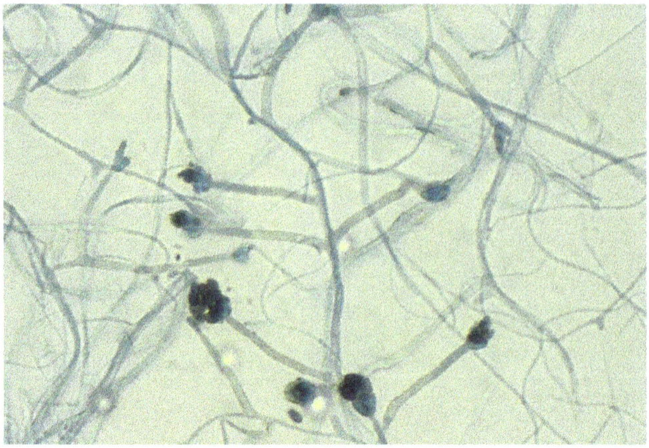

Fig. 9.27.11 Typical branching.

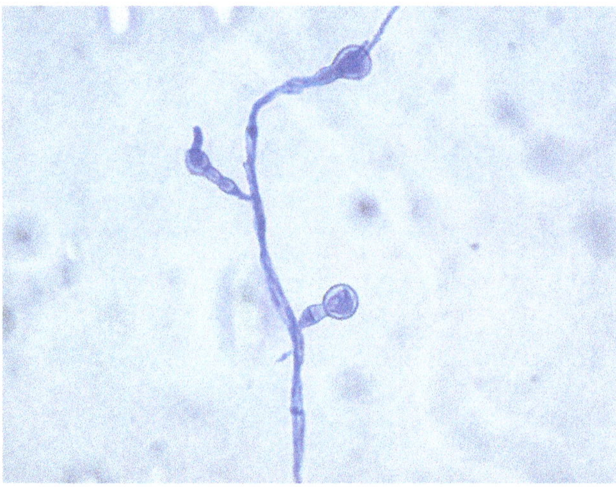

Fig. 9.27.12 Rare chlamydospore formation.

9.28 *Stemphylium* sp

Colonial morphology: colonies grow quickly and are brown to greenish-black or gray and rough, becoming fuzzy. The reverse is black.

Microscopic morphology: hyphae are hyaline to dark brown and septate. Conidiophores are perpendicular to the hyphae, solitary, long, rarely branched, and develop dark, terminal swelling. The conidia are large (12–20 × 15–30 μm), dark brown, oval or round, sometimes appearing as a "figure of 8" with cross-septations. The cross-septations (5–6) are vertical and horizontal, sometimes diagonal. The septations often appear in a checkerboard pattern. Hyphae become highly septate, dark, and form intercalary conidia in a "figure of 8", similar to *Alternaria*.

Stemphylium can be distinguished from *Ulocladium* which produces sympodial conidiophores, whereas *Stemphylium* does not. The conidia of *Stemphylium* are generally smooth while those of *Pithomyces* and *Epicoccum* are rough and bumpy.

Stemphylium

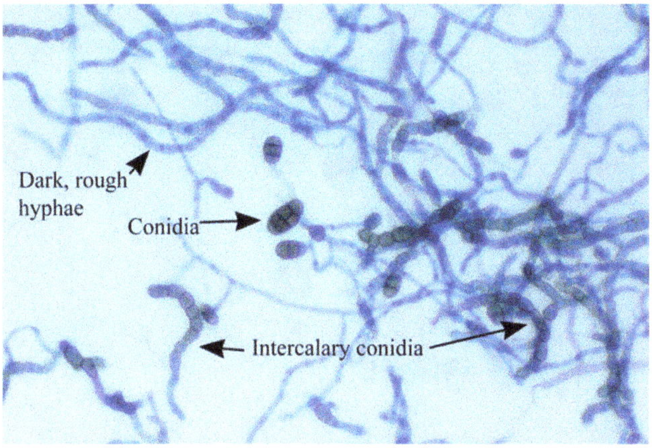

Fig. 9.28.1 *Stemphylium* basic morphology.

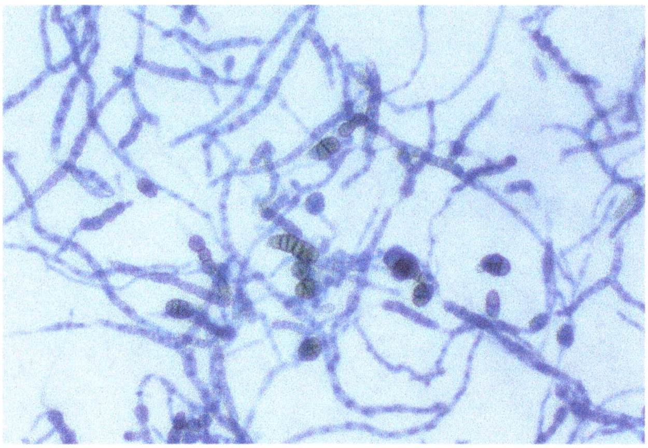

Fig. 9.28.2 Abundant intercalary conidia production.

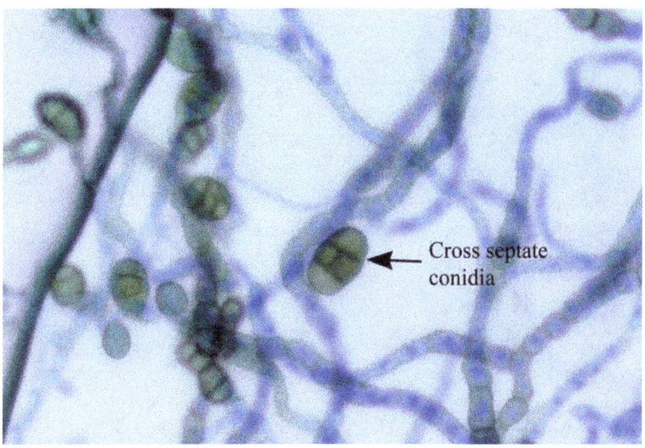

Fig. 9.28.3 Conidia.

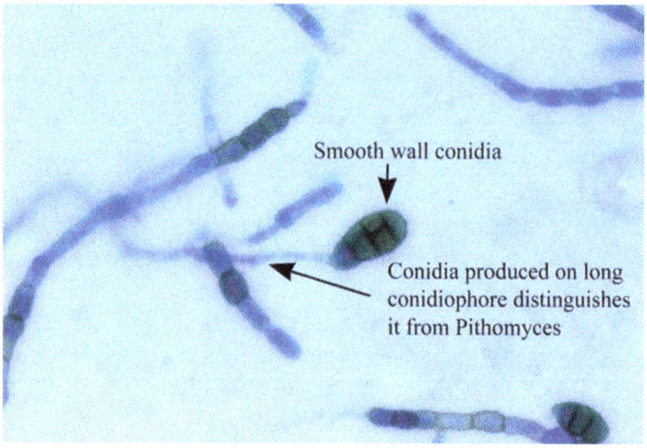

Fig. 9.28.4 Differentiation from *Pithomyces*.

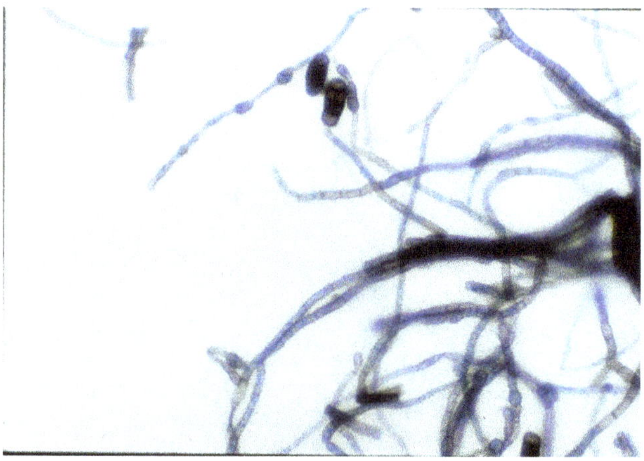

Fig. 9.28.5 Light stain showing dark color of hyphae and conidia.

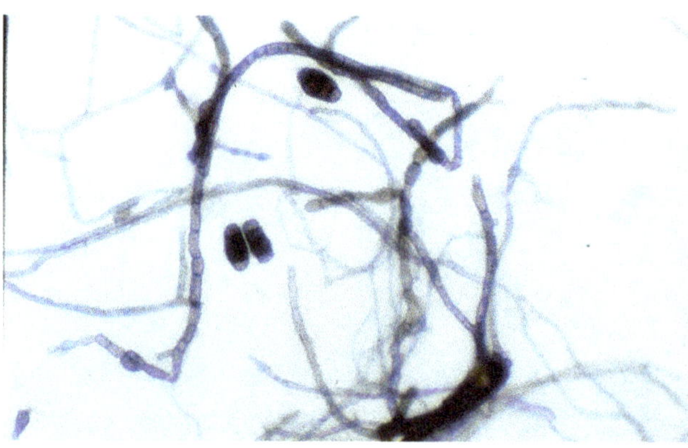

Fig. 9.28.6 Light stain showing dark morphology.

9.29 *Ulocladium* sp

Colonial morphology: colonies are gray or brown to black, fluffy and cottony, with white accents around the border of the colony. The reverse has a black center surrounded by a brown to orange border.

Microscopic morphology: hyphae are light to dark brown and septate. Individual, multicelled conidia (dictyoconidia) are formed through a pore (poroconidia) at the tip of the conidiophore. As more conidia (7–12 × 18–24 µm) are formed, they are deposited along the conidiophore in a zig-zag arrangement (sympodially). The oval or round conidia are narrow at the base (some flat on one end), rough and bumpy, and contain horizontal and vertical septations. Multiseptate conidia are not as often seen as with other genera that also produce dark brown septate conidia (i.e. *Stemphylium*, *Pithomyces*, and *Epicoccum*). *Ulocladium* dictyoconidia have a characteristic "y" septation that is unique, helping to identify it (Fig. 9.29.9). Older cultures may produce long, cigar-shaped conidia with three or more horizontal septations, resembling macroconidia of the dermatophytes.

Ulocladium

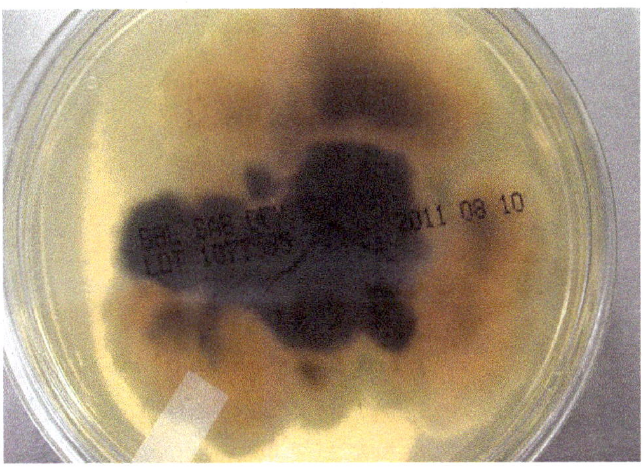

Fig. 9.29.1 *Ulocladium* sp (bottom).

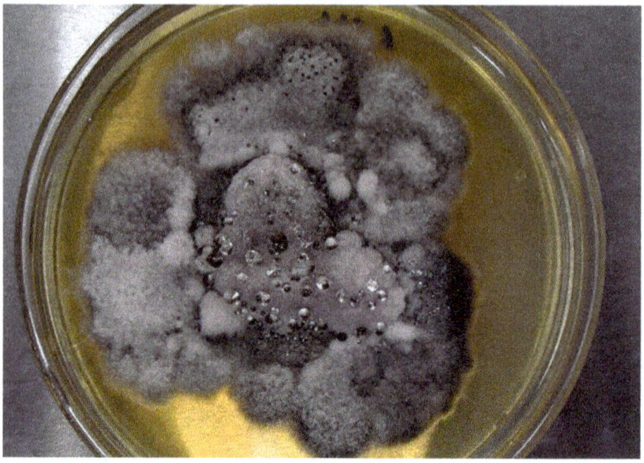

Fig. 9.29.2 *Ulocladium* sp (top).

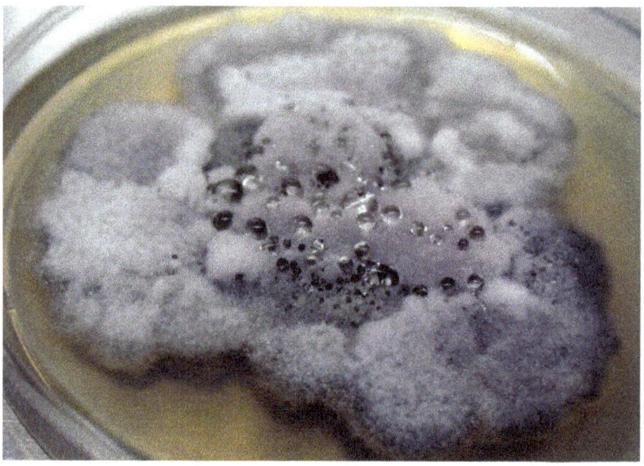

Fig. 9.29.3 *Ulocladium* sp (side).

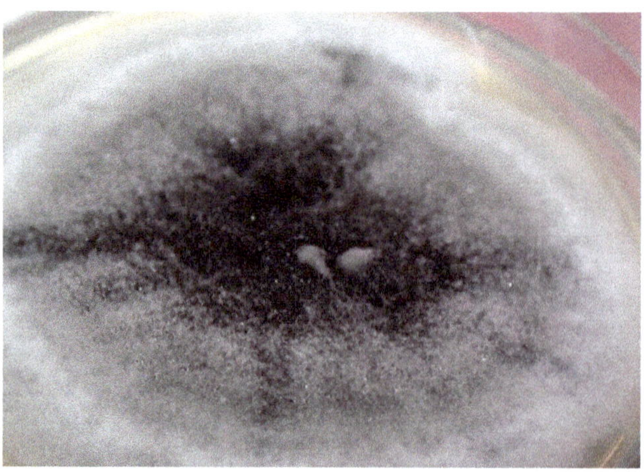

Fig. 9.29.4 *Ulocladium* sp (side).

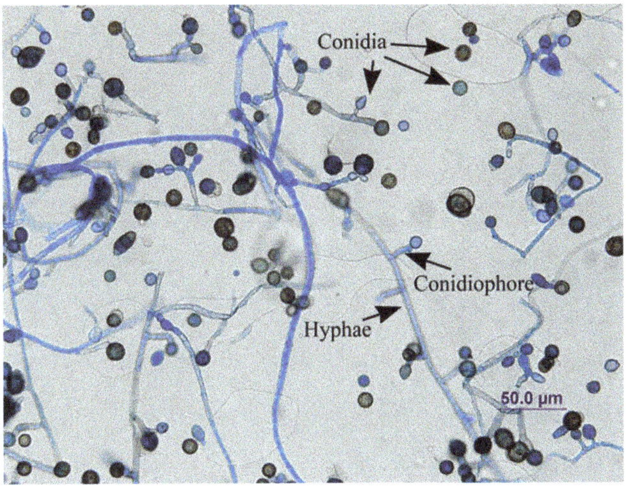

Fig. 9.29.5 *Ulocladium*, showing short clonidiophores and round conidia.

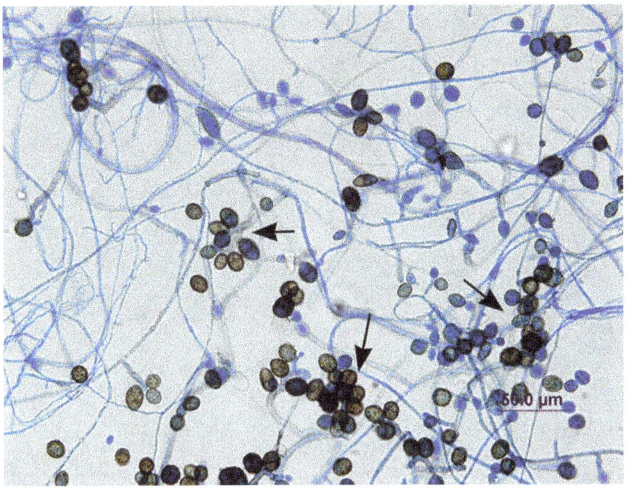

Fig. 9.29.6 Conidia accumulate along conidiophores.

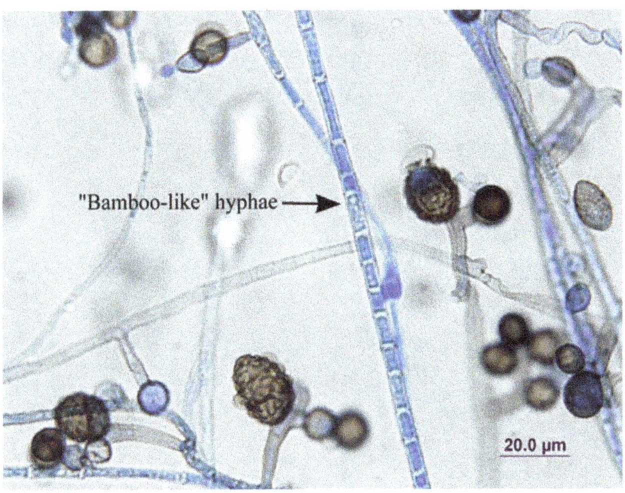

Fig. 9.29.7 Bamboo-like hyphae.

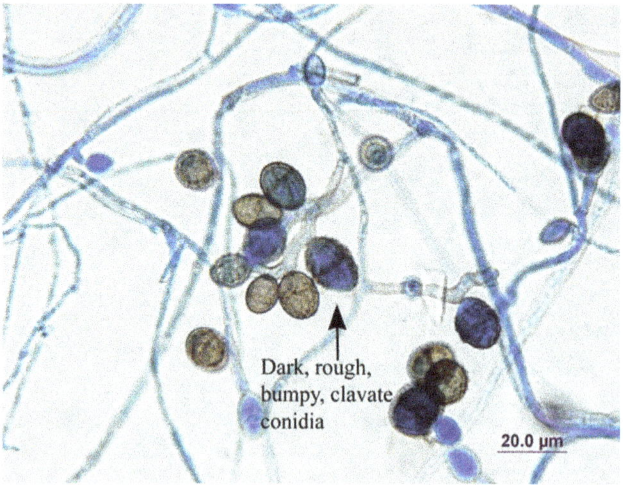

Fig. 9.29.8 *Ulocladium*, with rough, bumpy, clavate dictyoconidia.

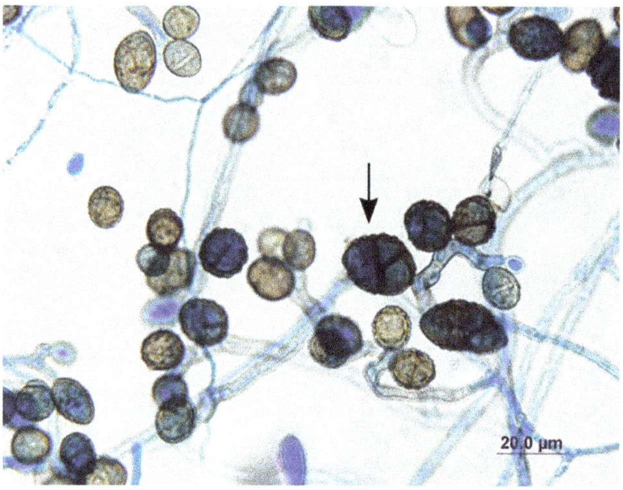

Fig. 9.29.9 Showing Y septation of dictyoconidia (right center).

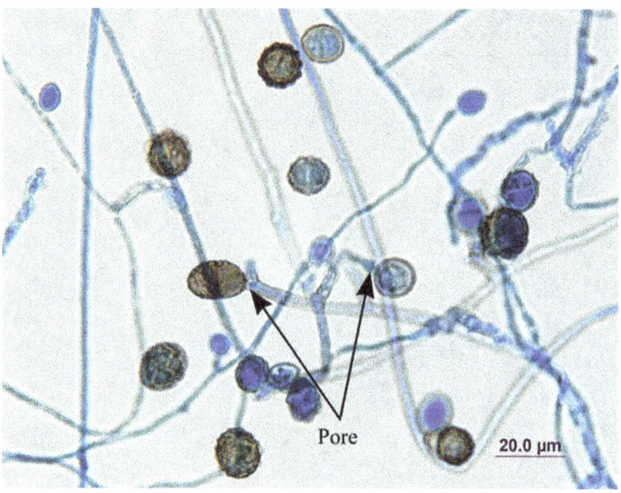

Fig. 9.29.10 Poroconidia.

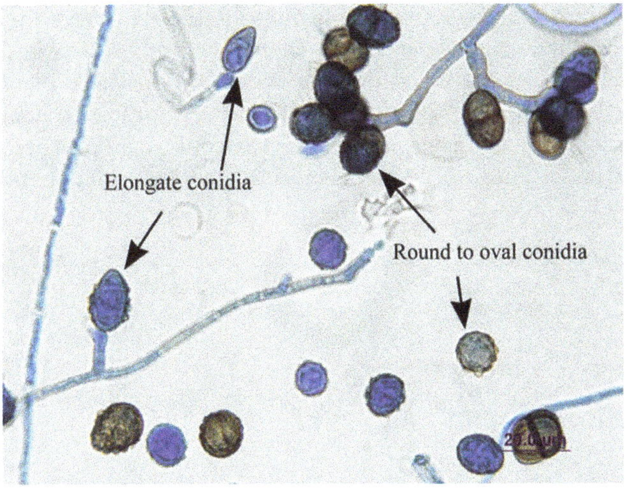

Fig. 9.29.11 Types of conidia.

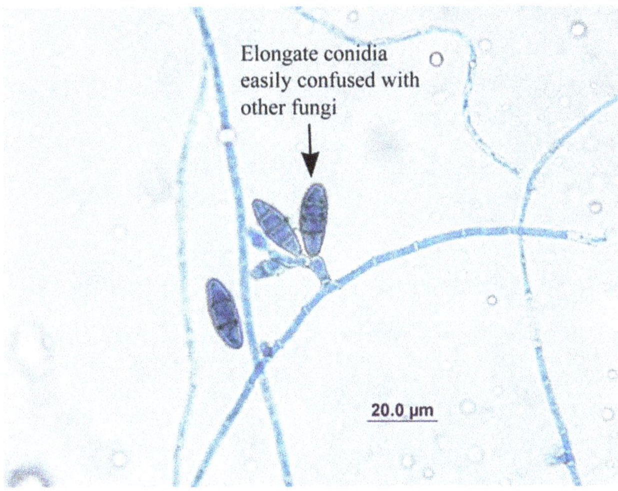

Fig. 9.29.12 Elongate conidia.

9.30 *Phoma* sp

Colonial morphology: on Sabouraud dextrose agar, colonies begin as grayish-white and become light buff and are fuzzy or velvety on the surface. The reverse color is buff. Colonial color varies on other media, most often dark to black, olive-gray or grayish-brown. A diffusible pigment, either reddish or brown, is sometimes released into the media. Colonies do not wrinkle, but may form concentric circles with darker shades of color in the center.

Microscopic morphology: hyphae are septate and hyaline. Conidiophores are not produced on hyphae but internal to the pycnidia. Asexual fruiting bodies, pycnidia, are frequently produced; they are flask shaped to round, and approximately 60–120 µm in diameter. Pycnidia are produced intercalary. Additional hyphae often extend from the pycnidia, but are sparse in comparison to *Chaetomium* which it closely resembles. Pycnidia must be distinguished from the perithecia of *Chaetomium* and the cleistothecia of *Pseudallescheria boydii*. Within pycnidia, conidia are abundantly produced and then released through an opening (ostiole) in the pycnidium. The conidia are oval and elongated, single celled and hyaline (2–6 μm). In some species, chlamydoconidia are seen resembling those of *Alternaria*.

Phoma spp culture photos

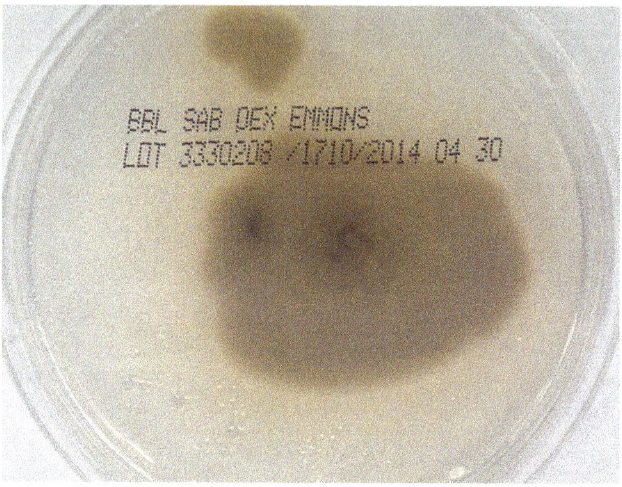

Fig. 9.30.1 *Phoma* on SABDX (bottom).

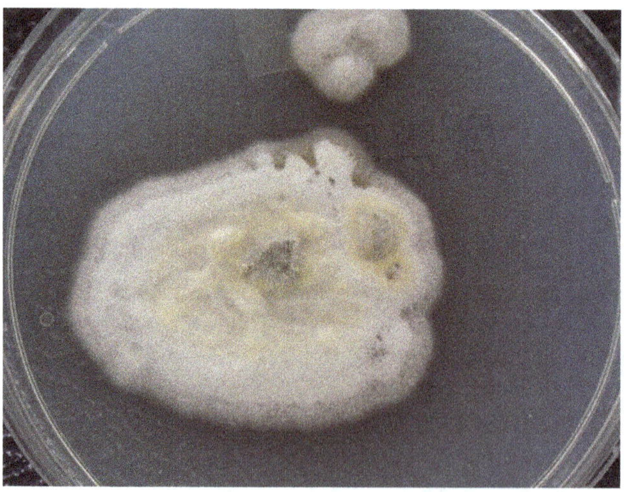

Fig. 9.30.2 *Phoma* on SABDX (top).

Fig. 9.30.3 *Phoma* on SABDX (side).

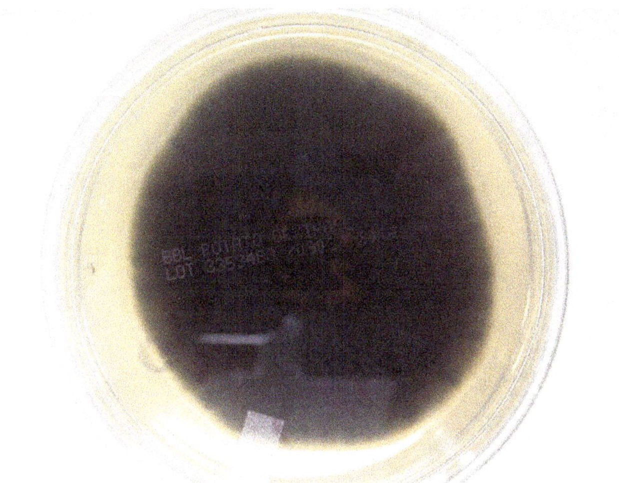

Fig. 9.30.4 *Phoma* on PDA (bottom).

Fig. 9.30.5 *Phoma* on PDA (top).

Fig. 9.30.6 *Phoma* on PDA (side).

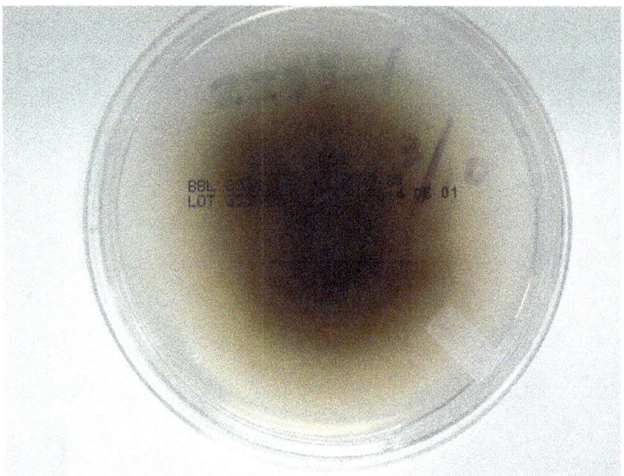

Fig. 9.30.7 *Phoma* on CMA (bottom).

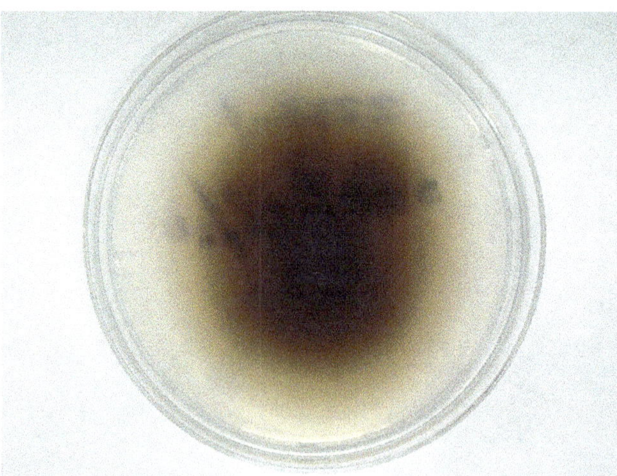

Fig. 9.30.8 *Phoma* on CMA (top).

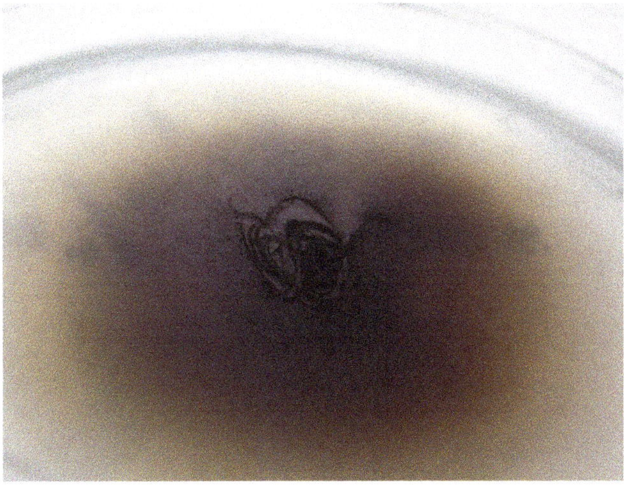

Fig. 9.30.9 *Phoma* on CMA (side).

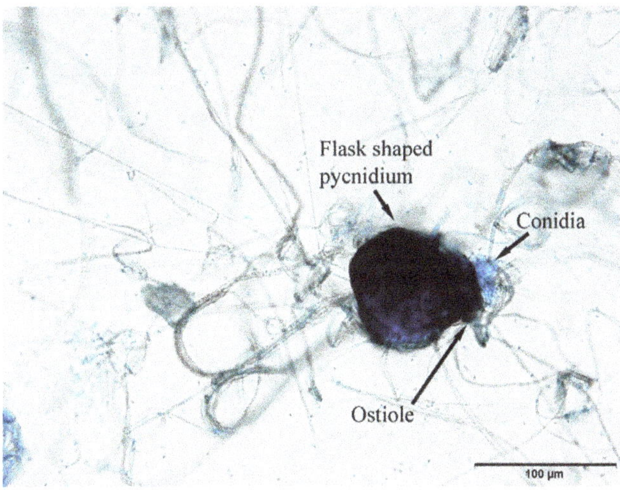

Fig. 9.30.10 Pycnidium with conidia.

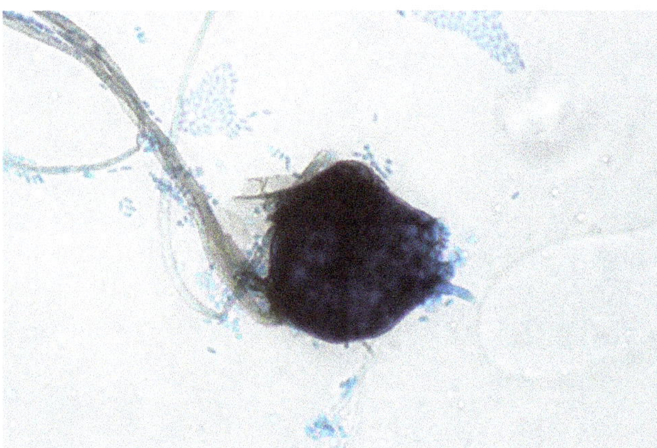

Fig. 9.30.11 *Phoma*, pycnidium.

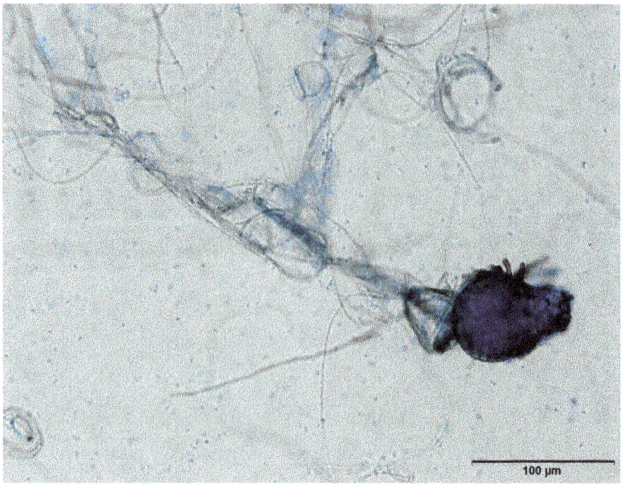

Fig. 9.30.12 *Phoma*, pycnidium.

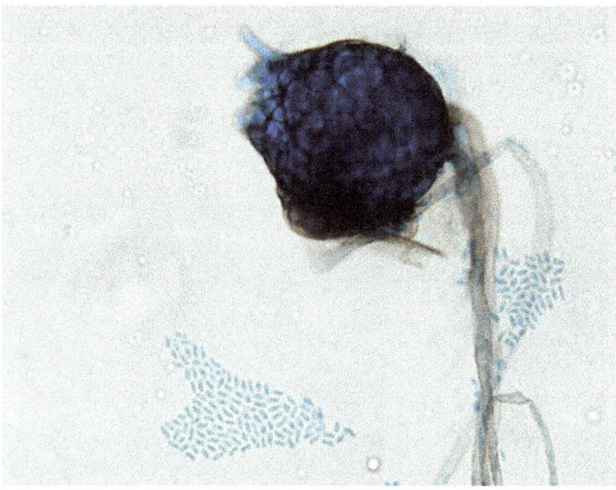

Fig. 9.30.13 *Phoma*, pycnidium.

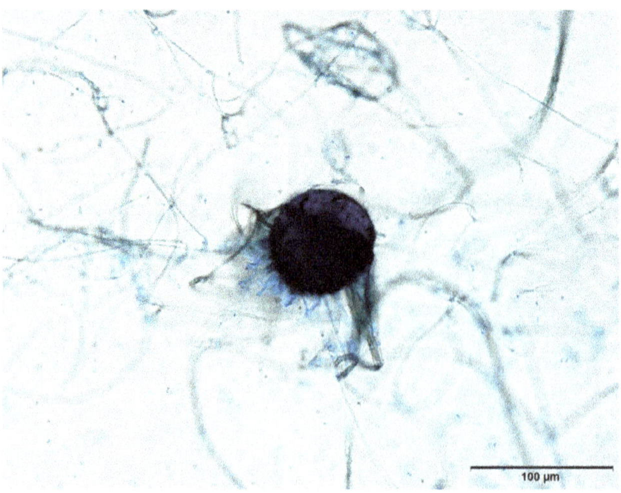

Fig. 9.30.14 Rounded pycnidia.

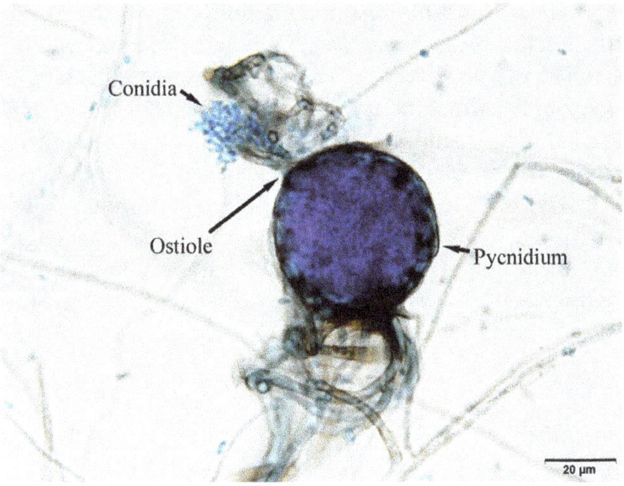

Fig. 9.30.15 Rounded pycnidia with conidia.

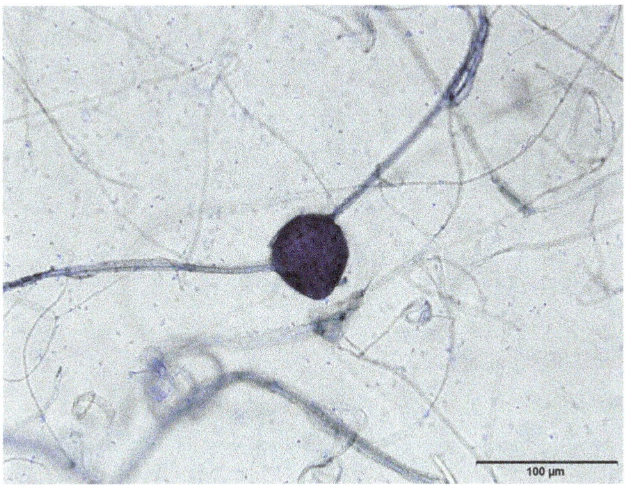

Fig. 9.30.16 Intercalary pycnidia.

9.31 *Rhinocladiella basitona*

Colonial morphology: colonies are first dark and smooth, then become wrinkled with a blue-gray to brown surface resembling *Cladosporium* sp. Mature colonies are 1.5–2.5 cm in diameter. The reverse is brown to black. Colonies produce submerged hyphae that pit colonies into the agar.

Microscopic morphology: hyphae are hyaline to dark and septate. Combinations of fine hyaline hyphae are mixed with dark hyphae composed of conidiogenous cells that are almost yeast-like (Fig. 9.31.8). Conidiogenous cells develop into new hyphae with rachiform conidiophores. Conidiophores have a basal septum with a slightly widened base that descends into the rachis (stem). A second septum may be evident prior to the rachis development (Fig. 9.31.11). The rachis is not distinctly raduliform (brush-like) as in *Ramichloridium*. Instead, the rachis produces only sympodial arranged conidia whereas *Ramichloridium* has both sympodial and opposing (non-sympodial) arrangements of conidia. At the point of conidial attachment to the rachis, some dark scarring occurs on the rachis (Fig. 9.31.14). The conidiophores (phialides) are 1.5–2 μm at the base, and can reach lengths of 65 μm, bearing 2–30 conidia. Verticillate hyphae (V-shaped whorls) are frequent (Fig. 9.31.16). Other conidiophores are abundantly produced perpendicular to the hyphae. Conidia are aseptate, smooth, pear shaped or tear shaped to elongated (1.5 × 4 μm). Conidia retain a darkened basal hilum that may not always be evident with light microscopy.

Rhinocladiella basitona

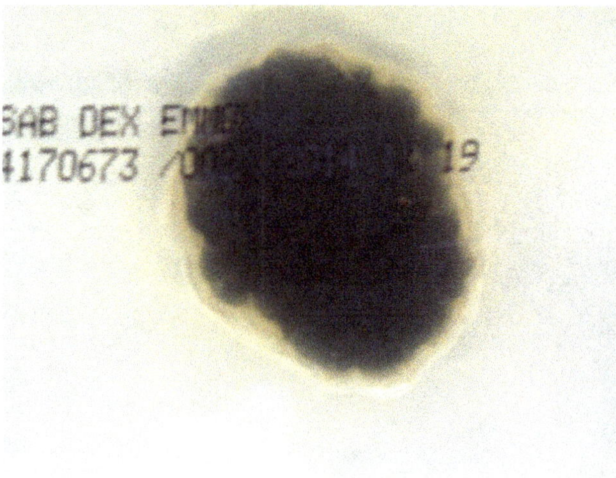

Fig. 9.31.1 *Rhinocladiella basitona* on SABDX (bottom).

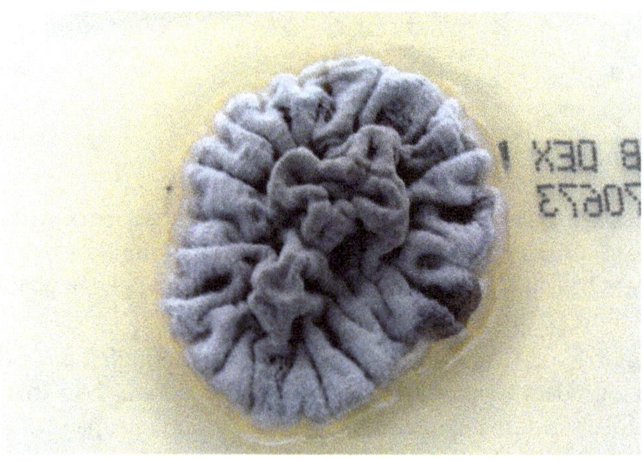

Fig. 9.31.2 *R. basitona* on SABDX (top).

Fig. 9.31.3 *R. basitona* on SABDX (side).

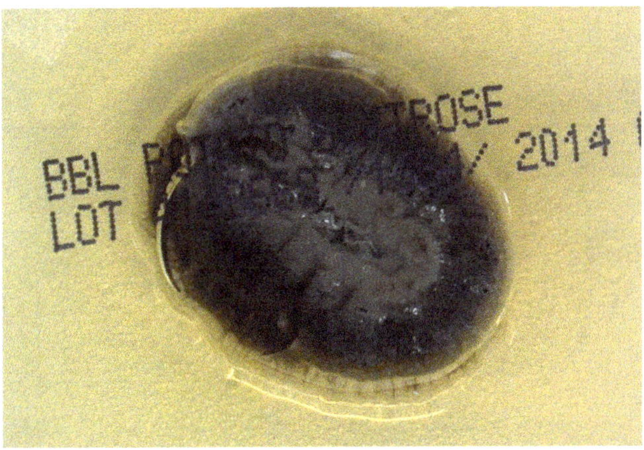

Fig. 9.31.4 *R. basitona* on PDA (bottom).

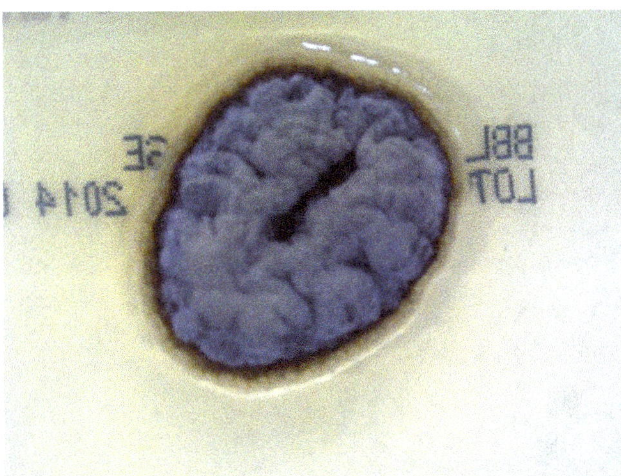

Fig. 9.31.5 *R. basitona* on PDA (top).

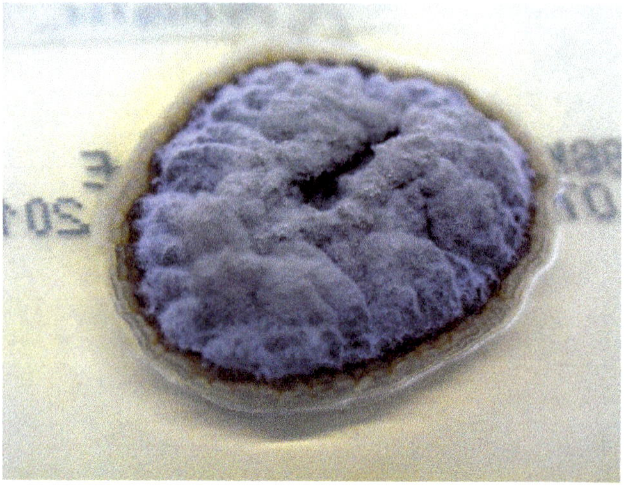

Fig. 9.31.6 *R. basitona* on PDA (side).

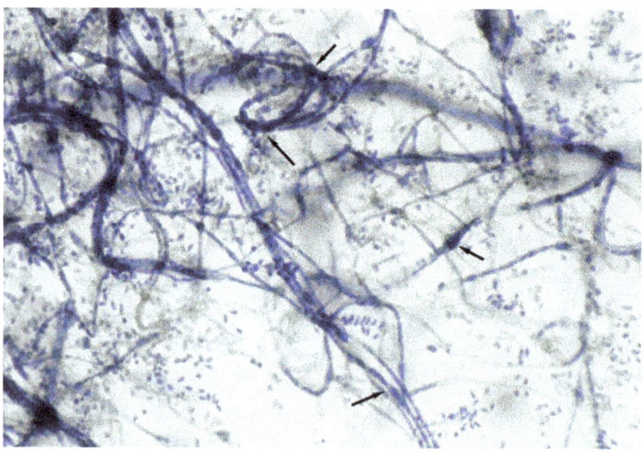

Fig. 9.31.7 Hyphae with intercalary swellings.

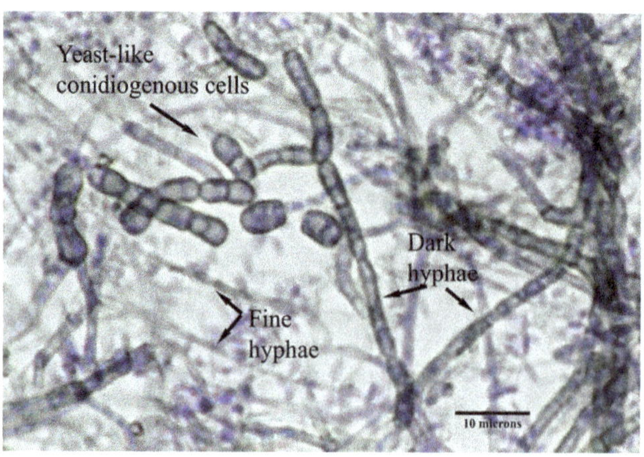

Fig. 9.31.8 Dark conidiogenous *Exophiala*-type budding cells.

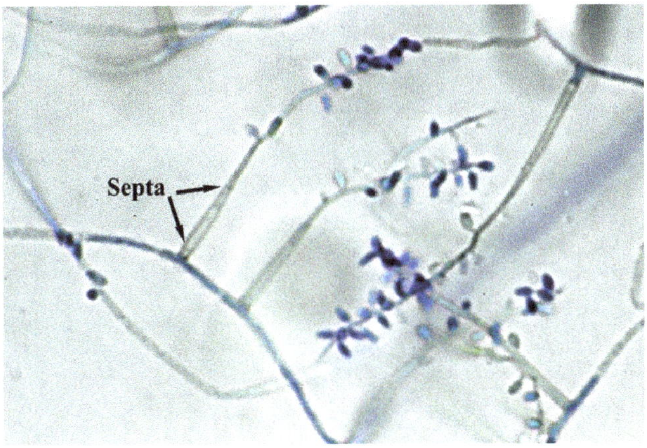

Fig. 9.31.9 Septate conidiophore.

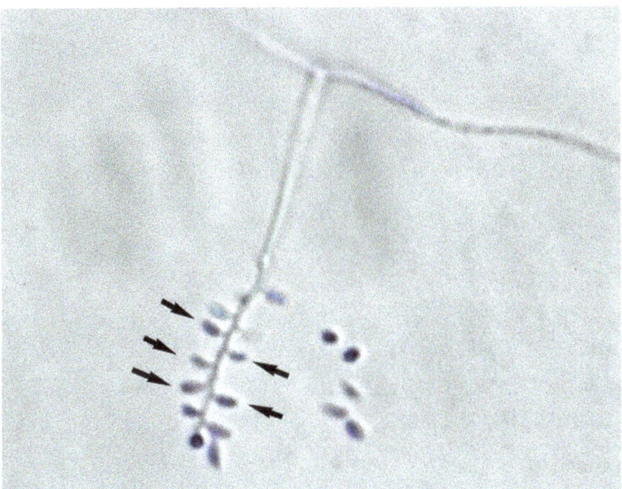

Fig. 9.31.10 Sympodial arrangement of conidia.

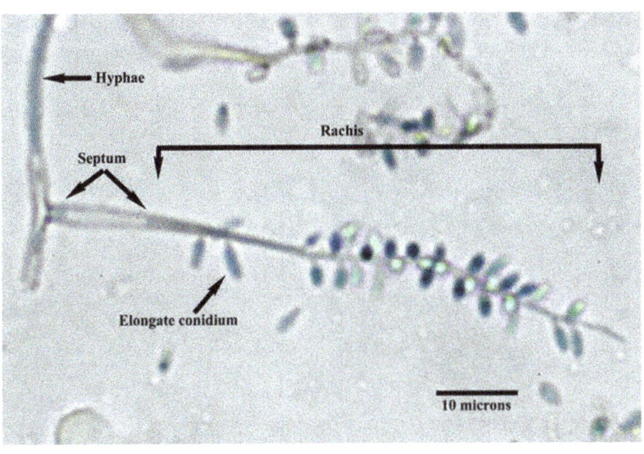

Fig. 9.31.11 Long rachis bearing up to 30 conidia.

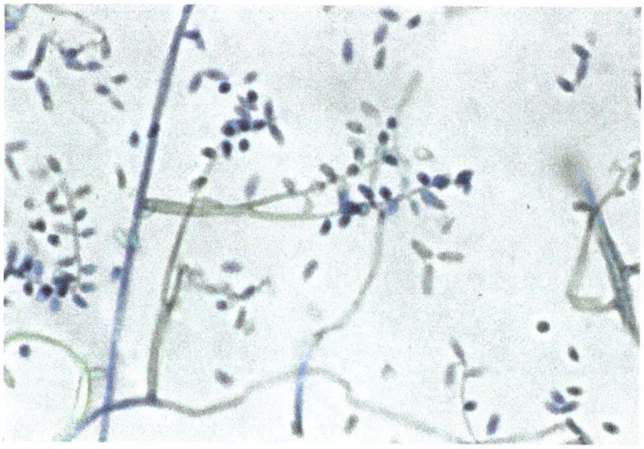

Fig. 9.31.12 Branching conidiophores.

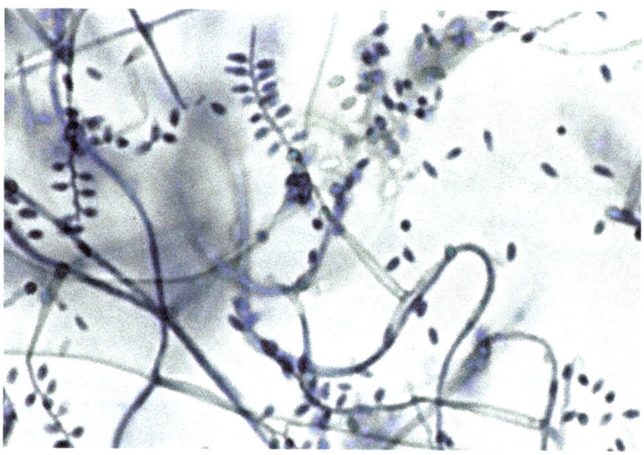

Fig. 9.31.13 Rachis with sympodial arranged conidia.

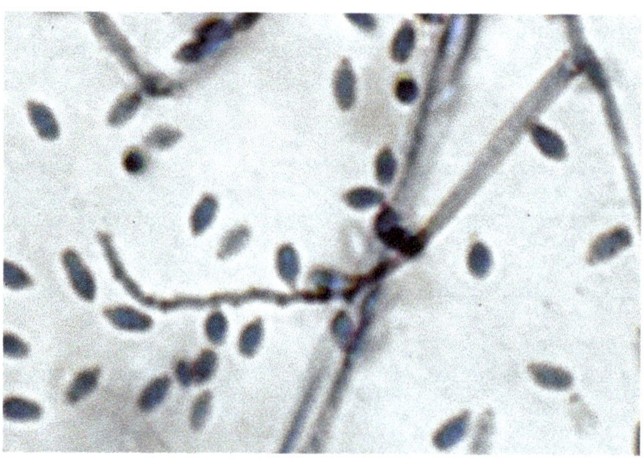

Fig. 9.31.14 Computer enhanced to show rachi scars and hilum of conidia.

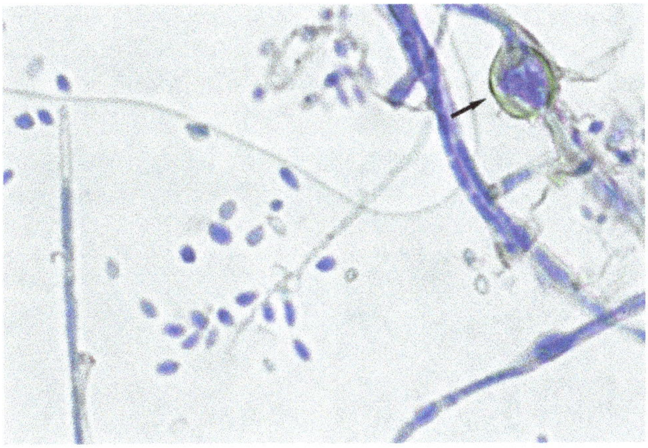

Fig. 9.31.15 Chlamydospores present.

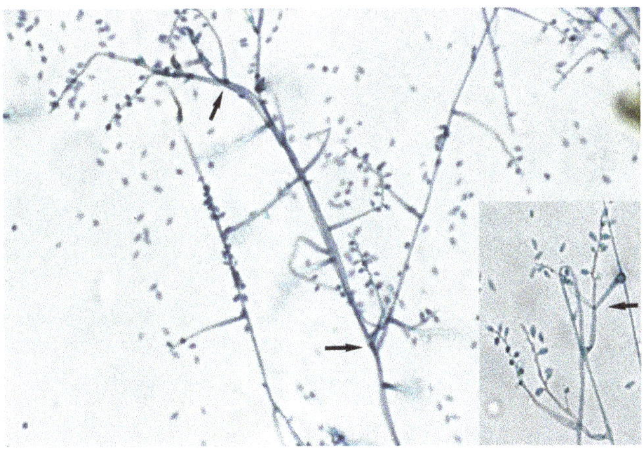

Fig. 9.31.16 Hyphal verticillate branching.

9.32 *Exophiala dermatitidis (Wangella dermatitidis)*

Colonial morphology: colonies appear brown-black on Sabouraud dextrose agar and black on Sabouraud brain heart infusion agar. Colonies resemble black yeast when grown on these two agars. Eventually, yeast-like colonies may develop dark, septate hyphae and become fluffy in appearance. Alternatively, colonies grown on potato dextrose agar are first brown and suede-like, with a velvety texture. Cultures grow at 42 °C.

Microscopic morphology: black colony yeast cells are first hyaline, oval, sometimes bud, and darken into thick brown dematiaceous yeast with thick cell walls. Yeast cells develop into dark brown mycelia that are irregular and give the appearance of chains of yeast cells strung together. Phialides often arise from terminal mycelia, but sometimes they branch centrally and direct from the mycelium. Phialides are sometimes flask shaped and annellide, but most often cylindrical without tapered ends, ending with terminal conidia. Conidia may be produced singly or in clusters (glioconidia) at the termini of the phialides or accumulate down their sides. Conidia are hyaline to light brown, one-celled (but look biclavate), round to oblong, 2.0–4.0 × 2.5–6.0 μm, and smooth walled. Large, thick-walled round chlamydospores (6–10 μm in diameter) are also produced at the apices of mycelia.

Exophiala dermatitidis

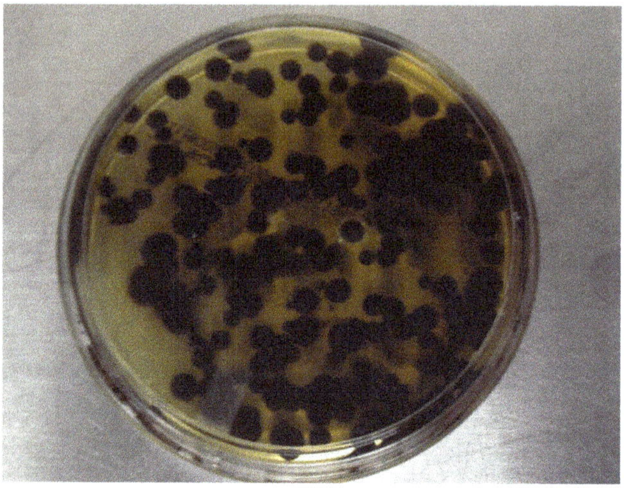

Fig. 9.32.1 *Exophiala dermatitidis* on Sabouraud BHI (bottom).

Fig. 9.32.2 *E. dermititidis* on Sabouraud BHI.

Fig. 9.32.3 *E. dermititidis* on Sabouraud BHI (side).

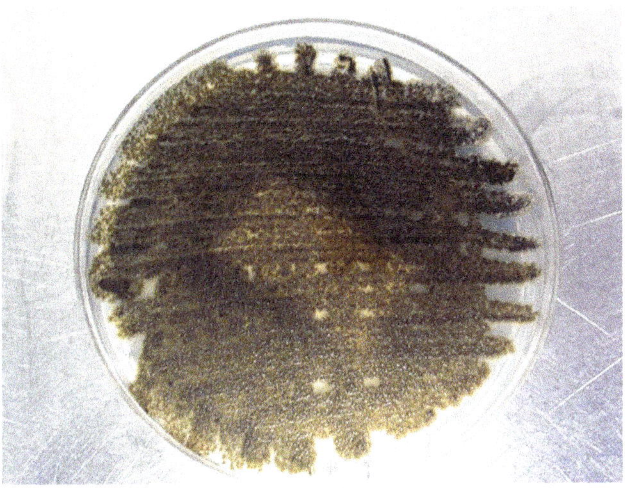

Fig. 9.32.4 *E. dermititidis* on SABDX.

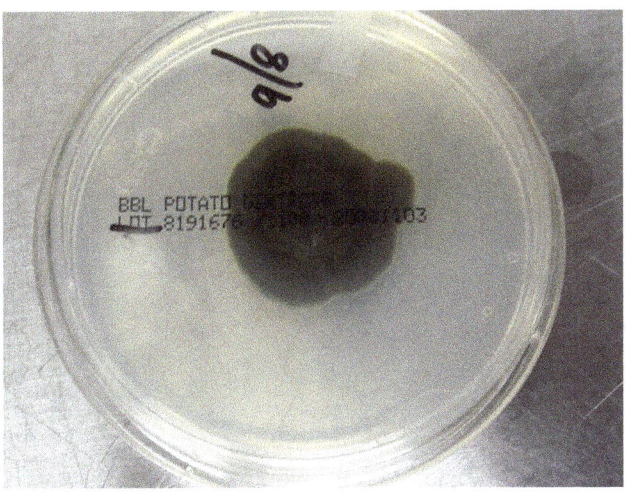

Fig. 9.32.5 *E. dermititidis* on PDA (bottom).

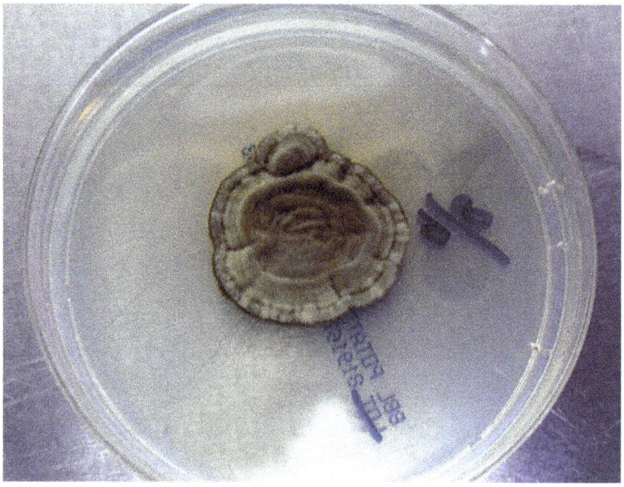

Fig. 9.32.6 *E. dermitititidis* on PDA.

Fig. 9.32.7 *E. dermitititidis* on PDA (side).

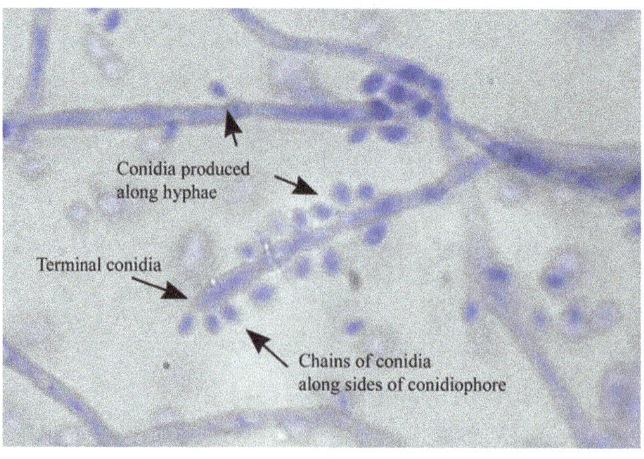

Fig. 9.32.8 Conidia production and arrangement.

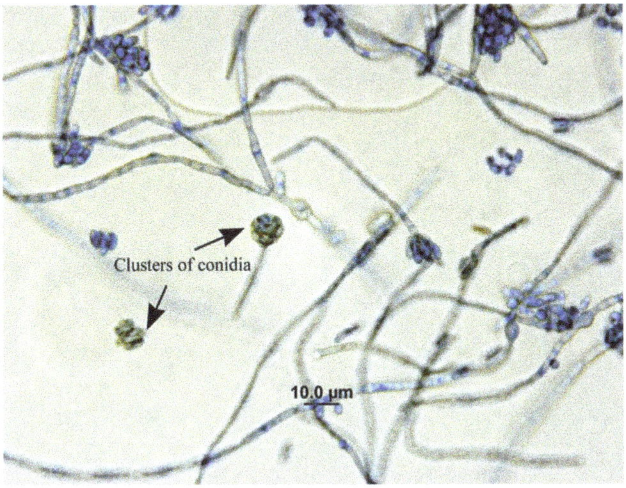

Fig. 9.32.9 Conidia production and arrangement.

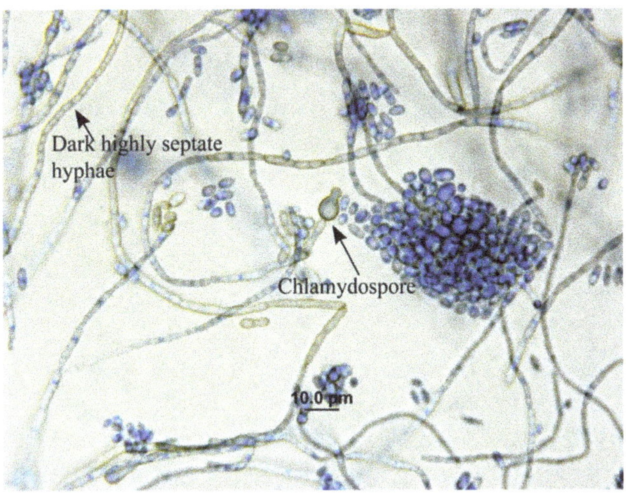

Fig. 9.32.10 Chlamydospores produced at ends of highly septate hyphae.

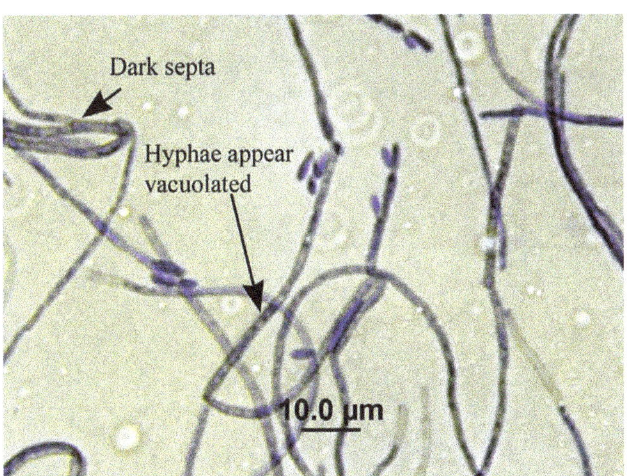

Fig. 9.32.11 Septate hyphae.

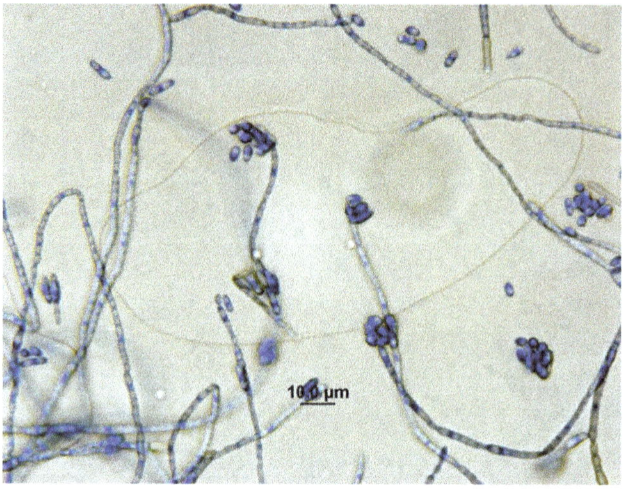

Fig. 9.32.12 *E. dermititidis*, morphology.

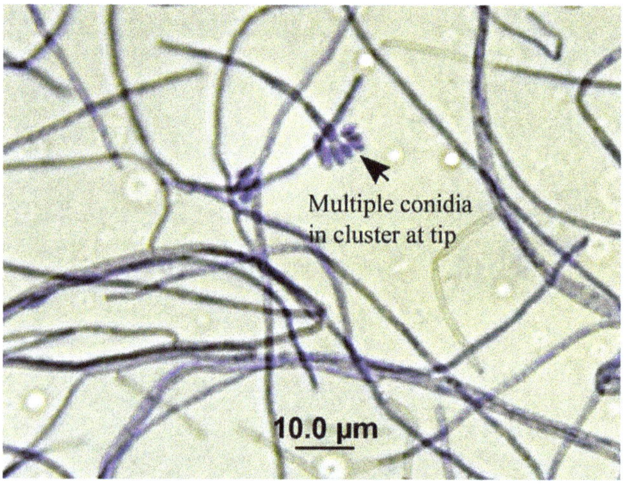

Fig. 9.32.13 *E. dermititidis*, conidia.

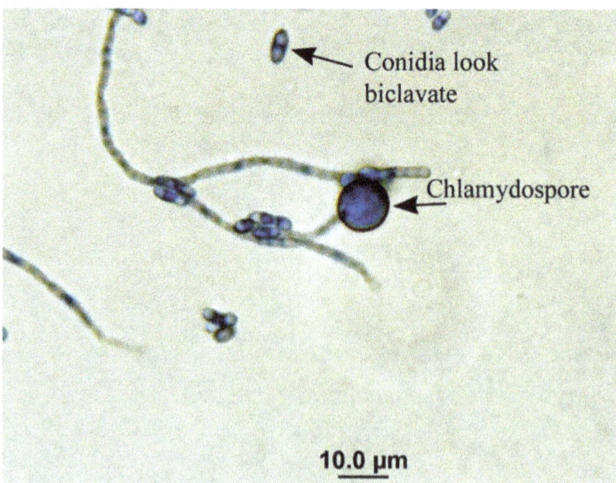

Fig. 9.32.14 Mature chlamydospore.

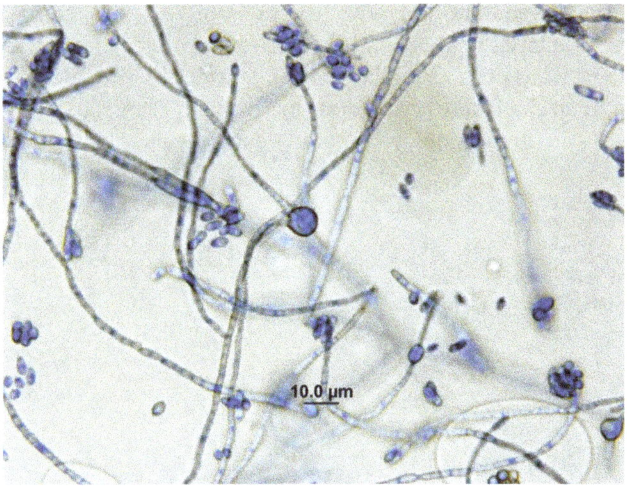

Fig. 9.32.15 Chlamydospore formation.

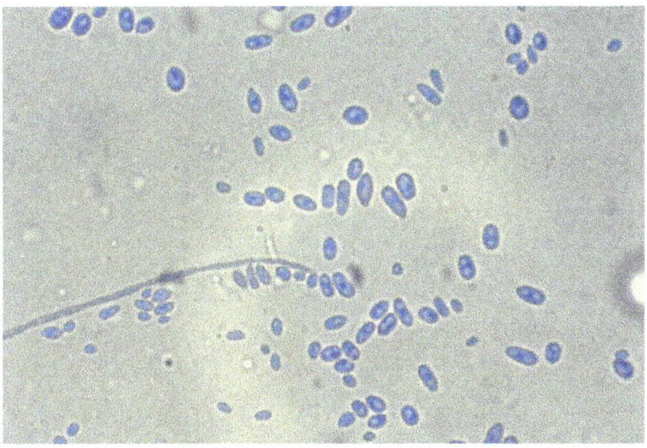

Fig. 9.32.16 Yeast-like cells.

9.33 *Exophiala jeanselmei*

Colonial morphology: most colonies are initially smooth, pine-green to black, mucoid and yeast-like, becoming raised, dome-shaped, dark green, and suede-like in texture and developing tufts of aerial mycelium with age. Reverse is light green-black. More recently recognized strains develop raised, fluffy colonies that show concentric rings of color varying from khaki, to brown, to white along the edges. Reverse is gray.

Microscopic morphology: microscopic morphology is complex for this organism. Young cultures show ellipsoidal yeast-like budding cells (unicellular or bicellular). Large germinating cells are scattered amongst budding yeast cells. These large broad cells develop hyphal shoots which later become septate hyphae. Conidiogenous cells form singly and perpendicularly to the axis of the hyphae (sometimes sprouting from detached conidia). They are flask shaped and may have divisions of 2–3 cells, sometimes branching before they become tapered at the end. Conidia are produced from the tapered, annellide end of conidiogenous cells. Conidia are hyaline, smooth, thin walled, and tear shaped, 2–5 × 1–3 μm in size, with inconspicuous basal scars. They cluster at the tip of the annellides or adhere down the side of the conidiogenous cell.

Exophiala jeanselmei

Fig. 9.33.1 *Exophiala jeanselmei* (bottom).

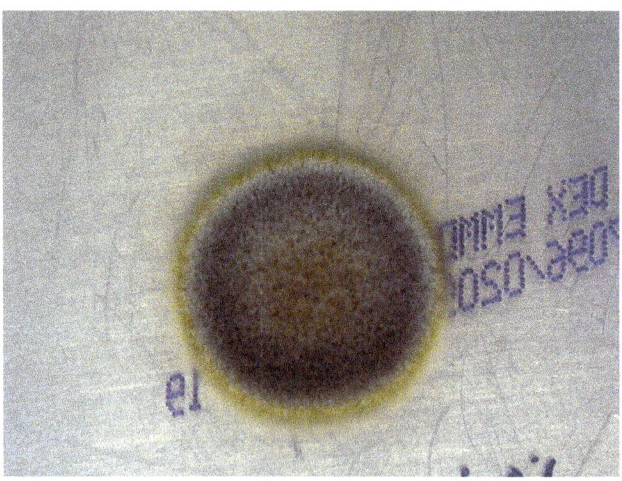

Fig. 9.33.2 *E. jeanselmei.*

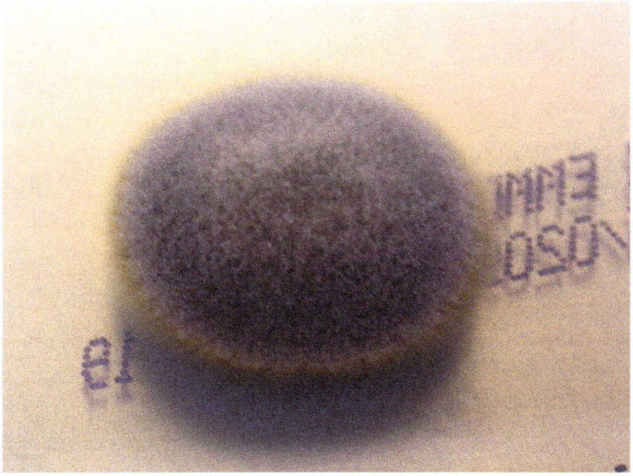

Fig. 9.33.3 *E. jeanselmei* (side).

Fig. 9.33.4 *Exophiala* sp, corn meal agar.

Fig. 9.33.5 *Exophiala* sp, corn meal agar.

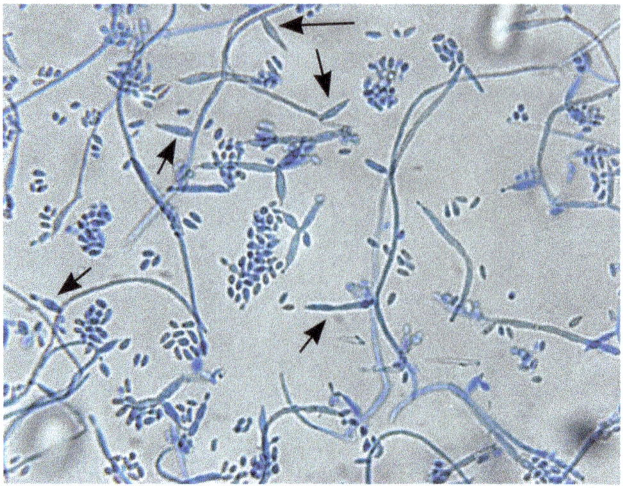

Fig. 9.33.6 Phialides form on hyphae.

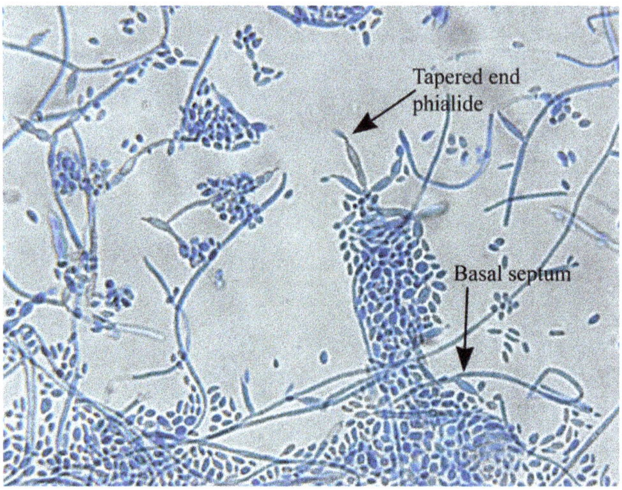

Fig. 9.33.7 *E. jeanselmei*, phialide morphology.

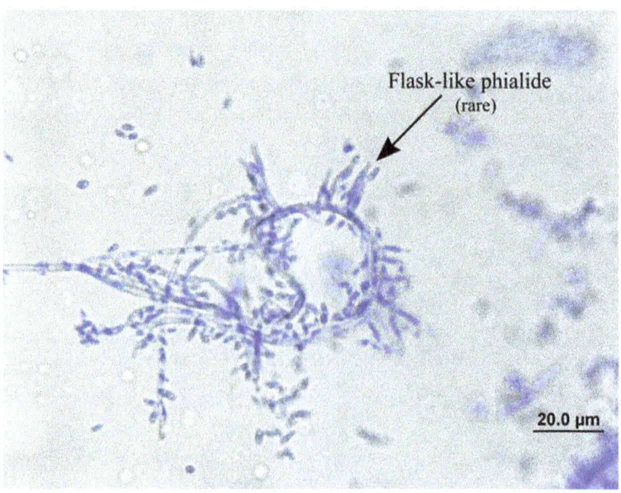

Fig. 9.33.8 PDA agar.

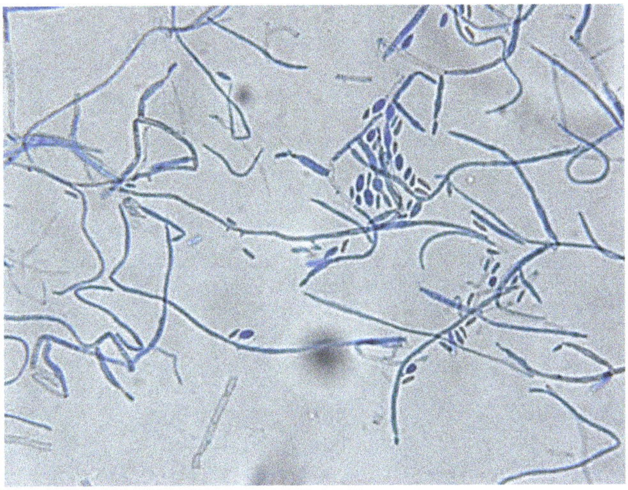

Fig. 9.33.9 *E. jeanselmei*, phialide morphology.

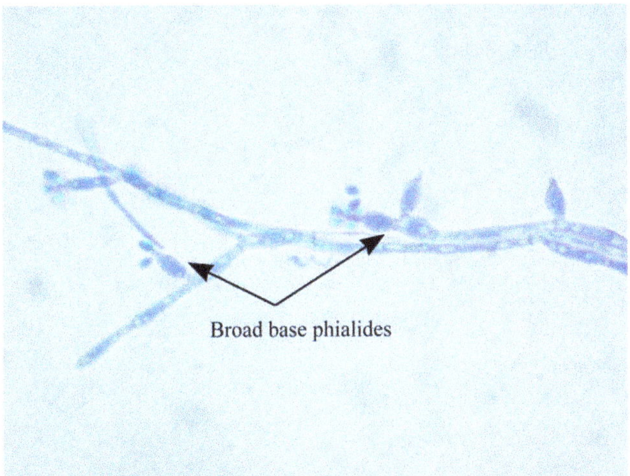

Fig. 9.33.10 *E. jeanselmei*, phialide morphology.

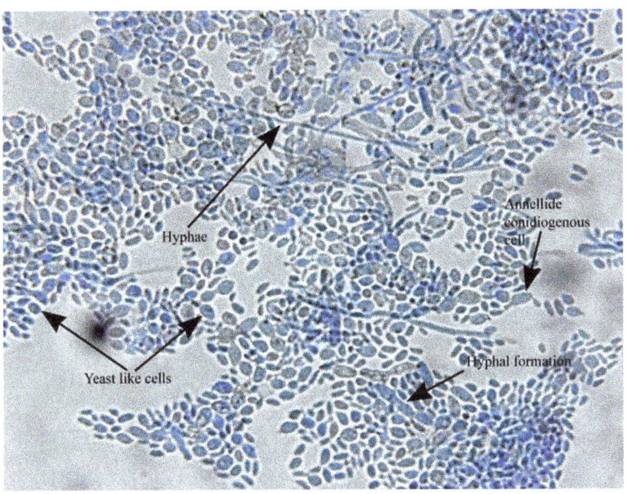

Fig. 9.33.11 Young culture with yeast-like cells.

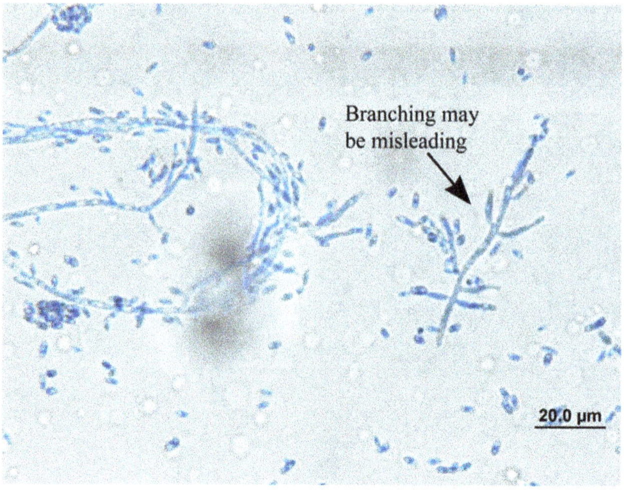

Fig. 9.33.12 *E. jeanselmei* can resemble *Trichoderma*.

Additional Reading

Blastomycosis

1 Bariola JR, Perry P, Pappas PG, *et al.* Blastomycosis of the central nervous system: a multicenter review of diagnosis and treatment in the modern era. *Clin Infect Dis* 2010; 50: 797–804. Available at: http://cid.oxfordjournals.org/content/50/6/797.full.pdf+html?sid=edc0f0e1-0cc5-4db6-9ca1-94b439d183ec

2 Bradsher RW Jr. Pulmonary blastomycosis. *Semin Resp Crit Care Med* 2008; 29: 174–181. Available at: www.thieme-connect.com/ejournals/pdf/10.1055/s-2008-1063856.pdf

3 Saccente M, Woods GL. Clinical and laboratory update on blastomycosis. *Clin Microbiol Rev* 2010; 23: 367–381. Available at: http://cmr.asm.org/content/23/2/367.full.pdf+html?sid=7bd53555-0a3e-4f7c-9411-43c969993d1a

4 Schwartz VJ. Intra-ocular blastomycosis. *Arch Ophthalmol* 1931; 5(4): 581–590. Available at: http://archopht.jamanetwork.com/article.aspx?articleid=609018

Histoplasmosis

5 Kauffman CA. Histoplasmosis: a clinical and laboratory update. *Clin Microbiol Rev* 2007; 20: 115–132. Available at: http://cmr.asm.org/content/20/1/115.full.pdf+html?sid=bec5a4be-6b56-49d4-957b-dd93fa49b795

6 Kurita N, Terao K, Brummer E, Ito E, Nishimura K, Miyaji M. Resistance of *Histoplasma capsulatum* to killing by human neutrophils. *Mycopathologia* 1991; 115: 207–213 (abstract only). Available at: http://link.springer.com/article/10.1007%2FBF00462229?LI=true#

7 McKinsey DS, McKinsey JP. Pulmonary histoplasmosis. *Semin Respir Crit Care Med* 2011; 32: 735–744. Available at: www.thieme-connect.com/ejournals/pdf/10.1055/s-0031-1295721.pdf

8 Swartzentruber S, L. Rhodes L, Kurkjian K, *et al.* Diagnosis of acute pulmonary histoplasmosis by antigen detection. *Clin Infect Dis* 2009; 49: 1878–1882. Available at: http://cid.oxfordjournals.org/content/49/12/1878.full.pdf+html?sid=42854d23-b0b1-46cc-aaf8-717b56bbd99e

9 Weinberg JM, Ali R, Badve S, Pelker RR. Musculoskeletal histoplasmosis: a case report and review of the literature. *J Bone Joint Surg Am* 2001; 83: 1718–1722 (abstract only). Available at: http://jbjs.org/article.aspx?articleid=29724

Atlas of Clinically Important Fungi, First Edition. Carmen V. Sciortino.

Coccidioidomycosis

10 Spinello IM, Munoz A, Johnson RH. Pulmonary coccidioidomycosis. *Semin Respir Crit Care Med* 2008; 29: 166–173. Available at: www.thieme-connect.com/ejournals/pdf/10.1055/s-2008-1063855.pdf

11 Thompson GR III. Pulmonary coccidioidomycosis. *Semin Respir Crit Care Med* 2011; 32: 754–763. Available at: www.thieme-connect.com/ejournals/pdf/10.1055/s-0031-1295723.pdf

Paracoccidioidomycosis

12 Flavio Queiroz-Telles F, Escuissato DL. Pulmonary paracoccidioidomycosis. *Semin Respir Crit Care Med* 2011; 32: 764–774. Available at: www.thieme-connect.com/ejournals/pdf/10.1055/s-0031-1295724.pdf

13 Restrepo A, Benard G, de Castro CC, Agudelo CA, Tobón AM. Pulmonary paracoccidioidomycosis. *Semin Respir Crit Care Med* 2008; 29: 182–197. Available at: www.thieme-connect.com/ejournals/pdf/10.1055/s-2008-1063857.pdf

Zygomycosis

14 Alvarez E, Sutton DA, Cano J, *et al* Spectrum of zygomycete species identified in clinically significant specimens in the United States. *J Clin Microbiol* 2009; 47: 1650–1656. Available at: http://jcm.asm.org/content/47/6/1650.full.pdf+html?sid=8ba1876f-29e0-4d4c-89ac-11fac64e30ed

15 Freifeld AG, Iwen PC. Zygomycosis. *Semin Respir Crit Care Med* 2004; 25: 221–231. Available at: www.thieme-connect.com/ejournals/pdf/10.1055/s-2004-824905.pdf

16 Kemna ME, Neri RC, Ali R, Salkin IF. *Cokeromyces recurvatus*, a mucoraceous zygomycete rarely isolated in clinical laboratories. *J Clin Microbiol* 1994; 32(3): 843. Available at: http://jcm.asm.org/content/32/3/843.full.pdf+html?sid=d69842cf-a32f-4ae5-980b-004746e98b2d

17 Khan ZU, Khoursheed M, Makar R, *et al. Basidiobolus ranarum* as an etiologic agent of gastrointestinal zygomycosis. *J Clin Microbiol* 2001; 39(6): 2360. http://jcm.asm.org/content/39/6/2360.full.pdf+html?sid=25a6df9f-2c71-40ac-8e47-ffa090dde7e2

18 Kimura M, Yaguchi T, Sutton DA, Fothergill AW, Thompson EH, Wickes BL. Disseminated human conidiobolomycosis due to *Conidiobolus lamprauges*. *J Clin Microbiol* 2011; 49: 752–756. Available at: http://jcm.asm.org/content/49/2/752.full.pdf+html?sid=af964437-9bdb-4c7b-a8dc-516f5a8536b1

19 Munipalli B, Rinaldi MG, Greenberg SB. *Cokeromyces recurvatus* isolated from pleural and peritoneal fluid: case report. *J Clin Microbiol* 1996; 34(10): 2601. Available at: http://jcm.asm.org/content/34/10/2601.full.pdf+html?sid=75a0c047-b0f1-4b5c-a0f0-4f2bb2868f06

20 Padhye AA, Ajello L. Simple method of inducing sporulation by *Apophysomyces elegans* and *Saksenaea vasiformis*. *J Clin Microbiol* 1988; 26(9): 1861. Available at: http://jcm.asm.org/content/26/9/1861.full.pdf+html?sid=b34a8add-41f3-4e67-bf34-0196ed77fefb

21 Pyrgos V, Shoham S, Walsh TJ. Pulmonary zygomycosis. *Semin Respir Crit Care Med* 2008; 29: 111–120. Available at: www.thieme-connect.com/ejournals/pdf/10.1055/s-2008-1063850.pdf

22 Ribes JA, Vanover-Sams CL, Baker DJ. Zygomycetes in human disease. *Clin Microbiol Rev* 2000; 13: 236–301. Available at: http://cmr.asm.org/content/13/2/236.full.pdf+html?sid=45942e0c-60f8-4e69-9354-970ae4ff8b4d

23 Spellberg B, Edwards J Jr, Ibrahim A. Novel perspectives on mucormycosis: pathophysiology, presentation, and management. *Clin Microbiol Rev* 2005; 18: 556–569. Available at: http://cmr.asm.org/content/18/3/556.full.pdf+html?sid=ee656f44-4a57-458c-9683-53216e167780

24 Weitzman I, Whittier S, Mckitrick JC, Della-Latta P. Zygospores: the last word in identification of rare or atypical zygomycetes isolated from clinical specimens. *J Clin Microbiol* 1995; 33: 781–783. Available at: http://jcm.asm.org/content/33/3/781.full.pdf+html?sid=750113e8-2ede-429c-a0fe-be4f534a0f89

25 Wieden MA, Steinbronn KK, Padhye AA, Ajello L, Chandler FW. Zygomycosis caused by *Apophysomyces elegans*. *J Clin Microbiol* 1985; 22(4): 522. Available at: http://jcm.asm.org/content/22/4/522.full.pdf+html?sid=f990238b-9fcd-4761-b6cc-75ab3beec1a4

Hyaline molds and yeasts

26 Burnham CD, Dunne WM. Answer to photo quiz: disseminated *Penicillium marneffei*. *J Clin Microbiol* 2011; 49: 3726. Available at: http://jcm.asm.org/content/49/10/3449.full.pdf+html?sid=2e07cb87-bde3-42be-b4fa-e2bd913993a3

27 Cristini A, Garcia-Hermoso D, Celard M, Albrand G, Lortholary O. Cerebral phaeohyphomycosis caused by *Rhinocladiella mackenziei* in a woman native to Afghanistan. *J Clin Microbiol* 2010; 48: 3451–3454. Available at: http://jcm.asm.org/content/48/9/3451.full.pdf+html?sid=cca2f945-53f4-4645-99f4-7e2b18af9bc7

28 Duong TA. Infection due to *Penicillium marneffei*, an emerging pathogen: review of 155 reported cases. *Clin Infect Dis* 1996; 23: 125–130. Available at: http://cid.oxfordjournals.org/content/23/1/125.full.pdf+html

29 Gordon RA, Sutton DA, Thompson EH, *et al.* Cutaneous phaeohyphomycosis caused by *Paraconiothyrium cyclothyrioides*. *J Clin Microbiol* 2012; 50(11): 3795. Available at: http://jcm.asm.org/content/50/11/3795.full.pdf+html?sid=2d594d9b-7faa-4c17-872e-5fa673ac2e97

30 Hayashi S, Naitoh K, Matsubara S, *et al.* Pulmonary colonization by *Chrysosporium zonatum* associated with allergic inflammation in an immunocompetent subject. *J Clin Microbiol* 2002; 40(3): 1113. Available at: http://jcm.asm.org/content/40/3/1113.full.pdf+html?sid=2242bcb3-2ed2-4630-9063-a194919f0efb

31 Houbraken J, Samson RA. Phylogeny of *Penicillium* and the segregation of *Trichocomaceae* into three families. *Studies Mycol* 2011; 70: 1–51. Available at: www.studiesinmycology.org/content/70/1/1.full.pdf+html

32 Houbraken J, Frisvad JC, Samson RA. Taxonomy of *Penicillium* section *Citrina*. *Studies Mycol* 2011; 70: 53–138. Available at: www.studiesinmycology.org/content/70/1/53.full.pdf+html

33 Houbraken J, Verweij PE, Anthonius JM, Rijs M, Borman AM, Samson RA. Identification of *Paecilomyces variotii* in clinical samples and settings. *J Clin Microbiol* 2010; 48(8): 2754. Available at: http://jcm.asm.org/content/48/8/2754.full.pdf+html?sid=1b8b4dbf-48a1-45ad-9064-f271c12f8316

34 Khan Z, Ahmad S, Al-Ghimlas F, *et al. Purpureocillium lilacinum* as a cause of cavitary pulmonary disease: a new clinical presentation and observations on atypical morphologic characteristics of the isolate. *J Clin Microbiol* 2012; 50(5): 1800. Available at: http://jcm.asm.org/content/50/5/1800.full.pdf+html?sid=bcbcdfbf-2a3e-4aef-a818-8cf7f9b22665

35 Krijgsheld P, Bleichrodt R, van Veluw GJ, *et al.* Development in *Aspergillus. Studies Mycol* 2012; 74:1–29. Available at: www.studiesinmycology.org/content/74/1/1.full.pdf+html

36 Kumar D, Sigler L, Gibas CFC, *et al. Graphium basitruncatum* fungemia in a patient with acute leukemia. *J Clin Microbiol* 2007; 45(5): 1644. Available at: http://jcm.asm.org/content/45/5/1644.full.pdf+html?sid=c130a1b0-90c3-4380-b2e3-30484bef7087

37 Nelson PE, Dignani MC, Anaissie EJ. Taxonomy, biology, and clinical aspects of *Fusarium* species. *Clin Microbiol Rev* 1994; 7(4): 479. Available at: http://cmr.asm.org/content/7/4/479.full.pdf+html?sid=8114f158-19ff-4ab1-af9a-db71652c81a2

38 Perdomo H, Sutton DA, García D, *et al.* Spectrum of clinically relevant *Acremonium* species in the United States. *J Clin Microbiol* 2011; 49(1): 243. Available at: http://jcm.asm.org/content/49/1/243.full.pdf+html?sid=a001b2f9-3c6d-48af-87ec-3d45e7abd7a5

39 Revankar SG, Sutton DA, Sanche S, *et al. Metarrhizium anisopliae* as a cause of sinusitis in immunocompetent hosts. *J Clin Microbiol* 1999; 37: 195–198. Available at: http://jcm.asm.org/content/37/1/195.full.pdf+html?sid=a7c9d2f3-b99b-4f29-b7a4-50cfc14dea29

40 Roilides E, Sigler L, Bibashi E, Katsifa H, Flaris N, Panteliadis C. Disseminated infection due to *Chrysosporium zonatum* in a patient with chronic granulomatous disease and review of non-*Aspergillus* fungal infections in patients with this disease. *J Clin Microbiol* 1999; 37(1): 18. Available at: http://jcm.asm.org/content/37/1/18.full.pdf+html?sid=5e1737b1-0a58-41a2-b55a-98254f0dab8e

41 Sutton DA, Slifkin M, Yakulis R, Rinaldi MG. U.S. case report of cerebral phaeohyphomycosis caused by *Ramichloridium obovoideum (R. mackenziei)*: criteria for identification, therapy, and review of other known dematiaceous neurotropic taxa. *J Clin Microbiol* 1998; 36(3): 708. Available at: http://jcm.asm.org/content/36/3/708.full.pdf+html?sid=29f68f88-7e49-4839-b7b3-25953ea3d82f

Dematiaceous fungi

42 Arzanlou M, Groenewald JZ, Gams W, Braun U, Shin HD, Crous PW. Phylogenetic and morphotaxonomic revision of *Ramichloridium* and allied genera. *Studies Mycol* 2007; 58: 57–93. Available at: www.studiesinmycology.org/content/58/1/57.full.pdf+html

43 Aveskamp MM, de Gruyter J, Woudenberg JHC, Verkley GJM, Crous PW. Highlights of the *Didymellaceae*: a polyphasic approach to characterise *Phoma* and related pleosporalean genera. *Studies Mycol* 2010; 65: 1–60. Available at: www.studiesinmycology.org/content/65/1/1.full.pdf+html

44 Badali H, Gueidan C, Najafzadeh MJ, Bonifaz A, Gerrits van den Ende AHG, de Hoog GS. Biodiversity of the genus *Cladophialophora*. *Studies Mycol* 2008; 61: 175–191. Available at: www.studiesinmycology.org/content/61/1/175.full.pdf+html

45 Bensch K, Braun U, Groenewald JZ, Crous PW. The genus *Cladosporium*. *Studies Mycol* 2012: 72: 1–401. Available at: www.studiesinmycology.org/content/72/1/1.full.pdf+html

46 Bensch K, Groenewald JZ, Dijksterhuis J, *et al.* Species and ecological diversity within the *Cladosporium cladosporioides* complex (*Davidiellaceae, Capnodiales*). *Studies Mycol* 2010; 67: 1–94. Available at: www.studiesinmycology.org/content/67/1/1.full.pdf+html

47 Bonifaz A, Badali H, de Hoog GS, *et al.* Tinea nigra by *Hortaea werneckii*, a report of 22 cases from Mexico. *Studies Mycol* 2008; 61: 77–82. Available at: www.studiesinmycology.org/content/61/1/77.full.pdf+html

48 Cortez KJ, Roilides E, Quiroz-Telles F, *et al.* Infections caused by *Scedosporium* spp. *Clin Microbiol Rev* 2008; 21: 157–197. Available at: http://cmr.asm.org/content/21/1/157.full.pdf+html?sid=6a18b23d-3627-4eba-b244-7a42aff37221

49 Crous PW, Braun U, Schubert K, Groenewald JZ. Delimiting *Cladosporium* from morphologically similar genera. *Studies Mycol* 2007; 58: 33–56. Available at: www.studiesinmycology.org/content/58/1/33.full.pdf+html

50 de Hoog GS, Nishikaku AS, Fernandez-Zeppenfeldt G, *et al.* Molecular analysis and pathogenicity of the *Cladophialophora carrionii* complex, with the description of a novel species. *Studies Mycol* 2007; 58: 219–234. Available at: www.studiesinmycology.org/content/58/1/219.full.pdf+html

51 de Hoog GS, Vicente V, Caligiorne RB, *et al.* Species diversity and polymorphism in the *Exophiala spinifera* clade containing opportunistic black yeast-like fungi. *J Clin Microbiol* 2003; 41: 4767–4778. Available at: http://jcm.asm.org/content/41/10/4767.full.pdf+html?sid=1f3e176f-9e9a-4aa6-a45a-7b1907afe1fd

52 Fothergill AW. Identification of dematiaceous fungi and their role in human disease. *Clin Infect Dis* 1996; 22(Suppl 2): S179–184. Available at: http://cid.oxfordjournals.org/content/22/Supplement_2/S179.full.pdf+html?sid=579e044f-09ce-460c-87a6-284345d18e9f

53 Perfect JR, Schell WA. The New fungal opportunists are coming. *Clin Infect Dis* 1996; 22(Suppl 2): S112–118. Available at: http://cid.oxfordjournals.org/content/22/Supplement_2/S112.full.pdf+html?sid=43865cf2-a16d-4499-8294-a2470924e1d3

54 Revankar SG, Sutton DA. Melanized fungi in human disease. *Clin Microbiol Rev* 2010; 23(4): 884. Available at: http://cmr.asm.org/content/23/4/884.full.pdf+html?sid=fb5938ad-542b-4c38-9fc6-b3dc4f6c0255

55 Schubert K, Groenewald JZ, Braun U, *et al.* Biodiversity in the *Cladosporium herbarum* complex (*Davidiellaceae, Capnodiales*), with standardisation of methods for *Cladosporium* taxonomy and diagnostics. *Studies Mycol* 2007; 58: 105–156. Available at: www.studiesinmycology.org/content/58/1/105.full.pdf+html

56 Shin JH, Lee SK, Suh SP, *et al.* Fatal *Hormonema dematioides* peritonitis in a patient on continuous ambulatory peritoneal dialysis: criteria for organism identification and review of other known fungal etiologic agents. *J Clin Microbiol* 1998; 36: 2157–2163. Available at: http://jcm.asm.org/content/36/7/2157.full.pdf+html?sid=d9803664-03e9-4baa-8dc9-6bc1001cef6c

57 Woo PCY, Lau SKP, Ngan AHY, Tse H, Tung ETK, Yuen KY. *Lasiodiplodia theobromae* pneumonia in a liver transplant recipient. *J Clin Microbiol* 2008; 46: 380–384. Available at: http://jcm.asm.org/content/46/1/380.full.pdf+html

58 Zalar P, de Hoog GS, Schroers HJ, Crous PW, Groenewald JZ, Gunde-Cimerman N. Phylogeny and ecology of the ubiquitous saprobe *Cladosporium sphaerospermum*, with descriptions of seven new species from hypersaline environments. *Studies Mycol* 2007; 58: 157–183. Available at: www.studiesinmycology.org/content/58/1/157.full.pdf+html

59 Zalar P, Gostinčar1 C, de Hoog GS, Uršič1 V, Sudhadham M, Gunde-Cimerman N. Redefinition of *Aureobasidium pullulans* and its varieties. *Studies Mycol* 2008; 6121–38. Available at: www.studiesinmycology.org/content/61/1/21.full.pdf+html

Pleosporales anamorphs: *Alternaria, Bipolaris, Coniothyrium, Curvularia, Drechslera, Exserohilum, Leptophoma, Phoma, Pithomyces, Stemphylium*

60 Zhang Y, Schoch CL, Fournier J, *et al.* Multi-locus phylogeny of *Pleosporales*: a taxonomic, ecological and evolutionary re-evaluation. *Studies Mycol* 2009; 64: 85–102. Available at: www.studiesinmycology.org/content/64/1/85.full.pdf+html

Dermatophytes

61 D'Antonio D, Romano F, Iacone A, *et al.* Onychomycosis caused by *Blastoschizomyces capitatus*. *J Clin Microbiol* 1999; 37: 2927–2930. Available at: http://jcm.asm.org/content/37/9/2927.full.pdf+html?sid=fe67f54d-e12f-4ed9-a9d1-750f355b1bf0

62 Elewski BE. Onychomycosis: pathogenesis, diagnosis, and management. *Clin Microbiol Rev* 1998; 11: 415–429. Available at: http://cmr.asm.org/content/11/3/415.full.pdf+html?sid=4965c2d6-554e-43d0-95b9-420b049758df

Yeasts

63 Colombo AL, Padovan ACB, Chaves GM. Current knowledge of *Trichosporon spp.* and trichosporonosis. *Clin Microbiol Rev* 2011; 24(4): 682. Available at: http://cmr.asm.org/content/24/4/682.full.pdf+html?sid=77e4c92d-b170-4438-8804-096451a921a3

64 Erke KH. Light microscopy of basidia, basidiospores, and nuclei in spores and hyphae of *Filobasidiella neoformans* (*Cryptococcus neoformans*). *J Bacteriol* 1976; 128: 445–455. Available at: http://jb.asm.org/content/128/1/445.full.pdf+html?sid=a0337fac-5900-469c-beca-488387ec55d6

65 Gaitanis G, Magiatis P, Hantschke M, Bassukas ID, Velegraki A. The *Malassezia* genus in skin and systemic diseases. *Clin Microbiol Rev* 2012; 25(1):106. Available at: http://cmr.asm.org/content/25/1/106.full.pdf+html?sid=133e965a-ada9-46ed-8f3c-bc6b96252935

66 Stringer JR. *Pneumocystis carinii*: what is it, exactly? *Clin Microbiol Rev* 1996; 9: 489–498. Available at: http://cmr.asm.org/content/9/4/489.full.pdf+html?sid=9ebcb12d-131f-474b-8068-2ebdb625dc1c

67 Lass-Flörl C, Mayr A. Human protothecosis. *Clin Microbiol Rev* 2007; 20(2): 230. Available at: http://cmr.asm.org/content/20/2/230.full.pdf+html

Identification techniques

68 Guarner J, Brandt ME. Histopathologic diagnosis of fungal infections in the 21st century. *Clin Microbiol Rev* 2011; 24(2): 247. Available at: http://cmr.asm.org/content/24/2/247.full.pdf+html?sid=51e6d209-7550-45a5-b1b1-cffc84b521f4

69 Woo PCY, Ngan AHY, Chui HK, Lau SKP, Yuen KY. Agar block smear preparation: a novel method of slide 1 preparation for preservation of native fungal structures for microscopic examination and long-term storage. *J Clin Microbiol* 2010; 48(9): 3053. Available at: http://jcm.asm.org/content/48/9/3053.full.pdf+html?sid=ed489508-19d4-4bb4-be11-7d4c4c28bae0

Index

Printed in the USA
CPSIA information can be obtained
at www.ICGtesting.com
CBHW080923190824
13269CB00005B/79

9 781119 069669